Analysis for Healthcare Diagnostics and Theranostics

University of Edinburgh,
United Kingdom
6–8 September 2010

FARADAY DISCUSSIONS
Volume 149, 2011

RSC Publishing

Faraday Discussions (Print ISSN 1359-6640, Electronic ISSN 1364-5498) is published 4 times a year by the Royal Society of Chemistry, Thomas Graham House, Science Park, Milton Road, Cambridge, UK CB4 0WF. Volume 149 ISBN-13: 978 1 84973 2345

2011 annual subscription price: print+electronic £669, US $1,247; electronic only £602, US $1,122. Customers in Canada will be subject to a surcharge to cover GST. Customers in the EU subscribing to the electronic version only will be charged VAT. All orders, with cheques made payable to the Royal Society of Chemistry, should be sent to RSC Distribution Services, c/o Portland Customer Services, Commerce Way, Colchester, Essex, UK CO2 8HP.
Tel +44 (0) 1206 226050;
E-mail sales@rscdistribution.org

If you take an institutional subscription to any RSC journal you are entitled to free, site-wide web access to that journal. You can arrange access *via* Internet Protocol (IP) address at www.rsc.org/ip. Customers should make payments by cheque in sterling payable on a UK clearing bank or in US dollars payable on a US clearing bank. Periodicals postage is paid at Rahway, NJ and at additional mailing offices. Airfreight and mailing in the USA by Mercury Airfreight International Ltd., 365 Blair Road, Avenel, NJ 07001, USA.

US Postmaster: send address changes to *Faraday Discussions*, c/o Mercury Airfreight International Ltd., 365 Blair Road, Avenel, NJ 07001. All despatches outside the UK by Consolidated Airfreight.

PRINTED IN THE UK

Faraday Discussions documents a long-established series of *Faraday Discussion* meetings which provide a unique international forum for the exchange of views and newly acquired results in developing areas of physical chemistry, biophysical chemistry and chemical physics.

Analysis for Healthcare Diagnostics and Theranostics

Faraday Discussions

www.rsc.org/faraday_d

A General Discussion on Analysis for Healthcare Diagnostics and Theranostics was held at the University of Edinburgh, United Kingdom on 6th, 7th and 8th September 2010.

CONTENTS

ISSN 1359-6640; ISBN 978-1-84973-234-5

Cover
Image reproduced by permission of Dr. Harald Peter, Division of Pathway Medicine, The University of Edinburgh.

INTRODUCTORY LECTURE

PAPERS AND DISCUSSIONS

CONCLUDING REMARKS

ADDITIONAL INFORMATION

Spiers Memorial Lecture
Breeding and building molecular spies

X. Shu,[a] V. Lev-Ram,[a] E. S. Olson,[a] T. A. Aguilera,[a] T. Jiang,[a] M. Whitney,[a] J. L. Crisp,[a] P. Steinbach,[a] T. Deerinck,[b] M. H. Ellisman,[b] L. G. Ellies,[c] Q. T. Nguyen[d] and R. Y. Tsien[a]

Received 13th December 2010, Accepted 14th December 2010
DOI: 10.1039/c0fd90021d

To circumvent the limited spatial resolution of fluorescent protein imaging, we are developing genetically encoded tags for electron microscopy (EM).

Arabidopsis phototropin, a photoreceptor containing flavin mononucleotide (FMN), can be engineered into a small (106-residue) singlet oxygen generator ("miniSOG"), which efficiently generates $^1O_2$ upon blue light illumination. $^1O_2$ polymerizes diaminobenzidine into an osmiophilic deposit, enabling correlative EM with nanometer spatial resolution. In an initial biological application, EM shows that the closely related cell-adhesion molecules SynCAM1 and SynCAM2, separately fused to miniSOG, predominantly localize respectively to the pre-synaptic and post-synaptic sides of mammalian CNS synapses. MiniSOG may do for EM what the green fluorescent protein (GFP) did for optical microscopy.

For clinical applications, we need synthetic molecules with novel amplifying mechanisms for homing to diseased tissues. Activatable cell penetrating peptides (ACPPs) are polycationic cell penetrating peptides (CPPs) whose cellular uptake is minimized by a polyanionic inhibitory domain and then restored upon proteolysis of the peptide linker connecting the polyanionic and polycationic domains. Local activity of proteases able to cut the linker causes amplified retention in tissues and uptake into cells. ACPPs on dendrimers labelled with Cy5 and Gd-DOTA enable whole body magnetic resonance imaging[1] followed by fluorescence-guided surgery. Such fluorescence guidance improves tumor-free survival in two animal models.[2] Thrombin-cleavable ACPPs accumulate in atherosclerotic plaques, and their labelling intensity seems to correlate with progression towards rupture. Separately, we have developed fluorescent peptides that light up peripheral nerves to show surgeons where not to cut.

References

1 E. Olson, *Proc. Natl. Acad. Sci. U. S. A.*, 2010, **107**(9), 4311.
2 Q. Nguyen, *Proc. Natl. Acad. Sci. U. S. A.*, 2010, **107**(9), 4317.

[a] *HHMI and Depts. Pharmacology, Univ. California, San Diego, USA*
[b] *HHMI and Depts. Neuroscience, Univ. California, San Diego, USA*
[c] *HHMI and Depts. Pathology, Univ. California, San Diego, USA*
[d] *HHMI and Depts. Surgery, Univ. California, San Diego, USA*

Amplified detection of DNA hybridization using post-labelling with a biotin-modified intercalator

Magdalena Gębala,[a] Gerhard Hartwich[b] and Wolfgang Schuhmann*[a]

Received 18th May 2010, Accepted 22nd July 2010
DOI: 10.1039/c005365a

A 32-electrode microelectrode array modified with a self-assembled monolayer of a thiolated DNA capture strand and 11-mercapto-1-undecanol was used for the detection of multi-resistant *Staphylococcus aureus* (MRSA) upon hybridization of the complementary target DNA. In the proposed assay strategy the obtained double-stranded DNA (dsDNA) is at first non-covalently labeled by intercalation of a proflavine derivative which is functionalized *via* a flexible spacer with biotin moieties. Subsequent to this ępost-labelling a avidin/alkaline phosphatase conjugate is bound to the biotin moieties thus introducing a reporter group at sites bearing dsDNA. Hybridization and hence the presence of MRSA DNA is detected *via* oxidation of *p*-aminophenol enzymatically generated from *p*-aminophenylphosphate. The assay strategy was carefully evaluated using ferrocene-modified target strands. An increase in sensitivity of the proposed label-free DNA assays based on a careful design of the sensing interface and the implemented enzymatic amplification was achieved.

Introduction

The diagnosis of genetic diseases or infectious agents, gene expression or drug screening is increasingly based on the use of microchip technology. DNA microarrays offer an enormous potential for rapid complex analysis of nucleic acid samples.[1–5] Specific microarray fabrication methods include micro-jet deposition of pre-synthesized oligonucleotides[6] as well as light-directed[7] or electrochemical oligonucleotide synthesis.[8] DNA microarrays are manufactured on glass, silicon or plastic supports comprising thousands of typically 10–100 μm diameter individually addressable electrodes.[9] Affymetrix pioneered the field of high-density DNA microarrays[10,11] using on-chip oligonucleotides synthesis and fluorescence read-out of complementary DNA hybridization. This technology has been successfully adapted for—among others—screening of breast cancer and identification of bacterial pathogens.[12] Another commercially available high-density array system, NimbleGen, with a similar read-out principle has been launched by Roche.[13] The CombiMatrix Custom Array, which consists of 12 544 individually addressable 40 μm diameter platinum electrodes and an overall analysis time of 60 s, is an example for a system with electrochemical read-out.[14,15] Moreover, 'lab-on-a-chip' devices which integrate the sample preparation including sample collection, DNA extraction, amplification and DNA array detection on a microfluidic platform have been proposed and are increasingly used.[16] All mentioned microarray systems, however, involve covalent labelling of the target DNA molecules with either dyes or enzymes. The used labels significantly enhance the sensitivity of the detection

[a]*Analytische Chemie – Elektroanalytik & Sensorik, Ruhr-Universität Bochum, Universitätsstr. 150, D-44780 Bochum, Germany. E-mail: wolfgang.schuhmann@rub.de; Fax: + 49-234-3214638; Tel: +49-234-3226200*
[b]*FRIZBiochem Gesellschaft für Bioanalytik mbH, Floriansbogen 2-4, 82061 Neuried, Germany*

schemes, however, at cost of simplicity in performing the assay or preparation of the sample.

Recently, we proposed an intrinsically label-free strategy for DNA detection based on a selective post-modification of double-stranded DNA (dsDNA) by means of intercalation of a specifically designed proflavine derivative bearing biotin moieties attached *via* a flexible spacer.[17] Here, this concept is extended to the detection of multi-resistant *Staphylococcus aureus* (MRSA) using a microelectrode array (MEA) consisting of 32 individually addressable square gold electrodes.

Experimental

Materials and reagents

Capture-probe DNA and complementary DNA were provided by FRIZ Biochem (Neuried, Germany) The sequence of the specific L-Uni-POG (MRSA) capture probe is GAA ACG TGC TGT GGG GAt tta ttt att tta tttt–(DTPA)$_3$OH–3′ (small letters indicate the linker region). The complementary target DNA was covalently modified with four ferrocene units at the 5′-end thus allowing for a direct voltammetric evaluation of the quality of the spotting procedure, the amount of DNA strands at the individual gold electrodes and the degree of intercalation. The 3′-5′-sequence of L-Uni signal oligo is 3′-CTT TGC ACG ACA CCC CT–[Y–FcAc]$_4$ with Y representing an unilink-C$_6$-aminomodifier and FcAc the ferrocene label.

All aqueous solutions were made with purified water from a Milli-Q system (Millipore). Ethanol and $NaH_2PO_4 \cdot 2H_2O$, were from Riedel-de-Haën (Seelze, Germany). $Na_2PO_4 \cdot 2H_2O$, Na_2SO_4, $NaHCO_3$, Na_2CO_3 were from J. T. Baker (Deventer, The Netherlands). The dibiotinylated proflavine derivative (*N*3,*N*3′-acridine-3,6-diylbis{*N*1-[15-oxo-19-(2-oxohexahydro-1*H*-thieno[3,4-d]imidazol-4-yl)-4,7,10-trioxa-14-azanonadec-1-yl]succinamide}) was synthesized as previously reported.[17] Milk powder was from Carl Roth, (Karlsruhe, Germany). *p*-Aminophenyl phosphate (*p*-APP) was from Biosynth AG (Staad, Switzerland), and the avidin/alkaline phosphatase conjugate dissolved in 0.05 M Tris buffer, pH 8.0, containing 1% bovine serum albumin, 1 mM $MgCl_2$ and 15 mM sodium azide was from Sigma (Deisenhofen, Germany). All reagents were of analytical grade and used as received.

Hybridization and intercalation procedure

The 32-electrode (70 μm × 70 μm) MEA chip was bonded to a so called "dipstick" that can be immersed into conventional PCR Eppendorf tubes. The microarrays were modified with the thiol-tethered specific DNA capture probes by spotting in the desired pattern according to a procedure established previously.[18] Pre-spotted MEAs were hybridized by immersing the dipsticks into a PCR Eppendorf tube containing a buffered solution (10 mM phosphate buffer, 1 M Na_2SO_4, pH 7.0) of the ferrocene-modified target DNA at pre-defined concentrations. Hybridization was carried out for 30 min at ambient temperature. Afterwards, the MEAs were rinsed with phosphate buffer and characterized by means of cyclic voltammetry (CV). The surface coverage (*G*) of dsDNA strands which are modified with ferrocene moieties was evaluated by the integration of the anodic peak ($i = dQ/dT$) obtained at a comparatively slow scan rate (500 mV s^{-1}) and using Faraday's law ($\Gamma = Q/nFA$) where Q is the background-corrected charge under the anodic redox wave in the cyclic voltammogram, n is the number of electrons involved in the redox process (in this case 4 since there are four ferrocene (Fc) moieties bound at each hybridized target DNA strand), F is the Faraday constant, and A is the real area of the electrode, calculated from the geometric area multiplied by the roughness factor which was determined to be 1.2. The roughness factor of the electrodes was determined from the ratio of the real area A_{real} to the geometric area $A_{geometric}$ of the electrode. A_{real} is the microscopic surface area obtained from the integration of the gold oxide reduction peak and dividing it by 482 mC cm^{-2}.[19]

Intercalation was accomplished by incubation of the hybridized MEAs for 10 min at ambient temperature in 100 μM of a dibiotinylated proflavine solution prepared in the same buffer which was used for hybridization. Afterwards the MEAs were thoroughly rinsed with phosphate buffer and again characterized by means of CV to estimate the degree of intercalation.

Blocking of DNA-modified MEA surfaces against nonspecific adsorption and modification with an avidin/alkaline phosphatase conjugate

Avidin and streptavidin exhibit a strong tendency for nonspecific adsorption on surfaces. This leads to a low signal-to-noise ratio of assays based on biotin–avidin interactions at surfaces. In order to prevent the undesired adsorption of the avidin/alkaline phosphatase conjugate or the target DNA strands at the MEA surface, the dipsticks were incubated in milk powder solution (1 mg mL^{-1}) in 500 mM carbonate buffer (pH 9.8) for 10 min at ambient temperature. Afterwards, they were thoroughly rinsed with 500 mM carbonate buffer and subsequently incubated with the solution containing the avidin/alkaline phosphatase conjugate (avidin/alkaline phosphatase). The commercial avidin/alkaline phosphatase solution was diluted by 1 : 1000 with carbonate buffer (500 mM; pH 9) and the MEAs were immersed in this solution for 10 min before thoroughly rinsing them with carbonate buffer.

Electrochemical measurements

All electrochemical measurements were performed in a three-electrode set-up with the on-chip counter and pseudo-reference gold electrodes. Dissolved 100 μM [Os(bipy)$_2$Cl$_2$] was used as an internal Nernstian reference system. A parallel 32-channel potentiostat (Dotria, FrizBiochem, Neuried, Germany) was used to individually polarize each of the 32 working electrodes of the MEA in parallel and accomplish the electrochemical measurements. CV was performed, in most cases, in a potential range from −300 to +300 mV *vs.* the internal reference at a scan rate of 500 mV s^{-1} in phosphate buffer (1 M Na_2SO_4, pH 7.0).

Amperometry was carried out in 100 mM carbonate buffer containing 250 mM Na_2SO_4, pH 9.0. The microelectrodes were polarized at a potential of +300 mV *vs.* the potential of the [Os(bipy)$_2$Cl$_2$] redox couple. The raw data were obtained as ASCII files and analyzed with ORIGIN (8.0, Microcal). The peak potential, E_p, and the peak current, i_p, of the Faradaic processes were determined from the cyclic voltammograms after subtraction of the charging current, i_c.

Results and discussion

Fig. 1 shows the 32-electrode array mounted on a dipstick and outlines the proposed assay principle. For the assay development, a ferrocene-modified complementary target DNA strand was used which allows additionally characterizing the DNA hybridization as shown in Fig. 1C. Moreover, as will be shown below, the electron-transfer between the Fc-modified DNA and the electrode surface is attenuated upon intercalation which allows evaluation of the proposed post-modification using the biotinylated intercalator. The fact that 4 Fc units are attached to one target DNA strand in close proximity to each other supposes that all Fc units are able to undergo electron-transfer reactions with the neighbouring Fc moiety. Thus, it can be assumed that all Fc units are oxidized and reduced at the same formal potential behaving like a virtual 4 electron-transfer redox species.

After hybridization with the Fc-modified target DNA and post-labelling of the dsDNA with the biotin-modified intercalator, a reporter system such as an avidin/alkaline phosphatase conjugate (avidin/alkaline phosphatase) is specifically bound to the dsDNA *via* the introduced biotin units. Hybridization and hence the amount of surface assembled Fc units can be determined by means of CV. Since the target

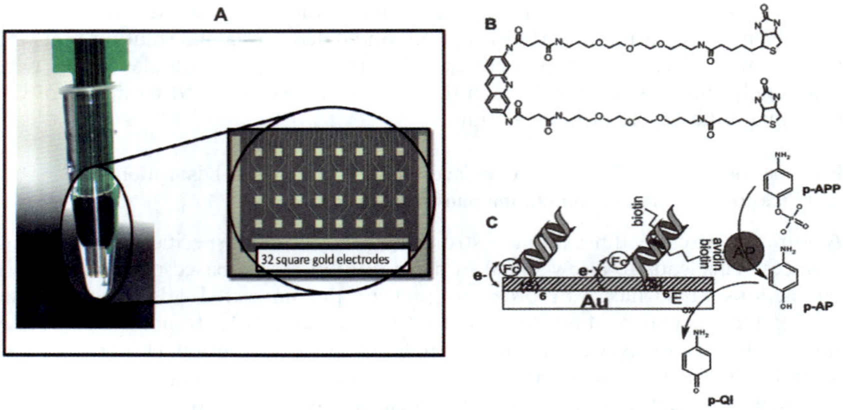

Fig. 1 (A) Photography of the dipstick-type MEA (left) consisting of 32 square gold microelectrodes. (B) Structure of the dibiotinylated proflavine derivative. (C) Schematic representation of the proposed assay strategy. Hybridization and intercalation of the dibiotinylated intercalator is detected on the one hand by the redox process of the ferrocene-label at the target DNA strand as well as by oxidation of *p*-AP formed in the enzyme-catalyzed hydrolysis of *p*-APP.

DNA strands were designed in such a way, that the Fc moieties are located at the 5′-end and hence in close proximity to the electrode surface, a fast electron transfer is expected for the DNA-bound Fc units. If the intercalation of the dibiotinylated proflavine derivative into the dsDNA is inducing significant changes in the structure of the dsDNA a modulation of the electron-transfer rate for the Fc redox reaction can be detected. The complementary binding of the enzymatic reporter system allows the amplified detection of successful hybridization by oxidation of *p*-aminophenol (*p*-AP) formed by the enzyme-catalyzed hydrolysis of *p*-aminophenyl phosphate (*p*-APP). Due to the comparatively large distance between the individual electrodes of the MEA it is assumed that the crosstalk between the electrodes caused by the diffusion of enzymatically generated *p*-AP remains negligible.

The spotting pattern is shown in Fig. 2. The grey sites indicate capture-probe-modified electrodes while white sites were functionalized with 11-mercapto-1-undecanol alone. Overall, 11 microelectrodes of the 32 electrodes on the MEA were finally modified with the L-Uni-POG capture strands. In order to force the DNA strands to stand up,[20] the electrodes were incubated in a solution containing 11-mercapto-1-undecanol. Thus, 21 electrodes were modified only with a self-assembled monolayer of 11-mercapto-1-undecanol.

25	26	27	28	29	30	31	32
L-Uni POG	empty	empty	L-Uni POG	empty	empty	L-Uni POG	empty
empty	empty	L-Uni POG	empty	empty	L-Uni POG	empty	empty
empty	L-Uni POG	Empty	empty	L-Uni POG	empty	empty	L-Uni POG
L-Uni POG	empty	Empty	L-Uni POG	empty	empty	L-Uni POG	empty
1	2	3	4	5	6	7	8

Fig. 2 Spotting pattern on the 32 electrodes of the MEA. Grey sites (L-Uni-POG) designate electrodes modified with the capture probe by needle spotting and subsequently incubated in a solution containing 11-mercapto-1-undecanol for filling the gaps between the DNA strands; white sites (empty) correspond to electrodes modified only with 11-mercapto-1-undecanol.

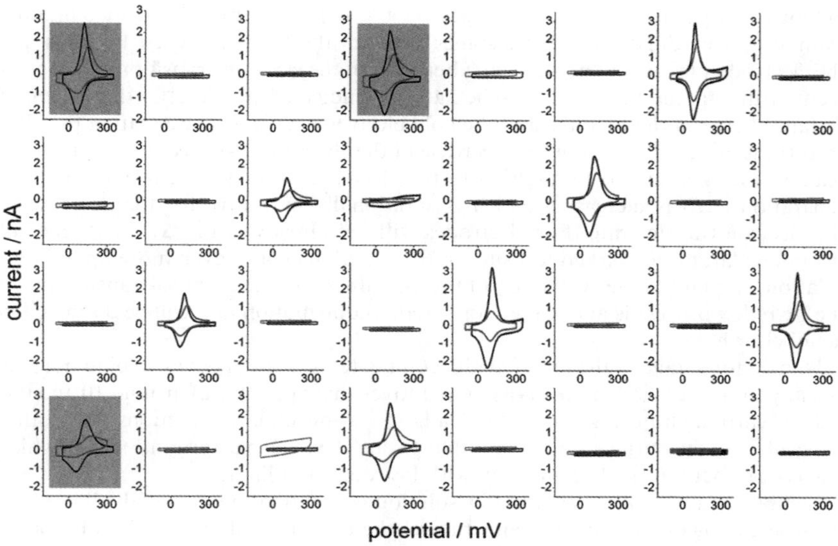

Fig. 3 Cyclic voltammograms at the MEA after hybridization with the Fc-labelled complementary target DNA (black lines) and after intercalation with the dibiotinylated proflavin derivative (grey lines). The capture probes were spotted following the layout shown in Fig. 2. The complementary target DNA was modified with four Fc moieties and hence the redox process of the DNA-bound Fc units is displayed. Scan rate: 500 mV s^{-1}. The potentials are *vs.* the internal $[Os(bipy)_2Cl_2]^{+/0}$ Nernstian reference. The supporting electrolyte was 10 mM phosphate buffer containing 1 M Na_2SO_4, pH 7.0.

Fig. 3 depicts the simultaneously recorded cyclic voltammograms of a MEA after hybridization and intercalation, respectively. The solid black curves correspond to the situation after hybridization of the capture probe with the complementary Fc-modified signalling oligonucleotides whereas the grey curves represent the response of the electrodes after intercalation with the dibiotinylated proflavine derivative. The cyclic voltammograms were recorded using a 32-fold parallel potentiostat. Since the MEAs are equipped with a gold pseudo-reference electrode a soluble Nernstian reference system, 50 nM $[Os(bipy)_2Cl_2]^{+/0}$, with an appropriate formal potential was used as an internal standard. The reversible redox process of the $[Os(bipy)_2Cl_2]^{+/0}$ couple occurs at a formal potential $E_0 \approx 0$ *vs.* Ag/AgCl.[18]

Two well-defined redox peaks of the DNA-bound Fc units are clearly visible in the cyclic voltammogram with current maxima at potentials of about 100 mV for the oxidation and 80 mV for the reduction process, respectively. The surface coverage was determined to be 7.9×10^{-11}–1.23×10^{-10} mol cm^{-2} which equals 4.74–7.4 $\times$ 10^{13} molecules cm^{-2}. The observed differences in the peak current values of the individual electrodes result from dissimilar loadings with the capture probe during the spotting process and hence differences in the amount of the Fc-modified target strands bound at the different electrodes of the MEA. The use of the Fc-modified target DNA strands allows an evaluation of the reproducibility and the accuracy of the spotting process by determining the charge transferred during the redox conversion of the DNA-bound Fc units. From the cyclic voltammograms in Fig. 3 it becomes evident, that occasionally during the spotting process a fraction of the capture probe solution is accidentally transferred onto test sites designated as *empty* (see *e.g.* electrode #20).

After intercalation of the dibiotinylated proflavine derivative into the formed Fc-modified dsDNA, a decrease of the Fc current signal is observed (see the differences between the black and grey cyclic voltammograms in Fig. 3). Intercalation is obviously imposing structural changes on the dsDNA which are concomitantly leading to a modulation of the distance between the DNA-attached Fc units and the electrode surface and hence to an alteration of the electron-transfer rate in agreement with Marcus theory.[21] As schematically indicated in Fig. 1C, the increased distance between the Fc units and the gold electrode surface is attenuating the electron-transfer rate[22,23] leading to a decrease of the charge transferred during the time-scale of the experiment. In addition to this decrease in the peak currents the peak separation (ΔE) is increasing which is additionally indicative for the decrease of the electron-transfer rate. For electrodes with an already larger ΔE in absence of the intercalator (see marked test sites in Fig. 3) ΔE increases even more upon intercalation. A quantitative evaluation of the modulation of the electron-transfer rate of the Fc redox process is above the scope of this contribution and will be discussed in detail elsewhere.

One disadvantage of the use of avidin conjugates is their tendency to bind nonspecifically to surfaces.[24] Avidin has a comparatively high pI value of around 10 and it is rich in carbohydrate residues. This leads to strong ionic interactions with other molecules at the surface and thus often to an elevated background signal which makes it difficult to realize an assay with low detection limit.

Incubation of MEAs with a buffer solution containing the avidin/alkaline phosphatase conjugate at ambient temperature (the commercially available solution of avidin/alkaline phosphatase was diluted 1 : 1000 in carbonate buffer, pH 9.8) for 10 min results in a pronounced increase of the anodic current upon addition of *p*-APP at all test sites irrespective to their previous modification with dsDNA (data not shown). Obviously, under these conditions the avidin/alkaline phosphatase conjugate does not exclusively bind to those sites which are modified with the intercalated dsDNA but rather adsorbs nonspecifically at the exposed surfaces. In order to minimize the nonspecific adsorption of avidin, the MEAs were incubated with a 1 mg mL^{-1} milk powder solution in carbonate buffer, pH 9.8, for 10 min and subsequently thoroughly rinsed with 500 mM carbonate buffer. After this, incubation with the diluted avidin/alkaline phosphatase conjugate was performed followed by an intensive rinsing with buffer solution. Constant-potential amperometry was carried out in 100 mM carbonate buffer containing 250 mM Na_2SO_4, pH 9.0, at 300 mV *vs.* the internal $[Os(bipy)_2Cl_2]^{+/0}$ reference. After a stable background current was attained, the alkaline phosphatase substrate, 100 mM *p*-APP, was added. Fig. 4 shows the simultaneously obtained chronoamperograms of the 32 electrodes of the MEA. As expected, the electrodes which were modified with the dsDNA and hence with the biotin groups after intercalation of the dibiotinylated proflavine derivative showed an increasing anodic current upon addition of *p*-APP. The test sites which were only modified with mercaptoundecanol did not show the increase in current. These results reveal the effectiveness of the milk powder in blocking the array surface against nonspecific adsorption.

The amperometric response of the individual electrodes of the MEA depends on the initial amount of DNA at the electrode and on the degree of intercalation. This information can be easily extracted from the Fc redox signal as described above. For example, electrodes #1 and #10 show a similar Fc signal with a peak current of about 1.75 nA. However, the degree of intercalation is significantly different. The suppression of the current upon intercalation is more noticeable for electrode #1 (Fig. 5).

Since the suppression of the Fc redox reaction is caused by the distortion of the DNA structure upon intercalation, it is assumed that a stronger modulation of the Fc redox reaction is due to a more efficient intercalation process. Consequently, test sites with a more efficient intercalation are exposing more biotin units which may lead to a more efficient complementary binding of the avidin/alkaline phosphatase conjugate. Hence, electrode #1 shows also a higher amperometric response upon addition of *p*-APP. The highest anodic current was detected for electrode #13. This electrode carries the highest amount of DNA as calculated from the

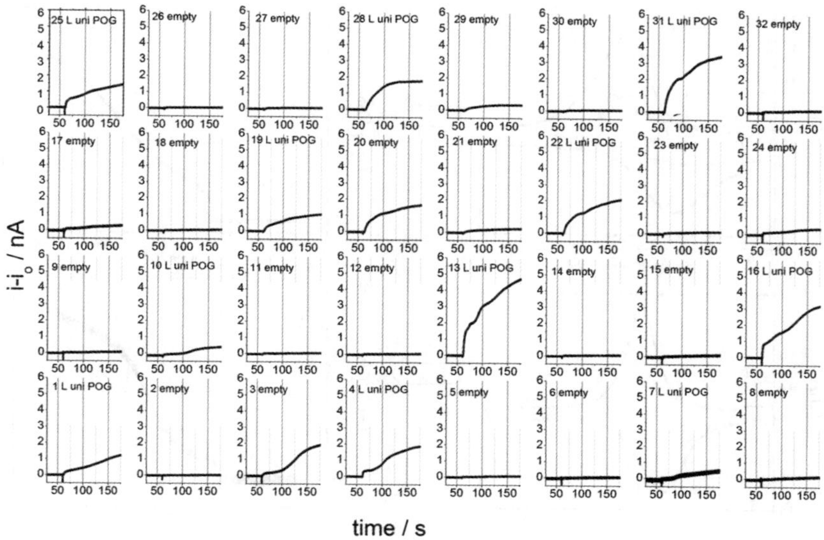

Fig. 4 Constant-potential chronoamperograms after complementary binding of the avidin/alkaline phosphatase conjugate to the Fc-modified dsDNA *via* the biotinylated intercalator. The measurements were recorded in 100 mM carbonate buffer containing 250 mM Na_2SO_4, pH 9.0 at a potential of +300 mV *vs.* the internal $[Os(bipy)_2Cl_2]^{+/0}$ Nernstian reference. 100 mM *p*-APP was added after 60 s after a stable background current was obtained.

charge of the Fc redox reaction. Moreover, the peak current decreased by 73% upon intercalation.

To evaluate the sensitivity of the proposed assay concept, a number of MEAs was prepared with the same spotting layout and the same amount of immobilized capture probes. The MEAs were hybridized with decreasing concentrations of the complementary target DNA starting from 100 nM, 1 nM, 100 pM, to 10 pM. After incubation with the biotinylated intercalator, blocking with milk powder solution and complementary binding of the avidin/alkaline phosphatase conjugate they were investigated by parallel chronoamperometry after addition of *p*-APP. The post-labelling of the dsDNA modified test sites with the dibiotinylated proflavine derivative allows the detection of DNA hybridization at concentrations at which the Fc signal can barely be distinguished from the capacitive current. For example, the CVs and the chronoamperograms of the electrodes #4 and #10 of a MEA hybridized in 100 pM of the target DNA after addition of *p*-APP are shown in Fig. 6. The Fc redox waves are barely noticeable. After addition of *p*-APP—in contrast to the expected steady-state current—a peak-type current response is detected indicative for the hydrolysis of *p*-APP and the oxidation of enzymatically produced *p*-AP.

Surprisingly, an increase in the *p*-AP oxidation current with decreasing amount of dsDNA on the surface of the individual test sites is observed concomitantly with a transition from the expected increase of the current to a steady-state value to a peak-type current response. The average peak currents are increasing from 2.1 ± 1.4 nA for hybridization with 100 nM target DNA to 10 ± 5 nA for 1 nM target DNA to 12 ± 5 nA for 100 pM target DNA to 19.9 ± 6 nA for a MEA hybridized with 10 pM target DNA.

Assuming simple Michaelis–Menten kinetics[25,26] in homogeneous solution a higher signal is expected for an increasing enzyme concentration if the substrate concentration is not limiting for the enzyme-catalyzed reaction. Thus, a current increase reaching finally a steady-state value would be expected. Obviously, the situation at the

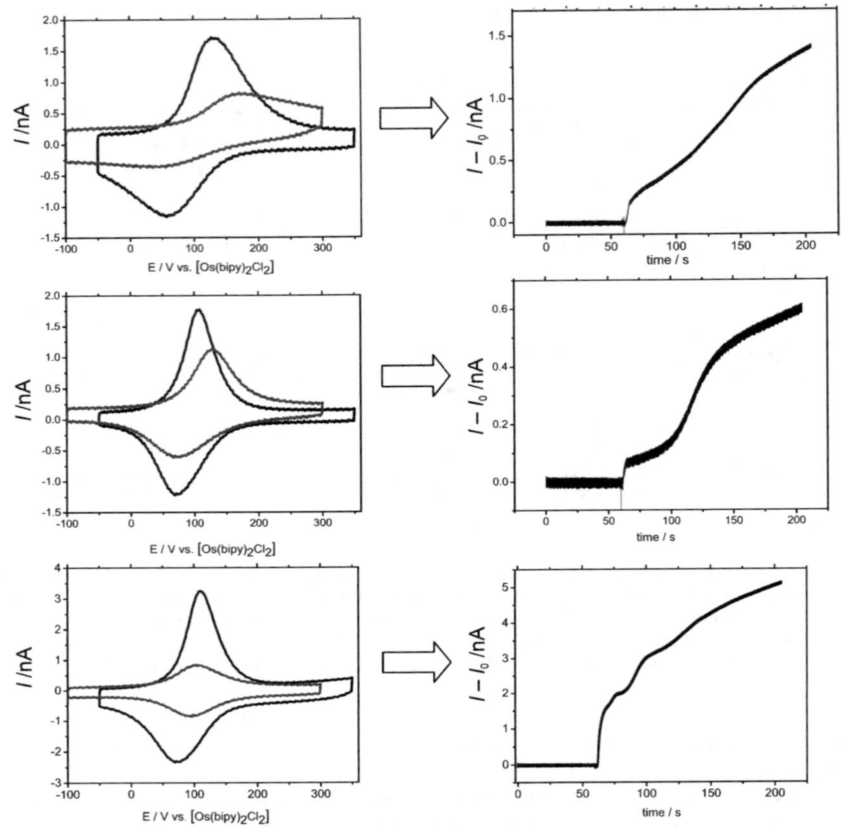

Fig. 5 Left: cyclic voltammograms of the Fc redox reaction at the dsDNA-modified test sites #1 (on top), #10 (middle) and #13 (bottom) before (black lines) and after intercalation with the dibiotinylated proflavine derivative (grey lines). Right: Corresponding chronoamperograms from the same test sites after complementary binding of the avidin/alkaline phosphatase conjugate and upon addition of 100 mM *p*-APP at $t = 60$ s.

electrode/electrolyte interface significantly deviates from the homogeneous case (see Fig. 7). With increasing concentration of the target DNA the number of the formed dsDNA duplexes and hence the number of intercalator molecules bound at the test site increase as well. The locally higher concentration of biotin groups allows binding of a higher number of avidin/alkaline phosphatase conjugates at the interface. Additionally, since avidin has four biotin binding sites, multiple and hence much stronger binding of an avidin molecule may occur at sites with a densely packed dsDNA layer. Taking into account the sequential building up of the interface architecture and the molecular weight of avidin of about 60 000 D, the comparatively smaller alkaline phosphatase will be located on top of an avidin layer being exposed towards the solution. After addition of the substrate *p*-APP the enzyme generated *p*-AP has hence a higher probability to escape towards the bulk of the solution than to diffuse through the densely packed DNA/protein layer towards the underlying electrode surface. Consequently, at high dsDNA concentrations the current signal caused by oxidation of *p*-AP is slowly increasing reflecting the restricted access to the electrode surface. However, the concentration of enzymatically generated *p*-AP is continuously increasing at the site of the enzyme leading to an increasingly steeper concentration gradient of *p*-AP towards the electrode. As a consequence, the current signal is slowly increasing for a surface with high concentrations of dsDNA. In

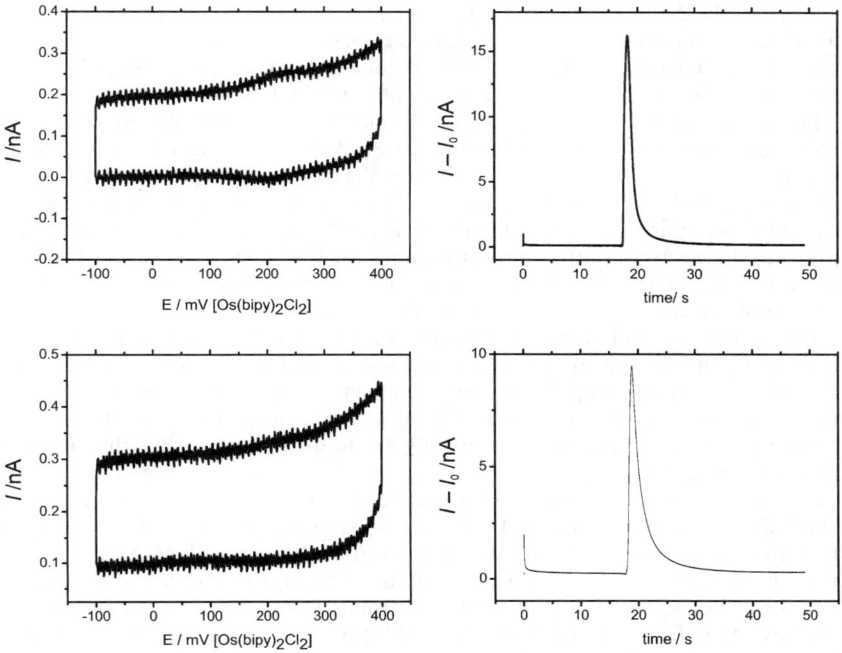

Fig. 6 Left: cyclic voltammograms of test sites #4 (top) and #10 (bottom) from a MEA which was hybridized with 100 pM of the target DNA. Right: Chronoamperograms at +300 mV for the same electrodes after intercalation and conjugation of avidin/alkaline phosphatase upon addition of *p*-APP at $t = 17$ s.

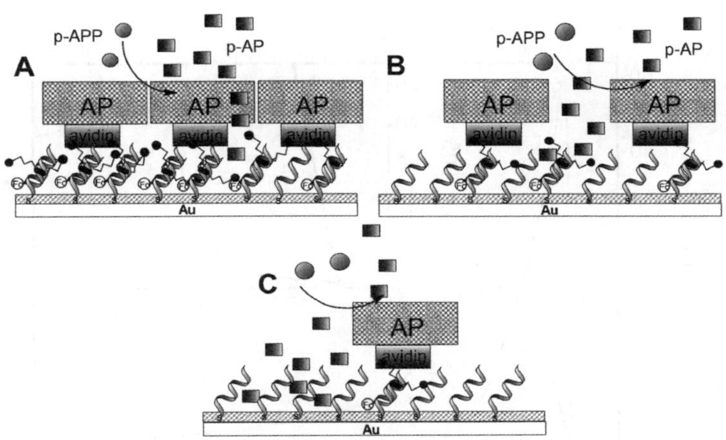

Fig. 7 Schematica representation of the interface architecture in dependence on the concentration of the target DNA during hybridization. (A) high target DNA concentration (> 100 pM). The protein layer is blocking the access of the *p*-AP to the electrode leading to a slow current increase. (B) smaller target DNA concentration (< 100 pM). The electrode becomes more easily accessible for *p*-AP. C) small Target DNA concentration (10 pM). The enzymatically generated *p*-AP as well as the protons generated during the oxidation of *p*-AP to *p*-QI can freely diffuse through the DNA/protein layer.

contrast, if the dsDNA concentration is small the number of intercalator molecules and hence the number of biotin binding sites is small as well. The DNA/protein layer is less dense allowing a fast diffusional access of enzymatically generated *p*-AP to the electrode surface and hence a fast increase of the oxidation current.

It is known that alkaline phosphatase is inhibited by phosphate ions which are liberated during the hydrolysis of the substrate *p*-APP[27–29] and the inhibitor constant K_i is about 1 μM. Depending on the source of the enzyme, phosphate may inhibit to 74% isoenzyme I, 15% isoenzyme II, and 2% isoenzyme III and in most cases it is a competitive inhibition.[30] Additionally, when *p*-AP is oxidized to *p*-quinoneimine (*p*-QI) at the electrode surface protons are released leading to a local acidification of the solution. The optimal pH value for alkaline phosphatase is 8–10.[27,31] Thus, the rate of the enzymatic reaction is on the one hand decreased by the generation of the competitive inhibitor phosphate and on the other hand by the local change of the pH value caused by the electrochemical oxidation of *p*-AP. Assuming that the high concentration of *p*-AP efficiently competes with the inhibitor for the active sites inside the enzyme molecules, the observed switching off of the enzyme-catalyzed reaction is supposed to be mainly due to the local change of the pH-value during *p*-AP oxidation.

Similar arguments explain the observed results shown in Fig. 8. In the case of a high dsDNA concentration and hence a comparatively dense DNA/protein layer the oxidation rate of *p*-AP is slow due to the restricted access of the *p*-AP to the electrode surface. The rate of proton formation is consequently rather slow and the impact on the local acidification of the electrolyte solution can be partly compensated by the buffer solution. Thus, the "switching off" of the enzyme is slow and the current decays slowly in this case. For a smaller target DNA concentration the *p*-AP oxidation current is higher due to the facilitated access of *p*-AP to the electrode. Simultaneously the local pH change is faster. This causes a decrease in enzyme activity leading to a fast decay of the *p*-AP oxidation current. However, due to the fact that the current is decaying to zero additional until now unknown factors have to be considered. Most likely, the oxidation of *p*-AP leads to the formation of a polyphenol-type polymer at the locally decreased pH value which is causing a blocking of the electrode surface. By comparing the current responses from the two MEAs being hybridized with 100 pM (MEA100) and 10 pM (MEA10) of target DNA,

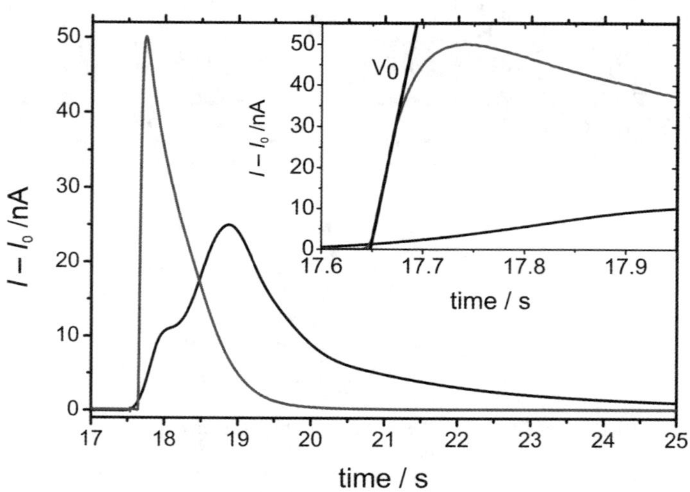

Fig. 8 A comparison of the amperometric responses for MEAs hybridized at concentration of 100 pM (black line; MEA100) and 10 pM (grey line; MEA10) upon additon of *p*-APP at t = 17.5 s.

respectively (Fig. 8), this hypothesis is supported. The black curve represents the result from MEA100 and the grey curve that from MEA10. Assuming a reaction in homogeneous solution, one can determine the initial velocity ($V_0 = k_{el}P$) which ascribes the catalytic rate by linear fitting of the *i vs. t* curves at the beginning of the reaction where k_{el} is the electrochemical rate constant and *P* is the product of the enzymatic reaction. The initial velocity was determined to be 0.16 nA s^{-1} for the MEA hybridized with 100 nM of target DNA, 28.8 nA s^{-1} for MEA100 and 1210 nA s^{-1} for MEA10. The current at MEA10 does not only rise much faster but also decays faster most likely due to the local change of the pH-value invoked by the oxidation of
p-AP.

Obviously, the sensor response depends on the interplay between the enzyme kinetics, the mass transport of *p*-AP to and protons from the electrode surface. The molecular architecture as schematically shown in Fig. 7 is responsible for the observed high sensitivity of the proposed assay concept. The dilution of the sensing monolayer by having a smaller amount of hybridized probes improves the communication between the electrode surface and the enzymatically generated *p*-AP.

Conclusions

A dibiotinylated proflavine derivative was used for selective post-labelling of dsDNA at individual electrodes of a 32-electrode MEA. Thiolated capture-DNA strands complementary to DNA from multi-resistant *Staphylococcus aureus* (MRSA) were spotted on the individual electrodes. The synthesized complementary target oligonucleotide strands were modified with four Fc moieties, which provides the basis for direct information about surface coverage with dsDNA molecules and the degree of intercalation. It was observed that the structural changes imposed on the dsDNA upon intercalation lead to an increase in the average distance of the Fc moieties to the electrode surface and hence to a decrease in the electron-transfer rate. After blocking the surface against nonspecific adsorption the selective binding of an avidin/alkaline phosphatase conjugate at the intercalator-modified dsDNA provides the basis for the high-sensitive detection of DNA hybridization. The interface architecture and the specific properties of the alkaline phosphatase lead to a peak type current response at small target DNA concentrations which are suggested to be due to a blocking of the access of enzymatically generated *p*-AP through the DNA/protein layer at high dsDNA loading and the switching off of the enzyme activity due to local pH changes imposed by the oxidation of *p*-AP at the electrode surface. Most likely, the current decay can be due to a blocking of the electrode surface by polymers derived from p-AP oxidation products.

The proposed assay architecture is the basis for future work aiming on the development of a primarily label-free, high-sensitive, selective and fast assay principle for the detection of DNA hybridization which may be performed in a point-of-care diagnostic situation. Future work will be aimed at the development of fast assays for the determination of the type of bacterial species causing urinary tract infection. From the obtained results it becomes clearly evident that an in-depth understanding and the careful design of the interface architecture provide the basis for the successful development of DNA assays for potential applications.

Acknowledgements

The authors are grateful to the EU for financial support in the framework of the project "INTCHEM" (MEST-CT-2005-020681). M. G. thanks the Ruhr-University Research School (DFG GSC 98/1) and the research department interfacial systems chemistry (RD-IFSC) at the Ruhr-University Bochum for support.

References

1 D. Shalon, S. J. Smith and P. O. Brown, *Genome Res.*, 1996, **6**, 639–645.
2 P. O. Brown and D. Botstein, *Nat. Genet.*, 1999, **21**, 33–37.
3 J. DeRisi, L. Penland, P. O. Brown, M. L. Bittner, P. S. Meltzer, M. Ray, Y. D. Chen, Y. A. Su and J. M. Trent, *Nature Genetics*, 1996, **4**, 457–460.
4 M. Schena, D. Shalon, R. W. Davis and P. O. Brown, *Science*, 1995, **270**, 467–470.
5 M. Schena, D. Shalon, R. Heller, A. Chai, P. O. Brown and R. W. Davis, *Proc. Natl. Acad. Sci. U. S. A.*, 1996, **93**, 10614–10619.
6 T. R. Hughes, M. Mao, A. R. Jones, J. Burchard, M. J. Marton, K. W. Shannon, S. M. Lefkowitz, M. Ziman, J. M. Schelter, M. R. Meyer, S. Kobayashi, C. Davis, H. Dai, Y. D. He, S. B. Stephaniants, G. Cavet, W. L. Walker, A. West, E. Coffey, D. D. Shoemaker, R. Stoughton, A. P. Blanchard, S. H. Friend and P. S. Linsley, *Nat. Biotechnol.*, 2001, **19**, 342–347.
7 A. C. Pease, D. Solas, E. J. Sulliivan, M. T. Cronin, C. P. Holmes and S. P. Fodor, *Proc. Natl. Acad. Sci. U. S. A.*, 1994, **91**, 5022–5026.
8 M. J. Lodes, D. Suciu, M. Elliott, A. G. Stover, M. Ross, M. Caraballo, K. Dix, J. Crye, R. J. Webby, W. J. Lyon, D. L. Danley and A. McShea, *J. Clin. Microbiol.*, 2006, **44**, 1209–1218.
9 J. Wang, *Nucleic Acids Res.*, 2000, **28**, 3011–3016.
10 W. Gette and T. Krein, *Am. Lab*, 1997, **6**, 15.
11 R. J. Lipshutz, D. Morris, M. Chee, E. Hubbell, M. J. Kozal, N. Shah, N. Shen, R. Yang and S. P. Fodor, *Biotechniques*, 1995, **3**, 442–447.
12 Affymetrix http://www.affymetrix.com/estore/ (last visited 12.03.2010).
13 Roche, Nimblegen: http://www.nimblegen.com/ (last visited 12.03.2010).
14 K. Dill, A. Ghindilis and K. Schwarzkopf, *Lab Chip*, 2006, **6**, 1052–1055.
15 S. Roy, X. J. Chen, M. H. Li, Y. F. Peng, F. Anariba and Z. Q. Gao, *J. Am. Chem. Soc.*, 2009, **131**, 12211–12217.
16 L. C. Waters, S. Jacobson, N. Kroutchinina, J. Khandurina, R. Foote and J. M. Ramsey, *Anal. Chem.*, 1998, **70**, 158–162.
17 M. Gębala, L. Stoica, D. Guschin, L. Stratmann and W. Schuhmann, *Electrochem. Commun.*, 2010, **12**, 684–688.
18 P. Liepold, T. Kratzmueller, N. Persike, M. Bandilla, M. Hinz, H. Wieder, H. Hillebrandt, E. Ferrer and G. Hartwich, *Anal. Bioanal. Chem.*, 2008, **391**, 1759–1772.
19 J. C. Hoogvliet, M. Dijksma, B. Kamp and W. P. van Bennekom, *Anal. Chem.*, 2000, **72**, 2016–2021.
20 T. M. Herne and M. J. Tarlov, *J. Am. Chem. Soc.*, 1997, **119**, 8916–8920.
21 A. J. Bard and L. R. Faulkner, *Electrochemical methods. Fundamentals and applications*, 2001, 2nd edn, Wiley, New York.
22 C. Li, B. Ren, Y. Zhang, Z. Cheng, X. Liu and Z. Tong, *Langmuir*, 2008, **24**, 12911–12918.
23 A. M. Napper, H. Liu and D. H. Waldeck, *J. Phys. Chem. B*, 2001, **105**, 7699–7707.
24 G. T. Hermanson, *Bioconjugate techniques*, 2007, Academic Press, San Diego.
25 J. M. Berg, J. L. Tymoczko, L. Stryer and N. D. Clarke, *Biochemistry*, 2003, 5th edn, W. H. Freeman, New York.
26 B. Alberts, *Molecular biology of the cell*, 2002, 4th edn, Garland Science, New York.
27 W. W. Cleland and A. C. Hengge, *Chem. Rev.*, 2006, **106**, 3252–3278.
28 R. Q. Thompson, M. Porter and C. Stuver, *Anal. Chim. Acta*, 1993, **271**, 223–229.
29 R. Q. Thompson, G. C. Barone, H. B. Halsall and W. R. Heineman, *Anal. Bioanal. Chem.*, 1991, **1**, 90–95.
30 Brenda: http://www.brenda-enzymes.org/ (last visited 19.01.2010).
31 R. K. Morton, *Biochem. J.*, 1957, **65**, 674–682.

Development of reagents and assays for the detection of pathogenic *Burkholderia* species

Omar Qazi,[a] Mridula Rani,[a] Annie J. Gnanam,[a] Thomas W. Cullen,[b] Christopher M. Stead,[b] Haley Kensing,[a] Kate McCaul,[a] Sarah Ngugi,[c] Joann L. Prior,[c] Alexandria Lipka,[d] Judit M. Nagy,[e] Gregory C. Whitlock,[f] Barbara M. Judy,[g] Sarah V. Harding,[c] Richard W. Titball,[h] Sachdev S. Sidhu,[i] M. Stephen Trent,[b] G. Barrie Kitto,[aj] Alfredo Torres,[gk] D. Mark Estes,[gk] Brent Iverson,[aj] George Georgiou[al] and Katherine A. Brown*[adj]

Received 3rd June 2010, Accepted 19th July 2010
DOI: 10.1039/c005422b

Rapid detection of the category B biothreat agents *Burkholderia pseudomallei* and *Burkholderia mallei* in acute infections is critical to ensure that appropriate treatment is administered quickly to reduce an otherwise high probability of mortality (*ca.* 40% for *B. pseudomallei*). We are developing assays that can be used in clinical laboratories or security applications for the direct detection of surface-localized and secreted macromolecules produced by these organisms. We present our current medium-throughout approach for target selection and production of *Burkholderia* macromolecules and describe the generation of a Fab molecule targeted to the *B. mallei* BimA protein. We also present development of prototype assays for detecting *Burkholderia* species using anti-lipopolysaccharide antibodies.

[a]*Institute for Cellular and Molecular Biology, The University of Texas at Austin, 1 University Station A4800, Austin, Texas, 78712, USA. E-mail: kate01@mail.utexas.edu; Tel: +1 (512) 471 9483*
[b]*Section of Molecular Genetics and Microbiology, The University of Texas at Austin, Austin, Texas, 78712, USA*
[c]*Defence Science and Technology Laboratory, Porton Down, Salisbury, Wiltshire, SP4 0JQ, UK*
[d]*Department of Life Sciences, Imperial College London, Exhibition Road, London, SW7 2AZ, UK*
[e]*Institute for Biomedical Engineering and the Department of Chemistry, Imperial College London, Exhibition Road, London, SW7 2AZ, UK; Deceased*
[f]*Department of Clinical Laboratory Sciences, University of Texas Medical Branch, Galveston, Texas, 77555, USA*
[g]*Department of Pathology, University of Texas Medical Branch, Galveston, Texas, 77555, USA*
[h]*School of Biosciences, University of Exeter, Geoffrey Pope Building, EX4 4QD, UK*
[i]*Terence Donnelly Center for Cellular and Biomolecular Research, Banting and Best Department of Biomedical Research, University of Toronto, Toronto, Ontario, M5S 3E1, Canada*
[j]*Department of Chemistry and Biochemistry, University of Texas at Austin, Austin, Texas, 78712, USA*
[k]*Department of Microbiology and Immunology and the Sealy Center for Vaccine Development, University of Texas Medical Branch, Galveston, Texas, 77555, USA*
[l]*Department of Chemical Engineering and Biomedical Engineering, The University of Texas at Austin, Austin, Texas, 78712, USA*

1. Introduction

Development of diagnostic assays for the rapid identification of infectious agents has a number of challenges that are often dependent upon the nature of the agent being detected and upon the progression of the disease. In addition, access to equipment, materials, and trained personnel can differ widely between a developed and a developing country and these criteria will have an enormous influence upon the nature of an assay and the platform that can be used. In the case of acute infections, it is widely recognized that better patient outcomes are directly associated with shorter times required for pathogen identification no matter what the setting. In many clinical laboratories, and especially those with limited resources, slow multi-step culture-based assays are often the only reliable methods available.[1] Slow or inadequate diagnosis can prove costly in acute-care settings (*e.g.*, emergency rooms and intensive care units) in developed areas,[1] whereas in areas with limited resources, patient outcomes are likely to be more severely affected. Furthermore, in situations of a pandemic or biothreat exposure, rapid and accurate point-of-care testing is an essential element in the appropriate management of both patient care and the containment of disease spread.[2]

The highly pathogenic Gram-negative *Burkholderia* bacterial species are good examples of pathogens for which rapid and accurate identification is essential in improving the chances of surviving an acute infection. *Burkholderia pseudomallei* and *Burkholderia mallei* are category B biothreat agents and the causative agents of the diseases melioidosis and glanders, respectively.[3,4] Whilst melioidosis infections occur in a wide range of mammals (including humans) and birds,[5] glanders predominantly affects horses and other solipeds,[4,6] though rare human infections do occur.[7] These organisms can be cultured easily, and are endemic in wide regions of the world, though the majority of reported cases of human melioidosis infections are in Southeast Asia and Northern Australia.[8] Pustular skin lesions and ulcers can be observed at the point of infection; or fever and coughing if the organisms are inhaled. Pulmonary infections progress rapidly. Septicemic infections result in multi-organ abscesses that, in the case of certain strains of *B. pseudomallei*, include the central nervous system. In acute infections bacterial sepsis can manifest within only a few days. Treatment with the correct antibiotics is critical as these organisms are naturally highly drug resistant.[4,5] Death usually follows within a few days if intensive and prolonged antimicrobial treatment is not rapidly applied. However, patient survival levels are estimated at approximately 40–50% even with the administration of the correct antibiotics.[3,4] A further complication, more widely discussed for melioidosis, is that infections have variable presentations and often mimic other infectious diseases.[5] This makes diagnosis based on clinical symptoms difficult.

In clinical settings, melioidosis and glanders are most commonly diagnosed using serological assays in a range of formats in conjunction with culturing of clinical samples. For melioidosis, positive culture of *B. pseudomallei* is considered the "gold standard" diagnostic but results can often take days and delayed correct antibiotic treatment can be inadequate in preventing death if bacterial sepsis has developed.[9] Furthermore serodiagnosis, which commonly uses the indirect hemagglutination assay (IHA), based upon visual evidence of agglutination of bacterial cells,[10] although rapid has proved to be of only limited utility (around 50% accurate) for diagnosing acute melioidosis infections or infections in endemic regions.[9,11] For glanders, veterinary diagnosis also relies upon serodiagnostic methods, the most commonly used being the mallein test that assesses allergic hypersensitivity.[12] Although a number of others also exist, including a complement fixation test (CFT) which has demonstrated improved sensitivity in situations of low prevalence of the disease.[6,13] Even so, current serological methods to diagnose glanders suffer from a considerable number of false negative and false positive results, linked to the crude preparation of the *B. mallei* antigens used in these tests.[6]

There is a key clinical need to develop improved reagents that can be used for the detection of these highly pathogenic *Burkholderia* organisms in clinical samples. Whilst a number of molecular based assays using, for example, polymerase chain reactions (PCR) to amplify specific targets have been reported (see Suppiah *et al.*[14] and references therein), direct detection of antigens or whole organisms may also prove advantageous in terms of both cost and sensitivity in acute infections, particularly in developing countries.[9] Therefore we have adopted a strategy which involves the development of high affinity reagents, aptamers and antibodies, to detect surface-associated or secreted macromolecules from *Burkholderia* species.[15,16] Such reagents would have the flexibility to be used in both rapid readout assays to address clinical needs in endemic areas (*e.g.*, rural Southeast Asia), as well as in more sophisticated high throughput multiplexed platforms for security-based applications. In this paper we describe our current progress in diagnostic target selection and production, including the generation of an engineered Fab directed against a *B. mallei* protein, BimA. We also present data obtained from microscopy, ELISA and Luminex assays using anti-lipopolysaccharide (LPS) antibodies that demonstrate the potential for detecting *Burkholderia* antigens and whole cells.

2. Experimental

2.1 Recombinant protein expression and purification

Construct design was based upon bioinformatics analyses of target sequences to indicate the presence (or absence) of an N-terminal secretion sequence, transmembrane domains and homology to published crystal structures as previously described.[17] DNA sequences encoding BPSS0213 (residues 1–196; 24 kDa), BPSL1763 (residues 34–457; 47 kDa), BPSS0493 (residues 40–365; 39 kDa), and BPSS1992 (residues 28–644; 68 kDa) were amplified as PCR fragments using genomic *B. pseudomallei* K96243 DNA as a template. All fragments contained an in-frame N-terminal 6× histidine tag and were cloned into the pET15b expression vector (Novagen). BPSS1492 (*B. pseudomallei* BimA; residues 15–234; 37 kDa) was cloned into the pCRT7/NT-TOPO vector (Invitrogen) in frame with a N-terminal 6× histidine tag. DNA sequences for BimA (BMAA0749; residues 19–265; 26 kDa) and GroEL (BMA2001; residues 1–550; 59 kDa) were amplified as PCR fragments using genomic *B. mallei* ATCC 23344 DNA as a template. These fragments contained an in-frame C-terminal 6× histidine tag and were cloned into the pET28a (+) expression vector (Novagen).

All proteins were expressed from *Escherichia coli* hosts and purified following established procedures.[15,17] Protein concentrations were determined using the BCA kit (Pierce) using bovine serum albumin as a standard, and sample purity assessed by SDS-PAGE.

2.2 Fab isolation using phage display methodologies

Fab phage library. A synthetic Fab phage display library expressing a bivalent Fab through a disulfide linkage, was obtained from Dr S. Sidhu (University of Toronto, Canada). The library was constructed using a single highly stable Fab framework with diversity introduced only to the Complementary Determining Regions (CDRs).[18] The theoretical diversity of the library is 1×10^{29}, but actual diversity is 3×10^{10}. The phage library was amplified using standard protocols with M13KO7 (NEB, Ipswich, MA) as helper phage. Phage titer was determined by serial dilution of infected *E. oli* (XL1 Blue) plated on 2x YT agar supplemented with 50 μg ml^{-1} of Carbeniciilin (Cb). The phage titer was calculated from the number of transformed colonies and found to be around 10^{14} cfu/ml. Aliquots of the amplified phage were stored at −80 °C.

Phage panning using immunotubes. Purified recombinant *B. mallei* BimA was used at 100 nM concentration in the first round of panning to coat a NUNC maxisorp immunotube in 50 mM sodium carbonate buffer pH 9.6 overnight. In the second and third rounds of panning the concentration of BimA was 50 nM and 25 nM, respectively. The immunotube was blocked with 2% milk/PBS for 2 h at room temperature. A depletion step was performed before each panning round by incubating the phage library (10^{12}–10^{13} cfu) in 2% milk/PBS in a BSA (10 μg ml^{-1}) coated immunotube. Following the depletion step, the phage library in 2% milk PBS was added to the immunotube coated with BimA and incubated for 2 h at room temp. The immunotube was washed five times with PBST (PBS+ 0.05% Tween 20) and five times with PBS to remove the non-specifically bound phage (10 washes each for 2nd round and 15 washes each for 3rd round). Bound phage were eluted by incubating with 1 ml of 0.1 M triethylamine (TEA) for 15 min while rotating and were neutralized with 0.5 ml of 1 M Tris-HCl, pH 7.5. Next 0.75 ml of the eluted phage was used to infect 10 ml of *E. coli* XL1 Blue cells in mid-log phase for 30 min at 37 °C without shaking. Phage titers were determined as before, and the infected cells were then transferred to 40 ml 2x YT media with 50 μg ml^{-1} Carbencillin (Cb) and were grown overnight at 30 °C with shaking. The next day, cells were harvested by centrifugation and resuspended in 5 ml 2x YT/50 μg ml^{-1} Cb/15% glycerol, and stored at −80 °C. For phage amplification, cells were grown to an OD_{600} of 0.3 and infected with M13KO7 and grown overnight. The following day cells were harvested by centrifugation at 10 000 rpm, and the phage from the media supernatant were rescued by precipitation with ¼ the volume of PEG/NaCl. Following incubation on ice for 1 h, the phage precipitate was collected by centrifugation and resuspended in PBS. The library was panned for three rounds before screening phage for their ability to bind to BimA by ELISA.

Monoclonal phage ELISA. Monoclonal phage ELISAs were performed with individual clones to identify BimA-specific binders. Single colonies from round 3 titer plates were inoculated in 96-well microplates (Costar-3790 Corning, Lowell, MA) containing 200 μl of 2x YT media with 50 μg ml^{-1} Cb and grown overnight at 37 °C. 30 μl of the overnight cultures were used to inoculate fresh 96-well plates containing 200 μl of 2x YT/50 μg ml^{-1} Cb/M13KO7 (10^{10} phage per ml), and grown overnight at 30 °C for phage propagation. 50 μl of the culture supernatant containing the phage were used for the ELISAs. 100 μl of 20 μg ml^{-1} of BimA in PBS buffer, and bovine serum albumin (BSA) at 10 μg ml^{-1} as control was used to coat 96-well ELISA plates (Corning) overnight at 4 °C. Plates were blocked with 2% milk/PBS for 2 h at room temperature, and washed once with PBS. 50 μl of the culture supernatant containing phage diluted with 50 μl of 2% milk PBS was added to each well. Each phage solution was tested for binding to both BimA and to the negative control BSA. After 2 h incubation at room temperature, wells were washed 3 times in PBST and once with PBS. 100 μl of *anti*-M13 HRP antibody (Pharmacia) diluted 1 : 1000 in 2% milk PBS was added to each well, and incubated for 1 h at room temperature. Following incubation, wells were washed 3 times with PBST and once with PBS. For detection, 100 μl of TMB substrate (Dako, Carpinteria, CA) were added to each well followed by incubation for 20–30 min until the color developed. The reaction was stopped by the addition of 100 μl of 4.5 M H_2SO_4, and absorbance values were read at 450 nm using a microtiter plate reader.

Fab purification and ELISA. For purification of Fab E8 selected from the monoclonal phage ELISA, the phagemid was converted into the Fab expression vector by introducing a stop codon between the heavy chain and gene lll *via* Kunkel mutagenesis. Fab E8 protein was secreted from *E. coli* XL1 Blue strain in low-phosphate medium at 30 °C for 18 h, as described.[19] Fab protein was purified with protein A affinity chromatography.

For ELISAs, 50 μl of 20 μg ml^{-1} of BimA in PBS buffer, and Lysozyme at 20 μg ml^{-1} as control were coated onto 96-well ELISA plates (Corning) overnight at 4 °C. After blocking the wells with 2% milk/PBS for 2 h, serial dilutions of Fab E8 were added at a starting concentration of 2 μM and incubated at room temperature for 2 h. Following washing as already described, 50 μl of an *anti*-human kappa HRP-conjugated antibody (Sigma) at a 1 : 5000 dilution were added. Following incubation for 1 h, wells were washed and detected by the addition of TMB substrate as described before.

2.3 LPS purification

Burkholderia thailandensis E264 LPS was purified using a modified hot phenol extraction method.[20] Briefly, bacterial cell pellets were lysed for 16 h at 4 °C using 15 000 U of lyzozyme per mg of bacteria. DNase I and RNase A were added at 20 μg ml^{-1} each and stirred at room temperature for a further 16 h, prior to digestion with 50 μg ml^{-1} Proteinase K for 6 h. An equal volume of 90% (v/v) aqueous phenol was added and the bacterial suspension heated to 70 °C for 30 min with vigorous stirring. The mixture was dialysed against dH_2O for 3 days, with frequent changes, prior to lyophilization. The dry material was subsequently digested with 50 μg ml^{-1} Proteinase K for 4 h at 45 °C and ultracentrifuged at 100 000 *g* for 3 h. Gel-like pellets were then resuspended in dH_2O and lyophilized. The dry LPS was weighed, resuspended in dH_2O, aliquoted and stored at −80 °C. The quality of the purified LPS was verified using SDS-PAGE and silver-staining (Bio-Rad Silver Stain Plus kit).

2.4 Live cell imaging

Anti-*B. mallei* LPS monoclonal antibodies (clone 3D11) were purchased from AbCam (10011) and anti-*B. pseudomallei* LPS monoclonal antibodies (clone CC6)[21] were provided by Dstl (Porton Down, UK). Binding of these Mabs to *B. thailandensis* LPS was confirmed by Western Blotting with a rabbit anti-mouse IgG (whole molecule) peroxidase-conjugated antibody and SuperSignal West Pico Chemiluminescent Substrate (ThermoPierce).

Cultures of *Burkholderia thailandensis* E264 (ATCC 700388) or *Pseudomonas aeruginosa* PA01_LAC (ATCC 47805) were grown at 37 °C, 250 rpm shaking in LB-broth. When an OD_{600} between 0.4–0.8 was obtained, cells were harvested by centrifugation, washed in PBS, resuspended in PBS/1% BSA and 0.3 OD units of bacteria added to sterile 1.5 ml microfuge tubes. The bacteria were pelleted by centrifugation, resuspended in 200 μl of PBS/1% BSA containing either 2.0 μg ml^{-1} anti-*B. mallei* LPS Mab or 4.0 μg ml^{-1} anti-*B. pseudomallei* LPS Mab, and then incubated for 45 min at 25 °C. Cells were washed by three rounds of centrifugation followed by resuspension in PBS/1% BSA. Bacterial pellets were then resuspended in 200 μl of anti-mouse IgG (whole molecule) FITC-conjugated antibodies (Sigma), diluted 1 in 100, and incubated for 30 min in the dark at 25 °C. Bacteria were washed and resuspended in 200 μl of PBS/1% BSA and 10 μl added to poly-L-lysine-coated microscope slides (Polysciences, Inc.) which were sealed with coverslips. Slides were viewed and fluorescence intensity assessed using a Nikon Eclipse 80i fluorescent microscope.

2.5 ELISA-detection of *Burkholderia* LPS and cells

Sample preparation. *B. thailandensis* E264 (ATCC 700388) and *P. aeruginosa* PA01_LAC (ATCC 47805) were grown in Luria Bertani (LB) broth and manipulated using standard microbiological protocols. Overnight cultures and exponential phase bacteria were harvested by centrifugation and washed in PBS. Cells were resuspended to 20 OD units per ml and either used as live cells, or were heat-killed at 80 °C for 2 h. Colony counts were also made indicating that 1 OD unit per ml of

B. thailandensis contained approximately 5 × 10^8 colony forming units (CFU)/ml. Heat-killed *B. pseudomallei* K96243 was prepared under biosafety level 3 containment at the University of Texas Medical Branch (Galveston, Texas) and supplied at a concentration of 1 × 10^{10} CFU/ml.

B. thailandensis E264 LPS was prepared as described above and phenol-extracted LPS from *P. aeruginosa* serotype 10 was purchased from Sigma (L9143).

Biotinylation of LPS and Mab. Anti-*B. mallei* LPS Mab was biotinylated using the EZ-link micro sulfo-NHS-LC-biotinylation kit (Thermo-Pierce), following the recommended protocol using 200 μg of Mab. Biotinylation of anti-*B. mallei* LPS Mab was confirmed by Western blotting in conjunction with the Streptavidin-peroxidase conjugate (Sigma) and SuperSignal West Pico Chemiluminescent Substrate.

Direct detection using *Burkholderia* Mabs. Maxisorp microplates were coated overnight with serially diluted or static concentrations of LPS or heat-killed bacteria in carbonate buffer at 4 °C. After blocking with 3% skimmed milk powder in PBS, washed plates were incubated for 1 h at 25 °C with 0.7 μg ml^{-1} anti-*B. mallei* LPS Mab, 1.6 μg ml^{-1} anti-*B. pseudomallei* LPS Mab or serial dilutions of either antibody in PBST. After washing in PBST, antibody binding was measured by incubation with a rabbit anti-mouse IgG (whole molecule) peroxidase-conjugated antibody (Sigma) diluted 1 in 2500, followed by development with 1-step Ultra TMB ELISA reagent (Pierce). After the addition of 2 M H_2SO_4, the plates were read at 450 nm.

Sandwich-capture ELISA. Maxisorp plates were coated overnight at 4 °C with 100 μl of between 2 and 10 μg ml^{-1} of anti-*B. mallei* LPS Mab in carbonate buffer. After washing and blocking with 3% skimmed milk powder in PBS, plates were incubated for 1 h at 25 °C with dilutions of purified LPS, heat-killed, or live *B. thailandensis* or *P. aeruginosa* (grown to OD_{600} = 0.4–0.8) bacteria. Biotinylated anti-*B. mallei* LPS Mab in PBST was added to washed plates at a concentration of 0.7 μg ml^{-1} and further incubated. The plates were washed in PBST and then incubated for 1 h at 25 °C with a Streptavidin-peroxidase polymer conjugate. Plates were washed prior to the addition of 1-step Ultra TMB ELISA reagent. Plates were read at 450 nm after the addition of 2M H_2SO_4 to stop the reaction. Reference-corrected data were analysed and plotted with GraFit v 7.0.0 (www.erithacus.co.uk).

2.6 Luminex assays

5 × 10^6 MicroPlex microspheres (region 17; LC10-017-01) were coupled with 75 μg of anti-*B. mallei* LPS Mab using the standard two-step carbodiimide coupling protocol from Luminex. Briefly, washed microspheres are activated by Sulfo-NHS and EDC prior to incubation with Mab in 50 mM MES, pH 5.0 for 1 h at 25 °C. Coupled microspheres were diluted to 40 000 beads per μl in 0.2% skimmed milk/PBS/0.8% polyvinylpyrrolidone (PVP) and 50 μl (2000 microsphere) distributed into the appropriate wells of a pre-wetted 1.2-μm filter plate (Millipore, MABVN1250). For the indirect detection assay, dilutions of LPS or heat-killed bacteria in 0.2% skimmed milk/PBS/0.8% PVP were added in 50 μl volumes and incubated for 1 h at 25 °C with 300 rpm shaking. Supernatants were aspirated by vacuum manifold and washed twice with 0.2% skimmed milk/PBS/0.8% PVP before adding 100 μl of biotinylated anti-B. mallei LPS Mab at a concentration of 4 ug ml^{-1}. After a 45 min incubation at 25 °C, 300 rpm, plates were washed by vacuum manifold and 100 ul Streptavidin R-phycoerythrin conjugate (SA-PE; Invitrogen, S866) at a final concentration of 4 μg μl^{-1}. Plates were incubated at 25 °C, 300 rpm for 30 min before aspiration by vacuum manifold and washing. Microspheres were resuspended in 100 μl of 0.2% skimmed milk/PBS/0.8% PVP and samples analyzed using a calibrated Luminex 200 IS System, with the doublet discriminator gate set to 7500 and 20 000. Reference-corrected data were analysed and plotted with GraFit v 7.0.0 (www.erithacus.co.uk).

3. Results and discussion

3.1 Target production

A key element of our current strategy to develop assays for the detection of pathogenic *Burkholderia* species requires the identification and purification of surface-associated or secreted proteins or other macromolecules produced by these organisms. These proteins and other macromolecules are used both for the development of molecular recognition reagents (antibody molecules or aptamers) directed against proteins and for validation and optimisation of diagnostic assays. Although the genome sequences for *B. pseudomallei*,[22] *B. mallei*,[23] and *Burkholderia thailandensis*[24] (a lowly virulent related organism) are available, selection of targets simply based upon bioinformatic prediction is not sufficient as macromolecule production varies depending upon the environment and growth stages of these organisms. To improve our choice of targets we used available experimental evidence of target production and bioinformatic analyses which suggest that protein targets are surface-localized or secreted,[16] and sequences from full-length open reading frames or domains which are likely to express high levels of soluble material that can be readily purified.

Examples of purified targets that form part of our current strategy are shown in Fig. 1. The two *B. mallei* proteins, full-length GroEL (Fig. 1A) and truncated BimA (lacking its C-terminal transmembrane domain) (Fig. 1B) were selected for a variety of reasons. The heat shock protein GroEL from *B. mallei* has been shown to be released in cell culture and has also been shown to be highly immunogenic in mice and human infections.[25] The homolog in *B. pseudomallei* which is 99% identical is also highly immunogenic[26,27] and has been observed in proteomic analyses of cell cultures.[28,29] In this case we anticipate a high level of cross-reactivity of recognition reagents to both proteins. The actin-binding protein BimA, in comparison, differs considerably in its size and composition in the amino terminal region of the protein that is exposed at the bacterial surface between *Burkholderia* species.[30] These

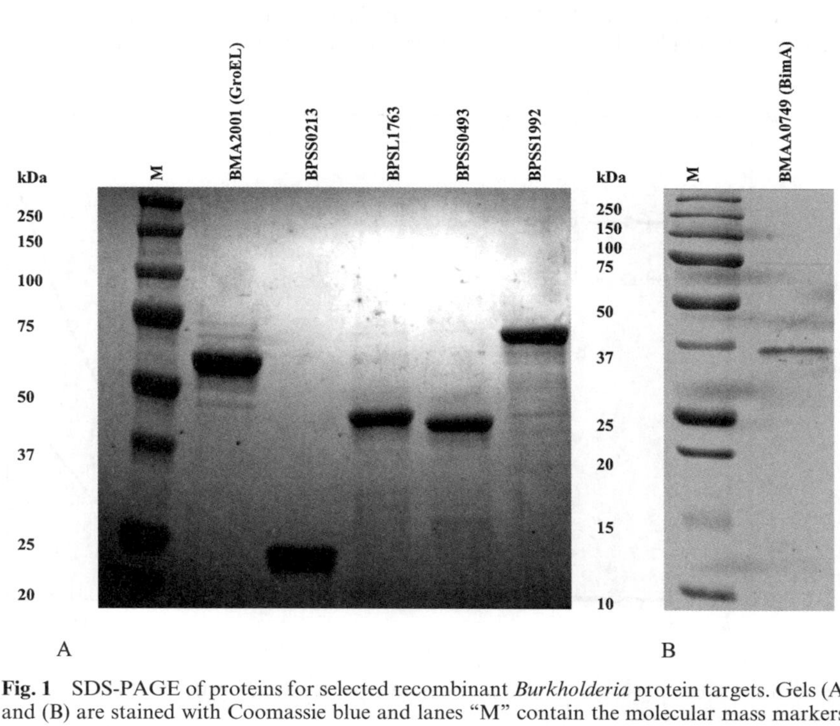

Fig. 1 SDS-PAGE of proteins for selected recombinant *Burkholderia* protein targets. Gels (A) and (B) are stained with Coomassie blue and lanes "M" contain the molecular mass markers.

sequence differences have been used to develop polymerase chain reaction (PCR) diagnostic primers to distinguish between *B. pseudomallei* and *B. mallei*,[31] although certain *B. pseudomallei* strains in Australia have been shown to possess *B. mallei*-type amino-terminal sequences that bring this diagnostic into question.[32] More interestingly, although BimA is known to function intracellularly,[30,33] it has recently been shown to be a protective antigen against *B. mallei* infection,[17] suggesting that it may be present at sufficient quantities on the cell surface in an extracellular environment, identifying it as a potential diagnostic target. Examples of other *B. pseudomallei* targets are also shown in Fig. 1A. These have been shown to be produced in comparative published[29] or in-house proteomic analyses with *B. thailandensis*. These data also provide a means of developing and optimizing methodologies for antigen detection under biosafety level 2 laboratory using *B. thailandensis* as a surrogate.

3.2 Generation of an anti-BimA Fab

The synthetic antibody library provided by Dr S. Sidhu (University of Toronto) was used for phage display biopanning experiments. The library diversity is $>3 \times 10^{10}$ members and was panned against *B. mallei* BimA. After three rounds of panning, enrichment of Fabs was observed. Monoclonal phage ELISA for 40 clones each from the round 3 populations was performed. 13 out of 40 clones screened for BimA showed specific binding, with some clones exhibiting a greater than 10-fold ELISA signal compared with the BSA control. Fab E8 was selected from the screen based on its ELISA signal over background. For purification of Fab, the phagemid was converted to an expression vector through the introduction of a stop codon upstream of gene lll. ELISA data demonstrate that purified Fab E8 binds to BimA with micromolar affinity in a concentration-dependent manner (Fig. 2). Fab E8 also demonstrates some affinity for *B. pseudomallei* BimA (BPSS1492), though reduced in comparison with the *B. mallei* protein (Fig. 2). This result is consistent with the observation that there is considerable sequence diversity between *B. mallei* and *B. pseudomallei* BimA proteins in their surface-exposed amino terminal regions, although these proteins have been shown to be functionally

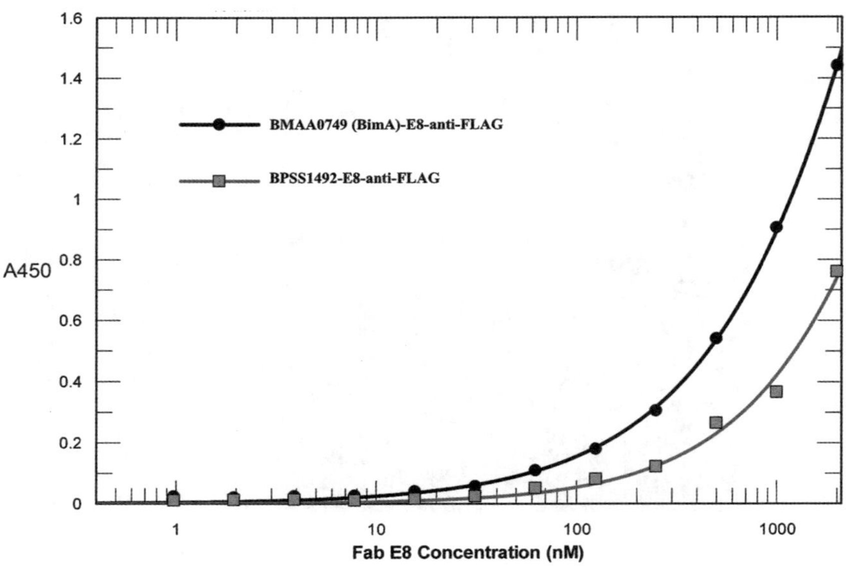

Fig. 2 ELISA binding curves of Fab E8 to BimA from *B. mallei* (BMAA0749) and from *B. pseudomallei* (BPSS1492).

complementary.[30] In addition, some structural similarity clearly exists between these BimA proteins and further optimization of Fabs has the potential to yield molecular recognition reagents which either bind to common structural epitopes, or bind to different epitopes for discrimination purposes.

3.3 Detection of *Burkholderia* LPS

In the search for potential targets that could be used in the development of diagnostic reagents, cell surface molecules such as LPS offer interesting options. In *B. pseudomallei* and *B. mallei* LPS and capsular polysaccharide are among the few genuine virulence factors that have been identified.[21,34] LPS is a tripartite molecule located in the outer cell wall of Gram-negative organisms including *Burkholderia* species. LPS is composed of a membrane-bound lipid A moiety (often responsible for the toxic effects of LPS), an oligosaccharide core (typically linking the lipid A moiety using ketooctulosonate), and a surface-exposed O-antigen made of repeating oligosaccharide units that can vary between bacterial species and strains. Mass spectrometric analyses of *B. pseudomallei* and *B. mallei* O-antigens indicate that these moieties are heteropolymers composed of repeating units of D-glucose and L-talose but differ in acetylation patterns of the talose residue.[35] Similarly, mass spectrometric analyses of the lipid portion of LPS show that both *B. pseudomallei* and *B. mallei* express very similar lipid A structures. However, *B. pseudomallei* lipid A is predicted to be penta-acylated[36] while *B. mallei* LPS is predicted to be a heterogeneous mixture of tetra- and penta-acylated species.[34]

As part of our strategy to develop protocols for detecting live or killed *Burkholderia* cells, we are using *B. thailandensis* LPS as a reagent for protocol development. This LPS has a lipid A moiety which is predicted to be similar to *B. pseudomallei*[36] and in previous studies has also been shown to be cross-reactive to *B. pseudomallei* and *B. mallei* immune sera by Western blot.[20,37] *B. thailandensis* LPS was purified from cell pellets of *B. thailandensis* following a modified hot phenol extraction protocol.[20,21] Purified *B. thailandensis* and commercially supplied *P. aeruginosa* LPS were visualized by SDS-PAGE followed by silver staining (data not shown). Both LPSs displayed a characteristic ladder banding pattern, caused by varying numbers of O-antigen units attached to the lipid A and core oligosaccharides.

Two mouse monoclonal antibodies (Mabs), directed against LPS from either *B. pseudomallei* or *B. mallei*, were initially investigated for their ability to bind to *B. thailandensis* LPS. Purified *B. thailandensis* LPS was separated by electrophoresis, transferred to nitrocellulose and probed with either anti-*B. pseudomallei* or anti-*B. mallei* LPS Mabs. Western blots (data not shown) revealed that both Mabs specifically bound to *B. thailandensis* LPS O-antigen repeating units ("ladder pattern") but not to the Lipid A band or to purified LPS from *E. coli* 0111:B4, which was included as a negative control.

For assay development, purified *B. thailandensis* LPS was initially immobilized directly onto immuno plates and binding of Mabs detected by ELISA. Sigmoidal binding curves were observed, with the anti-*B. pseudomallei* Mab showing higher affinity binding to purified LPS compared with the anti-*B. mallei* Mab (Fig. 3A). Based upon these results we adapted our methods to use the commercially available anti-*B. mallei* Mab in a Luminex-based indirect binding assay to detect purified *B. thailandensis* LPS. Luminex microspheres were first coupled with the anti-*B. mallei* LPS Mab. *B. thailandensis* and *P. aeruginosa* LPSs were then mixed with the coupled microspheres and binding detected using a biotinylated anti-*B. mallei* LPS Mab in conjunction with a streptavidin-phycoerethrin conjugate (SA-PE). Microspheres were analyzed on a Luminex 200 instrument. Binding of LPS from *B. thailandensis* but not *P. aeruginosa* LPS was detected (Fig. 3B), displaying dose-dependent binding. Initially, unacceptably high background binding was observed when 1% BSA/PBS was used as dilution buffer. However, this problem was eliminated by the use of 0.2% milk/PBS/0.8% PVP as a blocking buffer, with non-specific binding not exceeding 6 MFI units.

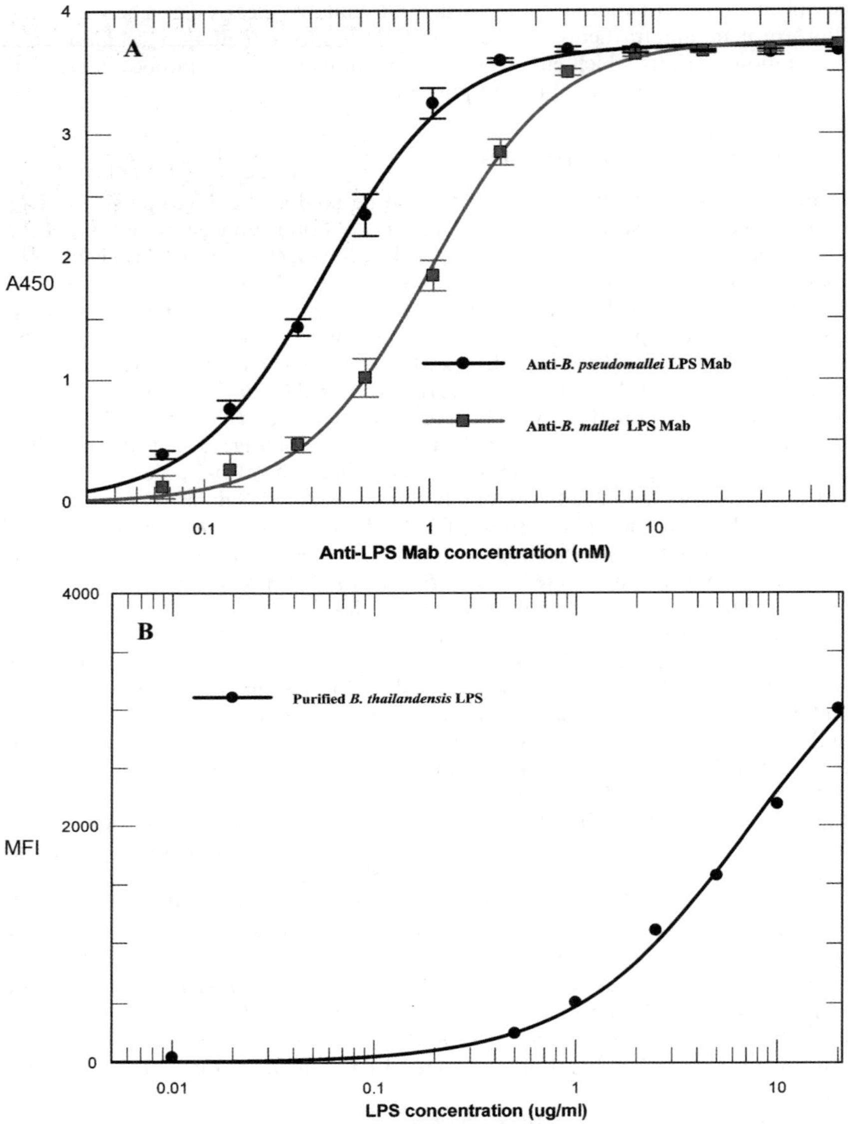

Fig. 3 Binding of anti-*B. mallei* LPS antibodies to LPS from *B. thailandensis*. Purified *B. thailandensis* LPS was incubated with (A) varying concentrations of anti-*B. mallei* LPS Mab or anti-*B. pseudomallei* Mab and binding detected by ELISA or (B) Luminex microspheres coupled to anti-*B. mallei* LPS Mab were incubated with purified *B. thailandensis* LPS and binding assayed using a Luminex 200 instrument (MFI, median fluorescence intensity).

3.4 Detection of *Burkholderia* cells

To confirm that the anti-LPS *Burkholderia* Mabs are viable reagents for the detection of live cells, exponential phase *B. thailandensis* or *P. aeruginosa* cells were incubated with each of the anti-LPS Mabs, followed by an anti-mouse IgG FITC-conjugated antibody. Cells were mounted onto poly-L-lysine-coated microscope slides and bacteria visualized by fluorescence microscopy (Fig. 4). Under white light, *B. thailandensis* rods were clearly visible at 100× magnification (Fig. 4A). Furthermore, when excited by the mercury vapour lamp, samples of *B. thailandensis*

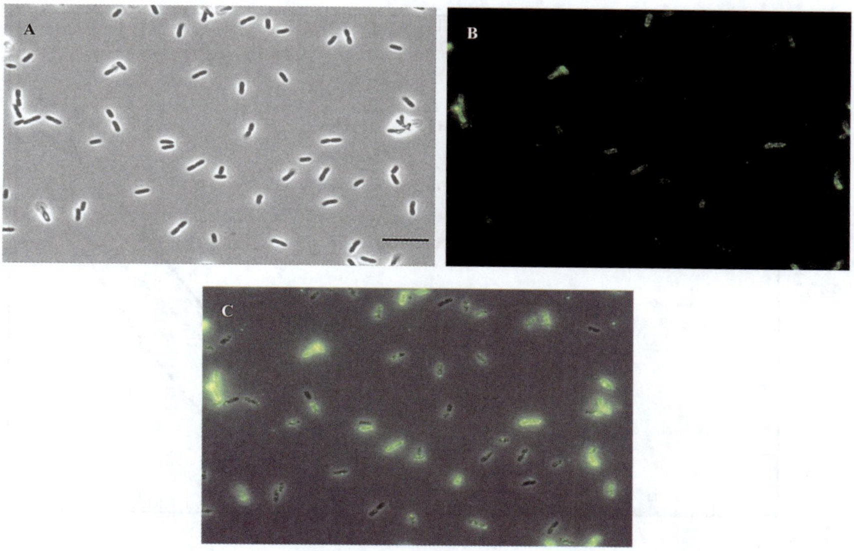

Fig. 4 Live cell imaging of *B. thailandensis*. Log phase *B. thailandensis* bacteria were probed with a mouse anti-*B.mallei* LPS Mab in combination with a rabbit anti-mouse whole IgG FITC-conjugated antibody. Binding was visualized at 100× magnification: (A) phase contrast image; (B) fluorescent image; (C) merged image. Scale bar represents 100 μm.

incubated with either of the anti-LPS Mabs displayed punctate green staining on the cell surface (Fig. 4B), whereas *P. aeruginosa* showed no staining at all (data not shown). Merging the phase and fluorescence images demonstrates that not all bacterial cells bind the Mabs equally (Fig. 4C). A proportion of the bacteria present did not stick to the surface of the microscope slides, and fluorescently-labelled bacteria could be seen to swim across the plane of vision. This observation suggests that these Mabs can be used to capture whole live cells in the timescale of this experiment.

Anti-*B. mallei* LPS Mab was also used to detect heat-killed and live *Burkholderia* species cells. Anti-*B. mallei* LPS Mab was coated onto immuno plates and heat-killed and exponential phase live *B. thailandensis* bacteria were subsequently detected by ELISA with an optimum Mab coating at 5 μg ml^{-1} (Fig. 5A). Comparison of data from the ELISA binding curves at 2 μg ml^{-1} Mab shows that the heat-killed bacteria bound with higher affinity compared to live cells. Binding of the anti-*B. mallei* LPS Mab to heat-killed *B. pseudomallei* bacteria was also shown to be similar to that of heat-killed *B. thailandensis* by direct detection ELISA using the anti-*B. mallei* LPS Mab (data not shown). A Luminex indirect detection assay was subsequently carried out on heat-killed *B. thailandensis*, demonstrating bacteria can also be detected using the anti-*B. mallei* LPS Mab with this system (Fig. 5B). The lower limit of detection for both the ELISA and Luminex assays is approximately 0.01 OD units per ml, which equates to approximately 5×10^5 bacteria.

4. Conclusions

In this study we have described how an integrated medium-throughput approach is being used to develop assays for the direct detection of cells of pathogenic *Burkholderia* species. It is important to note that the methodology presented here establishes a process to create and optimize diagnostic assays under BSL 2 conditions as a means of enabling more rapid translation of methods to BSL3 conditions. We have presented rapid and robust methods for the purification of milligram quantities of soluble proteins and LPS that have been identified to be surface-localized or secreted by *B. pseudomallei* and *B. mallei*. We have obtained a Fab using phage display

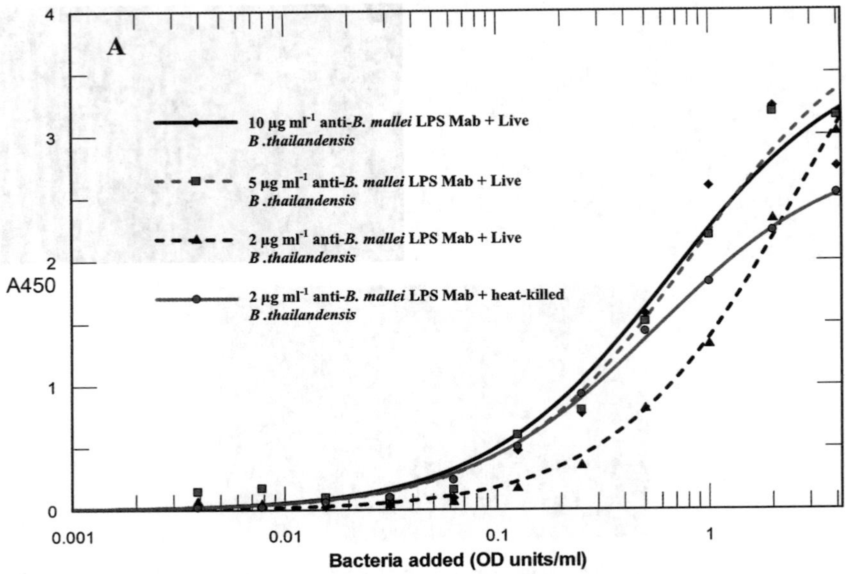

Fig. 5 Binding of anti-*B. mallei* LPS antibodies to live or heat-killed *B. thailandensis.* Microplates coated with various concentrations of anti-*B. mallei* LPS Mab were mixed with dilutions of live or heat-killed *B. thailandensis* and developed by sandwich ELISA (A) and (B) Luminex microspheres coupled to anti-*B. mallei* LPS Mab were incubated with heat-killed *B. thailandensis* and observed with a Luminex 200 instrument.

methods against the *B. mallei* BimA protein, a surface exposed protein which has a key role in infection[30] and is an antigen which confers protection against glanders.[17] Using *B. thailandensis* LPS and cells as a surrogate for the highly pathogenic *Burkholderia* species, we have demonstrated that anti-LPS *Burkholderia* Mabs can be used to detect LPS and cells in ELISA and Luminex diagnostic formats, though admittedly at high cell titres.

There is, however, considerable scope for improvement and optimization of the procedures presented here. For example, although we have developed reliable and robust methods for cloning, expression and purification of *Burkholderia* proteins, correct target selection remains difficult. However, target selection may be guided by efforts such as recently published *B. pseudomallei* protein microarray studies that have identified abundant/immunoreactive proteins which are expressed in an active infection.[27] We have shown proof-of-principle for detection of cells but our efforts are now being focused upon improving both sensitivity and selectivity of diagnostic reagents. Optimization of the assay conditions are underway and include altering experimental conditions such as the addition of PVP to reduce non-specific binding in Luminex assays.[38] It is also interesting to note that the Luminex system has been used to detect killed whole cells of other Gram-negative bacteria at levels of 2.5–500 organisms per ml in a previously published study using capture and detection antibodies directed against bacterial protein antigens.[39] However, detection of live organisms remains largely unexplored and clearly demands an integrated approach like ours to exploit the potential of using high affinity recognition reagents to identify microorganisms in clinical samples with multiplex-capable platforms such as Luminex.

Acknowledgements

This work was supported by NIH grants 1U01AI078008-1, U54AI057156 and RO1AI064184, and funding from TI-3D (Welch Foundation) and Dstl, Porton Down, UK. We gratefully acknowledge the technical assistance and advice we received from Tonniele Naeher, LaChanda Dunlap and Bi Linton from the Luminex Corporation (Austin, Texas, USA).

References

1 S. Yang and R. E. Rothman, *Lancet Infect. Dis.*, 2004, **4**, 337–348.

2 R. F. Louie, T. Kitano, T. K. Brock, R. Derlet and G. J. Kost, *Disaster Med. Public Health Prep.*, 2009, **3**(Suppl. 2), S193–S202.

3 W. J. Wiersinga, T. van der Poll, N. J. White, N. P. Day and S. J. Peacock, *Infect. Immun.*, 2007, **75**, 3074–3079.

4 G. C. Whitlock, D. M. Estes and A. G. Torres, *FEMS Microbiol. Lett.*, 2007, **277**, 115–122.

5 N. J. White, *Lancet*, 2003, **361**, 1715–1722.

6 H. Neubauer, L. D. Sprague, R. Zacharia, H. Tomaso, S. Al Dahouk, R. Wernery, U. Wernery and H. C. Scholz, *J. Vet. Med. B Infect. Dis. Vet. Public Health*, 2005, **52**, 307–310.

7 A. Srinivasan, C. N. Kraus, D. DeShazer, P. M. Becker, J. D. Dick, L. Spacek, J. G. Bartlett, W. R. Byrne and D. L. Thomas, *N. Engl. J. Med.*, 2001, **345**, 256–258.

8 A. C. Cheng and B. J. Currie, *Clin. Microbiol. Rev.*, 2005, **18**, 383–416.

9 S. Peacock, *Curr. Opin. Infect. Dis.*, 2006, **19**, 421–428.

10 S. Ileri, *Br. Vet. J.*, 1965, **121**, 164–170.

11 A. C. Cheng, M. O'Brien, K. Freeman, G. Lum and B. J. Currie, *Am. J. Trop. Med. Hyg.*, 2006, **74**, 330–334.

12 J. B. Derbyshire, *Can. Vet. J.*, 2002, **43**, 722–726.

13 L. D. Sprague, R. Zachariah, H. Neubauer, R. Wernery, M. Joseph, H. C. Scholz and U. Wernery, *BMC Vet. Res.*, 2009, **5**, 32.

14 J. Suppiah, J. S. Thimma, S. H. Cheah and J. Vadivelu, *FEMS Microbiol. Lett.*, 2010, **306**, 9–14.

15 A. J. Gnanam, B. Hall, X. Shen, S. Piasecki, A. Vernados, E. E. Galyov, S. J. Smither, G. B. Kitto, R. W. Titball, A. D. Ellington and K. A. Brown, *Trans. R. Soc. Trop. Med. Hyg.*, 2008, **102**(Suppl. 1), S55–57.

16 D. B. Thompson, K. Crandall, S. V. Harding, S. J. Smither, G. B. Kitto, R. W. Titball and K. A. Brown, *Trans. R. Soc. Trop. Med. Hyg.*, 2008, **102**(Suppl. 1), S61–65.

17 G. C. Whitlock, A. Deeraksa, O. Qazi, B. M. Judy, K. Taylor, K. L. Propst, A. J. Duffy, K. Johnson, G. B. Kitto, K. A. Brown, S. W. Dow, A. G. Torres and D. M. Estes, *Proc. Vaccinol.*, 2010, **2**, 71–75.

18 F. A. Fellouse, K. Esaki, S. Birtalan, D. Raptis, V. J. Cancasci, A. Koide, P. Jhurani, M. Vasser, C. weismann, A. A. Kossiakoff, S. Koide and S. S. Sidhu, *J. Mol. Biol.*, 2007, **373**, 924–940.
19 C. V. Lee, W.-C. Liang, M. S. Dennis, C. Eigenbrot, S. S. Sidhu and G. Fuh, *J. Mol. Biol.*, 2004, **340**, 1073–1093.
20 O. Qazi, J. L. Prior, B. M. Judy, G. C. Whitlock, G. B. Kitto, A. G. Torres, D. M. Estes and K. A. Brown, *Trans. R. Soc. Trop. Med. Hyg.*, 2008, **102**(Suppl. 1), S58–60.
21 M. Nelson, J. L. Prior, M. S. Lever, H. E. Jones, T. P. Atkins and R. W. Titball, *J. Med. Microbiol.*, 2004, **53**, 1177–1182.
22 M. T. Holden, *et al.*, *Proc. Natl. Acad. Sci. U. S. A.*, 2004, **101**, 14240–14245.
23 W. C. Nierman, *et al.*, *Proc. Natl. Acad. Sci. U. S. A.*, 2004, **101**, 14246–14251.
24 Y. Yu, K. S. Kim, H. H. Chua, C. H. Lin, S. H. Sim, D. Lin, A. Derr, R. Engels, D. DeShazer, B. Birren, W. C. Nierman and P. Tan, *BMC Microbiol.*, 2006, **6**, 46.
25 K. Amemiya, J. L. Meyers, D. DeShazer, R. N. Riggins, S. Halasohoris, M. England, W. Ribot, S. L. Norris and D. M. Waag, *Diagn. Microbiol. Infect. Dis.*, 2007, **59**, 137–147.
26 P. C. Y. Woo, P. K. L. Leung, S. S. Y. Wong, P.-K. Ho and K.-Y. Yuen, *Clin. Diagn. Lab. Immunol.*, 2001, **8**, 832–826.
27 P. L. Felgner, *et al.*, *Proc. Natl. Acad. Sci. USA*, 2009, **106**, 13499–13504.
28 S. V. Harding, M. Sarkar-Tyson, S. J. Smither, T. P. Atkins, P. C. Oyston, K. A. Brown, Y. Liu, R. Wait and R. W. Titball, *Vaccine*, 2007, **25**, 2664–2672.
29 P. Wongtrakoongate, N. Mongkoldhumrongkui, S. Chaijan, S. Kamchonwongpaisan and S. Tungpradabkul, *Mol. Cell. Probes*, 2007, **21**, 81–91.
30 J. M. Stevens, R. L. Ulrich, L. A. Taylor, M. W. Wood, D. DeShazer, M. P. Stevens and E. E. Galyov, *J. Bacteriol.*, 2005, **187**, 7857–7862.
31 R. L. Ulrich, M. P. Ulrich, M. A. Schell, H. S. Kim and D. DeShazer, *Diagn. Microbiol. Infect. Dis.*, 2006, **55**, 37–45.
32 C. Sittidet, J. M. Stevens, N. Chantratita, B. J. Currie, S. J. Peacock, S. Korbsrisate and M. P. Stevens, *J. Clin. Microbiol.*, 2008, **46**, 2418–2422.
33 M. P. Stevens, J. M. Stevens, R. L. Jeng, L. A. Taylor, M. W. Wood, P. Hawes, P. Monaghan, M. D. Welch and E. E. Galyov, *Mol. Microbiol.*, 2005, **56**, 40–53.
34 P. J. Brett, M. N. Burtnick, D. S. Snyder, J. G. Shannon, P. Azadi and F. C. Gherardini, *Mol. Microbiol.*, 2007, **63**, 379–390.
35 M. N. Burtnick, P. J. Brett and D. E. Woods, *Methods Mol. Biol.*, 2002, **184**, 849–852.
36 V. Novem, G. Shui, D. Wang, A. K. Bendt, S. H. Sim, Y. Liu, T. W. Thong, S. P. Sivalingam, E. E. Ooi, M. R. Wenk and G. Tan, *Clin. Vaccine Immunol.*, 2009, **16**, 1420–1428.
37 N. Anuntagool, P. Intachote, V. Wuthiekanun, N. J. White and S. Sirsinha, *Clin. Diagn. Lab. Immunol.*, 1998, **5**, 225–229.
38 T. Waterboer, P. Sehr and M. Pawlita, *J. Immunol. Methods*, 2006, **309**, 200–204.
39 S. A. Dunbar, C. A. Vander Zee, K. G. Oliver, K. L. Karem and J. W. Jacobson, *J. Microbiol. Methods*, 2003, **53**, 245–352.

Generic protease detection technology for monitoring periodontal disease

Xinwei Zheng,[a] Joseph P. Cook,[a] Michael Watkinson,[b] Shoufeng Yang,[a] Ian Douglas,[c] Andrew Rawlinson[c] and Steffi Krause*[a]

Received 18th May 2010, Accepted 21st July 2010
DOI: 10.1039/c005364c

Periodontal diseases are inflammatory conditions that affect the supporting tissues of teeth and can lead to destruction of the bone support and ultimately tooth loss if untreated. Progression of periodontitis is usually site specific but not uniform, and currently there are no accurate clinical methods for distinguishing sites where there is active disease progression from sites that are quiescent. Consequently, unnecessary and costly treatment of periodontal sites that are not progressing may occur. Three proteases have been identified as suitable markers for distinguishing sites with active disease progression and quiescent sites: human neutrophil elastase, cathepsin G and MMP8. Generic sensor materials for the detection of these three proteases have been developed based on thin dextran hydrogel films cross-linked with peptides. Degradation of the hydrogel films was monitored using impedance measurements. The target proteases were detected in the clinically relevant range within a time frame of 3 min. Good specificity for different proteases was achieved by choosing appropriate peptide cross-linkers.

Introduction

Periodontal diseases are a group of related inflammatory conditions affecting the supporting tissues of teeth. Inflammation of the gums (gingivae) is caused by the local accumulation of an oral biofilm, dental plaque. In its most severe form the disease can lead to destruction of the tooth supporting tissues, including resorption of alveolar bone and loss of teeth. There are several distinct clinical forms of periodontitis but chronic periodontitis, most frequently found in adults, is the most common. Disease progression is usually site specific but is not uniform and it is difficult to distinguish clinically those sites that are progressing from those that are inflamed but not progressing. As a result a considerable level of unnecessary treatment may occur of periodontal sites that are not progressing with consequent drain on resources. A non-invasive technique for more accurately identifying periods of acute inflammatory destruction would be of major benefit, which could be fulfilled by an accurate biosensor.

Both early in the pathogenesis of periodontitis and as acute exacerbations during its chronic course, neutrophils (PMNL) migrate to the site[1] and there is a fluid exudate (gingival crevicular fluid (GCF)) into the "pocket" between the tooth and the gingival tissue. Analysis of this fluid provides a convenient, non-invasive means

[a]*School of Engineering and Materials Science, Queen Mary University of London, London, E1 4NS, UK. E-mail: s.krause@qmul.ac.uk; Fax: +44 20 89819804; Tel: +44 20 78823747*
[b]*School of Biological and Chemical Sciences, Queen Mary University of London, London, E1 4NS, UK*
[c]*School of Clinical Dentistry, Claremont Crescent, Sheffield, S10 2TA, UK*

of monitoring host or bacterial molecules. The PMNLs arriving in the periodontal pocket attempt to control the plaque biofilm by degranulation and release of lytic enzymes, which include the serine proteases (elastase, cathepsin G, proteinases-3), and the matrix metalloproteinases, MMP8 (collagenase) and MMP9 (gelatinase-B). Elastase activity has been detected in the GCF and is considered to be a useful quantitative measure of gingival inflammation.[2] In addition, cathepsin G and MMP8 have been correlated with disease severity[3] but our work has shown that although MMP9 is present in an activated form in 40% of sites, it did not correlate with the clinical status of the sites. In contrast, elastase levels correlate with the degree of loss of attachment between tooth and supporting tissue and its levels mirror that of the neutrophil MMP8 in GCF.[4] These also correlate with gingival and bleeding indices.[5] In addition, elastase has been shown to damage epithelial cells directly.[6] The major factor regulating elastase activity in the oral cavity and upper respiratory tract is thought to be secretory leukoproteinase inhibitor (SLPI)[7] but this, like α1-antitrypsin, can be inactivated in the presence of free radicals such as those produced by the neutrophils.

From the proteases mentioned above, we have identified human neutrophil elastase (HNE), cathepsin G, and MMP8 as suitable markers for disease progression and therefore as target proteases for our detection technology. Some 0.05 to ~30 U ml^{-1} of HNE, and up to 2.4 mU ml^{-1} cathepsin-G have been measured in GFC.[8] Activities of up to 0.7 mU ml^{-1} have been measured for MMP8.[9] A point of care system for the diagnosis of periodontitis should be able to provide a result within a few minutes, while the patient is in the dental chair. Consequently, the work presented in this paper is directed at the development of a disposable biosensor for the above mentioned target proteases that would provide accurate chair side evaluation of protease levels, which in turn would immediately inform treatment decisions.

Protease sensors based on the degradation of thin films

Disposable sensors based on the degradation of thin films have shown great promise for the detection of proteases.[10–18] In some cases, natural substrates of the target proteases were coated onto a variety of different transducers. For example, Saum *et al.* monitored collagenase activity by depositing its natural substrate gelatin onto interdigitated electrodes.[19] Collagenase concentrations as low as 0.2 mg ml^{-1} were detected; however, only low electrolyte concentrations could be used for reliable impedance measurements due to the hydrophilic nature of the films. Millington *et al.* detected trypsin using holograms based on gelatin.[10] Hanumegowda *et al.* employed an optical microsphere resonator coated with BSA to detect trypsin.[12] Ionescu *et al.* introduced an amperometric sensor coated with gelatine for the detection of trypsin.[13] Other natural protease substrates used in this sensor format included fibrinogen used for the detection of human neutrophil elastase by quartz crystal microbalance (QCM) and BSA cross-linked acrylamide degraded by pepsin.[18]

All of the protease sensors mentioned above are based on the measurement of the disappearance of the film from the transducer surface. Other sensors based on the enzymatic degradation of protein films have detected the formation of amino acids by optical means.[15,20]

A drawback of using natural enzyme substrates in the detection of proteases is a lack of specificity. Proteins contain a large number of different cleavage sites and are therefore likely to be degraded by a number of different proteases. Furthermore, the use of proteins makes the fine tuning of film properties for maximising the sensor response and reducing undesirable effects, such as non-specific binding, very difficult.

Greater specificity can be achieved using synthetic protease substrates.[16,17] For example poly(ester amide) films based on bis(L-phenylalanine)-α,ω-alkylene diesters with a single cleavage site were used for the detection of α-chymotrypsin. Film degradation was monitored using surface plasmon resonance,[16,17] quartz crystal microbalance[21] and impedance measurements.[21,22] The rate of degradation was found to be dependent on the enzyme concentration. The hydrophobic nature of the film caused a severe reduction in the sensor signal in the presence of BSA.[16]

A more generic approach has recently been described by Stair *et al.*[14] Hydrogel films were formed by cross-linking oxidised dextran with two different peptides: AAPVAAK, preferentially cleaved by HNE, and AAPFFK, preferentially cleaved by cathepsin G. Degradation of the hydrogel films was monitored using a quartz crystal microbalance. Both hydrogel films showed good sensitivity towards their respective proteases, with little cross-sensitivity for the other enzyme. Moreover, the approach is generic as specificity for different proteases was achieved by choosing an appropriate peptide sequence as cross-linker. In addition, hydrogels are less likely to suffer from the effects of non-specific binding.[16] A drawback of the use of peptide cross-linked dextran hydrogel films was that the degradation of the films in the presence of its target protease was preceded by a long induction period ranging from 1.5 min to 15 min depending on the protease activity rendering these films unsuitable for a fast chair side diagnostic.

In this study, a new coating method for peptide cross-linked dextran hydrogels has been explored to increase their response rate to the target proteases. Film degradation has been monitored using impedance measurements as this allows the use of cheap, disposable sensor substrates in any ultimate point of care device.

Experimental

Materials

Polished, gold-coated QCM crystals (10 MHz) were purchased from ICM (Oklahoma City, USA). Silicon with 20 nm thermal oxide and 30 nm CVD nitride was purchased from Si-Mat (Landsberg, Germany). 96% alumina (0.5 mm thick) was purchased from Laser cutting-ceramic Ltd. (Sheffield, UK).

Dextran ($M_r = 2 \times 10^6$ Da) was purchased from Fluka. Peptide sequences AAPVAAK, AAPFFK, GPQGIWGQK (95% purity) were purchased from Genscript (New Jersey, USA). Each peptide sequence was dissolved in water to a concentration of 100 mg ml^{-1} and filtered through a PL-HCO3 MP solid phase extraction (SPE) tube purchased from Polymer Laboratories (Massachusetts, USA) to neutralise the TFA salts and freeze dried before use. Human neutrophil elastase (HNE) and cathepsin G were purchased from Elastin Products (Missouri, USA). Matrix metalloproteinase 8 (MMP 8) was purchased from Calbiochem (Darmstadt, Germany).

Photoresist S1813 and developer 351 were purchased from Chestech Ltd (Rugby, UK). Other chemicals and reagents were obtained from commercial sources (Sigma-Aldrich, Fisher Scientific) and used without further purification. Ultra pure water was obtained from an ELGA Purelab Ultra system.

The following buffer solutions were prepared:

- pH 8.0 phosphate buffer (10 mM);
- pH 7.5 charge transfer buffer: pH 7.5 phosphate buffered saline (10 mM) containing 5 mM potassium ferricyanide, 5 mM of potassium ferrocyanide and 140 mM sodium chloride;
- pH 5 charge transfer buffer: pH 5 sodium acetate buffer (1 mM) containing 5 mM potassium ferricyanide, 5 mM of potassium ferrocyanide and 140 mM sodium chloride.

Fabrication of electrodes

Chromium/gold interdigitated electrodes were patterned onto silicon substrates with a SiO_2/Si_3N_4 insulator and 96% alumina substrates. S1813 photoresist was spin-coated using CHEMAT Technology spin coater kw-4A and exposed with a Karl

Suss MicroTec MJB3 mask aligner following a standard procedure. The exposed photoresist was left in chlorobenzene for 10 min to harden the film surface. The photoresist was then developed. Some 30 nm chromium and 150 nm gold were deposited by thermal evaporation using an Edwards Coating System E306A, followed by lift-off in acetone. Interdigitated electrodes with 10 μm line width and 10 μm line spacing were patterned on the silicon substrates. Interdigitated electrodes with 50 μm line width and 300 μm spacing were patterned on the alumina substrate.

Oxidation of dextran

Dextran was oxidised adapting a procedure described by Ruys *et al.*[23] Sodium periodate (1.41 g, 6.6 mmol) was added cautiously to a stirred dextran (0.48 g, 3 mmol repeat units) solution in 7 ml de-ionized water. The mixture was left for 24 h in the dark. The oxidized dextran solution was then dialyzed against 1 l water using Spectra/Por Float-a-Lyser (MWCO 500) from Spectrum Laboratories Inc for 2 days. (California, USA). The bulk water was changed three times during dialysis. After freeze drying, a white fluffy solid was obtained (0.32 g, 83%), which was stored at room temperature. The degree of oxidation of dextran was confirmed to be 100% by titration as described previously,[14] *i.e.* each glucose unit was modified with two aldehyde groups resulting in a decrease of molecular weight of the repeat units from 162 g mol^{-1} to 130 g mol^{-1} during oxidation. δ_H (270 MHz, D_2O, Me_4Si) 3.80 (b, CH_2) 4.02 (b, CH), 5.10 (b, CH), 5.39 (b, CHOH), 8.38 (s, CHO). IR: (cm^{-1}) 3375 (OH), 2934 (CH), 1637(CHO), 1342, 1103, 1021.

Film preparation

Hydrogel films were prepared following two different procedures. For comparison with previous results, gold coated quartz crystals were coated by spin-coating a solution of oxidised dextran and peptide following the procedure described by Stair *et al.*[14] To overcome the problems encountered with these films, a new coating procedure was developed where oxidised dextran and peptide were deposited successively.

The interdigitated electrodes/quartz crystals were immersed in piranha solution (3 : 1 v/v concentrated H_2SO_4 and 30% H_2O_2) for one minute. After rinsing with water, the electrodes were transferred into a 10 mM solution of 4-mercapto-1-butanol in 4 : 1 v/v ethanol and water and left for one hour to form a self-assembled monolayer. Afterwards, the electrodes were rinsed with water and dried with nitrogen. A solution of oxidized dextran (6.5 mg, 0.05 mmol repeat units) in pH 8.0 phosphate buffer (0.5 ml) was then spin-coated at 3000 rpm for 25 s onto the surface of the electrode. The coated dextran layer was left in a sample box at room temperature overnight to dry. A solution of AAPVAAK (3.1 mg, 0.005 mmol) in pH 8.0 phosphate buffer (50 μl) was prepared. Some 200 μl of methanol was then added and mixed with a vortex mixer. The solution of AAPVAAK was then drop coated onto the surface of the dextran layer. The amount of AAPVAAK solution for drop coating was dependent on the area of the electrode. Specifically, 0.5 μl of AAPVAAK solution was drop coated onto a 3 mm × 3 mm electrode area. Film curing was finished by incubating the film at 30 °C for 1 day. The thickness of the film on the silicon substrate was measured to be 200–300 nm thick using a Dektak^{3}ST surface profiler. The thickness of the film on the alumina surface could not be measured due to the surface roughness of the substrate. The average grain size of alumina was estimated to be about 2 μm using an SEM.

AAPVAAK cross-linked hydrogel films were used for the detection of HNE, AAPFFK cross-linked films were used for the detection of cathepsin G. These peptide sequences were previously used by Stair *et al.*[14] MMP8 activated by *p*-aminophenylmercuric acetate using organomercurial activation protocol was detected using GPQGIWGQK as the cross-linker. This peptide sequence was obtained by adding lysine to a peptide that was shown to be a preferred substrate for MMP8 previously[24] to obtain an amine functionality at the C terminal of the peptide. Cross-linking with AAPFFK and GPQGIWGQK was carried out using the same molar concentration as described for AAPVAAK above. Initial investigations were carried out using the AAPVAAK cross-linked hydrogel because of the low cost of HNE compared to the other enzymes.

Film degradation experiments

The hydrogel coated interdigitated electrodes were inserted into a customized cell with an o-ring to insulate the contacts from the solution and a magnetic stirrer bar. Some 950 µl of pH 7.5 charge transfer buffer was then added to the cell at room temperature. The charge transfer buffer was constantly stirred and the impedance measured with time in a frequency range from 10 Hz to 50 kHz using an Autolab PGSTAT10 with FRA2 (Windsor Scientific, UK). All impedance measurements were carried out between the two electrodes of the interdigitated structure with an ac voltage amplitude of 10 mV and at a dc potential of 0 V. The impedance usually became stable after 15 min, and then 50 µl of different concentration enzyme solution in pH 5 charge transfer buffer was added into the pH 7.5 charge transfer buffer in the cell. After the addition of enzyme, an impedance change was generally observed in less than 30 s. A control experiment was carried out by replacing the enzyme solution with pH 5 charge transfer buffer.

For measurements in the presence of 2% BSA, 250 µl of 8% BSA solution in pH 7.5 charge transfer buffer was added to 700 µl of pH 7.5 charge transfer buffer followed by the addition of 50 µl of enzyme solution in pH 5 charge transfer buffer after the 15 min stabilisation period.

Measurements in small sample volumes

Capillary fill devices were constructed on top of the hydrogel coated IDEs on 96% alumina after coating the hydrogel as described above. A single-sided adhesive tape was attached to the hydrogel-coated electrode surface as an insulator that also formed a well around the electrode followed by a double-sided adhesive as a spacer and a hydrophilic lid manufactured by Adhesives Research Ireland Ltd. A schematic of the device is shown in Fig. 1. The volume of the capillary fill device was about 1.5 µl. After filling the device with test solution, it was sealed from both sides of the capillary channel using silicone grease to prevent evaporation of the

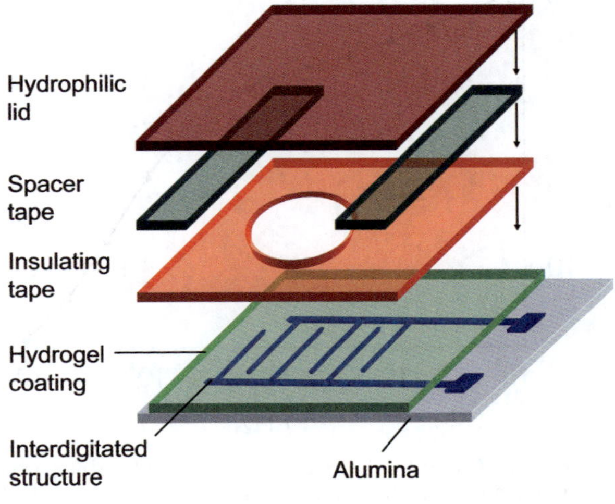

Fig. 1 Schematic of capillary fill device.

solution. Impedance changes were monitored using an Autolab PGSTAT10 with FRA2 at 100 Hz.

Standard characterisation techniques

^{1}H and ^{13}C NMR spectra were recorded on a Jeol Ex270 instrument at room temperature. Infrared spectra were recorded on a Bruker Tensor 37 FT-IR system with an ATR diamond press. Scanning electron microscope images were obtained from FEI Inspect F instrument with acceleration voltages of 10 keV. Optical microscope image was obtained from OLYMPUS BX60 microscope.

Results and discussion

The impedance change during the enzyme induced degradation of the peptide cross-linked hydrogel films was initially investigated using gold-coated quartz crystals as substrates for the hydrogel to allow comparison with previous data obtained with quartz crystal microbalance measurements.[14] Impedance spectra were measured in a two electrode arrangement using a platinum electrode as the counter electrode. No impedance change was observed during the degradation of an AAPVAAK cross-linked hydrogel film by HNE in phosphate buffered saline. This was attributed to the hydrophilicity of the film resulting in a significant uptake of electrolyte and therefore a low film resistance. In the presence of a charge transfer reagent (ferricyanide/ferrocyanide), a significant change in impedance was observed at the low frequency end of the spectrum (Fig. 2) due to the change in the charge transfer resistance during the degradation of the film. Hence subsequent degradation experiments were carried out at frequencies between 30 Hz and 100 Hz.

Two-step coating of hydrogel films

Hydrogel films produced using a single step method where a solution containing both oxidised dextran and the peptide was spin-coated onto a quartz crystal showed a relatively slow response preceded by an induction period when measured by QCM measurements.[14] A similar result was obtained when measuring the degradation of these hydrogel films by impedance measurements (Fig. 3, curve A). The impedance was stable in pH 7.5 charge transfer buffer, it increased briefly upon addition of HNE and then decreased slowly over a long period of time.

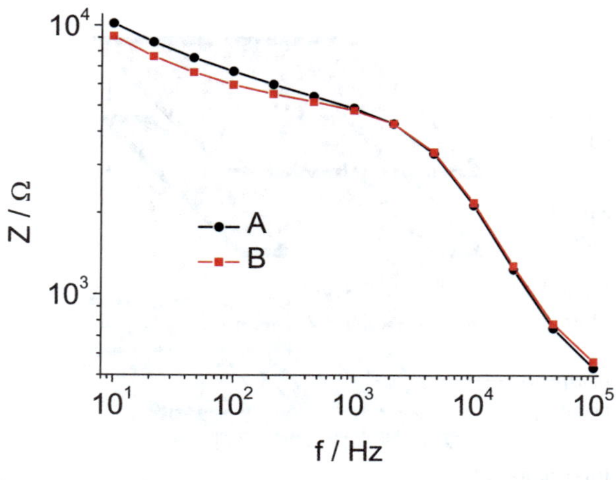

Fig. 2 Impedance spectra of AAPVAAK cross-linked hydrogel films prepared by a one-step coating method before (A) and after (B) the addition of 10 U ml^{-1} HNE to pH 7.5 charge transfer buffer.

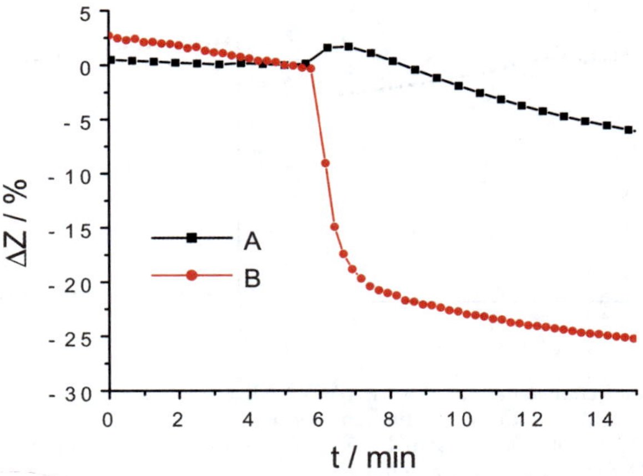

Fig. 3 Impedance change measured before and after addition of 10 U ml^{-1} HNE to pH 7.5 charge transfer buffer. Curve A shows the response of an AAPVAAK cross-linked hydrogel film produced by a single step coating method on a QCM crystal. Curve B shows the response of an AAPVAAK cross-linked hydrogel film produced by a two step coating method on a QCM crystal. HNE was added at t = 5 min.

For the production of the final clinical device, the hydrogel films will be deposited using standard technologies such as drop coating or inkjet printing. As pre-mixing of the amine containing peptide and the oxidised dextran in solution prior to coating would lead to cross-linking, albeit at a slow rate, we did not anticipate being able to use this approach with any standard printing method. Hence a two-step coating method was developed in which the dextran and the peptide were deposited in two consecutive steps. Using this two step coating strategy, the cross-linking condensation reaction would only happen on the surface of the substrate instead of in nozzles or ink cartridges. Adopting this approach had the attendant advantage of producing a highly cross-linked surface layer on the dextran which provided a sufficiently rapid response to the enzyme. Presumably once the cross-links thus formed in the surface layer were broken by HNE, the underlying non-cross-linked dextran dissolved rapidly resulting in a large sensor response (Fig. 3, curve B).

The effect of non-specific binding

Further development work was carried out using hydrogel-coated IDEs on insulator/silicon and alumina substrates. Cr/Au IDEs were obtained by microfabrication. It is anticipated that the microfabricated electrodes will be replaced by screen printed electrodes on alumina in the future to provide low-cost mass-produced electrodes. Response curves to different activities of HNE using IDEs on a silicon substrate are shown in Fig. 4a. The impedance response is comparable to that observed with two opposing electrodes (Fig. 3, curve B). The rate of degradation was found to be strongly activity dependent.

Samples of GCF are expected to contain protein concentrations between 2% and 4%. The effect of BSA on the rate of degradation was therefore investigated. Fig. 4 shows the comparison of the impedance change before and after the addition of three different activities of HNE in the absence and in the presence of 2% BSA. In the absence of BSA (Fig. 4a), the impedance change after the addition of 10 U ml^{-1} was about 18% in 3 min. The addition of BSA resulted in slower degradation (Fig. 4b). The impedance change in 3 min was 13% in the presence of 2% BSA. The reduction in the degradation rate was mainly attributed to an increase in the viscosity of the solution. The response in the presence of BSA remained sufficient

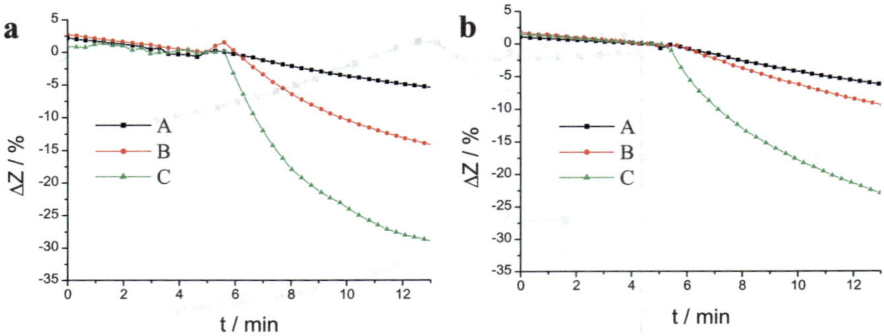

Fig. 4 Impedance change measured with AAPVAAK cross-linked hydrogel coated IDEs on silicon before and after addition of different concentrations of HNE (0.1 U ml^{-1} (A), 1 U mL^{-1} (B) and 10 U ml^{-1} (C)) to a pH 7.5 charge transfer buffer containing (a) 0% BSA, (b) 2% BSA. HNE was added at t = 5 min.

over the entire concentration range of HNE for obtaining adequate sensor signals in the 3 min timeframe required by the dental application.

The effect of the monolayer on the mechanism of degradation

If not specified, the gold electrodes on all substrates were modified with a hydroxyl terminated monolayer prior to hydrogel deposition to aid the spreading of the solution on the surface. To study the effect of the self-assembled monolayer on the degradation of hydrogel films, the gold IDEs were modified using two different thiols – one with a hydroxyl terminal group, which provided a hydrophilic surface but did not react with the oxidised dextran, and a second one with an amine terminal group that provided a hydrophilic surface, which was also capable of reacting with the aldehyde groups of the oxidised dextran. The response curves to both types of hydrogel films are shown in Fig. 5. Both films appeared stable in pH 7.5 charge transfer buffer. Upon addition of HNE, the hydrogel film on a hydroxyl terminated monolayer displayed a decrease in impedance caused by a dissolution of the film, while the hydrogel film formed on the amine terminated monolayer showed an increase in impedance indicative of a swelling rather than a dissolution process as

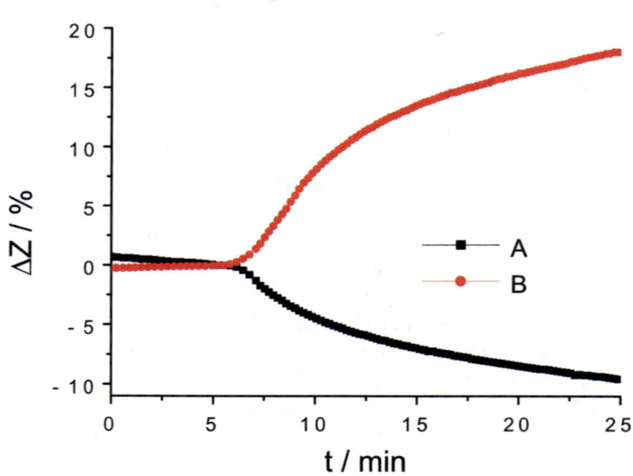

Fig. 5 Impedance change measured with an AAPVAAK cross-linked hydrogel coated on IDEs on 96% alumina modified with 4-mercapto-1-butanol as monolayer (A) and 2-aminoethanethiol as monolayer (B) before and after addition of 10 U ml^{-1} HNE to pH 7.5 charge transfer buffer. HNE was added at t = 6 min.

the dextran film was covalently bound to the surface after the cleavage of cross-links due to the enzymatic reaction.

The use of an amine terminated monolayer may be preferable in applications where greater mechanical stability is required, *e.g.* if the sensor were required to be placed into a flowing medium, and where the release of degradation products of the films is undesirable, *e.g.* for *in vivo* applications.

Sensor response to different proteases

Three host proteases, HNE, cathepsin G and MMP8, have been identified as suitable markers for monitoring periodontal disease (*vide supra*). Oxidised dextran films were thus cross-linked with three different peptides that each contained cleavage sites for one of the three target enzymes. The activity dependent response of the sensor materials to the different proteases was investigated. The response of AAPVAAK cross-linked films to different activities of HNE was discussed above (Fig. 4). Fig. 6a and 7a show the impedance change during degradation of dextran hydrogel films cross-linked with AAPFFK and GPQGIWGQK under the same experimental conditions. The film impedance for AAPFFK and GPQGIWGQK cross-linked films increased over time in the charge transfer buffer. This can be attributed to continued swelling of the films. Addition of the proteases caused the films to degrade for all three hydrogel/enzyme systems. At high enzyme activities, this resulted in a decrease of impedance; at low activities, this resulted in a reduced rate of impedance increase. Calibration curves for cathepsin G (Fig. 6b) and MMP8 (Fig. 7b) were obtained by determining the rate of degradation within the first 100 s of enzyme addition.

Measurement of protease activities in small volumes

A critical requirement of clinical measurements of GCF is the size of the sample volumes of 1–5 μl. To enable measurements in such small sample volumes, capillary fill devices were constructed on top of the hydrogel coated IDEs on alumina substrates. In contrast to the results shown above, there was no stabilisation period in PBS prior to the addition of enzyme solution, but the test solutions were added to the device directly. Sample response curves for different HNE activities are shown in Fig. 8. The response to pH 7.5 charge transfer buffer shows a small decrease in impedance followed by a slow increase in the impedance with time. It is assumed that the decrease in impedance was caused by the hydration of the hydrogel film. The continued swelling of the film resulted in an increase of its thickness and therefore an increase in the impedance. The addition of HNE to the capillary fill device

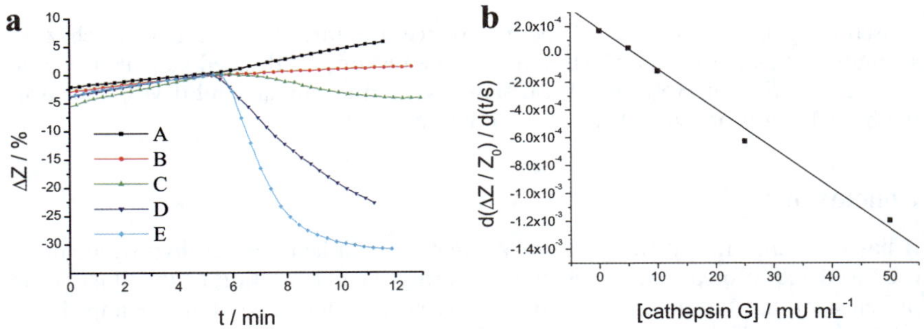

Fig. 6 (a) Impedance change measured with AAPFFK cross-linked hydrogel coated IDEs on alumina before and after addition of different activities of cathepsin G to pH 7.5 charge transfer buffer: 0 mU ml^{-1} (A), 5 mU ml^{-1} (B), 10 mU ml^{-1} (C), 25 mU ml^{-1} (D), 50 mU ml^{-1} (E); (b) calibration curve showing the rates of impedance change for the first 100 s after the addition of cathepsin G *versus* cathepsin G activity. Points represent the average of two individual measurements at each activity.

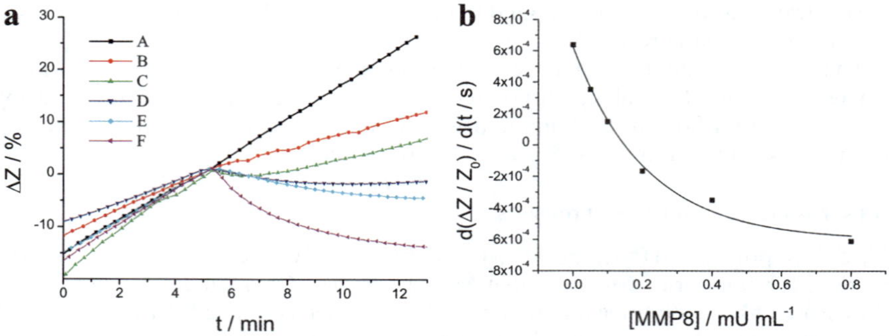

Fig. 7 (a) Impedance change measured with GPQGIWGQK cross-linked hydrogel coated IDEs on alumina before and after addition of different activities of MMP8 to a pH 7.5 charge transfer buffer: 0 mU mL^{-1} (A), 0.05 mU ml^{-1} (B), 0.1 mU ml^{-1} (C), 0.2 mU ml^{-1} (D), 0.4 mU ml^{-1} (E), 0.8 mU ml^{-1} (F); (b) calibration curve showing the rates of impedance change for the first 100 s after the addition of MMP8 *versus* MMP8 activity. Points represent the average of two individual measurements at each activity.

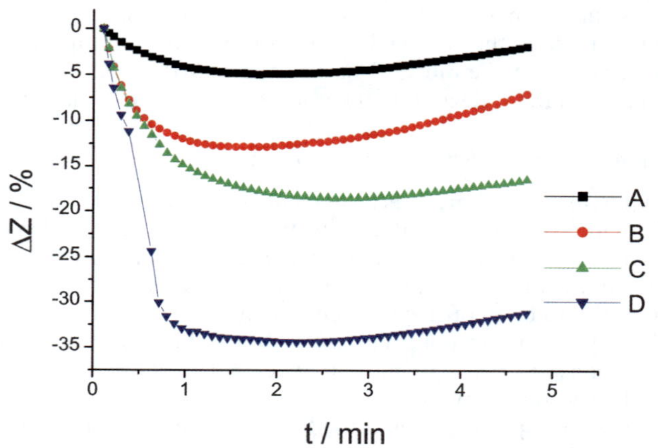

Fig. 8 Impedance change after filling capillary fill device on IDEs coated with an AAPVAAK cross-linked dextran hydrogel with pH 7.5 charge transfer buffer containing different activities of HNE: 0 U ml^{-1} (A), 5 U ml^{-1} (B), 10 U ml^{-1} (C) and 30 U ml^{-1} (D).

caused a significant decrease of the impedance, the rate and extent of which were strongly dependent on the HNE activity. The response is deemed sufficiently large for the detection of HNE in clinical samples, *i.e.* the capillary fill device will form the basis for a prototype of the clinical device.

Conclusions

It has been shown that thin films of peptide cross-linked dextran hydrogel can be used to detect protease activities by monitoring the degradation of the films in the presence of the target protease and a charge transfer reagent using impedance measurements. The sensor materials are generic – specificity for a particular protease can be achieved by employing an appropriate peptide sequence as a cross-linker. To date, three markers for periodontal disease, HNE, cathepsin G and MMP8, have been detected using this sensor system. Preliminary results have shown that measurements in small sample volumes can be facilitated by using a sealed capillary fill device providing the basis for a prototype device for clinical applications.

Depending on the self-assembled monolayer used to modify the electrode surface, hydrogel films were either shown to dissolve or to swell upon interaction with the target protease causing either a decrease or an increase in the impedance. Dissolution was achieved using a chemically inert monolayer with hydroxyl terminal groups, while swelling was observed when using an amine terminated monolayer, which bound the oxidised dextran covalently to the sensor surface. As a fairly high-molecular weight dextran was utilised, it can be assumed that most of the dextran molecules present in the film were bound to the surface in this manner. A surface-bound film may have advantages for applications where greater mechanical stability is required.

Measurements of the sensor response in the presence of BSA showed that background protein slowed down the sensor response. This was attributed to an increase in the viscosity of the solution and possibly to an interaction between the target protease and the dissolved protein.

Acknowledgements

The authors are grateful to the Technology Strategy Board (TP/8/BIO/6/I/Q0020H) for funding this project.

References

1 T. Sorsa, Y. L. Ding, T. Ingman, T. Salo, U. Westerlund, M. Haapasalo, H. Tschesche and Y. T. Konttinen, *J. Clin. Periodontol.*, 1995, **22**, 709–717.
2 J. M. Herrmann, J. R. Gonzales, R. H. Boedeker, J. Vonholdt and J. Meyle, *J. Clin. Periodontol.*, 2001, **28**, 31–37.
3 J. M. Mailhot, J. Potempa, S. H. Stein, J. Travis, J. D. Sterrett, P. J. Hanes and C. M. Russell, *J. Clin. Periodontol.*, 1998, **25**, 578–584.
4 B. M. Eley and S. W. Cox, *Br. Dent. J.*, 1998, **184**, 323–328.
5 H. Y. Chen, S. W. Cox, B. M. Eley, P. Mantyla, H. Ronka and T. Sorsa, *J. Clin. Periodontol.*, 2000, **27**, 366–369.
6 T. J. Venaille, G. Ryan and B. W. S. Robinson, *Respir. Med.*, 1998, **92**, 233–240.
7 A. Gervaix, R. Thompson, J. G. Bieth, H. P. Nick and S. Suter, *Eur. Respir. J.*, 1992, **5**, 566–575.
8 T. Tervahartiala, Y. T. Konttinen, T. Ingman, R. HayrinenImmonen, Y. Ding and T. Sorsa, *J. Clin. Periodontol.*, 1996, **23**, 68–75.
9 M. Wikstrom, J. Potempa, A. Polanowski, J. Travis and S. Renvert, *J. Periodontol.*, 1994, **65**, 47–55.
10 R. B. Millington, A. G. Mayes, J. Blyth and C. R. Lowe, *Sens. Actuators, B*, 1996, **33**, 55–59.
11 A. G. E. Saum, R. H. Cumming and F. J. Rowell, *Biosens. Bioelectron.*, 2000, **15**, 305–313.
12 N. M. Hanumegowda, I. M. White, H. Oveys and X. D. Fan, *Sens. Lett.*, 2005, **3**, 315–319.
13 R. E. Ionescu, S. Cosnier and R. S. Marks, *Anal. Chem.*, 2006, **78**, 6327–6331.
14 J. L. Stair, M. Watkinson and S. Krause, *Biosens. Bioelectron.*, 2009, **24**, 2113–2118.
15 X. D. Lou, L. Y. Zhang, J. G. Qin and Z. Li, *Langmuir*, 2010, **26**, 1566–1569.
16 C. Sumner, A. Sabot, K. Turner and S. Krause, *Anal. Chem.*, 2000, **72**, 5225–5232.
17 C. Sumner, S. Krause, A. Sabot, K. Turner and C. J. McNeil, *Biosens. Bioelectron.*, 2001, **16**, 709–714.
18 S. Krause, C. J. McNeil, C. Fernández-Sánchez and A. Sabot, in *Encyclopedia of sensors*, ed. C. Grimes, American Scientific Publishers, 2006, pp. 289–306.
19 A. G. E. Saum, R. H. Cumming and F. J. Rowell, *Biosens. Bioelectron.*, 1998, **13**, 511–518.
20 M. M. Orosco, C. Pacholski, G. M. Miskelly and M. J. Sailor, *Adv. Mater.*, 2006, **18**, 1393.
21 A. Sabot and S. Krause, *Anal. Chem.*, 2002, **74**, 3304–3311.
22 Y. Zhou, S. Jiang, S. Krause and J. N. Chazalviel, *Anal. Chem.*, 2007, **79**, 8974–8978.
23 L. Ruys, J. Vermeersch, E. Schacht, E. Goethals, P. Gyselinck, P. Braeckman and R. Van Severen, *Acta Pharm. Technol.*, 1983, **29**, 105–112.
24 S. Netzelarnett, G. Fields, H. Birkedalhansen and H. E. Vanwart, *J. Biol. Chem.*, 1991, **266**, 6747–6755.

[illegible]

Acknowledgements

[illegible]

References

[illegible]

Nucleic acid aptamers: ideal reagents for point-of-care diagnostics?

Anthony E. G. Cass* and Yangyang Zhang

Received 7th July 2010, Accepted 23rd July 2010
DOI: 10.1039/c005487a

Nucleic acid aptamers have many of the properties that make for effective reagents in point-of-care diagnostic devices and whilst superficially similar to antibodies as affinity reagents the scope for engineering them to fit this role is considerable. Synthesis of aptamers allows for the incorporation of functionality for both immobilisation and electrochemical signalling in a way that is compatible with the 'strip' type sensors familiar in enzyme sensors, such as those for glucose. Control of the structure of DNA aptamers through Watson–Crick base pairing allows for different electrochemical assay formats, whilst ferrocenes provide a versatile redox label and insights into the interactions between the aptamer and its target are obtained through both cyclic and square-wave voltammetries. Square-wave voltammetry in particular demonstrates good analytical utility. Two different approaches were used to assemble aptamer/redox probe structures on the surface of gold electrodes and both showed "signal on" behaviour (*i.e.* the current increases in the presence of analyte) although the two different methods appear to give quite distinct surface coatings.

1 Introduction

The use of point-of-care testing (POCT) has expanded greatly over the past 20 years driven by a combination of clinical need and technology development with the sector now accounting for about 15% by value of the US clinical diagnostics market.[1] More significantly it is the major growth driver in this industry showing a compound annual growth rate of around 7% compared to essentially a static position in laboratory-based diagnostic instruments and reagents. The expectation is that POCT will continue to grow as a consequence of three convergent trends; an ageing population with the concomitant shift from acute to chronic illness, the need to contain healthcare costs by shifting more of the delivery to home and primary care settings and finally the acceptance of the effectiveness of personalised therapy rather than the "one size fits all" model.

Amongst the different POCT formats, affinity techniques form a large and diverse group with target analytes ranging from small molecules such as drugs to large protein biomarkers associated with particular disease states.[2] Traditionally affinity-based methods rely on the specificity and high affinity of antibodies to detect the target analyte; specificity because clinical samples are complex with many potential interferents present and high affinity to give the requisite detection limits for trace biomarkers

Despite their well established position in POCT devices, there is a rapidly growing interest in alternative affinity reagents and amongst these are the class of molecules known as aptamers. Whilst an aptamer can be DNA-, RNA-or protein-based they share the common characteristics of small size and their selection, *in vitro*, from large

Institute of Biomedical Engineering, Imperial College London, UK. E-mail: t.cass@imperial.ac.uk; Fax: +02075945196; Tel: +02075955195

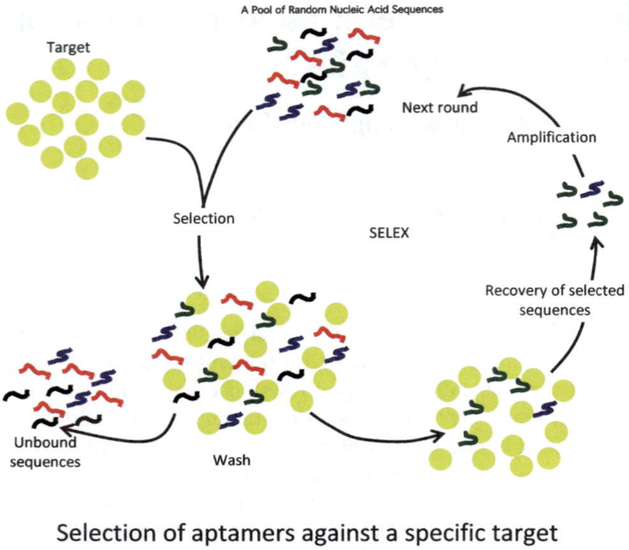

Fig. 1 The systematic evolution of ligands by exponential enrichment (SELEX) process to isolate nucleic acid aptamers from a large random sequence pool.

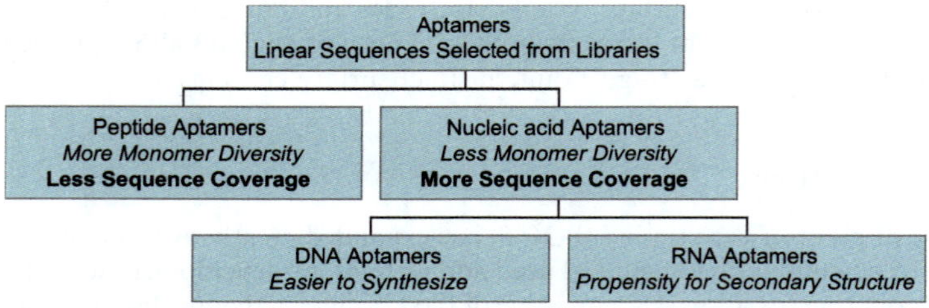

Fig. 2 A classification scheme for aptamers highlighting some of their characteristic properties.

random sequence libraries.[3] The selection of sequences with the desired specificity and affinity *in vitro* follows from successive rounds of binding and amplification (Fig. 1) and once the most suitable sequence has been found the molecule can be synthesised (nucleic acids and peptides) or expressed (proteins). Fig. 2 illustrates a classification of aptamers and in this paper we will be concerned solely with nucleic acid aptamers.

In drawing up a "wish list" of desirable properties for POCT affinity reagents nucleic acid aptamers fulfil most of the requirements including:

- Generic physical and chemical properties
- Readily obtainable
- Traceable
- Stable
- Reproducible
- Controlled affinity and specificity
- Specific chemical modification
- Can be produced to any target molecule
- Flexible signal transduction schemes
- Straightforward modular design rules
- Low molecular weight

however they should not be seen simply as replacement molecules for existing antibody-based devices, but rather should be the basis of new device designs and formats that exploit the unique properties associated with their molecular structures. In particular these properties include:

• The capacity for modification at the 5′ and 3′ ends with functional groups to allow the elaboration of additional chemistries for immobilisation and the attachment of signalling groups.

• Well understood rules ('Watson–Crick base pairing') for the interaction between nucleic acid strands; both inter- and intra-molecular

• Often large structural changes associated with target binding ('induced fit')

• Robust structures

In the light of the above advantages it is unsurprising that aptamers have become increasingly used in analytical applications in general and in biosensors in particular.[4–6] It can therefore only be a matter of time and development before they begin to form the basis of POCT devices.

Although they have been combined with many different "read-out" mechanisms the combination of aptamers with electrochemical signal transduction is an especially well matched format. Aptamers are readily immobilised on electrode surfaces, redox-active moieties can be attached either covalently or non-covalently to DNA and the small size of aptamers means that such attached redox probes can be close enough to the electrode surface to undergo electron exchange.

Aptamer-based electrochemical sensors can be classified into either "signal on" or "signal off" formats, in the former the current increases when target analyte binds whilst in the latter it decreases. Although the exact design varies we can associate "signal on" electrochemical aptamer sensors ('aptasensors') with the redox reporter moving closer to the electrode surface and "signal off" aptasensors with it moving further away.

Examples of "signal on" sensors include the work of Kim *et al.*[7]who took advantage of the blocking of the electrode to ferricyanide by the negatively charged DNA backbone of an oxytetracycline binding aptamer. When the aptamer bound its ligand a conformational change 'unblocked' the electrode resulting in an increased current. Rodriguez *et al.* used a similar unblocking concept with a lysozyme binding aptamer but measured the change in accessibility by electrochemical impedance spectroscopy. In a different approach Lai *et al.*[8] took advantage of a conformation-switching aptamer that bound platelet-derived growth factor and in doing so moved a redox label (methylene blue) nearer to a gold electrode to which the aptamer was tethered. Another alternative redox label in "signal on" aptamer sensing was that of Radi *et al.*[9] who used a ferrocene-labelled aptamer, again thiol tethered to gold where the conformational switch in the presence of thrombin moved the redox label close to the surface.

Amongst the signal off sensors are those described by Lu *et al.*[10] who hybridised a short, ferrocene-labelled oligonucleotide to an immobilised thrombin binding aptamer. In the presence of the aptamer the labelled oligonucleotide was displaced with a subsequent decrease in the current. Xiao *et al.*[11] used a different approach with a single-stranded methylene blue labelled thrombin binding aptamer where the conformation switching G-quadruplex formation resulted in a reduction in current as the methylene blue label was unable to approach the electrode closely enough to undergo a fast electron exchange.

Lysozyme is a biomarker for bacterial infections in a variety of bodily fluids including serum, saliva, CSF and wound exudate and lysozyme binding electrochemical aptasensors have been described by several groups in a variety of formats including differential pulse[12] and stripping voltammmetry of the protein,[13] as well as cyclic voltammetry,[14,15] electrogenerated chemiluminescence[16] and layer-by-layer voltammetry[17] of labelled aptamers.

2 Experimental

Synthetic oligonucleotides were supplied by Thermo-Fisher and calculations of hybridisation-free energies/melting temperatures used the Vector NTi programe (Invitrogen). 3′-labelling of oligonucleotides was through a two-step reaction; ferrocene carboxylic acid was converted to an active ester through reaction with HOBt (hydroxybenzotriazole) in dichloromethane in the presence of solid 1-ethyl-3-dimethyl-aminopropyl carbodiimide (EDC). The product of the reaction was then added to 3′-aminohexyl modified oligonucleotide in bicarbonate buffer pH 9, using a 10-fold excess of the ferrocene derivative. Modified oligonucleotides were then purified by a sequence of precipitation and dialysis steps and the modification confirmed by mass spectrometry. The concentration of single-stranded DNA was estimated using OliGreen (Invitrogen Ltd).

Immobilisation of the aptamers onto a gold electrode (250 μm diameter, cleaned by polishing with alumina and then electrochemically) used either a one-step procedure with the apatmer prehybridised to a short oligonucleotide in solution before adding it to the electrode, or a two-step procedure where the apatmer was first immobilised and then hybridised to the short oligo. In both cases the electrode was incubated with the aptamer for 24 h at room temperature and after washing in 10 mM Tris buffer pH 7.4 treated with a solution of a low-molecular-weight thiol such as mercaptosuccinic acid for 30 min to block unoccupied binding sites on the gold surface.

Cyclic and square wave voltammetries were performed using a CHI 103a potentiostat. Cyclic voltammograms were acquired at scan rates between 5 and 100 mV s^{-1} and square-wave voltammagrams at frequencies between 10 and 1000 s^{-1} with a step potential of 25 mV and an incremental potential of 2 mV. All potentials are with respect to a silver/silver chloride reference electrode and used a three-electrode configuration.

Surface plasmon resonance (SPR) measurements were made on a BiaCore instrument using bare gold chips with a layer of adsorbed thiol modified aptamer.

Hen egg white lysozyme was from Sigma-Aldrich and all titrations were performed using a phosphate buffered saline background prepared as a 1× PBS solution.

3 Results and discussion

A variety of lysozyme binding aptamers (both DNA and RNA) have been described in the literature and we have chosen to use the sequence described by Wang and co-workers in their electrochemical impedance spectroscopy based sensors.[18] There is some controversy surrounding this aptamer as its sequence was derived from that of an RNA aptamer that bound lysozyme, wherein the DNA sequence was derived from simply reverse transcribing the published RNA sequence.[19] As the conclusion that this DNA aptamer bound lysozyme was based solely on the electrochemical data we first of all sought to confirm this by an independent technique and used SPR to this effect and Fig. 3 shows the results of immobilising the Wang aptamer (carrying a 5′-hexane thiol linker) on a gold chip and titrating with lysozyme. Fitting the data to a Langmuir isotherm yielded a dissociation constant of 125 nM (about 6 times larger than that of the original RNA aptamer, albeit measured using a different technique).

Based on these results 2 short oligonucleotides were synthesised with 5′-aminohexyl groups, one designed to hybridise close to the 5′ (surface bound) end of the aptamer and the other designed to bind to the distal (3′) end. Both oligonucleotides had calculated affinities sufficiently high to ensure stability at room temperature, whilst being low enough to be displaced by lysozyme. The sequences and locations of the aptamers and probes are shown in Fig. 4. As nothing is known of the structure of the aptamer/lysozyme complex it is not possible to predict accurately which

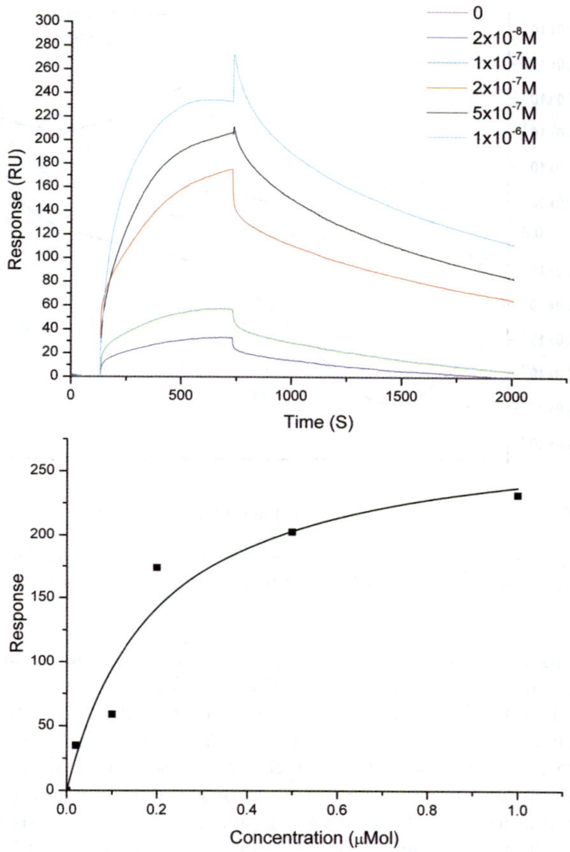

Fig. 3 (a) SPR response curves for lysozyme binding to a DNA aptamer. (b) The resulting binding isotherm.

```
                              FcNH.C6.TTCGTACG

GATTCCATTGAGACGTGAGAAATCGGGACTACTTAAGCATCTA.C6SH
3'                                           5'


FcNH.C6.AGTAACTAGA

   GATTCCATTGAGACGTGAGAAATCGGGACTACTTAAGCATCTA.C6SH
   3'                                           5'
```

Fig. 4 The sequences of a lysozyme binding DNA aptamer and two short oligonucleotide redox probes.

nucleotides are involved in the binding and hence the optimal location of the redox probe oligonucleotide.

Two approaches to assembling the aptamer-probe sequences were used with both aptamers; in one the hybridisation was performed in solution and the resulting duplexes adsorbed to the electrodes. In the second approach the apatmer is adsorbed

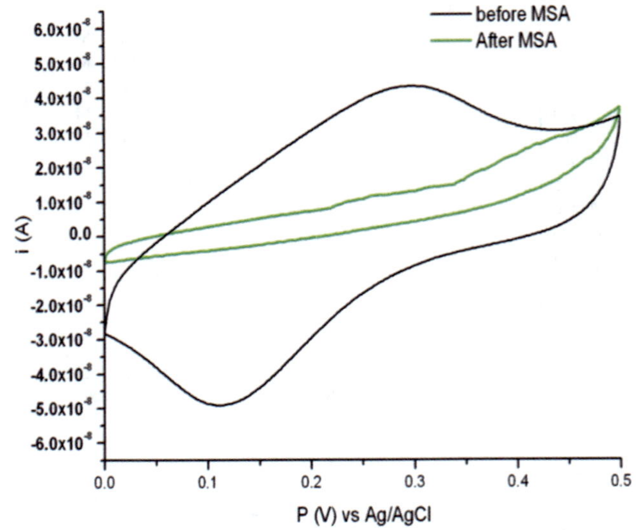

Fig. 5 Cyclic voltammogram of the aptamer/probe hybrid on a gold electrode in the absence (black line) and presence (green line) of mercaptosuccinic acid as a blocking agent.

and then hybridised to the probe. In both cases exposed gold surfaces were then 'back-filled with mercaptosuccinic acid (MSA). In the case where the redox probe is hybridised close to the surface (*i.e.* at the 5′-end of the aptamer) the effect of the mercaptosuccinic acid is to block electron transfer from the ferrocene to the surface (Fig. 5). This phenomenon is well known with diffusible redox couples and we assume in this case that the ferrocene is behaving in a similar fashion. The alternative explanation is that the MSA is displacing the aptamer from the surface, although as we shall see in the next paragraph this explanation is difficult to reconcile with the restoration of the current in the presence of lysozyme.

When an electrode prepared by the one-step method (*i.e.* hybridisation of the probe in solution and then adsorption of the duplex on the surface) is titrated with lysozyme and the resulting square-wave voltammograms (SWVs) measured (Fig. 6a) it can be seen that there is an increase in current, followed by a subsequent decrease (Fig. 6b). This reduced signal at high dose is reminiscent of the 'hook effect' seen in sandwich immunoassays and attributed in that case to a high loading of capture antibody resulting in a large amount of bound antigen that in turn inhibits the binding of the signalling antibody. In the case here of the apatamer/probe system we assume that lysozyme binding doesn't displace the probe but rather changes the structure of the aptamer and/or shields the charges on the MSA to allow electron transfer to occur. When the notional amount of probe that is added to the aptamer in solution is increased from a 1 : 1 stoichiometry to a 2 : 1 stoichiometry then the maximum current roughly doubles (Fig. 7a) and the "hook effect" disappears (Fig. 7b). We can attribute both these observations to a greater ratio of the probe to aptamer on the surface.

However inspection of the maximum currents at both low and high probe loading suggests that the surface organisation my be more complex that a simple cartoon would suggest. The electrode diameter is 250 μm and assuming a roughness factor of around 3 for planar gold electrodes then at a typical close packed density of 10^{12} molecules cm^{-2} we would expect around 10^9 molecules on the surface. The SWV allows a calculation of the total charge passed during an experiment and using the Faraday constant (10^5 C mol^{-1}) this can be converted to the number of molecules oxidised or reduced and when this calculation is carried out there is a factor of nearly

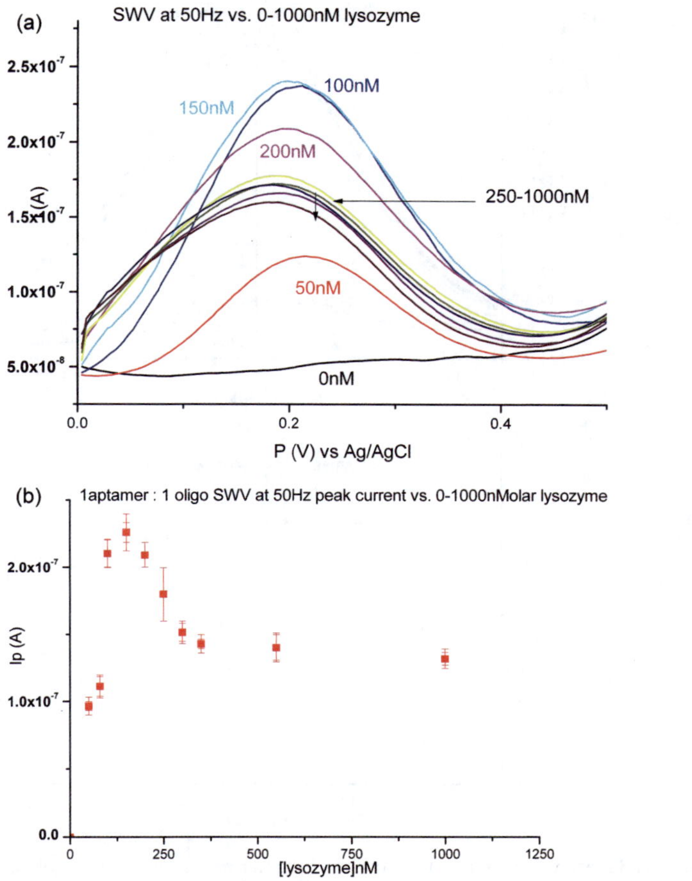

Fig. 6 (a) Square-wave voltammograms of an aptamer/redox probe hybrid immobilised on a gold electrode. (b) The corresponding binding isotherm.

10^3 between the number calculated from the charge and the number expected from a close packed monolayer, suggesting that the aptamer/probe layer is many molecules thick. If this is indeed the case then the probes must also be in electrochemically connected ('wired') to each other in the presence of lysozyme. Further work on the nature of the aptamer and aptamer/probe layers will require additional surface analytical tools.

Fig. 6a and 7a also show a concentration-dependent change in peak potential, and this is most clearcut in Fig. 7a. The shift in peak potential to more cathodic values (Fig. 8) is consistent with the ferrocene becoming easier to oxidise and this can be understood if the local environment around the redox centre is negatively charged as this would serve to stabilise the ferrocenium form. This is further evidence that the binding of lysozyme results in a significant reorganization of the aptamer/probe structure. Fitting both the current- and potential-dependent changes to a hyperbolic binding curve yields apparent dissociation constants of around 200 nM not dissimilar to that determined by SPR (125 nM).

A final point to notice with the one-step aptasensor is that this was initially designed as a "signal off" displacement sensor where the lysozyme resulted in a loss in current due to displacement of the redox probe. From an analytical perspective the behaviour of this probe is quite different and potentially more robust in giving

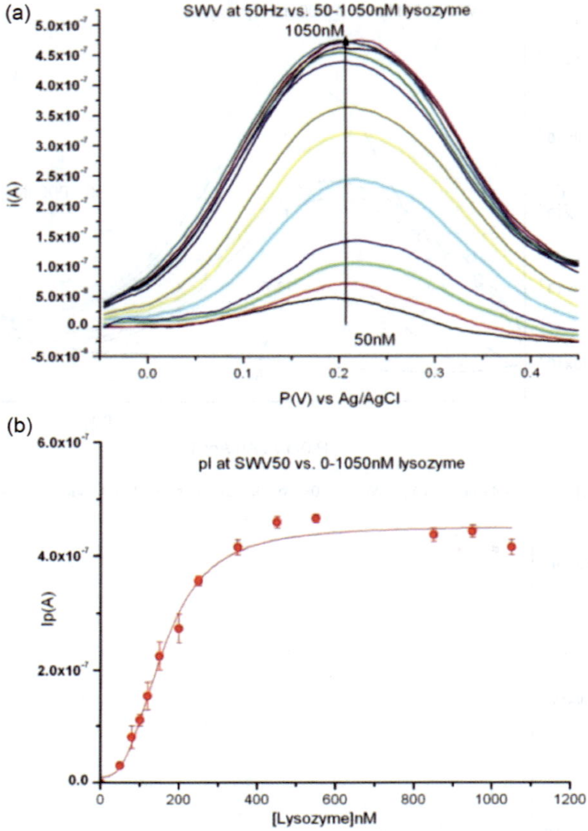

Fig. 7 (a) Square-wave voltammograms of a aptamer/probe hybrid on a gold electrode in the presence of increasing concentrations of lysozyme. (b) Corresponding dose–response curve.

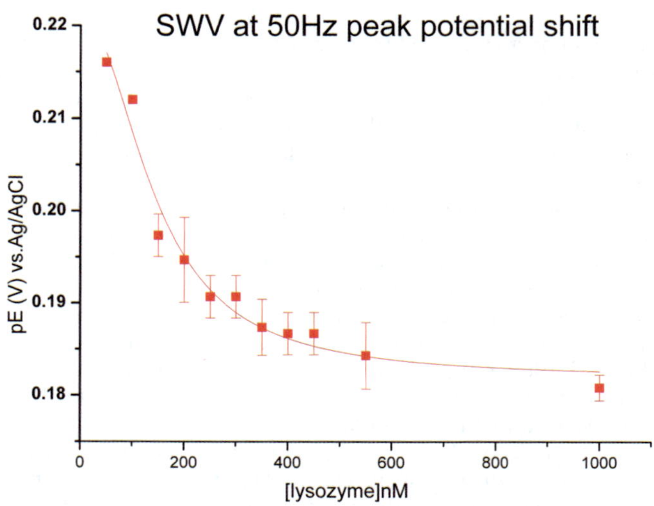

Fig. 8 Shift in peak potential as a function of lysozyme concentration.

a "signal on" response to analyte. Although we do not have a definitive explanation for this observation it is likely that the lysozyme, rather than displacing the labelled oligonucleotide causes a condensation of the multilayered aptamer/probe layer on the electrode that results in a decrease in the average distance of the ferrocene reporter from the surface.

Although the response of the aptasensor as shown in Fig. 7 is encouraging, the apparent complexity of the surface structure led us to consider a two-step approach to assembling the aptamer/probe complex on the electrode. In the two-step method the aptamer is first bound to the gold surface through the 5′-thiol and then the redox probe is subsequently hybridised to the aptamer. Again the surface is back-filled with MSA and a lysozyme titration carried out. As shown in Fig. 9a this is again a "signal on" response and shows the "hook effect" seen with the one-step electrode preparation (Fig. 9b). However the maximum current is around 2 orders of magnitude smaller than reported in Fig. 5 consistent with coverage on the electrode closer to a monolayer. Once again the peak potential shifts with lysozyme concentration (Fig. 10), and there is considerable scatter on the data although the potentials are

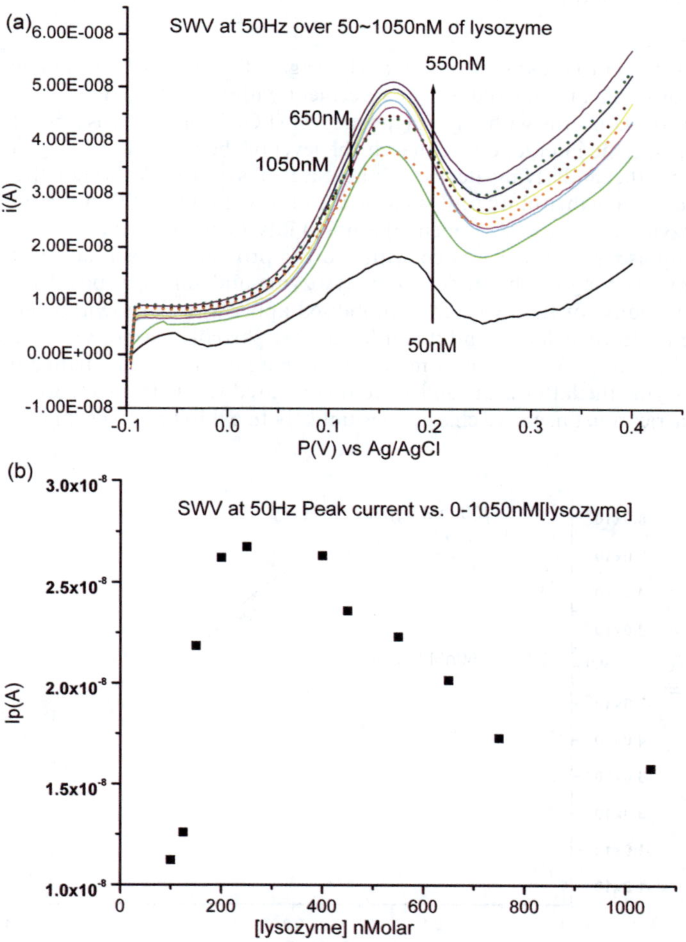

Fig. 9 (a) Square-wave voltammograms of an aptamer/probe electrode prepared by a two-step hybridisation reaction. (b) The corresponding dose response curves.

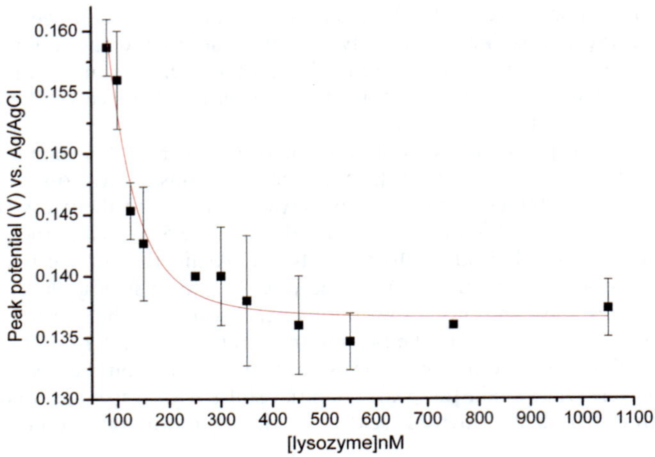

Fig. 10 Peak potential shift as a function of lysozyme concentration as for a two-step preparation of the aptamer/probe surface.

slightly different to those seen in Fig. 8. Here again the fitting of the data gives dissociation constants around 100 nM in agreement with the SPR data.

One of the concerns with any diagnostic test (POCT or otherwise) is interference from non-specific interactions, either at the level of the molecular recognition reaction or on the underlying device in the case of sensors. Although the aptamer described has a high affinity for lysozyme the observation that DNA is a poly(anion) and lysozyme a poly(cation) opens the possibility of non-specificity and so to test this the aptasensor was evaluated with 2 other proteins; trypsin and lactalbumin. The former is roughly the same size as lysozyme and similarly positively charged and the response of the two-step immobilised aptasensor is shown in Fig. 11. The response is clearly evident, and it can be at least partially suppressed by increasing the ionic strength. Moreover, comparing the magnitude of the change in current with lysozyme the latter is around 10-fold compared to the former's 3-fold. Lactalbumin carries a net negative charge so is unlikely to exhibit non-specific interactions

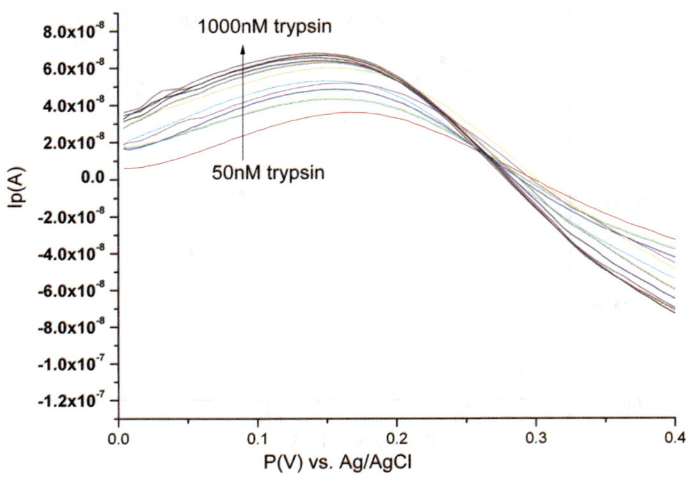

Fig. 11 Response of the aptamer probe sensor to trypsin.

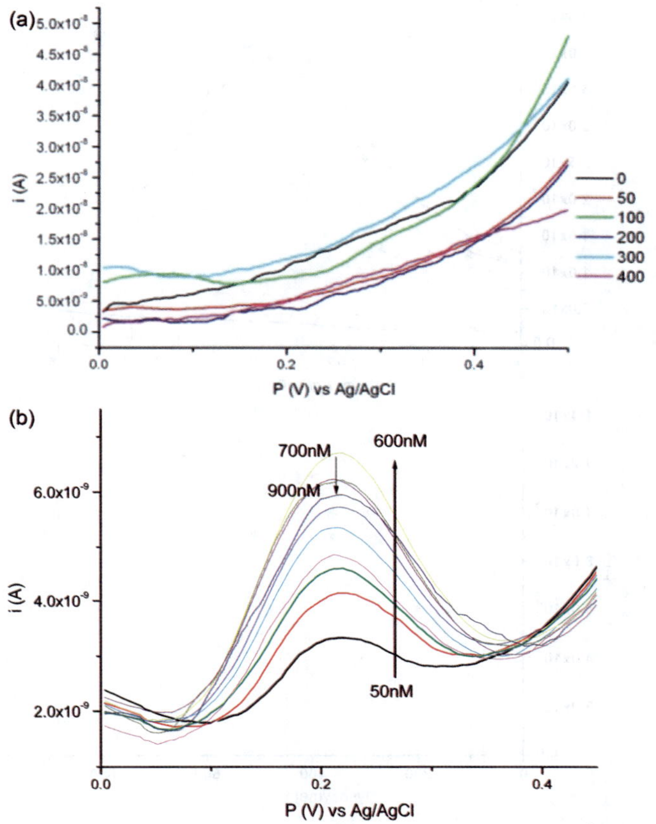

Fig. 12 (a) Response of the aptamer/probe sensor to lactalbumin. (b) Subsequent response to lysozyme.

with the aptamer, it is however quite hydrophobic and therefore can be used to probe non-specific surface interactions. As Fig. 12a a shows there is little effect of lactalbumin on producing a background signal and it appears not to interfere with the subsequent lysozyme dependent response (Fig. 12b).

In order to explore the effect of the location of the redox probe oligonucleotide on the aptasensor a second ferrocene labelled was hybridised to a region of the aptamer distal from the 5′-thiol group (and hence presumably the surface). A one-step attachment was used (*i.e.* the redox probe hybridised to the aptamer in solution and the duplex then adsorbed on the electrode surface). At a notional 1 : 1 stoichiometry the same "signal on" behaviour is seen (Fig. 13a) with currents consistent with multilayer formation and a cathodic shift in peak potential. In this case also a hook effect is seen in the dose response curve and the estimated dissociation constant is again around 200 nM (Fig. 13b). Finally the effect of the surface blocker was briefly investigated in this system, using either MSA or the nonpolar butanethiol. The results were broadly similar with the exception that with the uncharged blocker no hook effect was observed.

4 Summary and conclusions

Currently POCT uses one of two main technologies, either electrochemical 'strips' or lateral flow with a visual read-out. Both are mature and are at the end of their core patent life. The former can give quantitative numerical data but has been restricted

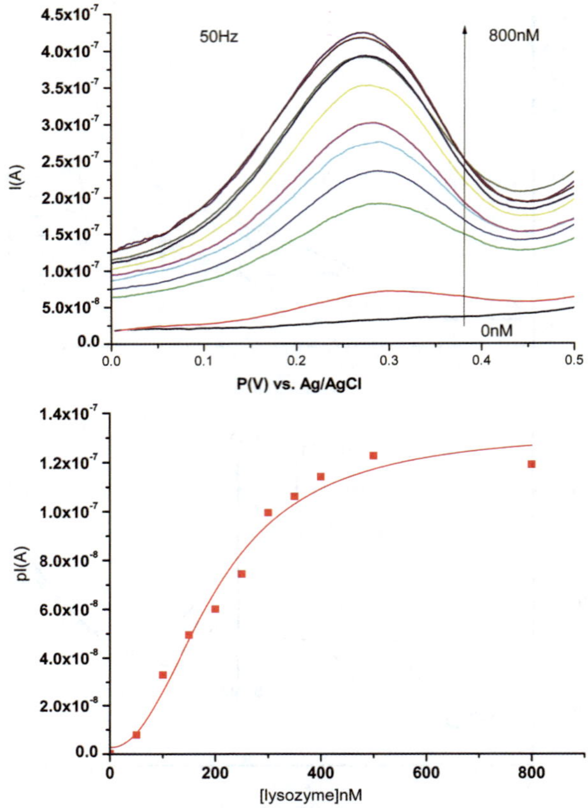

Fig. 13 (a) Square-wave voltammograms of a distal aptamer/probe hybrid. (b) Resulting dose–response curves.

to enzyme-based reactions of metabolites whilst the latter is a qualitative antibody based test for both large (protein) and small analytes.

Developing a new generation of electrochemical sensors that move beyond the enzymes/metabolites combination promises to expand this technology into a host of new analytes that will be critical in delivering the low cost, decentralized monitoring of chronic disease that will be the norm this century. Currently many molecular recognition/instrumentation approaches are being developed in both academic and commercial laboratories and it is the contention of this paper that electroactive nucleic acid aptamers ('aptrodes') offer many significant advantages in this regard.

Using relatively straightforward electrochemistry the binding of the target transduces into a change in current and the sensitivity of the aptrode is determined by the aptamer affinity as expected. Contrary to expectations however, was the observation that the response was a "signal on" one when what was expected from a simple displacement model was a "signal off" response. It appears from this unexpected result, in conjunction with the larger than anticipated currents, that the surface of the aptrode has a more complex structure than a simple cartoon would suggest.

References

1 J. F. Dooley, *Point of Care: The Journal of Near-Patient Testing & Technology*, 2009, **8**, 154–156.

2 S. E. F. Melanson, *Point of Care: The Journal of Near-Patient Testing & Technology*, 2009, **8**, 166–170.
3 G. Mayer, *Angew. Chem., Int. Ed.*, 2009, **48**, 2672–2689.
4 W. Mok and Y. Li, *Sensors*, 2008, **8**, 7050–7083.
5 J.-O. Lee, H.-M. So, E.-K. Jeon, H. Chang, K. Won and Y. H. Kim, *Anal. Bioanal. Chem.*, 2008, **390**, 1023–1032.
6 I. Willner and M. Zayats, *Angew. Chem., Int. Ed.*, 2007, **46**, 6408.
7 Y. S. Kim, J. H. Niazi and M. B. Gu, *Anal. Chim. Acta*, 2009, **634**, 250–254.
8 R. Y. Lai, K. W. Plaxco and A. J. Heeger, *Anal. Chem.*, 2007, **79**, 229–233.
9 A. Radi, J. Sanchez, E. Baldrich and C. O'Sullivan, *J. Am. Chem. Soc.*, 2006, **128**, 117–124.
10 Y. Lu, N. Zhu, P. Yu and L. Mao, *Analyst*, 2008, **133**, 1256–1260.
11 Y. Xiao, A. Lubin, A. Heeger and K. Plaxco, *Angew. Chem., Int. Ed.*, 2005, **44**, 5456–5459.
12 A. Erdem, H. Karadeniz, G. Mayer, M. Famulok and A. Caliskan, *Electroanalysis*, 2009, **21**, 1278–1284.
13 A. N. Kawde, M. C. Rodriguez, T. M. H. Lee and J. Wang, *Electrochem. Commun.*, 2005, **7**, 537–540.
14 A. K. H. Cheng, B. Ge and H.-Z. Yu, *Anal. Chem.*, 2007, **79**, 5158–5164.
15 C. Y. Deng, J. H. Chen, L. H. Nie, Z. Nie and S. Z. Yao, *Anal. Chem.*, 2009, **81**, 9972–9978.
16 J. G. Bai, H. Wei, B. L. Li, L. H. Song, L. Y. Fang, Z. Z. Lv, W. H. Zhou and E. K. Wang, *Chem.–Asian J.*, 2008, **3**, 1935–1941.
17 Y. Du, C. G. Chen, B. L. Li, M. Zhou, E. K. Wang and S. J. Dong, *Biosens. Bioelectron.*, 2010, **25**, 1902–1907.
18 M. C. Rodriguez, A.-N. Kawde and J. Wang, *Chem. Commun.*, 2005, 4267.
19 E. J. Cho, J.-W. Lee and A. D. Ellington, *Annu. Rev. Anal. Chem.*, 2009, **2**, 241–264.

General discussion

Professor Graham opened the discussion of the paper by Professor Schuhmann: The conclusion mentions label-free detection and I was interested in your view of this in light of the use of an intercalator and the ferrocene, although I now realise the ferrocene was playing a different role.

Professor Schuhmann responded: The electroanalysis of the introduced ferrocene at the target DNA strands provides information about the quality of the spotting procedure and the amount of hybridized DNA molecules at rather high concentrations of target DNA. Even more importantly we could examine the degree of intercalation *via* the modulation of the electron-transfer kinetics of the DNA-bound ferrocene units. However, it was not intended to use a ferrocene-modified DNA in the final DNA assay.

We call the assay principle essentially label-free because no binding of a reporter group to either the capture or the target DNA strand has to be done prior to the hybridization process. The specificity of the developed intercalator allows for a selective post-labelling of exclusively double-stranded DNA and is hence seen as a generic principle for the read-out of hybridization events.

Professor Graham then asked: The signal amplification is interesting and I was wondering if the limit of detection would be competitive with PCR-based assays using target amplification?

Professor Schuhmann replied: In PCR-based assays an exponential amplification of the DNA is achieved which can be adjusted to the sensitivity of the later read-out process. The proposed assay including enzyme amplification relies on the available DNA strands. We presently assume that the detection limit will be sufficient for DNA or RNA sequences which are highly expressed (*e.g.* S16 RNA). This may allow the detection of RNA or DNA without PCR directly from the sample.

Dr Zhou asked: The use of biotin-modified intercalator coupled with avidin–enzyme conjugate amplification appears to be very powerful. Is the intercalation reaction sequence specific? How many intercalator molecules can bind to one DNA duplex?

Professor Schuhmann responded: We have evaluated the intercalator (proflavin) mainly with respect of its selectivity to intercalate into dsDNA. See our previous manuscripts[1] for the evaluation of the properties of the intercalator with respect to unspecific adsorption at the SAM interface, as well as its inability to bind to ssDNA. The intercalator binds specifically at G–C base pairs. It can be assumed that several intercalator molecules are bound by the dsDNA. However, due to the large size of the avidin/alkaline phosphatase conjugate this will not lead to an increased immobilized enzyme activity at the monolayer.

1 (*a*) M. Gębala and W. Schuhmann, Controlled orientation of DNA in a binary SAM as a key for the successful determination of DNA hybridization by means of electrochemical impedance spectroscopy, *ChemPhysChem*, 2010, **11**, 2887–2895; (*b*) M. Gębala, L. Stoica, D. Guschin, L. Stratmann, G. Hartwich and W. Schuhmann, A biotinylated intercalator for selective post-labeling of double-stranded DNA as a basis for high-sensitive DNA assays, *Electrochem. Commun.*, 2010 **12**, 684–688; (*c*) M. Gębala, L. Stoica, S. Neugebauer and W. Schuhmann, Label-free detection of DNA hybridization in the presence of intercalators using electrochemical impedance spectroscopy, *Electroanalysis*, 2009, **21**, 325–331.

Dr Davies asked: In a biological sample, *e.g.* blood, there may be inhibitors of alkaline phosphatase and variable biotin (which sits around 100 to 400 pM), biocytin and catabolite levels which may cause assay variability. In addition, the use of an enzyme substrate complicates the assay, as does the enzyme, which may adsorb to a small extent and contribute background signal. Having shown that the four ferrocene units are in electronic contact, and so all are capable of rapid electron delivery and removal from the electrode, could you simplify the assay by using a much larger ferrocene label to replace the biotin on the proflavine compound?

Professor Schuhmann answered: As shown, the redox conversion of the DNA-bound ferrocene units does not ensure a sufficient amplification of the current signal, because only a limited number of electrons can be exchanged during the time of potential cycling.

In order to detect low amounts of hybridized DNA, a system with a amplification factor is required. Assuming that an enzyme may have a turnover of 1000 substrate molecules per second, one enzyme molecule is providing the current of 4000 ferrocene units during 1 s potential cycling, or even more for slower potential scans. We additionally do not foresee a significant impact of the sample components. After hybridization the sensing surface can be thoroughly washed before incubation with first the intercalator and second the avidin/alkaline phosphatase conjugate. Thus, the enzyme will not come in contact with sample components.

Professor Gooding remarked: In the system you have presented, the constraints for the DNA recognition interface appear to be quite different to other DNA sensors as you do not need high hybridisation efficiencies for a high signal, what you need is space for the avidin/alkaline phosphatase conjugate to bind. So it seems that a low surface density of probe DNA might be better. Have you looked at ways to reduce probe DNA density on the surface to give optimal avidin/alkaline phosphatase binding?

Professor Schuhmann responded: A specific study addressing the density of the grafted DNA molecules for optimal binding of the avidin/alkaline phosphatase conjugate has not been carried out yet. In the presented system always the same predefined concentration of the capture probes was spotted on each chip. Hybridization was then carried out in solutions that had different concentrations of the target DNA. In order to achieve a certain correlation between the target DNA concentration and the signal, we assumed that we have to offer sufficient capture strands at the interface. However, even a small fraction of hybridized DNA should lead to a measurable current response.

We have observed that for those chips that were hybridized with a lower amount of target DNA, the signal was higher. Avidin has 4 binding sites for biotin. Therefore, at interfaces which contain a high density of dsDNA the avidin may cross-bind DNA molecules *via* biotin moieties at the intercalator. This would lead to lower amplification of the signal. Based on these observations, we agree that a sufficiently low loading of the capture DNA strands should be envisaged for optimal sensitivity.

Professor Cass said: Do you have an idea of what proportion of the probe molecules are actually available for hybridisation? How could this be determined?

Professor Schuhmann replied: Actually, we do not know about the hybridization efficiency in this specific case and we did not perform related experiments, which are quite difficult at the level of the DNA chips. For bigger electrode surfaces one can determine the amount of immobilized capture strands by using positively charged transition metal complexes (*e.g.* ruthenium(III) hexaamine) which can be electrostatically bound at the phosphate backbones of DNA molecules (capture probe). By means of voltammetry of these complexes one can estimate the amount of DNA strands being grafted at an electrode surface, as presented by Steel *et al.*

Subsequently, one needs a signal from dsDNA (after the hybridization event). This can be obtained from the redox reaction of the ferrocene moieties attached to the target strand or by the method mentioned above. Finally, the relation between both signals for ssDNA and dsDNA has to be calculated to determine the efficiency of the hybridization event.

Dr Mount commented: What is the separation of the electrodes in your MEA? Assuming diffusion lengths of the order of the square root of $4Dt$, I calculate diffusion over tens of micrometres in a few seconds with a reasonable value of $D = 5 \times 10^{-6}$ cm^2 s^{-1}. Why therefore does cross-talk from the product from neighbouring electrodes not affect, for example, the chronoamperograms shown in Fig. 4 of your paper?

Professor Schuhmann responded: The separation of the electrodes in the array is 125 μm (edge to edge). Since the enzyme is immobilized in a monolayer on top of the electrode with a distance of only some nm the probability that the enzymatically produced product is diffusing toward the own electrode. The concentration of the enzymatically produced *p*-aminophenol is very tiny and will lead only to a tiny increase of the background current over time.

Professor Beynon opened the discussion of the paper by Dr Brown: The assays have to be delivered in the context of the biological samples that need to be sampled and assayed. I was wondering what volume of biological matrix would be needed for confident detection levels. I also wonder about the extent to which such assays survive in ‘dirty’ biological matrices.

Dr Brown answered: The volume of biological matrix is going to vary depending upon what is sampled (blood, sputum, urine) and at what stage in the infection the samples are taken. At this stage we have shown that aptamers/antibodies can be used to detect our targets in “dirty” biological matrices but we clearly need to further optimize our assays and improve the affinity of our detection reagents to increase the sensitivity the current detection levels. We also intend to use Luminex’s new MAGPIX® system (http://www.luminexcorp.com/magpix) that includes the use of magnetic beads and enhanced imaging technology, which should improve the capabilities of our current assays for detection of our targets under challenging conditions.

Professor Vadgama commented: Monoclonal antibodies (MAbs) are not necessarily better than polyclonal for affinity, but you do seem to have isolated MAbs in a nice way that demonstrate a very high affinity.

However, whatever the affinity, you need to locate the target in a complex sample—that constitutes a major sampling problem.

Dr Brown responded: We are currently focusing upon identifying abundant targets that are present in complex samples from acute infections. Our strategy for detection aims to use a “cocktail” of high affinity MAbs, FAbs and aptamers to achieve our objective of developing robust, sensitive assays for detecting these pathogens. Using this strategy we hope to achieve assays with as good or better target affinities compared to those obtained with polyclonal antibodies but with improved selectivities.

Professor Vadgama the remarked: I think you do need to allocate specific effort on how you might focus on the specific targets in a large sample volume and achieve the required extraction. Special attention needs to be given over to sampling strategies.

Dr Brown responded: We agree with this comment. It is also clear that our sampling strategies must be developed with input from clinicians who diagnose and treat these diseases in endemic regions of the world.

Dr Campbell asked: Since your target market is the developing world, and given that "paper microfluidics" (by George Whitesides, Paul Yager and others) seems to promise extremely low cost tests, what will the cost of your test be?

Dr Brown responded: Our assays are at an early stage of development and our work is mainly focused upon identifying the correct targets and producing recognition reagents. In theory, our reagents could be used on a variety of platforms and this is clearly an area we will be exploring with our clinical colleagues who reside in endemic areas. The cost of these assays is clearly an important factor which will be taken into account as part of our ongoing development strategy.

Dr Ko Ferrigno said: Luminex is an expensive platform, but it has been taken up by diagnostic labs. What are its advantages, and what do we have to do to compete with or improve upon this technology?

Dr Brown answered: I cannot comment on Luminex's competitive aspects from a commercial standpoint, although there were a number of factors that influenced our selection of this system for our assay development. Our work is focused upon developing the molecular recognition reagents against two biothreat agents and not development of a new platform. We decided to work with Luminex because: (1) the company has already established the use of their platform in biodefense applications; (2) assays can be multiplexed; (3) detection is rapid; (4) the use of antibodies/aptamers on their beads is already established; (5) efforts exist for developing assays for protein and whole cell detection in complex samples; and (6) magnetic beads can be used to enhance capture of targets in complex samples. Also, working with Luminex also represented a good strategic decision for us as Luminex's R&D facility is only a short distance from our laboratory in Austin, Texas, where the bulk of the research is being undertaken.

Professor Vadgama said: Luminex appears to have an analytical advantage because it does not use a fixed surface, but displaceable beads, possibly less prone to fouling; this is coupled with its configuration to give high density arrays. In electrochemistry or standard waveguide optics, the problem relates to the question of what else is in the sample to cause fouling at the sensing interface, or to give a false signal and so reduce sensitivity. PCR amplification of the target is not easy to avoid in such analysis because no matter how much you amplify the signalling cascade *after* the initial complementary strand binding event, this initial affinity is finite. So a lack of a forward reaction at ultra-low target concentrations does not allow for the initial reactive encounter.

Professor Beynon commented: This is more of a general comment, but as a protein chemist, I have a sense of increasing separation between the analytical capabilities in the protein and nucleic acid molecular domains. My high throughput sequencing colleagues can now achieve a complete genome sequence in the time it takes us to run a few proteomics samples. Proteomics (at lest delivered in LCMS paradigms) is serial and slow, limited by three factors: intrinsic sample complexity, the serial nature of a 2 h chromatographic delivery to the mass spectrometer and the complication of analysing low abundance proteins in the background of abundant proteins. I could foresee a future in which deep transcript profiling, for example, might be used to define putative protein changes that are then validated by proteomics, a far cry from using proteomics as a discovery, primary tool.

Professor Gooding opened the discussion of the paper by Dr Krause: What are the important variables in determining the reproducibility of the protease sensor? For example, how important is the thickness of the hydrogel film?

Dr Krause answered: The cleaning of the electrode surface, the quality of the self-assembled monolayer, the thickness of the dextran and peptide layers and the temperature and humidity during curing have all been found to have a strong influence on the reproducibility of the sensor. Thicker films were shown to degrade more slowly. When the dextran film was too thick, no degradation was observed. We assumed that in this case, the dextran was not cross-linked throughout the film causing the hydrogel to lift off upon the addition of buffer.

Professor Tsien remarked: How do you get samples out of a patient's mouth? What about the concern that protease activities may vary with location, perhaps being highest at a deep site of inflammation and much lower in the general saliva? In that case, it would be important to sample fluid from the gingival pocket. Furthermore, many of these proteases often are neutralized by endogenous inhibitor proteins, *e.g.* α1-antitrypsin for elastase, and tissue inhibitors of metalloproteinases (TIMPs) for MMPs. Do such inhibitors affect your measurements?

Dr Krause answered: Fluid will only be sampled from the gingival pocket. It is envisaged that a capillary tube or irrigation needle will be inserted into the gingival pocket. Gingival crevicular fluid (GCF) will be withdrawn either by capillary fill or using a syringe attached to the capillary or irrigation needle. The information obtained from the fluid will be specific to the gingival pocket investigated and will be used to inform treatment of that pocket. Selection of the protease activities being sought is based on a fairly extensive literature in which these enzymes are found in an active form in GCF. The activity of free protease is most relevant to determine disease progression, *i.e.* inhibitors should not affect the measurement.

Dr Tsaloglou communicated: Use of methacrylate (PMMA) capillaries may require surface treatment for optimisation due to the tendency of polymers, and specifically PMMA, to adsorb proteins and nucleic acids.[1] Because of its acrylate groups, PMMA offers excellent potential of chemical surface modification for specific target attachment (DNA, enzymes) in micro-analytical devices.[2] Especially for dental applications, PMMA is typically used as the base material of dentures. Göcke *et al.*[3] studied salivary pellicle formation of PMMA denture bases. Salivary peroxidases and lysozyme adsorbed onto the PMMA and formed a layer of positively charged molecules on which further salivary molecules could adsorb. Since periodontal disease monitoring in this paper is based on protease detection, potential adsorption of the target elastase, cathepsin G and MMP8 on the walls of the capillaries could introduce significant bias to the measurement of the sensor.

1 S. Ghosal, *Anal. Chem.*, 2002, **74**, 771–775, DOI: 10.1021/ac010571p.
2 A. C. Henry, T. J. Tutt, M. Galloway, Y. Y. Davidson, C. S. McWhorter, S. A. Soper and R. L. McCarley, *Anal. Chem.*, 2000, **72**, 5331–5337, DOI: 10.1021/ac000685l.
3 R. Göcke, F. Gerath and H. von Schwanewede, *Clin. Oral Invest.*, 2002, **6**, 227–235, DOI: 10.1007/s00784-002-0176-1.

Dr Krause communicated in response: Thank you for pointing out this potential problem. We will compare the sensor response obtained by directly adding enzyme solutions to the sensor and by picking up and depositing enzyme solutions onto the sensor using a PMMA capillary to study the effect of protein adsorption in the capillary. The clinical sample will be in contact with the PMMA capillary for only 30 s rather than the 30 min described in the paper by Göcke *et al.* We expect the effect of protein adsorption to be less pronounced in that relatively short time period.

However, if enzyme activity is significantly reduced due to the solution passing through a PMMA capillary, we will consider a surface modification.

Professor Tsien asked: How selective for HNE, cathepsin G, and MMP8 are the respective substrates used in this work? For example, how much discrimination against other MMPs does GPQGIWGQK show?

Dr Krause replied: It was shown previously that there is negligible cross-sensitivity between HNE and cathepsin G.[1] GPQGIWGQK has been shown to have high substrate specificity when compared to MMP1.[2] Catalytic data for the cleavage of GPQGIWGQK by other MMPs are not available. The sensitivity of hydrogel films cross-linked with GPQGIWGQK to MMP9, which is the protease most likely to interfere with the measurement, will be tested and the peptide sequence modified if required.

1 J. L. Stair, M. Watkinson and S. Krause, *Biosens. Bioelectron.*, 2008, **24**, 2113, DOI: 10.1016/j.bios.2008.11.002.
2 S. Netzel-Arnett, G. B. Fields, H. Birkedal-Hansen, H. E. Van Wart and G. Fields, *J. Biol. Chem.*, 1991, **266** , 6747–6755.

Professor Schuhmann said: How does the spacing of the interdigitated electrode relates to the film thickness of the polymer film for getting the best and fastest response?

Dr Krause answered: As shown in our paper (Fig. 3 and 4), responses obtained with opposing electrodes and with interdigitated electrodes of 10 μm line spacing were very similar. With a hydrogel film thickness of 200–300 nm, the electrode spacing was considerably bigger than the film thickness. The current is therefore expected to cross the hydrogel layer and to flow predominantly through the solution to bridge the gap between the interdigitated electrodes, as is the case with opposing electrodes. It could be speculated that the resistance measured at high frequencies would also be sensitive to the dissolution/swelling of the film if the electrode spacing was similar to the film thickness. The rate and magnitude of the change in the charge transfer resistance that was measured at low frequencies in our work should, however, not be affected by the electrode spacing.

Dr Ko Ferrigno remarked: How long can the cross-linking peptide spacer be? Would there be scope for making a polymer that could shrink or swell following conformational changes in the cross-linking peptide.

Dr Krause responded: We have only tested peptide cross-linkers up to 9 amino acid residues long. We assume that the longer the cross-linker the more the properties of the hydrogel films will be dominated by the peptide. As we are interested in producing hydrogels with similar film properties, we have been trying to keep the peptide cross-linkers as short as possible. In principle it should be possible to make hydrogel films swell and shrink reversibly if a binding event changes the conformation of the peptide cross-linker, particularly if the conformational changes were caused by a small molecule that could easily penetrate the hydrogel matrix. I would expect this to work less well for large proteins that would only bind to the film surface. An added problem is that the binding of the protein itself would cause a change in impedance and be indistinguishable from the impedance change caused by the shrinking or swelling of the hydrogel.

Dr Mount asked: There seems to be significant variation in the background impedance signals, for example in Fig. 4–7 of your paper before protease addition ($t < 5$ s), with some rising and some falling. These points are typically quoted as being the average of two individual measurements. What is the variation in slope of these individual measurements, is there peptide sequence selectivity in this response and what is the physicochemical origin of this variation? Are the slopes constant over extended periods and how does this affect detection sensitivity, time to result and reproducibility?

Dr Krause answered: Only the points shown in the calibration curves in Fig. 6b and 7b are the average of two points. All the percentage impedance change *vs.* time graphs (Fig. 4, 5, 6a and 7a) show individual measurements. Curves presented in the same Figure were fabricated under identical conditions. The variation in slope before the addition of enzyme is therefore representative for the variability of the different groups of samples. The slopes are typically constant for up to 20 min. It is noticeable that the increase in impedance prior to enzyme addition is greater for GPQGIWGQK cross-linked films than that for AAPVAAK and AAPFFK cross-linked films. Due to the two glutamine residues, GPQGIWGQK is much more polar than the other two peptides. This could explain increased swelling and therefore an increase in impedance in buffer.

In a clinical device, the sample would be added directly to the sensor by means of a capillary fill channel (see Fig. 1 and 8 in our paper). The response obtained in a capillary fill device without prior stabilisation in buffer seems to be reproducibly faster than the response after a stabilisation period in buffer. We assume that the surface concentration of cleavage sites on a dry film is greater than on a film that has been allowed to swell leading to a faster response in a capillary fill device. The detection sensitivity and time to result are therefore improved when using a capillary fill device without a stabilisation period. This has been confirmed with recent measurements with cathepsin G in a capillary fill device (results not shown). The variability in the impedance change prior to enzyme addition indicates a poor reproducibility of the coating process and will also affect the reproducibility in the degradation rates measured at different protease activities. We expect to see a significant improvement when moving from drop-coating by hand to automated drop-coating.

Dr Mount said: At least the initial measurements use quartz crystals for comparison with previous work, however the mass change results readily available from QCM measurements are not shown. The explanation of impedance changes is typically in terms of such effects as film swelling and/or dissolution, the effects of which should be evident in mass change measurements. What do these measurements show and do they support these hypotheses? Why is impedance a better method than resonance mass measurement for protease detection?

Dr Krause answered: The results of a BVD fit of quartz crystal admittance spectra measured during the degradation of AAPVAAK cross-linked dextran hydrogel films deposited in a single step on a 4-mercapto-1-butanol modified electrode QCM crystal showed an initial increase of the resistance ΔR after enzyme addition indicating an increase in the viscoelasticity of the film due to swelling followed by a decrease in ΔR. The reactive inductance $\omega\Delta L$ and therefore the mass of the film initially remained constant after enzyme addition and only started to decrease when R decreased.[1] It was assumed that this behaviour could be interpreted as a competition of mass gain due to swelling and mass loss due to hydrogel dissolution. Fig. 1 shows the results of simultaneous quartz crystal admittance and impedance measurements before and during the degradation of an AAPVAAK cross-linked hydrogel film obtained by successive deposition of oxidised dextran and peptide on a 4-mercapto-1-butanol modified QCM crystal (see our paper). In contrast to films obtained by single step coating, the BVD fit of quartz crystal admittance spectra measured during the degradation of two-step coated films shows an instant decrease in both the mass ($\omega\Delta L$) and the viscoelasticity (ΔR) upon the addition of enzyme (Fig. 1(a)), which indicates a straight forward dissolution process

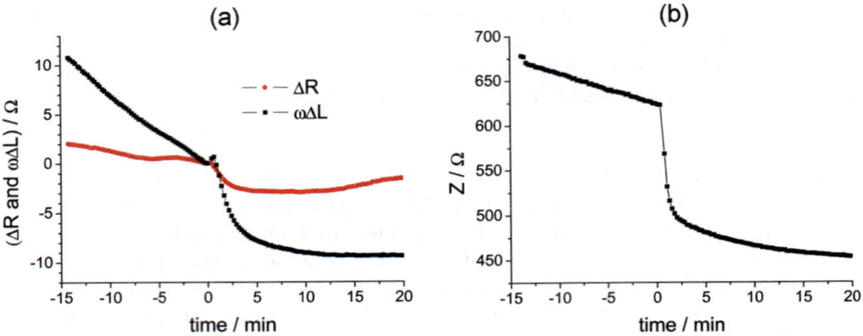

Fig. 1 Simultaneous QCM and impedance measurements of an AAPVAAK cross-linked hydrogel film produced by a two-step coating method on a 4-mercapto-1-butanol modified QCM crystal before and after addition of 10 U ml^{-1} HNE to pH 7.5 charge transfer buffer. (a) Results of a BVD fit; ΔR represents changes in the viscoelasticity of the film, $\omega \Delta L$ represents mass changes. (b) Impedance change at 100 Hz.

without significant swelling. This behaviour is mirrored by the impedance change shown in Fig. 1(b). The decrease of the impedance prior to the addition of enzyme also corresponds to a mass loss from the surface although this instability of the film seems to be much more pronounced in the mass change than the impedance change. The changes in parameters ΔR and $\omega \Delta L$ for films produced in two steps (Fig. 1(a)) were by about a factor of 10 smaller than the changes observed for films produced in a single step coating procedure.[1] There are several significant differences between the films obtained by the two different coating methods that could explain the differences in behaviour. Films obtained by spin-coating a solution containing both oxidised dextran and peptide contained a surfactant (Tween 20) and a lot more salt as they were coated from a PBS solution. They were also more likely to contain unreacted aldehyde groups that could bind the enzyme and therefore slow down degradation. Films obtained by successive deposition of oxidised dextran and peptide were likely to have unreacted peptide and buffer salts on the surface, which would dissolve prior to the addition of enzyme. Analogous measurements with a 2-aminoethanethiol modified surface were not carried out as, at that stage, quartz crystals had already been abandoned as sensor substrates because they were deemed too expensive for a disposable sensor. The ultimate aim of this work is a sensor array that can measure the activities of three proteases simultaneously. Impedance measurements have the advantage that arrays of screen printed interdigitated electrodes on alumina can be mass produced much more cheaply than the equivalent arrays of mass sensitive devices.

1 J. L. Stair, M. Watkinson and S. Krause, *Biosens. Bioelectron.*, 2008, **24**, 2113, DOI: 10.1016/j.bios.2008.11.002.

Professor Bradley asked: What are the practicalities of impedance measurements on more traditional reporters such as colour or fluorescence that could be measured in real-time? Could your gel be adapted to such a colour change system?

Dr Krause answered: Impedance measurements on traditional chromogenic substrates are not really practical as impedance is not sensitive to light, and cleaving off a small fluorescent molecule would result in a significantly smaller impedance response than the degradation of a hydrogel layer. Integrating chromogenic groups into the hydrogel would in no way enhance the impedance response. Traditional fluorescent assays for periodontal markers have not succeeded on the market as their response was too slow for a busy dental practice.

Professor Bradley said: You could use "colour" based reporters in real time *in vivo* in the patient's mouth – similar to classical nitrophenol based substrates.

Dr Krause replied: I am not certain how easy it would be to measure colour changes inside the periodontal pocket and in the presence of blood.

Dr Davies said: In Fig. 5 of your paper you have an increase in impedance on the amino-derivatised surface and a decrease on the hydroxyl functionalised electrodes. You attribute this to a covalently attached gel in the former and to dissolution of the gel in the latter. Could it be due to different behaviour of the ferri- or ferrocyanide at the two surfaces – their local pH would be different. Could Prussian blue be depositing on the amino surface? Do you get the same results with other redox couples?

In the case of the films here we have an entangled, high molecular weight dialdehyde dextran with protein covalently linked. The film swells by water uptake (osmotic) which causes the polymer chains to slide apart by a mechanism the polymer physics people know as reptation. I picture a physically "knotted" polymer with some permanent covalent links and the slightly branched dextran may also have some protein knobs on it that could also slow down the rate of expansion and dissolution of the film. Although it is possible that most of the film does not truly dissolve – the polymer mesh may remain attached to the amino substrate, or it may depart from the substrate as an intact mesh. This is probably a far more complex system than you believe it to be, and perhaps an understanding of some fairly easy-to-grasp water soluble polymer physics concepts can help us design experiments to better understand the system.

Dr Krause replied: Previously, we investigated the degradation of peptide cross-linked hydrogel films on hydroxyl terminated monolayers with QCM measurements, which clearly showed that the hydrogel dissolved from the surface.[1] It is unlikely that the ferri/ferrocyanide redox couple behaves significantly differently at the different monolayers. Prussian blue should not be formed as there was no dc potential applied, and no free ferrous or ferric ions were present. For the sensor to work without a reference electrode and without applied dc potential, equal amounts of oxidised and reduced charge transfer reagent have to be present. For many other redox couples (*e.g.* ruthenium hexamine) one of the oxidation states is too unstable to be used in this system. Attempts to use other charge transfer reagents have therefore failed.

We have recently carried out a set of experiments that confirms that the hydrogel does not dissolve when it is attached to the sensor surface. Interdigitated gold electrodes were screen printed on a 96% alumina substrate (Gwent Electronic Materials, UK). SU-8 2000.5 photoresist (Chestech Ltd., Rugby, UK) was patterned onto the sensor substrate covering the immediate area surrounding the electrode and insulating the tracks leading to the contact pads. Monolayer formation and AAPVAAK cross-linked hydrogel deposition was then carried out in the same manner as described in our paper. Films deposited onto an unmodified surface, on a 4-mercapto-1-butanol modified surface and on a 2-aminoethanethiol modified surface all showed an increase in the impedance upon the addition of human neutrophil elastase (Fig. 2). It is assumed that the hydrogel film was bonded to the SU-8 edge due to the reaction of epoxy groups in the SU-8 to the amine functional groups of the peptide. Therefore, after the peptide was cleaved, the films swelled rather than dissolved in the solution. This experiment clearly shows that the chemical nature of the electrode surface does not affect the direction of the impedance response. The only explanation for an increase in impedance is the presence of a film after degradation by the enzyme. Swelling alone does not explain the increase in impedance as a greater influx of water should also result in an increase in the ferri/ferrocyanide concentration in the film. There are three possible explanations for this behaviour:

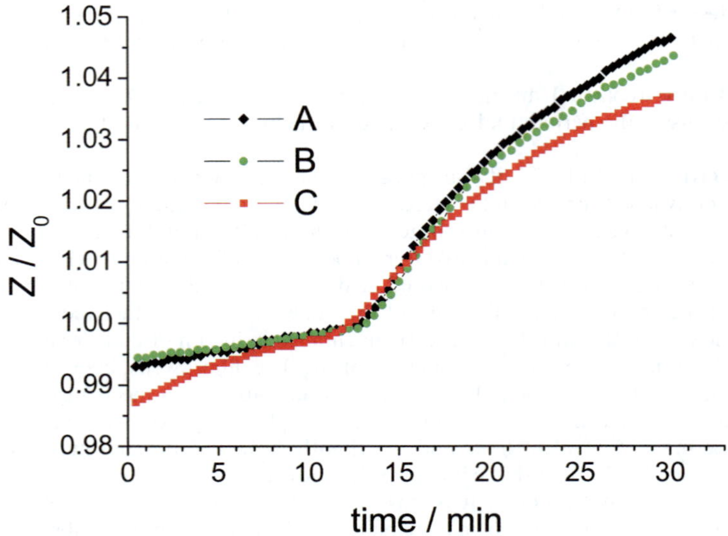

Fig. 2 Impedance change of AAPVAAK cross-linked hydrogel films deposited on interdigitated gold electrodes with different SAMs measured before and after addition of 10 U ml^{-1} HNE to pH 7.5 charge transfer buffer. A – hydroxyl-terminated monolayer, B – amino-terminated monolayer, C – unmodified (experiments performed by Li Chen, Queen Mary University London, UK).

(i) Only the surface of the film is 100% cross-linked. Degradation of the top layer of the film exposes unreacted aldehyde groups, which can react with the protein. Protein binding could reduce the concentration of ferri/ferrocyanide in the film.

(ii) If the electrode surface is initially incompletely covered with hydrogel, expansion of the film due to swelling could result in a greater surface coverage thereby impeding the access of charge transfer reagent to the surface.

(iii) Peptide cleavage results in the formation of carboxylic acid groups and amino groups. The negative charge of the carboxylate group at pH 7.4 could cause the ferri/ferrocyanide to be expelled. However, this effect would be counteracted by the formation of positively charged $-NH_3^+$ groups.

1 J. L. Stair, M. Watkinson and S. Krause, *Biosens. Bioelectron.*, 2008, **24**, 2113, DOI: 10.1016/j.bios.2008.11.002.

Professor Lu opened the discussion of the paper by Professor Cass: Fig. 12 of your paper showed impressive selectivity of the lysozyme aptamer for lysozyme over lactalbumin. I am wondering if there is any effect of lactalbumin on sensitivity and detection limit.

Professor Cass responded: We didn't observe any obvious effects.

Professor Schuhmann asked: For the post-assembling of the thiol monolayer you used the negatively charged thiol mercaptosuccinic acid? How do you consider the impact of this negatively charged surface on possible unspecific adsorption in real samples?

Professor Cass replied: Most proteins carry a net negative charge at physiological pH so the negatively charged surface should suppress non-specific adsorption. We have also tried neutral blocker thiols and the nature of the blocker appears to have some effect.

Dr Campbell commented: We have done some experiments where we found that we could achieve the best dose-response curves with an aptamer probe when the aptamer had been immobilised as a structured molecule (by adding the right amount of magnesium ions) rather than an unstructured molecule.[1] Have you tried immobilising your aptamers in a buffer that encourages them to adopt secondary/tertiary structure?

1 M. A. Ochsenkühn and C. J. Campbell, Probing biomolecular interactions using surface enhanced Raman spectroscopy: label-free protein detection using a G-quadruplex DNA aptamer, *Chem. Commun.*, 2010, **46**, 2799.

Professor Cass responded: All the selections are done in a magnesium-containing buffer and the binding reactions similarly. Typically the aptamer is melted by heating to 65 °C to disrupt any preformed structure and then the magnesium helps in the structuring as it binds. There is the danger that if the aptamer is pre-structured on the surface it may be in a kinetic or thermodynamic 'well' where it does not have the correct structure to bind the ligand. Although your work shows clearly that this is dependent on the aptamer/target combination.

Professor Graham remarked: Have you had any issues with degradation of the nucleic acid by nucleases?

Professor Cass answered: We haven't, although the sensors have not been used in complex samples. There doesn't seem to be a problem, at least over the timescale of the measurement, with adventitious nucleases.

Professor Tsien asked: Most of the tests in this paper seem to have been done on pure analytes. Wouldn't nucleases present in real-world, complex biological samples interfere strongly? How would one evolve DNA aptamers that are resistant to nucleases? The SELEX method can only work with unmodified DNA. If one takes a native sequence evolved by SELEX and naively converts nucleotides to unnatural analogs with significantly different charge or hydrogen-bonding potential, there would seem to be a considerable danger that the aptamer will change conformation or otherwise lose the binding selectivity.

Professor Cass replied: There is quite a large literature on successful modifications of aptamers that protect them from nucleases, examples include methyl or fluorine substituted bases.

Professor Lu said: It is quite interesting that, instead of a signal decrease due to lysozyme binding that displaces the short oligonucleotide containing the redox probe, you observed a signal increase, attributable to a conformational change that favors more efficient electron transfer. This "signal on" approach is better for point-of-care testing. I am wondering whether you have tried to apply this approach to a different aptamer for a different target.

Professor Cass replied: We haven't. Most of our subsequent work has been based on beacon-type aptamers as they appear better behaved.

Professor Vadgama commented: I would be hesitant to apply the 'hook effect' terminology here, as it is specifically used in the immunoassay field with particular connotations, notably antibody–antigen binding failure or conformational disruption of binding at very high target antigen concentration.

Professor Vadgama then asked: Have you considered using a probe added to the bulk solution for more conventional electrochemistry to assess if the 'hook' concentration effect is due to the surface being more masked, *e.g.* by a more compact DNA layer?

Professor Cass answered: No, but that would be worth trying.

Professor Gooding commented: One possible explanation for the decrease in electrochemical signal with high concentrations of lysozyme (the 'hook effect') relates to adsorbed proteins suppressing the electrochemistry of the surface-bound ferrocene. This has been seen by a number of workers and we have shown that we can modulate electrochemistry of a surface-bound ferrocene as an antibody binds and decreases the current or is displaced from a surface to enhance the current observed.[1] We hypothesise that the suppression of the ferrocene electrochemistry is because the ferrocene becomes immersed in a protein environment, which hinders the required counter ion from balancing the charge of the ferricinium ion formed during the oxidation process. So in the case of the lysozyme binding with the aptamer, perhaps the ferrocene is being immersed in a protein environment at higher lysozyme concentrations and suppressing the electrochemistry. The fact that the 'hook effect' is not observed when the ferrocene is at the distal end of the aptamer by support this suggestion.

1 G. Z. Liu, M. N. Paddon-Row and J. J. Gooding, Protein modulation of electrochemical signals: application to immunobiosensing for detecting small molecules, *Chem. Commun.*, 2008, 3870–3872.

Professor Cass replied: That is a good suggestion; it implies that at low fractional occupancy of the aptamer the ferrocenes on the lysozyme–aptamer complex behave as isolated moieties. As the fractional occupancy increases the probability of two aptamer lysozyme complexes being adjacent to one another also increases and that the ferrocene on aptamer is then immersed in the lysozyme on an adjacent aptamer.

Dr Davies commented: You were surprised to get a "signal-on" response – maybe the probes did not dissociate. How much dissociation would you expect from the probes during the time of the experiment? I guess you could get an idea of this using the BiaCore with gold or carboxydextran surfaces using immobilised target and a suitably labelled probe.

Could it be that the lysozyme binding site lay in the *ca.* 20 base region between the probes? Could one, or both, of the probes have occupied a small part of the lysozyme binding region so that lysozyme could still bind, but with reduced affinity? I would expect the influence of the probe to extend a little beyond the stiffer, double-helix region, as it would reduce the free volume "explorable" by the nearest neighbouring bases on the aptamer and interfere with binding to the protein surface. Is this type of binding interference seen with nucleic acid aptamers?

Professor Cass answered: The probes were designed to have a free energy of binding that was sufficient to make the duplex stable at room temperature but less than the measured (by SPR) free energy of lysozyme binding. Of course the free energy of hybridisation was calculated for the short oligonucleotide rather than the ferrocene derivative and if Professor Gooding's comment about the ferrocene becoming immersed in the protein holds in this case that would enhance the binding of the oligonucleotide to the aptamer.

Professor Nakano said: As noted in your talk, we often meet, in electrochemical biosensor systems, certain types of physicochemical phenomena that may be caused by molecular clouding. Such a particular effect plays a very important role in any biomolecular assemblies that are central in most biological processes. So, I think that the effect of molecular clouding should be solved and explained quantitatively to open up entirely new biosensor applications in the near future. Do you think what kind of opportunities will be given by taking advantages of electrochemical measurements?

Professor Cass replied: Molecular crowding is an important phenomenon *in vivo* where protein concentrations are typically much higher than *in vitro*. It has also been used to accelerate DNA hybridisation kinetics. Given that molecular crowding affects diffusion rates and amperometric electrochemical methods are sensitive to diffusion there is a link that should be exploitable.

Professor Nakano commented: With the molecular clouding manifests at electrodes whose surface are modified with biomolecular assemblies, electrode-reaction behavior tends to give certain kinds of discrepancies between theory and practice in most cases. For example, a surface-attached-redox system gives voltammetric peak broadenings, which are caused by specifically distributed rather than uniform kinetic properteis. Could you use any particular difference in electrode-reaction parameter distribution as a quantitative measure of molecular clouding?

Professor Cass responded: The exquisite sensitivity of voltammetric methods to the surface means it might be difficult to attribute observed behaviour to a single definitive cause.

Dr Dougan commented: In reference to your observation of the "hook" behaviour of your aptamer response, have you attempted to control the surface density of your aptamer assemblies and have you considered, instead of using a thiolated oligo, forming a SAM of thiolated active ester (such as thioctic acid NHS-ester) and reacting with an amino-modified oligonucleotide – as the competitive hydrolysis of the active ester would effectively "space" the oligonucleotides on the surface and, in addition, the coverage of the surface by a small molecule SAM may confer additional control to the surface chemistry?

Professor Cass answered: We haven't done that, we would however need an orthogonal chemistry for the ferrocene attachment as this currently uses the reaction of an amino-terminated oligonucleotide with ferrocene carboxylic acid.

Professor Cooks said: I wonder about the suitability of electrochemical methods for actual point of care analysis, which I take to include the measurement of specific compounds in complex biofluids. Are electrochemical methods generalizable to be up to this task? Do they have the sensitivity, specificity and speed to provide answers for particular compounds in blood, for example? How does one make rapid EC measurements on such samples?

Professor Cass replied: We know that there are endogenous reductants in biofluids that can give high background currents outside a given potential window and this dictates the choice of redox probe/substrate that can be used. Beyond that sensitivity and specificity are functions of the reagents and the surface chemistry.

Dr Mount enquired: Regarding the discussion of Fig. 6 of your paper, has it been excluded that the redox probe dissociates from the aptamer as originally planned but then binds either specifically or non-specifically to the lysozyme? Changing the stringency may also lead to greater discrimination between monolayer and multilayer formation and give further support to the mechanism postulated. Has this been tried?

Professor Cass responded: That's an interesting thought, effectively producing a sandwich. The stringency hasn't been systematically investigated by us except in so far as dissociating the lysozyme in a SPR experiment.

Professor Vadgama then opened a general discussion of the papers by Dr Krause and Professor Cass: It may be that there is a unifying aspect to the papers from Professor Cass and Dr Krause: this relates to Coulombic charge effects. Thus ssDNA has mutually repulsive negative charge and so some of the putative molecular collapse may take place when these are masked, *i.e.* with target binding. Alternatively, for the gel system, as presented by Dr Krause, charge shielding could alter gel swelling state and therefore serve as a reporter system.

Dr Krause replied: The cleavage of peptide bonds in the peptide cross-linked dextran hydrogel certainly introduces a lot of additional charges in the form of –COO– and $-NH_3^+$ groups, which are expected to cause swelling of the film, particularly when the film is bound covalently to the surface.

Professor Cooks asked: Should it not be standard practice to make molecular diagnostic measurements during anticancer drug infusion, for example, rather than simply examine vital signs as is now done? We need generalizable methods that can be applied to any molecules to make circumstance-appropriate *in situ* measurements.

Dr Davies communicated: This is an interesting idea, particularly if the patient is currently dosed on a body mass basis, with no regard is taken of the cancer mass or distribution and its responsiveness to the chemotherapy. With such treatments being given in a hospital there is potential to use sophisticated equipment to maintain the circulating level within the therapeutic window. There will be, however, a finite assay time which would need to be as short as possible to enable good feedback to the dosing mechanism. Ideally such an instrument would also be able to accept an input of cancer mass, distribution and perfusion of the tumor, as well as the tolerance of the patient's normal cells to the therapy, and could optimise the treatment regime. Sphere Medical (www.spheremedical.com) have a device for monitoring dosing of the intravenous anaesthetic Propofol. Their electrochemical sensor sits in a saline drip line and at intervals withdraws blood from the vein, performs a measurement and is then flushed with saline ready for another measurement. I believe that the analysed blood is washed back into the patient – a challenging protocol to get FDA approval. So there is activity in patient monitoring, and perhaps somewhere it is being looked at for improving chemotherapy.

Professor Cooks communicated in response: The case you cite is encouraging but the pace of progress is slow. To take the example you cite of anaesthetics in blood – this measurement was made by mass spectrometry as long ago as 1987.[1]

1 J. S. Brodbelt, R.G. Cooks, J.C. Tou and G.J. Kallos, In vivo mass spectrometric determination of organic compounds in blood with a membrane probe, *Anal. Chem.*, 1987, **59**, 454–458, DOI: 10.1021/ac00130a017.

Professor Vadgama said: The analytical turnaround for many drug assays is good enough for modulating anticancer treatment. Assays are done in the UK for a range of relevant drug agents, and also their metabolites where relevant.

Professor Cooks replied: One point of interest concerns the time scale of the analysis. My comments concerned the absence of measurements on oncology drugs during an infusion that might last a number of hours. I apologize for not making this clear. As I understand it, such measurements are not made in the UK either. However the question brings up another interesting point. In discussions with a number of oncologists in the US, I was given the information that even on a much longer time scale drug levels are not routinely measured. However, there seems to be agreement that such measurements are potentially extremely valuable, provided they have the analytical performance characteristics described above. There is a lack of monitoring except for vital signs monitoring during routine anticancer drug infusion.

Professor Vadgama replied: The technical point is correct, however, there is certainly availability of anticancer monitoring outwith the acute infusion procedure. This is used to help regulate therapeutic dose during chronic management. Sustained drug effects are clearly important as they determine the balance of side effects *vs.* efficacy. Effects of acute drug loading, *e.g.* transient highs in levels during infusions, may well be important, but need specific investigation to establish importance. Such monitoring, especially if it is to be continuous, will of course pose analytical challenges.

Peptide Aptamer Microarrays: Bridging the bio–detector interface

Qifeng Song,[a] Lukas Kurt Josef Stadler,[a] Jianhe Peng[b] and Paul Ko Ferrigno*[a]

Received 19th May 2010, Accepted 27th July 2010
DOI: 10.1039/c005376g

In the near future, personalised medicine and phase-0 trials will require that clinical practitioners move from the "one biomarker per disease" paradigm to the use of molecular signatures of disease for diagnosis and the prediction of a patient's response to treatment. These signatures will be composed of biomarkers specific to the disease, and will include over-expression of normal protein from a gene that does not carry a mutation; loss of expression of an essential protein; expression of a protein from a mutant gene; and metabolites whose levels are altered in disease. Surrogates for protein expression, such as alterations in the messenger RNA that encode for them, have already proved their value. The next challenge, then, in clinical biosensing is to enable the multiplexed detection of protein biomarkers, and perhaps the multiplexed detection of mixed biomarkers (metabolites, RNA and proteins) all in a single test. Given the plethora of available antibodies specific for biomarkers, why is this not already happening? We believe that the limitation lies in the nature of the antibody molecule itself, and especially the fact that antibodies have evolved to function in solution, while most diagnostic tests take place at a surface. We have accordingly turned to the design of alternative antibodies, and have identified a protein that appears to be unusually stable on surfaces. The new, non-antibody, scaffold protein is derived from human Stefin A, a natural inhibitor of the cathepsin family of proteases. We have engineered this protein so that it lacks natural binding partners, and introduced a series of new binding surfaces through randomisation or directed replacement of the surfaces used by Stefin A to bind to cathepsins. Our new probes show exquisite specificity and binding affinities comparable to antibodies, and can be used to probe biology in intact cells. More importantly, together and in collaboration with other groups in Chemistry or Engineering Departments, we have shown that these designer proteins can be used in optical detection of labelled target proteins from whole cell lysates in a highly multiplexed microarray format, as well as in label-free detection of unlabelled proteins using surface plasmon resonance, QCM, microcantilevers and using electrochemical assays on gold electrodes. We believe that the combination of optimised surface chemistry, robust and combinatorial designer biological probes and novel, robust and sensitive detection technologies will enable, in the near future, the introduction of multiplexed biomarker detection in the clinical setting, most likely in cancer where multiple biomarkers are known, but probes are still lacking.

[a]*Section of Experimental Therapeutics, Wellcome Trust Brenner Building, St James's University Hospital, Beckett Street, Leeds, LS9 7TF. E-mail: p.koferrigno@leeds.ac.uk*
[b]*Clinical and Biomedical Proteomics Group, Cancer Research UK Clinical Centre, Leeds Institute of Molecular Medicine, St James's University Hospital, Beckett Street, Leeds, LS9 7TF*

1 Introduction

1.1 Background

The advent of personalised medicine will require the introduction of increasingly sophisticated assays for the detection of markers of disease. Such assays need to be inexpensive, reliable (highly reproducible) and robust (useable with little training requirement). In some settings (the operating theatre, general practitioners' offices or outpatient visits), they will also need to be rapid. Clearly, there is a major opportunity for any technology that combines the low cost, high volume production of the semiconductor industry and the precision and sensitivity of biological detection. For these reasons, we have been collaborating with chemists, physicists and engineers for several years on the development of label-free assays that can be performed using electronic devices. Label-free assays have additional advantages in the avoidance of reagent costs, of lengthy labelling and sample processing steps, and in the immediacy of the read-out.

We will focus here on the multiplexed detection of proteins, with the rationale that proteins are the mediators of biological and pathological functions, and that the simultaneous detection of multiple proteins in a multiplexed format represents the future of clinical assays. Antibodies have long been the tools of choice for the study of biology at the molecular and cellular levels, for the diagnosis of disease and recently as therapeutics. Antibodies have many advantages, including high specificity and affinity, which can be further increased through multivalent recognition of repetitive epitopes, as occurs for example of the surface of many pathogens. However, they suffer from many limitations. Firstly, antibodies are traditionally raised in animal hosts. This raises ethical concerns, is expensive and has the inevitable technical limitation that not all targets are immunogenic, which can be for reasons as simple as evolutionary conservation with the host. Secondly, antibody molecules are complex. Thus, IgG comprises 4 polypeptide chains whose tertiary and quaternary structure require the formation of between 12 and 16 disulfide bonds as well as post-translational glycosylation.[1] This complexity means that *in vitro* expression of immunoglobulins is complicated, expensive and difficult to scale up. Thirdly, the immunoglobulin molecule comprises two moieties: the variable region that mediates recognition, and the constant region that mediates interactions with effector cells and molecules of the immune system. While the latter is useful in some applications (immunodetection, therapeutic elimination of disease-associated molecules or cells), it is counter-productive in others, including multiplexed assays where the requirement for glycosylation of the constant region limits high-throughput expression, and where the many natural interactors of the constant region may lead to high levels of "background" signal in diagnostic assays. Finally, the sheer size of antibody molecules may lead to steric hindrance and limit their utility in specific applications. Some of these limitations may be overcome by the use of single-chain antibodies, such as those from camelids[2,3] or engineered simplified versions of human or mouse antibodies[4–6] and, there have been many attempts to design or identify molecules with the desirable characteristics of antibodies (high specificity and affinity) while avoiding these drawbacks. All of these attempts use standard protein engineering technologies to create unique restriction sites in the open reading frame that encodes the scaffold. These restriction sites can then be used for the insertion of oligonucleotides encoding peptides designed to mediate a specific protein interaction, or of random oligonucleotides to create libraries that can be screened for binders to a specific target. Recent reviews have identified more than 40 engineered scaffolds and highlighted a high level of commercial as well as academic interest.[7,8] These scaffolds range from conserved protein domains (such as the Z-domain from protein A;[9] the IgG fold of fibronectin III;[10] and designer ankyrin repeat proteins or DARPins[11]) to intact proteins including bacterial thioredoxin[12,13] and lipocalins.[14] These scaffolds have been broadly shown to be useful, for example in novel imaging applications for cancer.[15] Although these efforts invariably are initiated in academe, they are rapidly acquired by companies (Wyeth acquired the thioredoxin scaffold) or spun out into companies (Affibody AG, Adnexus, Molecular Partners, and Pieris) that may themselves be acquired (*e.g.* Adnexus acquired by Bristol Myers Squibb), unfortunately limiting their application in basic biological research. Nonetheless, a great deal has been learned, leading to the development of increasingly versatile technologies. We have been applying these new tools in a range of label-free assays that are amenable to multiplexing, with a view to developing signatures of human disease that may be useful in personalised medicine.

The closest mature field of basic biological research is that of proteomics. Proteomics is the evolving science of exploring protein expression and interactions on the level of an entire cell, system or organism. The field is dominated by two technologies: mass spectrometry and two-dimensional gel electrophoresis.[16] Mass spectrometry enables top-down and bottom-up proteomics. Top-down proteomics allows the identification of intact protein expression signatures, while bottom-up proteomics sequences the enzymatic digested peptides before assembling them into proteins. Thus, mass spectrometry generally allows the identification of protein expression signatures, where the identity of the proteins present will not be accessible until a second, peptide identification run is performed. Nonetheless, mass spectrometry can be coupled with high-throughput approaches to lead to many crucial insights into protein expression on a global level. Similarly, two-dimensional gel electrophoresis, particularly when coupled with dual-labeling and high-throughput tandem mass spectrometry, can also lead to the identification of large numbers of proteins associated with a biological or pathological process. However, both approaches suffer from a series of technical limitations, some of which are overlapping. Thus highly charged, hydrophobic or rare proteins can often be problematic for both systems, particularly in top-down approaches. In bottom-up proteomics, identification of incorrectly annotated or missing peptides, or those that are post-translationally modified, can be a problem. Mass spectrometry also requires highly skilled personnel and expensive equipment. Alternative technologies are thus clearly needed. Most promising appear to be protein microarray approaches.[17] These take two forms: direct immobilization of many different proteins, so as to allow a direct assessment of whether interaction partners for each protein exist in a sample; and indirect approaches, where a large number of probes, each specific for a given protein, are immobilized and used to detect cognate partners in biological samples.

The availability and specificity of antibodies remain a restriction to large-scale proteomic investigations.[18] Compared with the number of the products of more than 22 700 human genes making up the whole human proteome, only a small percentage of antibodies are available. Although efforts aimed to increase the throughput of antibody generation have been made,[19] research laboratories are still limited to traditional hybridoma methods, and high-throughput technologies such as phage display can be limited by lengthy downstream manipulation.[20] Peptide aptamers are small proteins that comprise variable peptide sequences constrained in a neutral scaffold designed for use within eukaryotic cells. They are usually selected from large, randomised libraries for their ability to specifically bind a given protein. Because they bind targets with comparable affinity and specificity to antibodies, but are stable both in cells and, in some cases, on surfaces, they are an alternative to antibodies in both proteomic research and drug discovery. Several proteins including Thioredoxin A,[13] green fluorescent protein (GFP)[21] and Stefin A[22] have been engineered to accommodate peptides. This multiplicity alone is likely to be useful, as it has been shown that different scaffolds can display the same peptide differently.[23] Thus, it is predicted that any given scaffold will be unable to explore all of protein space. In addition, compared to antibodies, these engineered non-immunoglobulin binding protein scaffolds are expected to show increased robustness, to be useful when expressed in eukaryotic cells, and to circumvent limitations imposed by the immune system.

1.2 Recent work

In 2001, building on the state of the non-immunoglobulin scaffold field, we defined the requirements for a next generation scaffold, focusing on requirements for simplicity, robustness (to include both biophysical and biological stability), and biological neutrality, which we define as a scaffold with inherently low affinity for irrelevant proteins in the system to be studied.[22] This latter characteristic was desirable so as to maximize utility in cell biological applications, where the scaffold should (i) not itself affect the biology of the system and (ii) spend as little time as possible interacting with proteins in the milieu so as to maximize opportunities for specific interactions mediated by the engineered recognition surface. Biological inertness has the added virtue of minimizing background signal in diagnostic and detection assays, and is expected to increase the efficacy of eventual bio-therapeutics based on our scaffold. We identified 7 proteins that, on paper, matched our criteria but found that only one of these, human Stefin A (SteA), could be engineered for biological inertness without significant loss of structural stability.[22] SteA is a member of the cystatin super-family of protease inhibitors.[24] Cystatins function to inhibit the protease activity of extra-cellular proteases, whereas Stefins are generally cytoplasmic or lysosomal proteins whose role is to inhibit intracellular cathepsins.[24] Several solution and crystal structures of SteA alone and in complex with cathepsins are available.[25–28] Although we used the Stefin B/papain co-crystal structure[25] for our initial designs, crystal structures of the complex between human SteA and porcine cathepsin H[26] have since become available. The stefin B structure identifies three domains that mediate interaction with cathepsins: N-terminal residues, and two conserved hairpin loops that together comprise the so-called 'tripartite wedge'. This wedge is sufficient to mediate a binding affinity between Cystatin C and papain of 60 fM[22] while the interaction between SteA and cathepsins may be even tighter as interactions between the SteA wedge occur deeper within the active site of cathepsin H.[20] The role of SteA as a protease inhibitor, its known structure, ability to mediate high-affinity interactions and use of three interaction sites that appeared suitable for combinatorial engineering all make SteA an attractive scaffold candidate. This was confirmed through its resilience to engineering and lack of toxicity when expressed in cells.[17] Although the initial form of the scaffold, called STM (SteA Triple Mutant), was only used to present a single interaction surface at loop 2, subsequent engineering has shown that both loops[23] and more recently a combination of the two loops and the amino terminus[29] can be used to present multiple surfaces for interaction. Various versions of the scaffold have been used to demonstrate lack of affinity for the bulk of human proteins (low protein retention form human cell lysates on SPR chips coated with scaffold protein[25,26]); specific protein recognition in whole cell lysates[27–31] and replacement of antibodies for detection of proteins in western blotting and immunofluorescence.[32]

1.3 This work

In the present paper, our goal is to demonstrate that our biological probes, called here peptide aptamers, are amenable to high-throughput production protocols, can be used for robust protein detection in complex biological mixtures, and can be used in a multiplexed format for the detection of specific target proteins.

2 Methods

2.1 Plasmids and strains

The pET30 plasmid for hexa-histidine tagging and bacterial protein expression was from Novagen. All constructs were sequenced before use. XL10 gold ultra-competent bacterial cells used for cloning and BL21 (DE3) cells for protein expression were from Stratagene.

2.2 Construction of PA library

Double-stranded DNA encoding the 6- or 12-mer random amino acids sequence was generated as described.[13,22] Briefly, an annealed PCR substrate was generated by mixing together 100 μg primer and 50 μg of random oligonucleotide in a 800 μL reaction, incubating at 95 °C for 5 min then allowing the mix to slowly cool to room temperature. A single extension reaction was performed using 45 μL of 5 mM dNTPs and 20 μl Klenow polymerase. Nucleotide removal kits (Qiagen) were used to purify the resulting double-stranded DNA cassette. The double-stranded DNA cassette and pET30-STM or -SQM plasmid were separately digested with *Rsr*II. The cassette was purified using a nucleotide removal kit (Qiagen) while linearised pET30-STM or -SQM plasmid was treated with calf intestinal phosphatase to minimize re-ligation and purified by gel electrophoresis. To construct a library of peptide aptamers constrained in SQM, 4 ng of each double-stranded DNA cassette was ligated with 300 ng of the pET30-STM or -SQM vector at the appropriate site. Details of cloning sites for the different libraries are given in refs. 22 and 29.

2.3 Transformation

We used 150 ng DNA of the ligation to transform 100 μl XL10 Gold ultra competent cells (Stratagene) following the manufacturer's instructions. Single colonies grown on kanamycin selective plates were picked and cultured in 600 μL LB medium with 30 μg mL^{-1} kanamycin in deep well 96-well plates. To construct the expression clones, each plasmid was transformed into 50 μL BL21 DE3 competent cells by heat-shock at 42 °C for 30 s. After incubation with LB medium at 37 °C for 45 min, the cells were collected by centrifugation at 4000 rpm in a microfuge for one minute. The pelleted cells were resuspended with 30 μL sterilised water and plated onto LB (with 30 μg mL^{-1} kanamycin and 30 μg mL^{-1} chloramphanicol) plates.

2.4 Expression and purification of random peptide aptamers

A starter culture of expression clones was made by inoculation of single colonies into 600 μL LB medium supplemented with 30 μg mL^{-1} chloramphenicol and 30 μg mL^{-1} kanamycin in 96-well plates. Cells were allowed to grow overnight at 37 °C. Aliquots of each culture were diluted 10 fold into one millilitre LB with 30 μg mL^{-1} chloramphenicol and 30 μg mL^{-1} kanamycin in 96-well plates and grown to OD600 = 0.4. To induce the expression of each peptide aptamer, IPTG was added to a final concentration of 1 mM and the cells grown at 37 °C for a further 2 h. Cells were harvested at 4000 rpm in a Beckman centrifuge using the microplate attachment and at 4 °C for 5 min. The pellets can be stored at −20 °C or immediately subjected to the purification procedure.

The peptide aptamers were purified using a high-throughput method developed by Chen *et al.*[30] with a few modifications. The cells were resuspended in 100 μL lysis buffer (0.1M sodium phosphate with 300 mM NaCl, pH 7.4, 1 mM DTT, protease-inhibitor cocktail (Roche), 1 mM PMSF, 1 : 10 PopCulture (Novagen), benzonase (Novagen) at 10 units/mL) and incubated at room temperature for 20 min. Cell debris were removed by centrifugation at 4000 rpm at 4 °C for 10 min in an Eppendorf 5310R microfuge and the supernatants transferred to a clean 96-well plate followed by 20 μL Ni-NTA His-bind resin (Novagen) equilibrated in lysis buffer. The plates were sealed and incubated at 4 °C with shaking at 40 rpm for an hour. The Ni beads were pelleted at 2000 rpm at 4 °C for 2 min and the supernatant discarded by flipping over the plate. The resin was resuspended in 200 μL of pre-chilled wash buffer (0.1 M sodium phosphate buffer, 300 mM NaCl, 2 0mM imidazole, pH 7.4) and transferred to a filter plate (PALL) which was placed over a deep-well 96-well plate as a receiver. The stacked plates were centrifuged at 3000 rpm for 3 min at 4 °C and the wash procedure was repeated 6 times.

To elute the peptide aptamers, 50 μL elution buffer (0.1 M sodium phosphate buffer, pH 7.4, 250 mM imidazole, 300 mM NaCl, 30% glycerol, 0.05% Triton X-100) was added to each well. The filter plate was placed on top of a clean 96-well plate as a receiver plate and the stacked plates centrifuged at 3000 rpm at 4 °C for 3 min. The flow-through comprised the purified peptide aptamer solutions. Most peptide aptamers were well-expressed, yielding between 10 μg and 200 μg purified protein per mL of bacterial culture, with an average yield of 31 μg of peptide aptamer per mL of bacterial culture. Purity was greater than 90% as judged by Coomassie staining of multiple preparations analysed by SDS-PAGE and contaminant proteins were apparently uniform across the preparations analysed. Although future generations of this technology may require the spotting of equal amounts of peptide aptamer per spot, we decided to spot equal volumes of each peptide aptamer preparation, reasoning that impurities would yield uniform background signals at each spot on an array, and that, because they will be present in large excess over the target proteins, 20-fold variations in the number of molecules of peptide aptamer per spot would be acceptable.

2.5 Fabrication and probing of the peptide aptamer microarray

To prepare the peptide aptamer microarray, we employed a Qarray mini (Genetix) printer to spot all the peptide aptamers in duplicate onto Ni-NTA coated glass slides (XENOPORE) using split pins. It was not necessary to remove imidazole prior to spotting. The printed slides were allowed to sit at RT for one hour in the printer before storing at 4 °C.

To assess the quality of the PA microarray, we probed the slide with an anti-StefA antibody (Chemicon). Briefly, the microarray was blocked with 20 mL 3% (w/v) BSA in PBST (PBS with 0.1% [v/v] Tween 20) at room temperature for an hour with gentle shaking. The microarray was then incubated with 100 μL 0.2μg mL^{-1} rabbit anti-SteA antibody in blocking buffer covered with a cover slip (Thermo Fisher Scientific) in a humidity chamber. After washing three times with 20 mL PBST, the microarray was incubated with Cy3-labeled goat anti-rabbit IgG antibody (ZYMED) for one hour at room temperature in a humidified chamber with a cover slip on top of the slide. The slide was then washed 3 times with 20 mL PBST. After a further wash with distilled water for 5 min, the slide was dried by centrifuging at 1600 rpm in a microfuge at room temperature for 5 min and analysed with a standard DNA microarray scanner (Axon). Genepix 4.0 software was used to quantify the signal intensity for each spot. Only those spots with a signal-to-background ratio $\geq$3.0 in the anti-StefA probing were considered to represent peptide aptamers that had been successfully expressed and immobilised on the slide.

DG75 cells were grown in RPMI medium (Gibco) supplemented with 10% fetal bovine serum, penicillin and streptomycin to a density of one million per millilitre. The cells were collected, washed with pre-chilled PBS (pH 7.2) and lysed by sonication in lysis buffer (20 mM HEPES, pH 7.5, 150 mM KCl, 0.1 mM EDTA, 10% glycerol, 0.1% NP-40, 1 mM DTT, EDTA-free protease inhibitor cocktail (Roche), and 1 mM PMSF). The resulting cell lysate was centrifuged in a microfuge at 13 000 rpm for 20 min at 4 °C. The supernatant was transferred to pre-chilled clean vials and the total protein concentration determined by measuring the light absorption at 280 nm using a spectrophotometer (Nanodrop).

We used a Cy3 post-labelling reactive kit (GE Healthcare) to label the protein or lysate sample. Briefly, 7 nmol dye solution dissolved in DMSO was added to 10 μL protein or lysate (which was diluted to 2 mg total protein/mL with lysis buffer) premixed with 10 μL Na_2CO_3 (0.1M, pH 9.3). The reaction was allowed to proceed at 4 °C for 2 h. To remove excess unreacted dye molecules, labeled proteins were purified by use of a Biospin-6 column (Biorad) following the manufacturer's instructions.

To probe the peptide aptamer microarrays with the labeled lysate, the microarray was first blocked as described for the anti-SteA antibody-probing assay. The labeled lysate was diluted in 100 μL blocking buffer to a concentration of 40 μg protein/mL and applied to the slide, while the purified CDK2 probing varied as indicated. The slide was covered with a cover slip and placed in a humidified chamber to incubate at RT for an hour. The cover slip was removed and the microarray immersed in PBST to wash for 5 min with gentle shaking. The wash buffer was discarded and the wash repeated twice more. The microarray was washed with distilled water for 5 min, dried by centrifugation at 1600 rpm at room temperature for 5 min, and analysed with the Axon scanner. Two microarrays were used in parallel for each assay. In the Cy3-labeled DG75 lysate probing, we used those positive features in anti-SteA probing for data analysis, in which we set the cutoff at mean ± 1 standard deviation, that is, only those peptide aptamers binding to lysate proteins to give a signal intensity above the average value of all spots plus/minus the standard deviation value across all spots were considered reactive with the lysate.

3 Results and discussion

We envisage two applications for peptide aptamers in healthcare. In the first, rationally selected peptide aptamer ligands would be used to detect specific analytes in a biological or clinical sample. Here, we illustrate this idea with previously identified peptide aptamers that recognise human CDK2. In the second application, arrays of naive peptide aptamers selected randomly from large libraries can be arrayed on a microscope slide, and used to identify proteins altered in disease or following treatment in an open-ended, non-hypothesis driven fashion. To these ends, we have developed a proof-of-concept peptide aptamer microarray to demonstrate the multiplexed detection of human proteins. With a view to the future creation of highly multiplexed microarrays, we have developed an efficient workflow for the expression and purification of peptide aptamers in 96-well plate format† (Fig. 1, panel A). Briefly, oligonucleotides comprising a specified number of random codons flanked by restriction sites are ligated into the open reading frame of a constant scaffold protein, called SQM.[29] The resulting plasmid library is transformed into *E coli* BL21 cells optimised for protein expression and individual colonies picked into each well of a series of 96-deep-well plates. Starter cultures are diluted into fresh plates and the expression of peptide aptamers induced following conventional protocols. Peptide aptamers are purified from lysed cells using automation-compatible protocols (see Methods for details) on Ni-chelate beads, eluted and either spotted directly onto Ni-functionalised glass slides or stored in 30% glycerol at −80 °C until use. Peptide aptamer microarrays are created using standard DNA spotting approaches, using split pins and Ni-chelate coated microscope slides. Because the amount of matrix in each well of the purification plate is in excess over the amount of peptide aptamer present in the lysed culture, the amount of peptide aptamer purified from each well reflects the expression level of each peptide aptamer, and variations are clearly visible in the final array. The concentration of purified protein (comprising peptide aptamer and any contaminating *E. coli* proteins) in each preparation prior to spotting ranged from 0.006 to 0.2 mg mL^{-1}. Gel electrophoresis and western blotting revealed that the lower end of this range resulted from cultures that did not express any peptide aptamer (Fig. 1, panel B), which also explains unexpected absence of signal in some grids when we develop the array with a fluorescently-labelled anti-SteA antibody probe that should recognise every spotted protein (Fig. 1 panel B). Because of the relative uniformity in fluorescence intensity compared to the variable amounts of protein detectable by Coomassie

† Because the goal of this technology is to enable high-throughput production of peptide aptamers and peptide aptamer microarrays, we did not seek to optimise individual steps, except that we found it important to use an appropriate printing buffer and a high-quality (commercial grade rather than self-made) microscope slide surface. Further details are in the Methods section.

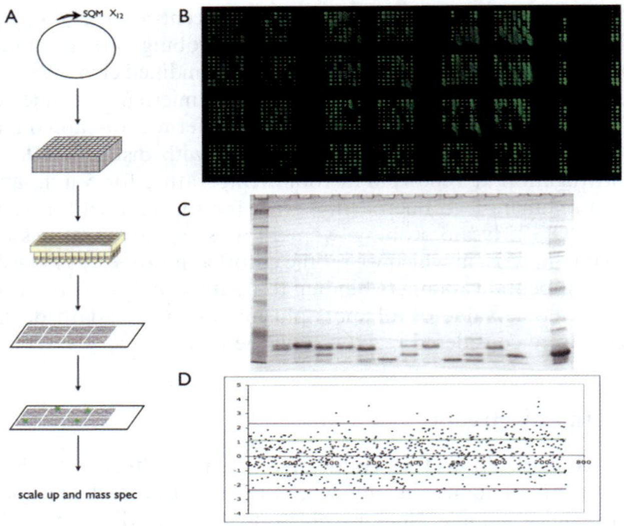

Fig. 1 Design and validation of peptide aptamer microarrays. Panel A: plasmids encoding random peptide aptamer libraries are expressed in and purified from *E. coli* in 96-well plates, and spotted on derivatised glass slides. Panel B: The resulting microarrays can be probed to assess the quality of protein production and printing. Panel C: Coomassie staining of SDS gels confirm that protein expression levels are responsible for the variations in spot intensity. Panel D: Quantification of spot intensity across the array confirm that the vast majority of peptide aptamers are equally well expressed and immobilised. See text for details.

staining of sampled preparations of expressed and purified peptide aptamers (Fig. 1, panel C), we suggest that excess peptide aptamer present in preparations of the most highly expressed peptide aptamers did not bind to the slide and was removed during wash steps. Each naïve peptide aptamer was printed in duplicate, and the achieved correlation efficiency of the signal intensity of the two spots representing one single peptide aptamer was as high as 0.996. Variability in spotting appears to contribute little as 94% of the measured fluorescence intensities across a single slide cluster within 2 standard deviations of the mean, and 68% within 1 sd of the mean (Fig. 1D). Given a reported spotting volume of 0.2 nL for split pins, we calculate that the maximal amount of protein immobilised ranged from 1.2 to 4 pg at each spot, of which peptide aptamer was 0–2.8 pg, corresponding to 0–0.19 fmoles of peptide aptamer. This value represents the calculated upper value for target protein that can be detected by our system.

We next asked whether we could use such arrays to detect specific binding partners from whole cell lysates (Fig. 2). We asked whether peptide aptamers with known specificity immobilized on similar arrays could be used to detect their target (human CDK2) in the same whole cell lysates. We used 4 peptide aptamers (pep2, pep6, pep9 and pep10m) that were derived from the original CDK2 binders described by Colas *et al.*[13,31] We have previously transferred the peptide recognition moiety of each peptide aptamer into the STM scaffold without apparent loss of reactivity.[22,33–35] Binding specificity was confirmed by showing that bacterially expressed peptide aptamers could be used to affinity-purify CDK2 from whole cell lysates (Fig. 2, panel H). As negative controls, we used the empty STM scaffold, the empty SQM scaffold and 64 different randomly selected peptide aptamers from the library of random peptide aptamers described above. In order to allow critical assessment of the performance of the arrays, negative controls were spotted at the maximal concentration, giving 0.2 fmoles of protein per spot, while the CDK2-binding peptide aptamers were spotted at concentrations spanning a range from 0 to 0.2 mg mL^{-1}

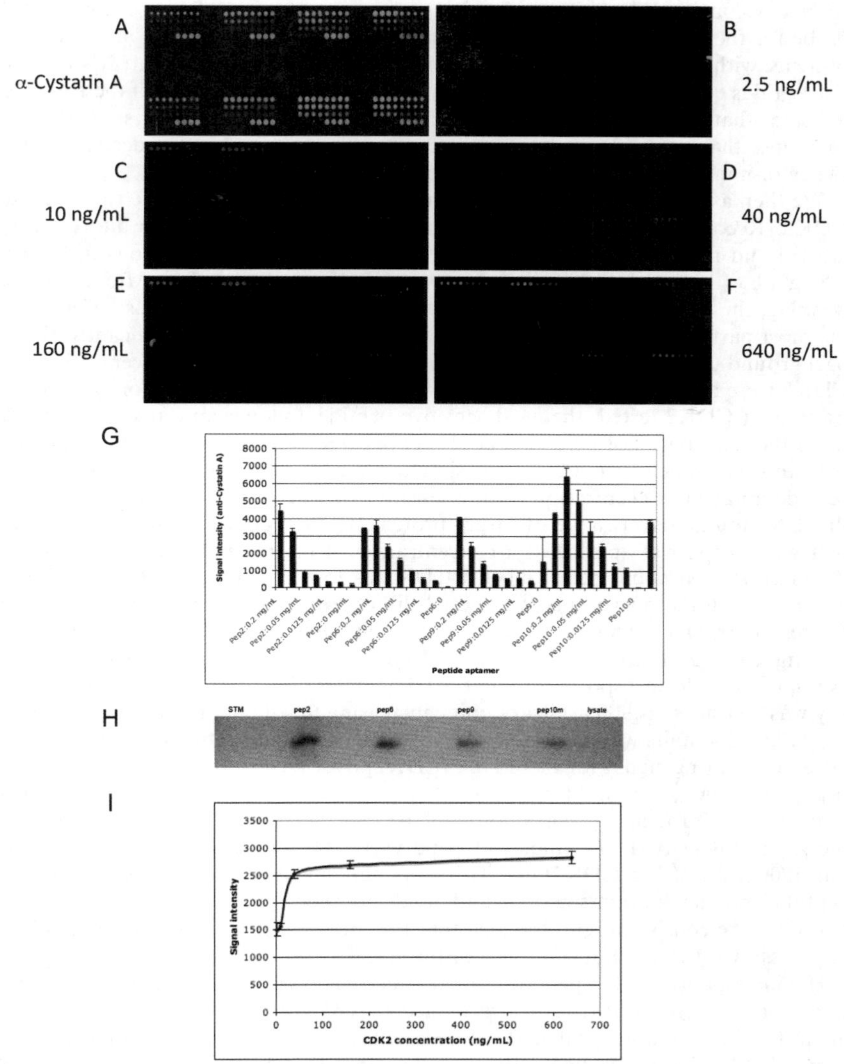

Fig. 2 Validation of peptide aptamer detection using purified target protein. A range of 4 peptide aptamers (pep2, pep6, pep9 and pep10m) specific for human CDK2 were immobilised from solutions of varying concentrations and challenged with varying concentrations of fluorescently labelled recombinant, active cyclin A-CDK2 from baculovirus. Panel A: quality control for the array, showing the dilution series for each peptide aptamer. The 4 bright spots underneath each grid represent empty scaffold protein. Panels B–F: a similar array to that shown in panel A was probed with 2.5 (panel B), 10 (panel C), 40 (panel D), 160 (panel E) or 640 (panel F) ng mL^{-1} CDK2. Panel G: Quantification of signal intensities across spots confirms that signal intensity correlates with the amount of immobilised peptide aptamer. Panel H: when immobilised on agarose beads, the four peptide aptamers used in panels A–F, but not the empty STM scaffold protein, are able to recruit CDK2 from whole cell lysates, confirming their reactivity and, to some extent, their specificity. Panel I: quantification of the data in panels B–F.

(Fig. 2, panel A). Each concentration of each peptide aptamer was spotted in 4 replicates per grid, and each grid is spotted in duplicate, so that each peptide aptamer spot is present 8 times on each array. Each control was spotted in 4 replicates and controls were evenly distributed across the 8 grids so as to allow the use of a large number of specificity controls. The quality of spotting was assessed using a fluorescently labeled probe for the S-tag that recognises the amino-terminal fusion portion of all spotted proteins, with fluorescence being measured using standard DNA microarray scanners. Each grid is qualitatively as expected (Fig. 2, panel A). Quantification of these images indicates that intensity increases with the amount of peptide aptamer spotted, with some loss that is readily explained by near-neighbour quenching or decreased efficiency of immobilisation at higher surface coverage (not shown).

We then asked whether the spotted peptide aptamers would be able to recognise CDK2. Recombinant human cyclin A/CDK2 from baculovirus was fluorescently labeled and incubated with microarrays at five different concentrations in buffer: 2.5 ng mL^{-1}, 10 ng mL^{-1}, 40 ng mL^{-1}, 160 ng mL^{-1} and 640 ng mL^{-1}. After extensive washing, the slides were analysed using a DNA microarray scanner as before. For the most part, signal intensity with 2.5 ng mL^{-1} CDK2 was not significantly above background (Fig. 2 panel B), but signal intensities for the higher concentrations of CDK2 were all detectable above background (Fig. 2, panels C–F). For all concentrations of CDK2 tested, the most sensitive peptide aptamer appeared to be pep2, while the sensitivities of pep9 and pep10m appeared similar and pep6 the weakest.

In this experiment, we are able to simultaneously assess the effect of increasing peptide aptamer concentration on the sensitivity of detection, and the linearity of the detection across a range of target protein (CDK2) concentrations. We found that when pep2 was spotted at a concentration of 0.1 mg mL^{-1} or higher (up to 0.2 mg mL^{-1}), it could detect labeled CDK2 as low as 2.5 ng mL^{-1}. In addition, when probing the microarray with serial dilutions of purified CDK2 protein, the fluorescent signal intensity changed correspondingly although the range tested (2.5 ng CDK2/mL to 160 ng CDK2/mL) was not linear (Fig. 2, panel I). For example, at the lowest spotted amount of 0.06 fmoles of pep2, the fluorescence intensity was measured as 900 arbitrary units, increasing to 4200 units with 0.1 fmoles of pep2 and 5200 units with 0.2 fmoles of pep2 (when probing with 60 ng CDK2/mL). When comparing signal for increasing [CDK2] over a constant amount of spotted peptide aptamer, we found that signal intensity increased with increasing concentration of CDK2 for any given amount of peptide aptamer spotted. For example, the signal intensity at 0.2 fmoles of pep2 was 2100 units with 20 ng CDK2/mL but 5200 with 60 ng CDK2/mL. This apparent linearity is conserved across all peptide aptamer amounts tested, and holds for pep2, pep6, pep9 and pep10m. Our data are consistent with the idea that the probes (peptide aptamers) would be in excess over the amount of CDK2 at the surface.

Having established that peptide aptamer microarrays can be used to detect known target proteins in purified form, we wished to ask whether they could be used to quantify a known target protein (CDK2) in cell lysates. Similar microarrays to those shown in Fig. 2 were probed with total, fluorescently labelled protein from DG75 cell lysates. As before, pep2 was the most sensitive binder, reproducibly giving signal with multiple samples (Fig. 3, panels A–D). To ask whether the arrays were able to detect any changes in CDK, signalling the cell lysates used in this experiment were treated either with a known CDK2 inhibitor, roscovitine[36] or with DMSO as carrier control. To confirm that the signals we measure are due to CDK2, we used a fluorescently labelled anti-CDK2 antibody to probe the arrays after the incubation and wash steps. Intriguingly, after both 24 h (Fig. 3, panels A and B) or 48 h (Fig. 3, panels C and D), roscovitine-treated cells showed a significantly decreased level of signal (Fig. 3, panel E). This could be due to an altered conformation of CDK2 with decreased affinity for the peptide aptamers, or to a decrease in CDK2 expression levels in treated cells. Western blotting showed that total CDK2 protein was indeed decreased in roscovitine-treated cells relative to tubulin as a control (Fig. 3, panel F), and that (as with the microarrays) the effect was slightly more pronounced at 24 than at 48 h. Comparing the fluorescence intensities obtained here to those obtained using purified CDK2 (Fig. 2), we calculate the amount of CDK2 in DMSO-treated cells to be 13×10^{-18} moles/cell, dropping to less than

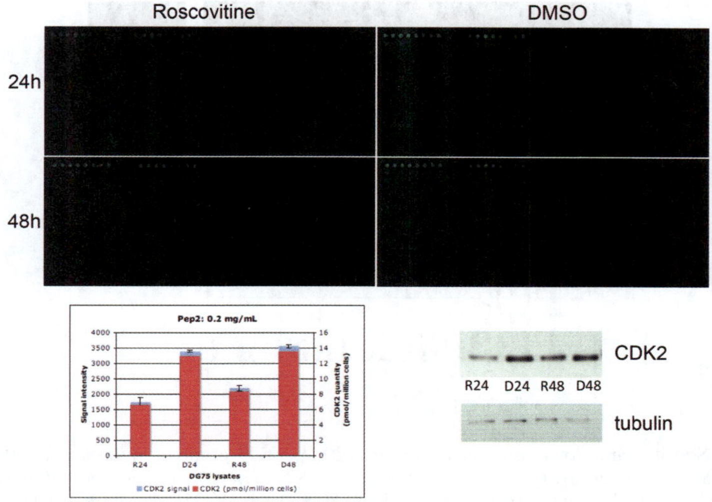

Fig. 3 CDK2-specific peptide aptamer microarrays reveal drug-induced protein changes. Similar peptide aptamer microarrays to those used in Fig. 3 were probed with proteins from lysed DG75 cells that had been treated for 24 or 48 h with roscovitine (R) or DMSO (D). Captured CDK2 was detected using a fluorescently labelled anti-CDK2 antibody. Quantification of the signals obtained using a microarray scanner is shown in the bar graph. The decreased levels of CDK2 revealed by the miroarray were confirmed using conventional Western blotting, using antibodies against CDK2 and tubulin.

7×10^{-18} moles/cell in roscovitine-treated cells after 24 h and 8×10^{-18} moles/cell after 48 h of roscovitine treatment.

Finally, having shown that peptide aptamer microarrays can be used to detect known proteins in cell lysates, we asked whether a naïve array could be used to detect unknown proteins in whole cell lysates. To this end, we incubated 100 μL of cell lysate corresponding to 4 μg of fluorescently labeled protein isolated from 8000 cells with microarrays identical to those described in Fig. 1. After extensive washing, the recruitment of fluorescently labelled proteins to immobilised peptide aptamers was measured using a microarray scanner (Fig. 4, panel A) and fluorescent signals were normalised to the amount of peptide aptamer immobilized at each pixel measured using the anti-StefA antibody (Fig. 4, panel D). Peptide aptamers pep2 and pep9 gave relatively strong signals, with ratios of 3.5 for the brightest pep2 spot (Fig. 4, panel C). Only one of the peptide aptamers from the random library, designated pep247, gave a significant signal (Fig. 4, panel B) corresponding to a ratio of 9.5. The lack of signal from the majority of the peptide aptamers spotted indicates that peptide aptamer microarrays are capable of highly specific protein detection even when challenged with complex biological samples such as whole cell lysates. In order to identify the protein(s) recognised by pep247, we immobilised the peptide aptamer on Ni-NTA beads as an affinity-purification column. Coomassie stained gels revealed 2 major bands that could be eluted from the column after incubation with whole cell lysates. We used pep247 to probe Western blots of cell lysates, and found that the peptide aptamer recognises a major band at 97 kDa. Mass spectrometry of the two bands revealed that one of these corresponded to the peptide aptamer itself, while the 97 kDa band recognised by pep247 comprised a mixture of two proteins of roughly equal molecular weight, nucleolin and PSF/SFPQ. Nucleolin and PSF/SFPQ both play roles in transcription and RNA transport, and we are unable to tell whether pep247 recognises a surface functionally conserved between both proteins, whether the two proteins may form a complex that has been purified by pep247 (which seems unlikely as they function in different transcriptional

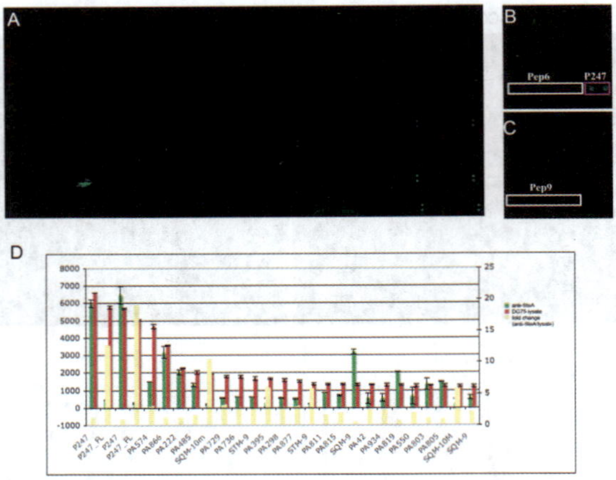

Fig. 4 Specific, unknown protein detection with peptide aptamer microarrays. Panel A. Peptide aptamer microarrays comprising in excess of 800 peptide aptamers spotted in duplicate were challenged with total labelled protein from 8000 lysed DG75 cells. Panel B. The peptide aptamer pair giving rise to the strongest signal is shown. Panel C: control peptide aptamers with known specificity are shown for comparison. Panel D: normalised fluorescent signal intensities for selected peptide aptamers across the array.

processes and localise to distinct subnuclear compartments[37,38]), or whether one of these proteins binds pep247 specifically and the second binds non-specifically to the complex. RNase treatment of the proteins bound to the pep247 column released both nucleolin and PSF/SFPQ, suggesting that pep247 may recognise an RNA that is specifically bound to these two proteins, but no other RNA-binding proteins, which would seem unlikely unless there was a functionally significant role for the complex.

4 Conclusions

Peptide aptamers based on the SQM scaffold are able to retain their specific recognition functions when immobilised on glass surfaces. They show low affinity for irrelevant proteins in whole cell lysates. They are amenable to high-throughput expression and multiplexed detection formats. Peptide aptamer microarrays can be used to detect specific proteins in cell lysates and are able to detect subtle changes in protein expression. They can be used to identify new proteins or perhaps even protein complexes from whole cell lysates. These data lay the foundations for a biomarker discovery platform using naïve peptide aptamer microarrays which will simultaneously yield a biomarker, a cognate probe and an assay. They also lay the foundation for the development of clinical assays using single or multiple immobilised peptide aptamers, although much work on the validation of these tools in clinical settings remains to be done. Finally, in combination with our published collaborative work on label-free detection, we suggest that peptide aptamers should be explored in parallel with antibodies when establishing assays using novel optical, electrical or M/NEMS (micro- or nano-electromechanical system) detection technologies.

References

1 G. M. Edelman and W. E. Gall, *Annu. Rev. Biochem.*, 1969, **38**, 415–466.
2 S. Muyldermans, T. N. Baral, V. C. Retamozzo, P. De Baetselier, E. De Genst, J. Kinne, H. Leonhardt, S. Magez, V. K. Nguyen, H. Revets, U. Rothbauer, B. Stijlemans,

S. Tillib, U. Wernery, L. Wyns, G. Hassanzadeh-Ghassabeh and D. Saerens, *Vet. Immunolo. Immunopathol.*, 2009, **128**, 178–183.
3 J. Wesolowski, V. Alzogaray, J. Reyelt, M. Unger, K. Juarez, M. Urrutia, A. Cauerhff, W. Danquah, B. Rissiek, F. Scheuplein, N. Schwarz, S. Adriouch, O. Boyer, M. Seman, A. Licea, D. V. Serreze, F. A. Goldbaum, F. Haag and F. Koch-Nolte, *Med. Microbiol. Immunol.*, 2009, **198**, 157–174.
4 E. S. Ward, D. Gussow, A. D. Griffiths, P. T. Jones and G. Winter, *Nature*, 1989, **341**, 544–546.
5 L. J. Holt, C. Herring, L. S. Jespers, B. P. Woolven and I. M. Tomlinson, *Trends Biotechnol.*, 2003, **21**, 484–490.
6 L. J. Holt, A. Basran, K. Jones, J. Chorlton, L. S. Jespers, N. D. Brewis and I. M. Tomlinson, *Protein Eng., Des. Sel.*, 2008, **21**, 283–288.
7 H. K. Binz, P. Amstutz and A. Pluckthun, *Nat. Biotechnol.*, 2005, **23**, 1257–1268.
8 M. Gebauer and A. Skerra, *Curr. Opin. Chem. Biol.*, 2009, **13**, 245–255.
9 K. Nord, J. Nilsson, B. Nilsson, M. Uhlen and P. A. Nygren, *Protein Eng., Des. Sel.*, 1995, **8**, 601–608.
10 A. Koide, C. W. Bailey, X. Huang and S. Koide, *J. Mol. Biol.*, 1998, **284**, 1141–1151.
11 H. K. Binz, M. T. Stumpp, P. Forrer, P. Amstutz and A. Pluckthun, *J. Mol. Biol.*, 2003, **332**, 489–503.
12 Z. Lu, K. S. Murray, V. Van Cleave, E. R. LaVallie, M. L. Stahl and J. M. McCoy, *Bio/ Technology*, 1995, **13**, 366–372.
13 P. Colas, B. Cohen, T. Jessen, I. Grishina, J. McCoy and R. Brent, *Nature*, 1996, **380**, 548–550.
14 G. Beste, F. S. Schmidt, T. Stibora and A. Skerra, *Proc. Natl. Acad. Sci. U. S. A.*, 1999, **96**, 1898–1903.
15 T. A. Tran, D. Rosik, L. Abrahmsen, M. Sandstrom, A. Sjoberg, H. Wallberg, S. Ahlgren, A. Orlova and V. Tolmachev, *Eur. J. Nucl. Med. Mol. Imaging*, 2009, **36**, 1864–1873.
16 P. R. Srinivas, B. S. Kramer and S. Srivastava, *Lancet Oncol.*, 2001, **2**, 698–704.
17 X. Yu, N. Schneiderhan-Marra and T. O. Joos, *Clin. Chem.*, 2010, **56**, 376–387.
18 H. T. Beernink and S. Nock, *Expert Rev. Proteomics*, 2005, **2**, 487–497.
19 M. Uhlen, E. Bjorling, C. Agaton, C. A. Szigyarto, B. Amini, E. Andersen, A. C. Andersson, P. Angelidou, A. Asplund, C. Asplund, L. Berglund, K. Bergstrom, H. Brumer, D. Cerjan, M. Ekstrom, A. Elobeid, C. Eriksson, L. Fagerberg, R. Falk, J. Fall, M. Forsberg, M. G. Bjorklund, K. Gumbel, A. Halimi, I. Hallin, C. Hamsten, M. Hansson, M. Hedhammar, G. Hercules, C. Kampf, K. Larsson, M. Lindskog, W. Lodewyckx, J. Lund, J. Lundeberg, K. Magnusson, E. Malm, P. Nilsson, J. Odling, P. Oksvold, I. Olsson, E. Oster, J. Ottosson, L. Paavilainen, A. Persson, R. Rimini, J. Rockberg, M. Runeson, A. Sivertsson, A. Skollermo, J. Steen, M. Stenvall, F. Sterky, S. Stromberg, M. Sundberg, H. Tegel, S. Tourle, E. Wahlund, A. Walden, J. Wan, H. Wernerus, J. Westberg, K. Wester, U. Wrethagen, L. L. Xu, S. Hober and F. Ponten, *Mol. Cell. Proteomics*, 2005, **4**, 1920–1932.
20 S. Hu, Y. Li, G. Liu, Q. Song, L. Wang, Y. Han, Y. Zhang, Y. Song, X. Yao, Y. Tao, H. Zeng, H. Yang, J. Wang, H. Zhu, Z. N. Chen and L. Wu, *Proteomics*, 2007, **7**, 2151–2161.
21 M. R. Abedi, G. Caponigro and A. Kamb, *Nucleic Acids Res.*, 1998, **26**, 623–630.
22 R. Woodman, J. T. Yeh, S. Laurenson and P. Ko Ferrigno, *J. Mol. Biol.*, 2005, **352**, 1118–1133.
23 B. Klevenz, K. Butz and F. Hoppe-Seyler, *Cell. Mol. Life Sci.*, 2002, **59**, 1993–1998.
24 D. Kordiš and V. Turk, *BMC Evol. Biol.*, 2009, **9**, 266.
25 M. T. Stubbs, B. Laber, W. Bode, R. Huber, R. Jerala, B. Lenarcic and V. Turk, *EMBO J.*, 1990, **9**, 1939–1947.
26 S. Jenko, I. Dolenc, G. Guncar, A. Dobersek, M. Podobnik and D. Turk, *J. Mol. Biol.*, 2003, **326**, 875–885.
27 J. R. Martin, C. J. Craven, R. Jerala, L. Kroon-Zitko, E. Zerovnik, V. Turk and J. P. Waltho, *J. Mol. Biol.*, 1995, **246**, 331–343.
28 J. R. Martin, R. Jerala, L. Kroon-Zitko, E. Zerovnik, V. Turk and J. P. Waltho, *Eur. J. Biochem.*, 1994, **225**, 1181–1194.
29 T. Hoffmann, L. K. Stadler, M. Busby, Q. Song, A. T. Buxton, S. D. Wagner, J. J. Davis and P. Ko Ferrigno, *Protein Eng., Des. Sel.*, 2010, **23**, 403–413.
30 C. S. Chen, E. Korobkova, H. Chen, J. Zhu, X. Jian, S. C. Tao, C. He and H. Zhu, *Nat. Methods*, 2008, **5**, 69–74.
31 B. A. Cohen, P. Colas and R. Brent, *Proc. Natl. Acad. Sci. U. S. A.*, 1998, **95**, 14272–14277.
32 Unpublished results.
33 D. Evans, S. Johnson, S. Laurenson, A. G. Davies, P. Ko Ferrigno and C. Walti, *J. Biol.*, 2008, **7**, 3.

34 J. J. Davis, J. Tkac, R. Humphreys, A. T. Buxton, T. A. Lee and P. Ko Ferrigno, *Anal. Chem.*, 2009, **81**, 3314–3320.
35 P. Estrela, D. Paul, Q. Song, L. K. Stadler, L. Wang, E. Huq, J. J. Davis, P. Ko Ferrigno and P. Migliorato, *Anal. Chem.*, 2010, **82**, 3531–3536.
36 L. Meijer, A. Borgne, O. Mulner, J. P. Chong, J. J. Blow, N. Inagaki, M. Inagaki, J. G. Delcros and J. P. Moulinoux, *Eur. J. Biochem.*, 1997, **243**, 527–536.
37 F. Mongelard and P. Bouvet, *Trends Cell Biol.*, 2007, **17**, 80–86.
38 C. S. Bond and A. H. Fox, *J. Cell Biol.*, 2009, **186**, 637–644.

Aptameric sensors based on structural change for diagnosis†

Koichi Abe, Daisuke Ogasawara, Wataru Yoshida, Koji Sode and Kazunori Ikebukuro*

Received 14th May 2010, Accepted 3rd July 2010
DOI: 10.1039/c005359g

Aptamers are nucleic acids that can bind to various molecules. Because they have some features that are lacking in antibodies, aptamers could serve as alternatives to antibodies. For the purpose of biosensing, we focused on aptamers that undergo structural changes on binding to their target molecules. We constructed an aptamer-based bound/free (B/F) separation system that uses a designed aptamer named the "capturable aptamer". The capturable aptamer changes its structure upon recognizing its target molecule thereby exposing a specific single-strand region. The oligonucleotide that is complementary to this exposed region, named the "capture DNA" is immobilized on a support. This design permits the exclusive capture by the capture DNA of the aptamer bound to its target, and subsequent removal of any unbound aptamer and contaminants by B/F separation. The removal of unbound contaminants or aptamers results in highly sensitive detection at similar levels to those achievable by sandwich-based immunoassay. We describe the construction of a thrombin-detection system by using a capturable aptamer, and we discuss the potential of capturable aptamers in clinical diagnostics.

Introduction

Aptamers are nucleic acids that bind to various molecular targets, and they are normally obtained by selection from libraries of nucleic acids through systematic evolution of ligands with exponential enrichment (SELEX).[1,2] Although SELEX is a powerful method for obtaining aptamers, it involves laborious experiments that generally require five to twenty rounds of screening. However, an automated SELEX process has been reported that is capable of producing aptamers against 120 targets per month.[3] Aptamers can therefore be developed more rapidly than antibodies. Furthermore, in addition to showing comparable affinities and specificities to antibodies, aptamers can be selected against not only proteins, but also against small molecules. Furthermore, aptamers can be synthesized and easily modified, and they are more stable than antibodies. Therefore, aptamers are expected to serve as alternative molecules to antibodies.

In comparison with proteins, nucleic acids carry many negative charges derived from their phosphate backbones. Whereas double-stranded DNA is stabilized by the presence of many Watson–Crick base pairs, the structures of single-stranded nucleic acids, including aptamers, are destabilized as result of the presence of fewer

Department of Biotechnology, Tokyo University of Agriculture and Technology, 2-24-16 Naka-cho, Koganei-shi, Tokyo 184-8588, Japan. E-mail: ikebu@cc.tuat.ac.jp; Fax: +81 42 388 7030; Tel: +81 42 388 7030

† Electronic supplementary information (ESI) available: Fig. S1 and Table S1. See DOI: 10.1039/c005359g

Watson–Crick base pairs than are present in double-stranded DNA. Single-stranded nucleic acids often exist as ensembles of conformational states distributed on various timescales.[4] Binding of a ligand can dramatically modify the energy landscape of the structures of aptamers, resulting in stabilization of one conformation or small subsets of conformations through additional binding energy. Therefore, aptamers have a tendency to undergo a change in conformation on binding to their target molecules.

Many aptameric sensors have been reported that rely on transduction of structural changes into detectable signals.[5–8] Such sensor systems can detect target molecules in homogeneous assays without bound/free (B/F) separation, which is ideal for simple and rapid assays of target molecules. Most sensors are based on fluorescence resonance energy transfer[9] or changes in the distance between an electrode and a molecule that reacts with the electrode after the aptamer has undergone a structural change.[6] We have reported a homogeneous detection system that uses an enzyme-inhibiting aptamer called an aptameric enzyme subunit (AES).[10–14] The AES consists of two aptamers, an enzyme-inhibiting aptamer and a target molecule-binding aptamer. Binding of the target molecule to the latter aptamer changes the enzyme-inhibitory activity of the former aptamer, allowing the AES to be used to measure the concentration of the target molecule through the resulting change in the activity of an enzyme. We have reported several detection systems based on the AES that can be used to assay DNA, small molecules, or proteins.[10–14] Because enzymes can amplify signals, the AES should be ideal for determining target molecule rapidly, easily, and with a high sensitivity.

However, for highly sensitive detection systems, B/F separation is important because it permits the removal of molecules that cause interference and because it allows labeling of the molecular-recognition elements by an enzyme to amplify any signals. Sandwich assays using two antibodies that bind to different regions of a target molecule simultaneously are commonly used for sensing of target molecules. We have reported a sandwich-based system for sensing thrombin by using two aptamers.[15] We used glucose dehydrogenase as the labeling enzyme, because this allows target molecules to be determined by using an electrochemical detection system. However, to construct a sandwich-based detection system, it is necessary to select two molecular recognition elements that can bind to different regions of the target molecule. The selection of two such antibodies is a laborious task, because in addition to having to bind to different regions simultaneously, the two antibodies also have to have low dissociation constants.

We have previously reported on two types of single aptamer-based B/F separation systems (Fig. 1).[16,17] The first type of system makes use of the features of aptamers whose structures are stabilized by binding to their target molecules.[16] Aptamers that are not bound to their target molecules can hybridize with capture DNA (CaDNA) immobilized on the support, whereas aptamers that are bound to their target molecules do not hybridize with the CaDNA [Fig. 1(a)]. The second system makes use of the structural changes that aptamers undergo on binding to their target molecules.[17] We added a sequence to an aptamer to form a designed aptamer with a new structure that we named a "capturable aptamer". Capturable aptamers do not hybridize with CaDNA in the absence of their target molecules, but can hybridize with CaDNA in the presence of their target molecules. In this case, the structure of a capturable aptamer in the presence of its target molecule changes to a different structure from that which is present in the absence of the target molecule [Fig. 1(b)]. Of these two types of single aptamer-based systems, the first can be easily designed, because it does not require any additional sequences, whereas the second system requires careful design of the additional sequence of the aptamer with structural prediction. On the other hand, the second system can eliminate many interfering compounds, whereas it is difficult for the first system to eliminate interfering compounds, because aptamers that bind to the target molecules are present in the supernatant. It is therefore necessary to select a particular system to suit the needs of each particular target molecule.

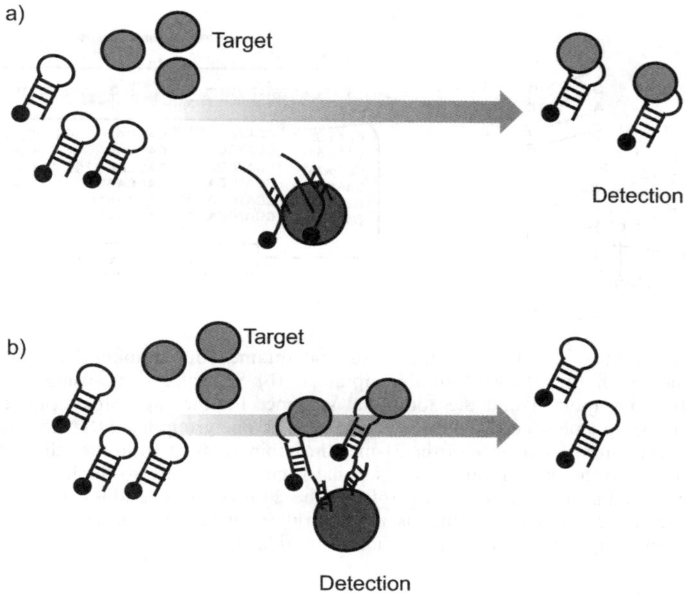

Fig. 1 The scheme for an aptamer-based bound/free separation system. (a) In the absence of a target protein, the aptamers are trapped by the immobilized beads containing CaDNA, whereas in the presence of the target protein, aptamers that bind to the target are not trapped. The target protein can therefore be detected by means of simple B/F separations and by measuring the fluorescence or enzymatic activity of the labeled aptamer in the supernatant. (b) The capturable aptamer undergoes a conformational change upon binding to the target molecule. This change induces the exposure of a partial single-strand that hybridizes with the CaDNA. Otherwise, the unbound capturable aptamer does not hybridize with the CaDNA and is removed by the bound/free separation.

We have previously reported a system for sensing mouse prion proteins that uses a capturable aptamer.[17] In the current study, we designed a capturable aptamer based on a thrombin aptamer to demonstrate the versatility of the capturable aptamer-based analytical system. The thrombin-binding aptamer, which has a G-quadruplex structure, was previously obtained as a 31-mer that has a greater inhibitory activity than a similar 15-mer thrombin aptamer.[18] The 31-mer thrombin aptamer contains a G-quadruplex motif and a duplex motif [Fig. 2(a)], but its structural properties have not yet been fully analyzed. We designed a capturable aptamer for thrombin detection, and we analyzed the structural changes in this aptamer to validate the system. We also discuss the potential of such systems in clinical diagnosis.

Experimental

Materials

All oligonucleotides were synthesized by Life technologies. The sequences of the oligonucleotides are listed in Table 1 and Table S1 of the ESI.† All aptamers were folded by heating at 95 °C for 5 min then gradually cooling to 25 °C over 30 min before use in the experiments.

Measurement of the T_m value for the thrombin aptamer

A solution of the 31-mer thrombin aptamer (1 μM) in binding buffer [20 mM Tris-HCl (pH 8.0), 5 mM KCl, 5 mM $MgCl_2$, 100 mM NaCl] was heated from 25 to 85 °C

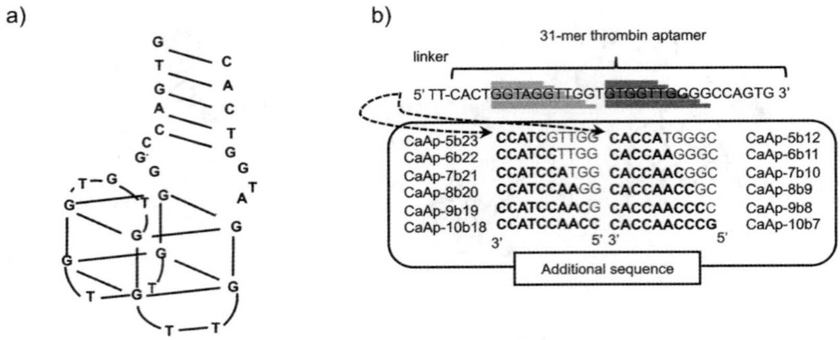

Fig. 2 The details of the design of the capturable aptamer for thrombin detection. (a) Predicted structure of the 31-mer thrombin aptamer. (b) Sequence relationship between the 31-mer thrombin aptamer and the additional sequence for the capturable aptamer design. The capturable aptamers were divided into two groups: one group has DNA that is complementary to the internal sequence of the 31-mer thrombin aptamer from the 5th to the 9th or 15th residues, and the other group has DNA that is complementary to the internal sequence of 31-mer thrombin aptamer from the 16th to the 20th or 26th residues. Each additional sequence has an extra sequence that is not hybridized with the internal sequence of the 31-mer thrombin aptamer to maintain sufficient hybridization with the CaDNA.

Table 1 Sequences of designed capturable aptamer and calculated T_m value

Name	Sequence	$T_m/°C$
CaAp-5b12	CCGGGTaccacTTCACT GGTAGGTTGGTGTGGT TGGGGCC AGTG[a]	59.3
CaAp-6b11	C CGGGaaccacTTCAC TGGTAGGTTGGTGTGGTT GGGGCCAGTG	59.3
CaAp-7b10	CCGGcaaccacTTCAC TGGTAGGTTGGTGTGGTTG GGGCCAGTG	62.7
CaAp-8b9	CCGccaaccacTTCACTGGTAGGT TGGTGTGGTTGG GGCCAGTG	68.4
CaAp-9b8	CGcccaaccacTTCACTGGTAGGT TGGTGTGGTTGGG GCCAGTG	73.0
CaAp-10b7	CccccaaccacTTCACTGGTAGGTT GGTGTGGTTGGGG CCAGTG	77.4
CaAp-5b23	TGGTTGctaccTTCACTGGTAG GTTGG TGTGGTTGGGGCCAGTG	54.8
CaAp-6b22	TGGTTcctaccTTCACTGGTAGG TTGGT GTGGTTGGGGCCAGTG	56.2
CaAp-7b21	TGGTacctaccTTCACTGGTAGGT TGG TGTGGTTGGGGCCAGTG	60.2
CaAp-8b20	TGGaacctaccTTCACTGGTAGGTT GG TGTGGTTGGGGCCAGTG	63.8
CaAp-9b19	TGcaacctaccTTCACTGGTAGGTTG GT GTGGTTGGGGCCAGTG	64.0
CaAp-10b18	TccaacctaccTTCACTGGTAGGTTGG TGT GGTTGGGGCCAGTG	67.0
TBA	CACTGGTAGGTTGGTGTGG TTGGGGCCAGTG	64.0[b]

[a] Small letters show additional DNA complementary to each underlined region of 31-mer thrombin binding aptamer (TBA). [b] The T_m value of the 31-mer TBA was measured by UV-melting assay.

at a rate of 1 °C min^{-1}, and its absorbance at 260 nm was measured at 1 °C intervals. The T_m value was calculated by means of non-linear fitting.

Electrophoresis mobility shift assay

The capturable aptamer (4 μM) and thrombin (4 μM) (Wako Pure Chemical Industries) were mixed in the binding buffer, and the mixture was incubated for 30 min at room temperature. The mixture was then subjected to polyacrylamide gel electrophoresis (PAGE). After PAGE, the gels were stained with Sybr Safe™ (Life technologies), and fluorescent images were analyzed and compared with the silver-stained images of thrombin. A 31-mer thrombin aptamer with an additional 11-mer consisting of thymine residues (C11c) at the 5′-terminus was also evaluated as a positive control.

Detection of thrombin

A 100-μL aliquot of CaDNA biotinylated at its 3′-end (750 μM) was introduced into each well of a React-Bind streptavidin-coated, high-binding-capacity, 96-well plate (Thermo Fisher Scientific Inc.), and the plate was incubated for 2 h at room temperature. Each well was then washed three times with binding buffer. Capturable aptamer labeled at its 5′-end with fluorescein was mixed with thrombin or bovine serum albumin (BSA) and, optionally, *Escherichia coli* homogenate (1 mg mL^{-1}) as a contaminant, and the resulting mixtures were incubated for 30 min at room temperature. The mixtures were added to the wells with the immobilized CaDNA, and incubated for 2 h at room temperature. After twenty washes with binding buffer, a 100 μL aliquot of binding buffer was added to each well, and the fluorescence intensity of the fluorescein-labeled capturable aptamer retained in each well was analyzed by using an Arvo MX 1420 Multilabel counter (Perkin Elmer). The limit of detection was calculated as three times the standard deviation of the blank.

Circular dichroism spectroscopy

The circular dichroism (CD) spectrum of the 31-mer thrombin aptamer and the capturable aptamer (20 μM) were measured in the presence or absence of a 100 mM solution of K^+ at wavelengths of 200–320 nm using a cuvette with a path length of 0.1 cm at 20 °C in a J-720 spectropolarimeter (JASCO).

Surface plasmon resonance measurements

Biotinylated aptamer (100 nM) was injected and immobilized on a streptavidin-modified sensor chip (SA chip; GE Healthcare). CaDNA, thrombin, or a mixture of thrombin and CaDNA were injected at a flow rate of 20 μL min^{-1}.

Results and discussion

Design of the capturable aptamer

For the detection of thrombin, we used a 31-mer thrombin aptamer. For regulation of the equilibrium between the two conformations of the capturable aptamer, it is important to know the T_m value of the aptamer. Although the typical secondary structure of DNA and RNA can be predicted by using predictive software such as mfold,[19] the structure of a G-quadruplex cannot be predicted, and the T_m values cannot be calculated. The 31-mer thrombin aptamer is composed of both a duplex and a G-quadruplex motif [Fig. 2(a)]. We therefore measured the T_m value of the aptamer by monitoring the temperature dependence of the absorbance of the molecule at 260 nm to identify the temperature at which breaking of the duplex structure occurred. The T_m value for the 31-mer thrombin aptamer was approximately 64 °C. We then designed capturable aptamers that had various T_m values, and formed various duplex structures with the 31-mer thrombin aptamer.

Fig. 2(b) shows the sequence relationships between the 31-mer thrombin aptamer and the designed capturable aptamers. The capturable aptamer should bind to the target molecule, but should not hybridize with the CaDNA in the absence of the target molecule. We therefore carefully designed additional sequences that induce the formation of various structures from the 31-mer thrombin aptamer in the absence of thrombin, without compromising its ability to bind to thrombin. We then designed two types of capturable aptamer and several CaDNA candidates. We set two different positions of the 31-mer thrombin aptamer with which additional DNA could hybridize. To regulate the stability of the structure in the absence of the target molecules, we designed capturable aptamers that had various lengths of additional DNA that were complementary to the internal sequence of the 31-mer thrombin aptamer. Each capturable aptamer can then form a different predicted structure, and Table 1 shows each calculated T_m value. The capturable aptamer has to be capable of hybridizing with the CaDNA after it has undergone a structural change on binding to thrombin. Because the lengths of some additional complementary DNA molecules of the capturable aptamer were insufficient to permit hybridization with CaDNAs, we added sequences to the capturable aptamers to maintain hybridization. All the resulting CaDNA molecules were 11-mers, and there was little difference in the T_m values for the various CaDNA groups.

Binding assay by electrophoresis mobility-shift assay

First, we evaluated the abilities of the aptamers to bind to thrombin by means of an electrophoresis mobility-shift assay (EMSA). In the group of aptamers that had DNA that was complementary to the internal sequence from the 5th to the 9th or the 14th residues of the 31-mer thrombin aptamer from the 3′ terminal (Fig. 3), CaAp-5b12 and CaAp-6b11 showed a significant band shift toward the position of thrombin, whereas the others in the group exhibited little band shift [Fig. 3(a)]. In the second group, CaAp-5b23, CaAp-6b22, and CaAp-7b21 all exhibited significant band shifts, whereas the others exhibited little band shift [Fig. 3(b)]. The EMSA study showed that aptamers whose T_m values are higher than that of the 31-mer thrombin aptamer showed little or no binding to thrombin, whereas aptamers whose T_m values were lower than that of the 31-mer thrombin aptamer became bound to thrombin. Capturable aptamers that have high T_m values are stabilized in a different structure from that of the 31-mer thrombin aptamer, which

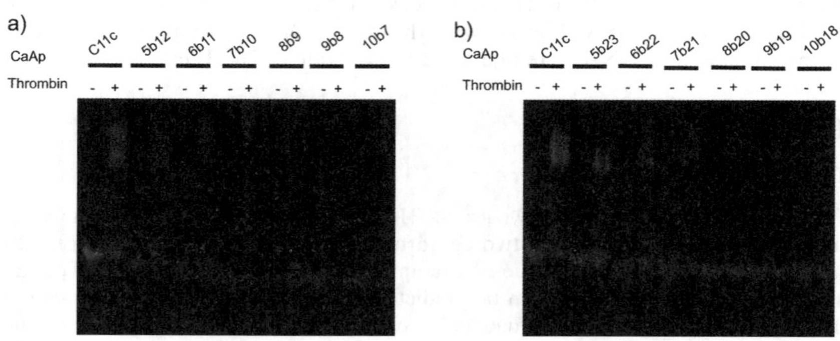

Fig. 3 The evaluation of the binding ability of each capturable aptamer candidate by means of EMSA. (a) and (b) are Sybr safe™-stained image of polyacrylamide gel. A mixture of thrombin (4 μM) and the capturable aptamer (4 μM) or the capturable aptamer (4 μM) was applied to each lane. C11c is a control capturable aptamer with an 11-mer of thiamine added to the 5′-position of the 31-mer thrombin aptamer.

shows little or no binding to thrombin. The capturable aptamers that have additional DNA complementary to the region from the 16th to the 20th or the 25th residues exhibited a greater tendency to retain their binding ability against thrombin in comparison with other capturable aptamers.

Thrombin detection

Firstly, we attempted to determine the correct concentrations of capturable aptamer and CaDNA for thrombin detection by using three capturable aptamer candidates: CaAp-6b22, CaAp-7b21, and CaAp-8b20. We evaluated various combinations of CaDNA (at concentrations of 375 nM or 750 nM) and the capturable aptamer (at concentrations of 100 nM or 500 nM), and we compared the fluorescence intensity of the capturable aptamer retained on each well in the presence or absence of thrombin. All the combinations showed higher fluorescence intensities in the presence of thrombin than in its absence (Fig. 4). For all the capturable aptamers, a combination of 750 nM CaDNA and 500 nM capturable aptamer produced the greatest difference between the fluorescence intensity in the presence of thrombin and that in its absence.

Next, we evaluated the dependence of the fluorescence intensity of the capturable aptamer on the thrombin concentration [Fig. 5(a) and (b)]. All the designed capturable aptamers showed higher fluorescence intensities in the presence of thrombin than in its absence. Each capturable aptamer showed a different background signal. Background signals were ascribed to instabilities in the structures of the capturable aptamers that allow them to bind to CaDNA in the absence of their target molecules. Although we considered that background signals depend on the calculated T_m value for each capturable aptamer, the order of the background signals was different from that of the calculated T_m value. This shows that the structural stability of aptamers is affected by slight changes in sequence, and that the prediction of this effect is difficult. Fig. 5(b) shows the absolute value of the increase in the signal above the background in the presence of thrombin. CaAp-8b20 showed the highest signals among all the capturable aptamers. The limit of detection was calculated to be 7.3 nM from the results shown in Fig. 5(c). This value is similar to that of a sandwich assay using two aptamers.[15] Whereas glucose dehydrogenase was used for signal amplification in the previous study, we did not use the enzyme in the current study. Because the capturable aptamer could easily be modified with the enzyme, this might permit the measurement of lower concentrations of thrombin than the use of a fluorescein-labeled capturable aptamer.

Next, we evaluated the detection of thrombin by using the capturable aptamer in a solution containing contaminants. Fig. 5(d) shows the result of for the detection of thrombin in the presence of *Escherichia coli* homogenate (1 mg mL^{-1}). Although the signal intensities were lower than those in buffer solution, we were still able to observe an increase in signal that depended on the concentration of thrombin but not that of BSA. The calculated limit of detection was 31 nM, so the capturable aptamer can be used to determine thrombin in solutions that contain contaminants.

Investigation of structural changes of the capturable aptamer

Next, we evaluated the structural changes of the capturable aptamer by measuring its CD spectrum. The 31-mer thrombin aptamer showed a typical CD spectrum of an antiparallel G-quadruplex structure with a positive band at 295 nm and a negative band at 265 nm [Fig. 6(a)].[20] On the other hand, CaAp-8b20 showed a spectrum that was unlike that of the 31-mer thrombin aptamer. In particular, the negative band at around 265 nm in the spectrum of the 31-mer thrombin aptamer was replaced by a positive band. These results show that additional sequences in the capturable aptamer can hybridize with the internal sequence of the 31-mer thrombin aptamer so that the structure of the capturable aptamer is different from that of the 31-mer

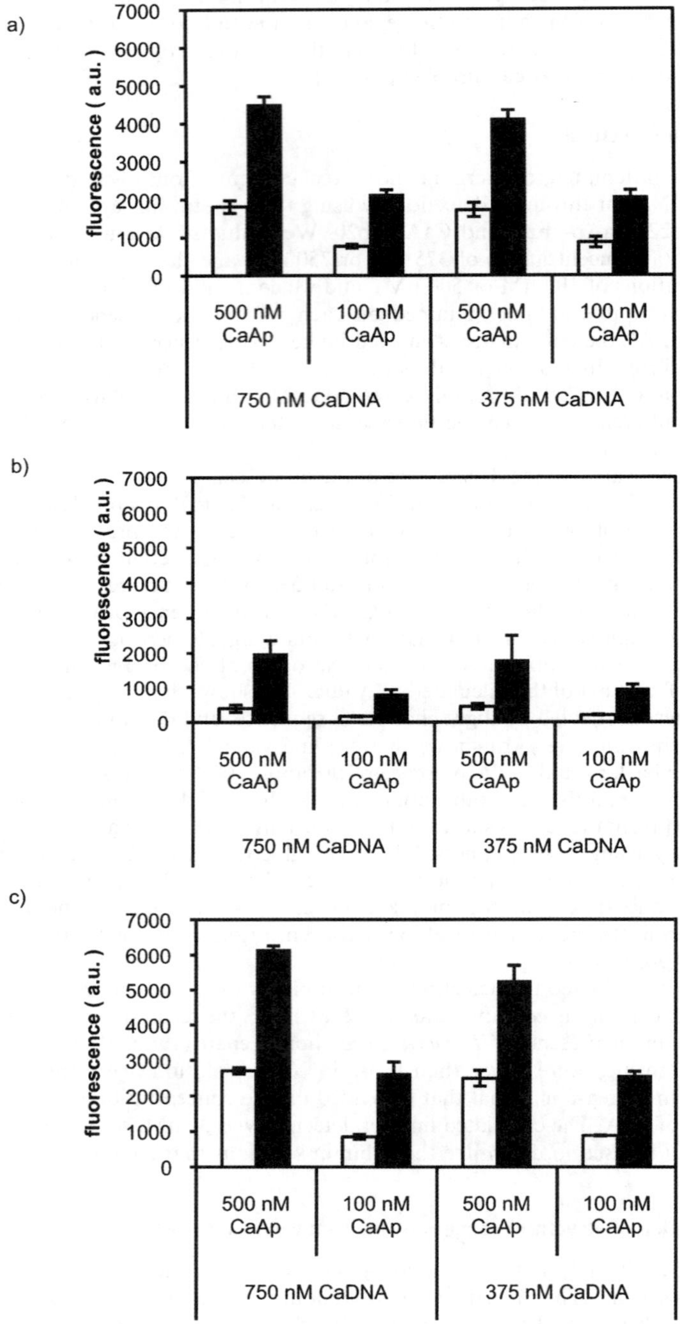

Fig. 4 Evaluation of the effects of the concentrations of the CaDNA and the capturable aptamer. Combinations of two concentrations of CaDNA (375 or 700 nM) and two concentrations of the capturable aptamer (100 nM or 500 nM) were evaluated for three capturable aptamer candidates: (a) CaAp-6b22, (b) CaAp-7b21, and (c) CaAp-8b20, in the presence of thrombin (200 nM) or without thrombin. Each graph shows the fluorescence intensity of the capturable aptamer retained on each well in the presence or absence of thrombin.

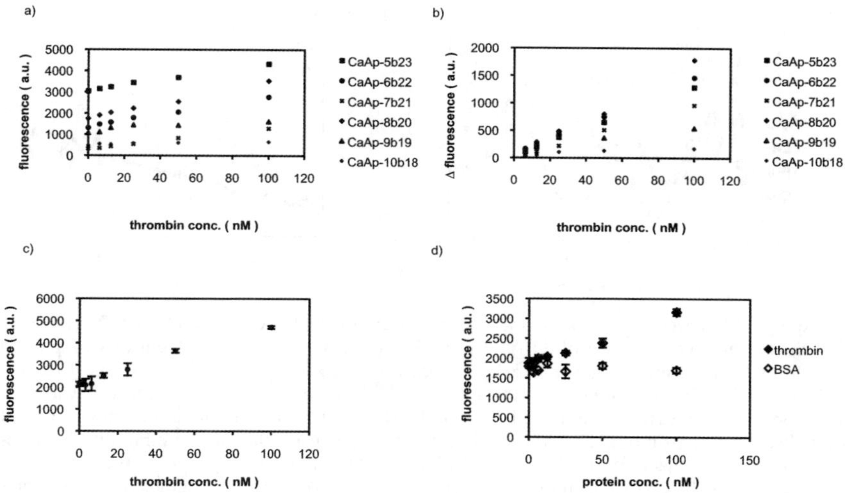

Fig. 5 Thrombin detection by bound/free separation using a capturable aptamer. (a) Dependence on the thrombin concentration (0, 3.1, 6.3, 12.5, 25, 50, or 100 nM) of the fluorescence intensity of the capturable aptamer candidates for each well containing immobilized CaDNA. (b) The change in fluorescence intensity, showing the increase in fluorescence intensity above the background level. (c) The dependence on the thrombin concentration of the fluorescence intensity of CaAp-8b20 retained on wells with immobilized CaDNA for CaAp-8b20. The detection procedure was carried out three times for each thrombin concentration (0, 3.1, 6.3, 12.5, 25, 50, and 100 nM). The error bars show the range of one standard deviation for three experimental measurements. (d) The fluorescence intensity of CaAp-8b20 retained on wells containing immobilized CaDNA for CaAp-8b20 depended on the thrombin concentration in solutions containing *Escherichia coli* homogenate. The detection procedure was carried out three times for each thrombin concentration (0, 3.1, 6.3, 12.5, 25, 50, or 100 nM). The error bars show the range of one standard deviation for three experimental measurements.

thrombin aptamer. However, the broadening of the peaks of the CD spectrum of CaAp-8b20 showed that the structure of the capturable aptamers in solution is heterogeneous, and some capturable aptamers in solution adopt the structure of the 31-mer thrombin aptamer.

To investigate the stability of the structure of the capturable aptamer, we added a high concentration of K^+ to induce structural changes in the G-quadruplex structure. K^+ stabilizes the G-quadruplex structure by insertion into the center of the G-quartet plane. However, the CD spectrum of CaAp-8b20 in binding buffer containing 100 mM K^+ was little different from that of CaAp-8b20 in the binding buffer (5 mM K^+). This shows that CaAp-8b20 forms a stable structure that is different to that of the 31-mer thrombin aptamer.

Because it is difficult to analyze structural changes in the capturable aptamer that accompany binding to thrombin unless a high concentration of thrombin is present, we evaluated the binding ability of CaAp-8b20 against CaDNA and thrombin. First, we measured the dissociation constant of CaAp-8b20 toward thrombin. CaAp-8b20 has a dissociation constant (K_d) of about 20 nM toward thrombin, which is similar to the K_d for the 31-mer thrombin aptamer (34 nM).[18] The effect of the additional sequence in the capturable aptamer is therefore small.

Next, we compared the surface plasmon resonance (SPR) signals of CaDNA with or without thrombin to evaluate the effect of binding of thrombin on CaDNA binding. Fig. S1 of the ESI†shows the effect of the concentration of CaDNA on the SPR signal in the presence of thrombin (50 nM) or in its absence. We observed an increase in the intensity of the signal with increasing CaDNA concentration in the absence of thrombin. On the other hand, the intensity of the SPR signal obtained on

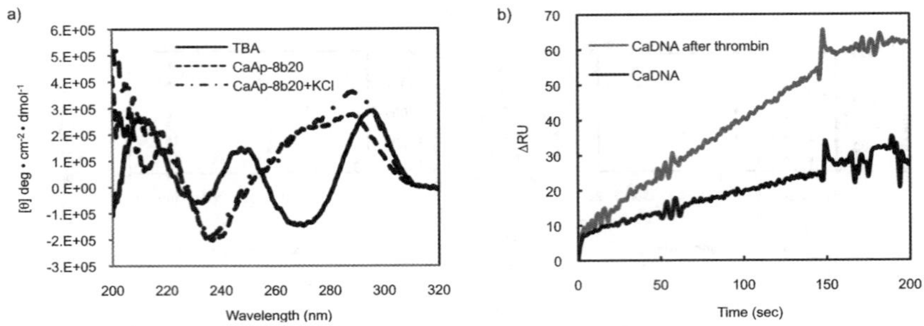

Fig. 6 (a) CD spectrum of the 31-mer thrombin aptamer (TBA) and CaAp-8b20. The solid line shows the CD spectrum of TBA. The dash line shows the CD spectrum of CaAp-8b20 in binding buffer containing 5 mM K^+. The dash-dot line shows the CD spectrum of CaAp-8b20 in binding buffer containing 100 mM K^+. (b) Comparison of SPR signals of CaDNA injection for an SA sensor chip with immobilized CaAp-8b20 that binds or does not bind with thrombin. Thrombin (10 nM) was injected into a sensor chip with immobilized CaAp-8b20. After the injection of thrombin, CaDNA (500 nM) was injected immediately. To compensate for the dissociation of thrombin during CaDNA injection, signals for the dissociation of thrombin were subtracted from the CaDNA injection signals after thrombin injection. The gray line shows the CaDNA signals after thrombin injection. The black line shows the CaDNA signals without injection of thrombin.

injection of a mixture of thrombin and CaDNA increased by more than that with the CaDNA alone. When 31-mer thrombin aptamer was immobilized on the SA sensor chip, CaDNA did not affect the SPR signal associated with thrombin injection (data not shown). We also evaluated CaDNA binding after thrombin binding. After thrombin injection, we immediately injected CaDNA without an intermediate washing step. Because thrombin dissociates from CaAp-8b20 during injection of CaDNA, we measured the SPR signal for thrombin dissociation, and we subtracted this signal from the CaDNA signal obtained by sequential injection of thrombin and CaDNA. Injection of CaDNA following injection of thrombin produced a greater increase than CaDNA injection alone [Fig. 6(b)]. These results show that thrombin binding promotes the binding of CaDNA.

The CD and SPR results, taken together, demonstrate that the capturable aptamer forms a stable structure that shows little or no binding to CaDNA in the absence of thrombin, and after binding of thrombin to the capturable aptamer, the structure of the capturable aptamer changes and binding of CaDNA to CaAp-8b20 is promoted.

Structural changes for biosensing

We have reported some biosensing systems that utilize structural changes of not only of aptamers, but also of periplasmic binding proteins, the structures of which change markedly in the presence of ligands.[10–14,16,17,21–23] Structural changes are useful features for the development of biosensors in cases where antibodies cannot be used. To construct a highly sensitive detection system, it is important to regulate the equilibrium of the structural changes. Two models have been proposed for changes in the structures of molecular-recognition elements that accompany their binding to target molecules: the pre-existing equilibrium model and the induced-fit model.[24] The pre-existing equilibrium model assumes the existence of an equilibrium state of more than two conformations, and that a conformation or several conformations within these structures undergo stabilization on binding of the target molecule. The induced-fit model assumes that the molecular conformation is normally stable and that it changes into another conformation only on binding of the target molecule. The mechanism for the structural change of maltose-binding protein may follow the induced-fit model,[25] whereas that of lysine/arginine/ornithine-binding protein may follow the pre-existing equilibrium model.[26] On the other hand, single-stranded nucleic acids can have an ensemble of conformations, and theophylline aptamer, a well-characterized RNA aptamer, forms multiple conformation in free RNA.[27] These ensembles of conformations of aptamers may include several conformations that are similar to the conformation that binds to the target molecule.

In biosensors that do not utilize structural changes in molecular-recognition elements, background signals are attributed to nonspecific binding to non-target molecules or carriers, whereas in biosensors that utilize structural change, the background signals are sensitive not only to nonspecific binding, but also to the equilibrium between conformers, particularly in the case of aptamers whose structural changes are based on the pre-existing equilibrium model. A structural change based on the induced-fit model would provide better suppression of background signals than one that based on pre-existing equilibrium model, because a structural change based on the pre-existing equilibrium model also involves some conformations that can generate signals. Because a high background signal attributed to equilibrium could result in an increase in noise in the absence of the target molecule, it is better to design aptamers that undergo structural changes in accord with the induced-fit model if they are to be used in a homogeneous detection system. Therefore, the design of a stable "unbound" form of a capturable aptamer to CaDNA in the absence of a target molecule and the design of its "bound" form to CaDNA in the presence of the target molecule are important. Additionally, because conformations of the capturable aptamer that can bind to CaDNA are unstable in the absence of the target molecule, these can be removed by means of a washing step. Therefore, the use of a capturable aptamer should minimize noise signals arising from equilibrium, and is expected to produce a greater signal-to-noise ratio.

In nature there are many natural aptameric sensors that use structural changes; these are named "riboswitches".[28] Riboswitches exist in untranslated regions or introns in mRNA, and they have an aptameric domain and a regulatory domain. They can regulate protein expression, transcription levels, or splicing with ligands.[29] The mechanism of a riboswitch that regulates translation is similar to that of a capturable aptamer. The structure of the riboswitch changes to a structure that can hybridize with CaDNA or an anti-Shine–Dalgarno sequence after binding of a target molecule. Recently, it has been reported that an engineered riboswitch showed a 58-fold promotion of protein expression in the presence of a target molecule in comparison with the case without the target molecule.[30] This paper demonstrates the potential of an aptameric sensor that uses structural change, and could help in designing highly sensitive aptameric sensors that show structural changes.

Capturable aptamers for biosensors

Because better biomarkers are urgently required to improve clinical diagnosis and drug development, many researchers are engaged in their discovery. Furthermore, the technologies of proteomics and metabolomics have facilitated the discovery of such biomarkers. To validate a biomarker and apply it to clinical diagnosis, it is necessary to develop appropriate molecular-recognition elements. As a result of the acceleration in the discovery of biomarkers, it has become necessary to reduce the timescale for the development of recognition elements. Although antibodies are widely used for *in vitro* diagnosis, the development of antibodies by means of immunization is a low-throughput technique. On the other hand, the development of aptamers, which is carried out by *in vitro* selection, can be automated and accelerated by means of the SELEX process. We have reported the selection of aptamers against unpurified samples, and we can obtain aptamers against unidentified biomarker candidate proteins that can be used to validate these biomarkers.[31] Furthermore, because aptamers can be evaluated by using microarray technology, the binding abilities of selected aptamers and their mutants can be evaluated comprehensively,[32–34] and the features of mutated aptamers can be improved by such a comprehensive evaluation. Therefore, the use of aptamers should facilitate the development of molecular-recognition elements for *in vitro* diagnosis. However, whereas antibodies that can recognize various positions can be obtained by immunization of various partial peptides of the target protein, the selection of aptamers that can bind to various regions of a protein is troublesome, because aptamers can recognize not only sequences of a target protein, but also structures of a target protein that cannot be found in partial peptides. Furthermore, the optimization of pairs of aptamers for high-sensitivity detection systems is a laborious task. In this study, we reported a sensing system that uses an aptamer-based B/F separation. Since this system does not require a pair of aptamers that bind to the target molecule simultaneously, it can facilitate the development of applications that use aptamers in clinical diagnosis.

Many small molecular biomarkers have recently been discovered by metabolomics techniques. Because blood, which is an attractive source of samples for clinical diagnosis, contains many small molecules that can cause interference, particularly in electrochemical detection systems, B/F separation is important if highly sensitive detection is to be achieved. However, it is difficult to analyze small molecules by means of a sandwich assay, because almost all small molecules have only a single binding site for molecular-recognition elements. Although many researchers have reported strategies for analyzing small molecules by using antibodies or aptamers, the techniques they employed required especial molecular-recognition elements or splitting of the aptamer or antibody.[35,36] Because capturable aptamers are designed on the basis of the conformation of the aptamers, a capturable aptamer can be used to analyze any molecule by means of B/F separation, which permits the removal of interfering molecules.

Conclusion

In summary, we constructed a thrombin-detection system that uses an aptamer-based B/F separation system. We previously reported the determination of a mouse prion protein by using a capturable aptamer. The prion aptamer is predicted to have a stem–loop structure that may, possibly, be regulated more easily than other structures. In this study, we succeeded in designing a capturable aptamer by using a thrombin-binding aptamer that has a G-quadruplex structure. Therefore capturable aptamers can be designed for various aptamers. Aptamers can be selected by *in vitro* selection, and by applying microarray technology, their sequence can be optimized more easily than that of antibodies. Despite their numerous advantages, aptamers have yet to be used for *in vitro* clinical diagnosis, whereas antibodies to many proteins have already been developed and used for *in vitro* diagnosis. Because sandwich assay-based immunoassay is an established and highly sensitive technique, it is difficult to substitute aptamers for antibodies in *in vitro* diagnosis. However, because there is a growth in the demand for biomarkers for use in drug development and *in vitro* diagnosis (possibly leading to personalized medicine), there is a need for more rapid screening of biomarkers, so the development of biosensors for newly identified biomarkers also needs to be expedited. Capturable aptamers could be developed more rapidly than sandwich assays because they required only an aptamer, and they permit the analysis of target molecules with a high sensitivity. We believe that the development of capturable aptamers could facilitate the use of aptamers for *in vitro* diagnosis.

Acknowledgements

This work was supported by the 2009 Industrial Technology Research Grant Program of the New Energy and Industrial Technology Development Organization (NEDO) of Japan.

References

1 C. Tuerk and L. Gold, *Science*, 1990, **249**, 505–510.
2 A. D. Ellington and J. W. Szostak, *Nature*, 1990, **346**, 818–822.
3 J. C. Cox and A. D. Ellington, *Bioorg. Med. Chem.*, 2001, **9**, 2525–2531.
4 H. M. Al-Hashimi and N. G. Walter, *Curr. Opin. Struct. Biol.*, 2008, **18**, 321–329.
5 M. Zayats, Y. Huang, R. Gill, C. A. Ma and I. Willner, *J. Am. Chem. Soc.*, 2006, **128**, 13666–13667.
6 A. E. Radi, J. L. Acero Sanchez, E. Baldrich and C. K. O'Sullivan, *J. Am. Chem. Soc.*, 2006, **128**, 117–124.
7 R. Yamamoto, T. Baba and P. K. Kumar, *Genes Cells*, 2000, **5**, 389–396.
8 N. Hamaguchi, A. Ellington and M. Stanton, *Anal. Biochem.*, 2001, **294**, 126–131.
9 N. Tuleuova, C. N. Jones, J. Yan, E. Ramanculov, Y. Yokobayashi and A. Revzin, *Anal. Chem.*, 2010, **82**, 1851–1857.
10 W. Yoshida, K. Sode and K. Ikebukuro, *Anal. Chem.*, 2006, **78**, 3296–3303.
11 W. Yoshida, K. Sode and K. Ikebukuro, *Biochem. Biophys. Res. Commun.*, 2006, **348**, 245–252.
12 K. Ikebukuro, W. Yoshida and K. Sode, *Biotechnol. Lett.*, 2008, **30**, 243–252.
13 W. Yoshida, K. Sode and K. Ikebukuro, *Biotechnol. Lett.*, 2008, **30**, 421–425.
14 W. Yoshida, E. Mochizuki, M. Takase, H. Hasegawa, Y. Morita, H. Yamazaki, K. Sode and K. Ikebukuro, *Biosens. Bioelectron.*, 2009, **24**, 1116–1120.
15 K. Ikebukuro, C. Kiyohara and K. Sode, *Biosens. Bioelectron.*, 2005, **20**, 2168–2172.
16 M. Fukasawa, W. Yoshida, H. Yamazaki, K. Sode and K. Ikebukuro, *Electroanalysis*, 2009, **21**, 1297–1302.
17 D. Ogasawara, N. S. Hachiya, K. Kaneko, K. Sode and K. Ikebukuro, *Biosens. Bioelectron.*, 2009, **24**, 1372–1376.
18 K. Ikebukuro, Y. Okumura, K. Sumikura and I. Karube, *Nucleic Acids Res.*, 2005, **33**, e108.
19 M. Zuker, *Nucleic Acids Res.*, 2003, **31**, 3406–3415.
20 J. Kypr, I. Kejnovska, D. Renciuk and M. Vorlickova, *Nucleic Acids Res.*, 2009, **37**, 1713–1725.
21 A. Taneoka, A. Sakaguchi-Mikami, T. Yamazaki, W. Tsugawa and K. Sode, *Biosens. Bioelectron.*, 2009, **25**, 76–81.
22 A. Sakaguchi-Mikami, A. Taneoka, R. Yamoto, S. Ferri and K. Sode, *Biotechnol. Lett.*, 2008, **30**, 1453–1460.
23 A. Sakaguchi, S. Ferri, W. Tsugawa and K. Sode, *Biosens. Bioelectron.*, 2007, **22**, 1933–1938.
24 C. S. Goh, D. Milburn and M. Gerstein, *Curr. Opin. Struct. Biol.*, 2004, **14**, 104–109.
25 O. Millet, R. P. Hudson and L. E. Kay, *Proc. Natl. Acad. Sci. U. S. A.*, 2003, **100**, 12700–12705.
26 B. H. Oh, J. Pandit, C. H. Kang, K. Nikaido, S. Gokcen, G. F. Ames and S. H. Kim, *J. Biol. Chem.*, 1993, **268**, 11348–11355.
27 S. W. Lee, L. Zhao, A. Pardi and T. Xia, *Biochemistry*, 2010, **49**, 2943–2951.
28 W. Winkler, A. Nahvi and R. R. Breaker, *Nature*, 2002, **419**, 952–956.
29 A. Roth and R. R. Breaker, *Annu. Rev. Biochem.*, 2009, **78**, 305–334.
30 N. Muranaka, V. Sharma, Y. Nomura and Y. Yokobayashi, *Nucleic Acids Res.*, 2009, **37**, e39.
31 T. Noma, K. Ikebukuro, K. Sode, T. Ohkubo, Y. Sakasegawa, N. Hachiya and K. Kaneko, *Biotechnol. Lett.*, 2006, **28**, 1377–1381.
32 M. Platt, W. Rowe, J. Knowles, P. J. Day and D. B. Kell, *Integr. Biol.*, 2009, **1**, 116–122.
33 N. O. Fischer, J. B. Tok and T. M. Tarasow, *PLoS One*, 2008, **3**, e2720.
34 E. Katilius, C. Flores and N. W. Woodbury, *Nucleic Acids Res.*, 2007, **35**, 7626–7635.
35 T. Sakata, M. Ihara, I. Makino, Y. Miyahara and H. Ueda, *Anal. Chem.*, 2009, **81**, 7532–7537.
36 X. Zuo, Y. Xiao and K. W. Plaxco, *J. Am. Chem. Soc.*, 2009, **131**, 6944–6945.

Aptameric sensors based on structural change for diagnosis

Koichi Abe, Daisuke Ogasawara, Wataru Yoshida, Koji Sode and Kazunori Ikebukuro

Faraday Discuss., 2011, **149**, 93–105 (DOI: 10.1039/c005359g). ***Amendment published 23rd December 2010***

Each sequence in Fig. 2b is missing a letter. The corrected version of Fig. 2 is given below:

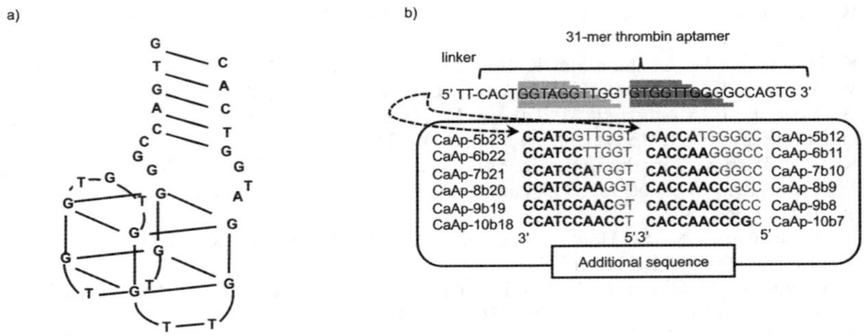

In the caption of Fig. 2: 'from the 5th to the 9th or 15th residues' should read 'from the 5th to the 9th or 14th residues' and 'from the 16th to the 20th or 26th residues' should read 'from the 16th to the 20th or 25th residues'.

On page 6, 2nd paragraph, line 3: 'from the 5th to the 9th or the 14th residues of the 31-mer thrombin from the 3′ terminal' should read 'from the 16th to the 20th or the 25th residues of the 31-mer thrombin from the 5′ terminal'.

The Royal Society of Chemistry apologises for these errors and any consequent inconvenience to authors and readers.

Additions and corrections can be viewed online by accessing the original article to which they apply.

Multi-modal molecular imaging approaches to detect primary cells in preclinical models†

Kevin Dhaliwal,*[a] Lois Alexander,[b] Geraldine Escher,[b] Asier Unciti-Broceta,[b] Maurits Jansen,[a] Neil Mcdonald,[a] Juan M. Cardenas-Maestre,[b] Rosario Sanchez-Martin,[b] John Simpson,[a] Chris Haslett,[a] and Mark Bradley*[b]

Received 1st June 2010, Accepted 5th July 2010
DOI: 10.1039/c005410k

The need to understand cellular trafficking *in vivo in situ* requires the development and application of novel methodologies for cellular labeling and cell tracking. Here we applied new technologies associated with advances in molecular imaging to demonstrate the feasibility of labeling primary immune cells. We demonstrate the utility of fluorescently tagged polystyrene microspheres, MRI susceptible emulsions and cell entry peptoids. The adaptation of these labeling agents will permit cell specific delivery, diagnostic sensing and the delivery of therapeutic agents to sites of inflammation and infection.

Introduction

Molecular imaging (MI) is a rapidly expanding field. It can be described as a broad technological approach by which molecular events and pathways can be observed and defined *in vivo in situ* (and in real-time) without perturbation and pathological sampling of the process. The technological platforms by which imaging contrast is generated and detected have been extensively reviewed and compared for sensitivity and depth penetration.[1] Modalities include computerised tomography (CT), magnetic resonance imaging (MRI), positron emission tomography (PET), single photon emission computerised tomography (SPECT) and optical molecular imaging (OMI).

The principal clinical MI modalities historically have been PET and SPECT.[2] MRI and CT have also been used for MI approaches but the paucity of specific molecular probes and low sensitivity have limited their clinical application.

OMI remains the most utilised modality in preclinical MI and is a rapidly advancing field with potential to revolutionise near patient testing and theranostics. The detection of light deep within living tissues has been made possible by the dramatic advances in both fluorophores and physical systems able to excite and detect photons.[3] Photons can be detected at both the macroscopic and microscopic level *in vivo in situ* using instruments that are now commercially available. Preclinically on a whole-body level, OMI platforms exist that are able to perform both epi and transillumination. These platforms employ varying forms of excitation sources coupled with extremely sensitive charge coupled device (CCD) detection systems.

[a]*MRC Centre for Inflammation Research, Queen's Medical Research Centre, University of Edinburgh, 47 Little France Crescent, Edinburgh, UK. E-mail: kdhaliwa@staffmail.ed.ac.uk*
[b]*School of Chemistry, Joseph Black Building, Kings Buildings, West Mains Road, Edinburgh, UK. E-mail: mark.bradley@ed.ac.uk*

† All animal experimentation had local ethical committee approval and Home Office licence authorisation.

The accurate detection of fluorophores within tissue are limited to approximately three centimetres for epiillumination platforms, whilst transillumination approaches may be extended deeper utilising near-infrared (NIR) fluorophores.[4]

However, the lack of clinical grade molecular probes have hindered translational implementation of OMI, highlighting the need for multidisciplinary interactions for the development of *chemical* probes in *biological* systems that can be detected by *physical* systems. There is thus a pressing need for the concerted global clinical development and application of MI methodologies and solutions to provide fundamental answers to biological processes that remain poorly defined.

The need to detect the recruitment and trafficking of cells is a major area of interest in preclinical and clinical arenas in the context of both developing a better understanding of the temporal kinetics of cell recruitment in inflammation and also in the application of cell-based therapies. Optical imaging approaches permit the rapid, economical and multiplexed ability to delineate such trafficking in a whole-body scenario in preclinical imaging and also potentially in macroscopic and microscopic resolution in man. The choice of fluorophore is dictated primarily by the detection system to be used. In the context of whole-body imaging, NIR fluorophores are used to enable tissue penetration and overcome the extensive tissue absorption of light that occurs with wavelengths less than 600 nm.

Requirements for labeling cells for *in vivo* imaging include: speed, low toxicity, uniform labeling and intracellular targeting to ensure no label is easily transferred between live cells. We describe in this discussion report varying approaches: using optically encoded polystyrene microspheres, dual-modality perfluoropolyether (PFPE) emulsions for MRI and cell entry peptoids. We present data demonstrating labeling of primary murine macrophages and *in vivo* imaging in preclinical models. Table 1 summarises the *in vitro* data of toxicity, speed of labeling and efficiency of the three approaches for labeling primary murine macrophages.

Microspheres as cell tracking reagents

We have previously described the synthesis of robust, biocompatible, functionalised cross-linked polystyrene microspheres of highly defined sizes that can be highly-functionalised, with a range of cargos, ranging from fluorophores and proteins to sensors and small interfering ribonucleic acid (siRNA).[5,6] We have reported their efficient entry into a range of cell lines and primary cells *in vitro.*[7] Here we report their synthesis and use in *ex vivo* labeling of primary murine leucocytes and subsequent adoptive transfer by intravenous administration into animals with established inflammation.

Briefly, 0.5 μm polystyrene microspheres synthesised as previously were coupled to in-house synthesised Cy5.5 *via* an aminohexanoic spacer.[7,8] Primary murine macrophages were isolated by obtaining femurs of mice and flushing the bones with ice-cold Hank's Balanced Salt Solution (HBSS), triturating through a 19 gauge needle and then growing the cells under adherent conditions at 37 °C/5% CO_2 for 5 days in macrophage colony-stimulating factor enriched media as previously described.[9] Day 5 murine macrophages were then exposed to microspheres (86 μg mL^{-1}) in complete buffer at 37 °C/5% CO_2 (Dulbelco's Modified Eagle's Medium (DMEM) plus 10% fetal calf serum (FCS))

Cell viability was assessed by propidium iodide incorporation measured on flow cytometry (FACScan, BD). Confocal analysis with co-staining for the nucleus (DAPI) and cell membrane (actin) (cells labeled in adherent conditions on coverslips, washed extensively and then fixed for 15 min followed by confocal imaging) ensured intracellular localisation of microspheres. Flow cytometry of macrophages (FACSdiva) in fluorescent quenching solution (0.4% trypan blue in HBSS) ensuring that only internalised microspheres could be detected, confirmed over 65% labeling after 4 h. Fig. 1 demonstrates the intracellular accumulation of microspheres in primary macrophages. Direct intratracheal intubation of mice and single lung

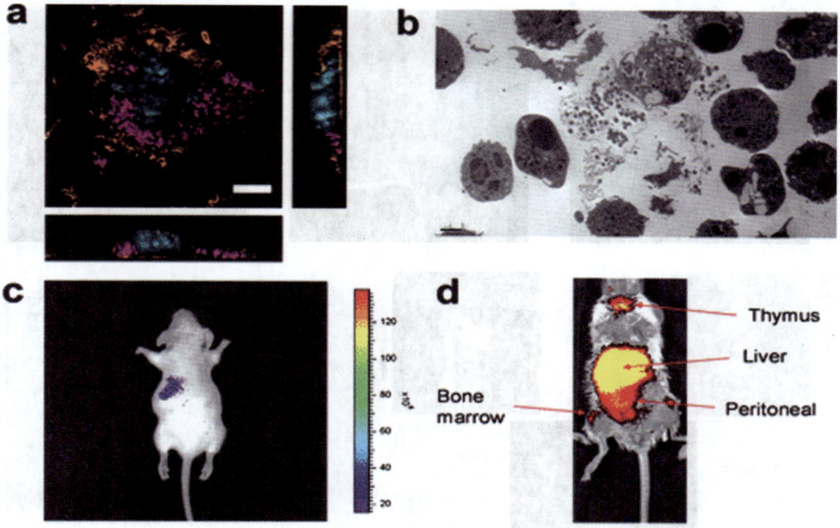

Fig. 1 Microsphere labeling of cells. (a) Merged confocal image of murine macrophages with microspheres (magenta), nuclei (DAPI, excitation 405 nm) and actin (phalloidin, excitation 543 nm). Scale bar 10 μm. (b) Apoptotic primary human neutrophils releasing intact microspheres. (c) Single lung instillation of 0.25×10^6 Cy 5.5 microspheres, imaged post instillation with a Xenogen IVIS Spectrum (excitation 675 nm, emission 720 nm). (d) 10×10^6 Cy7-microsphere labeled macrophages intravenously administered to mouse that had established peritoneal inflammation. Imaged with a Xenogen Spectrum (excitation 745 nm, emission 820 nm). Image shows cells recruited to liver and bone marrow but also to peritoneal space.

instillation of 0.5×10^6 labeled macrophages showed ability to detect microsphere labeled cells *in vivo in situ*. Extending the *in vivo* experiments, 10×10^6 Cy7 microsphere labeled cells were administered to mice that had an intraperitoneal injection of zymosan using methods described.[10] Mice were imaged the following day showing recruitment of cells to the peritoneal space. Control animals showed no peritoneal recruitment. Further, the ability of the microspheres to resist cellular degradation was tested by loading primary human neutrophils with microspheres and allowing the neutrophils to age overnight. Aged necrotic and apoptotic neutrophils were then fixed and imaged using scanning electron microscopy to demonstrate that the microspheres remained intact. Hence it is likely that any protracted imaging of cell fate *in vivo in situ* with fluorescently tagged microspheres may be difficult to interpret as cell-free microspheres may contribute to the signal/contrast. Nevertheless, the electron microscopy imaging confirmed that the polystyrene microspheres resist the proteolytic repertoire of the neutrophil, and suggest that further studies to use the neutrophil as a potential 'Trojan horse' are warranted.

Fluorescent perfluoropolyether nanoemulsions as MRI cell tracking agents

Cell tracking utilising MRI has received much interest with a number of iron-based contrast reagents.[11] Iron loading of cells with superparamagnetic iron oxide (SPIO) or ultrasmall superparamagnetic iron oxide (USPIO) results in signal dephasing due to magnetic field inhomogeneity leading to a signal 'void' or hypointensity on T2 and T2* weighted scans in the vicinity of cell localisation. An example of this is demonstrated in Fig. 2 where primary murine macrophages were isolated as described above and either:

(a) incubated with 100 μg ml^{-1} Feridex™ (SPIO) for 4 h in DMEM plus 10% FCS at 37 °C/5% CO_2; or

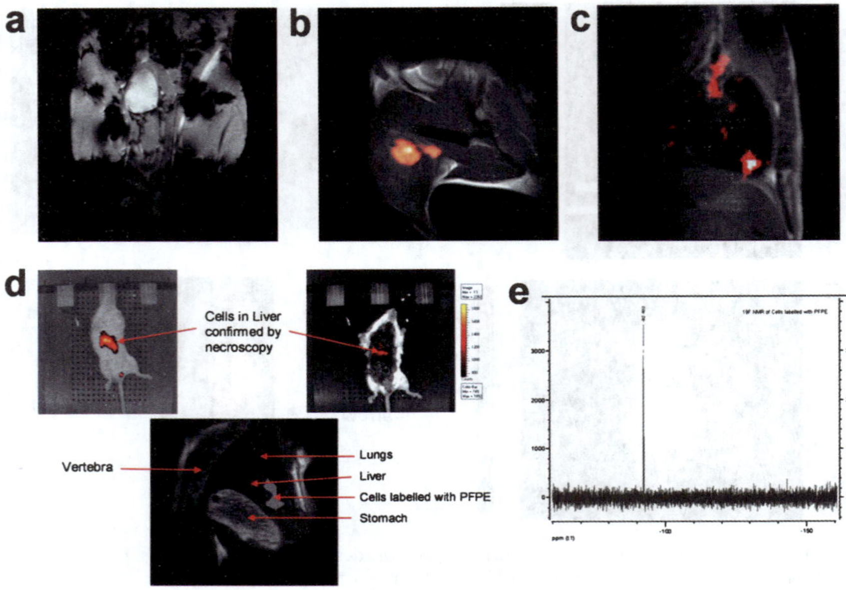

Fig. 2 MRI imaging of PFPE emulsions loaded into cells. (a) T2 weighted image of lower thighs of mice showing 'voids' of signal in the right thigh from 5×10^6 Cd11b Miltenyi Biotec microbead labeled macrophages and void in the left thigh from 5×10^6 Feridex labeled macrophages. (b) ^{19}F MRI imaging of 5×10^6 labeled macrophages injected directly into the the right thigh superimposed on a ^{1}H image for structure. (c) ^{19}F MRI of 5×10^6.macrophages labeled with PFPE directly instilled into the murine lung. (d) 5×10^6 Cy7-PFPE labeled macrophages homing to the murine liver. Optical images were acquired with a 3 s exposure with excitation filter 745 nm, emission filter 820 nm with a Xenogen Spectrum. MRI image of upper torso of the same mouse showing ^{19}F labeled cells in liver. Image acquisition was over one hour. (e) *In vitro* nuclear magnetic resonance (NMR) of labeled cells showing a single specific resonance for ^{19}F measured in a Brucker 400 MHz NMR.

(b) labeled with CD11b microbeads (Miltenyi Biotec) according to the manufacturers instructions.

5×10^6 cells were injected into each thigh of a mouse and imaged on a Varian 7 Tesla MRI DirectDrive system using a gradient echo scan (matrix size 256×256, FOV 40×40 mm, TR 423 ms, TE 10 ms, flip angle 30°, slice thickness 1 mm (30 slices), number of averages 2). Despite the localised injection of the cells, a considerable 'volume or blooming' effect was observed. This highlights some of the challenges in using paramagnetic iron for tracking cells with MRI using T2 weighted sequences in that the local signal void makes it impossible to perform quantification. Additionally, signal voids depend on the contrast that is inherently present in the tissue and inflammatory events may themselves cause signal dephasing due to iron-rich red cell extravasation.

One approach to overcome this problem is the ability to multiplex MRI using ^{1}H imaging for structural resolution alongside ^{19}F imaging for the localisation of labeled cells. This results in a positive contrast MRI cell labeling technique. We hence generated ^{19}F nanoemulsions using similar methods described previously.[12] PFPE emulsions were prepared using a 1 : 1 molar ratio of perfluoropolyethelene glycol (molecular weight 1750, Exfluor) and sterile filtered Pluronic (Sigma-Aldrich) with emulsification by sonication giving an average particle diameter of 150 nm (as measured by laser diffraction). In addition NIR fluorescent PFPE emulsion particles were prepared by mixing the PFPE emulsion with Cy7.

Primary murine macrophages (isolated as described above) were incubated (25μl mL^{-1} final concentration of ^{19}F nanoemulsions) for four hours in DMEM with 10% FCS at 37 °C/5% CO_2 prior to extensive washing. 5×10^6 cells were then injected into the thigh of a mouse. ^{19}F Images were acquired on a Varian 7T Direct-Drive MRI system using a 20 mm surface coil tuned to the ^{19}F resonance frequency. ^1H images were acquired with the same coil tuned to the ^1H resonance frequency and similar MRI parameters to serve as anatomical reference. Sequential images of the mouse were acquired using a respiratory-gated fast spin echo sequence (echo train length 8) with a matrix size of 256 × 256 and a field of view of 40 mm × 40 mm. The TR was 1000 ms and the TE was 10 ms. Slices were acquired with 2 mm thickness (8 slices) and the total number of averages was 512.

The ability to detect the cells within the thigh using this approach then led us to test if it would be possible to detect ^{19}F labeled primary macrophages in the lungs of a mouse. Cells were labeled as previously described and anaesthetised mice intubated with a 19 gauge needle. 5×10^6 cells were instilled in a 50 μl volume of PBS and the mouse allowed to recover prior to MRI imaging. Fig. 2 shows the distribution of labeled cells within the murine thorax. The integration time for scanning for ^{19}F labeled cells was over one hour. Imaging animals for this duration of time is fraught with anaesthetic concerns with the potential to modify the underlying inflammatory process. Hence we dual labeled our nanoemulsions by incorporating Cy7 during the emulsification process. Murine macrophages were injected intravenously into mice which allowed the dual detection with MRI and reflectance optical imaging and localisation of the cells within the liver. Optical imaging using the Xenogen Spectrum required only a one second exposure on large binning compared to over one hour with MRI, highlighting the sensitivity, speed and economical utility of preclinical optical imaging.

Cell entry peptoids as cell tracking agents

The description of HIV TAT as an efficient protein transduction domain heralded the development of peptide facilitated cell delivery.[13] Despite the widespread realisation of HIV TAT as an efficient means to potentiate cell entry, it remains only 50–60% efficient in delivering some NIR dyes into primary cells while also showing some cellular toxicity.[14] Hence based on our recent description of cell entry peptoids with nine lysine-containing repeating units as efficient tools for cell entry, we attached a NIR fluorophore to our nonalysine-like peptoid (NLLP) (Fig. 3).[15] (NLLPs were generated by using a highly-optimised microwave-based solid-phase synthesis route as detailed previously with the NIR dye coupled by using the *N*-hydroxysuccinimidyl ester of the Cy5.5 or Cy7 dye).[8] Cy5.5 and Cy7 tagged peptoids displayed no detectable toxicity at any of the concentrations used in this study and achieved rapid cell entry with 100% efficiency in primary murine macrophages. Washing and fixing at time points up to 30 min demonstrated efficient labeling within minutes. Intracellular incorporation was confirmed in all primary cells by confocal microscopy. Inflammatory cytokine liberation from primary murine macrophages loaded with Cy5.5-NLLP was not elevated compared to controls. Imaging of mice that had lipopolysaccharide (50 μg *Escherichia coli* LPS intratracheally instilled 24 h earlier) lung injury demonstrated recruitment of Cy7-NLLP labeled cells to the lung. The facile, non-toxic nature of this primary cell labeling technique for cell tracking and cell delivery opens up promising therapeutic applications.

Discussion

The ability to rapidly and safely deliver cargoes into cells and optically detect these cargoes will be of major interest in clinical applications to either use cells as 'Trojan horses', to deliver therapeutic payloads or to use cell recruitment and behaviour as key cellular 'sensors' of disease initiation, progression and resolution.

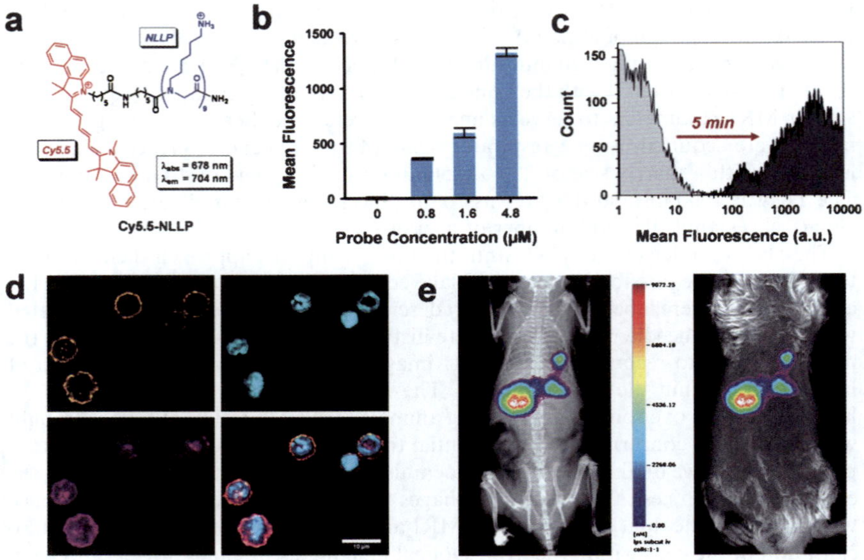

Fig. 3 (a) Chemical structure of the Cy5.5-NLLP. (b) Comparative uptake studies of Cy5.5-NLLP at various concentrations in primary murine macrophages as measured by flow cytometry analysis. (c) Representative cell population/fluorescence charts from primary murine macrophages: untreated cells (light grey) and cells after 5 min incubation with Cy5.5-NLLP (dark grey). (d) Confocal images of freshly-flushed murine marrow after 10 min incubation with 4.8 μM Cy5.5-NLLP: actin staining (543 nm excitation), DAPI nuclei staining (405 nm excitation), Cy5.5-NLLP labeling (633 nm excitation), and composite. (e) Tracking of Cy7-NLLP labelled murine macrophages to lipopolysaccharide induced pulmonary inflammation showing liver, spleen and lung recruitment. Image on left with X-ray superimposed on reflectance fluorescent images on Kodak Multispectral FX Pro (excitation 745 nm, emission 820 nm). Image on right showing the same mouse with just reflectance. Images show utility of using multimodal imaging with combined X-ray and fluorescent reflectance imaging to confirm localisation of cell recruitment.

The sensitivity, speed of imaging and anatomical resolution are all factors that need to be weighed up when determining the optimal modality for detection of cells *in vivo in situ*. Table 1 compares the three approaches used in this paper for tracking primary murine macrophages.

Microspheres represent a versatile platform for cell entry and cell tracking. Microspheres are easily functionalised and we have shown that they can be exploited to deliver proteins, siRNA, DNA and even pH and calcium sensors to the cell.[16] These remarkable facets of functional cellular delivery alongside other documented uses of microspheres as being scaffolds for the attachment and delivery of drugs such as antimicrobials emphasises their unique and effective versatility.[17]

^{19}F imaging permits positive contrast MRI and the ability to quantify labeled cells *in vivo in situ* and investigators have reported the ability to detect small numbers of cells using this methodology.[18] However, the current formulation of the nanoemulsions requires extended coincubation to gain intracellular entry. Further chemical modifications to improve cell entry and speed are required. Additionally ^{19}F MRI *in vivo in situ* requires prolonged image acquisition.

The NLLP represents the most efficient and versatile cell labeling tool discussed in this report. Five minutes incubation is all that is required for 100% cell labeling, with no toxicity demonstrated in cells or mice. The versatility of its chemical synthesis allows the decoration with a myriad of cargoes and opens up wide therapeutic possibilities that are currently being explored for theranostic applications *in vivo*.

Table 1 Summary of cell labeling characteristics of agents used to 'tag' primary murine macrophages. Data represents means and standard error of means in brackets derived from three independent experiments in triplicate for labeling of 5×10^6 primary murine macrophages. Significance ($p < 0.05$) by Student's t-test.

Cell labeling agent	% cell death[a]	% labeling[b]	TNF secretion[c]	Speed of imaging/s[d]
Feridex	22 (6.2)	45 (7.2)	38 (8.2)	180
CD11b microbeads	9 (3.4)	85 (9.4)	10 (3.5)	180
Cy7 PFPE	17 (4.3)	55 (6.3)	17 (4.2)	40 800
Cy 5.5 microspheres	6 (1.4)	68 (6.7)	ns	1
Cy 5.5 and Cy 7-peptoid	ns	100	ns	1

[a] Increase over control of propidium iodide incorporation 4 h after labelling. [b] Prussian blue staining for Feridex and Cd11b microbeads and fluorescence intensity increase on flow cytometry for Cy7 and Cy5.5 over control. [c] Cytokine in cell free supernatants at 16 h corrected for total protein and expressed as percentage increase over control. [d] 5×10^6 cells injected directly into thigh of mice and time in seconds to acquire images.

With regard to cell labeling, whilst adoptive transfer techniques such as the ones illustrated above represent one approach, a more robust and accurate method for 'sensing' inflammatory cell recruitment is the direct intravenous administration of cell specific tags. Developing these methodologies is the focus of ongoing work.

OMI is now well established in preclinical models as a rapid, cost effective platform to assess disease processes *in vivo in situ*. OMI has versatility over other modalities due to the range of possible fluorophores that can be spectrally discriminated. With the advent of human optical imaging becoming a reality utilising both microscopic devices such as confocal microendoscopy and hand-held or boom-operated devices for theatre, OMI has the potential to become a point of care MI modality.[19,20] Additionally, the expanding portfolio of 'smart' molecular chemical probes permits the multiplexed detection of disease processes.[21] The recent description of the ability to detect nuclear sources using CCD devices opens up intriguing and versatile applications.[22]

Acknowledgements

This work was supported by the MRC, EPSRC and BBSRC.

References

1 F. A. Jaffer and R. Weissleder, *JAMA, J. Am. Med. Assoc.*, 2005, **293**, 855.
2 G. D. Luker and D. Piwnica-Worms, *Acad. Radiol.*, 2001, **8**, 4.
3 S. A. Hilderbrand and R. Weissleder, *Curr. Opin. Chem. Biol.*, 2010, **14**, 71.
4 F. Leblond, S. C. Davis, P. A. Valdes and B. W. Pogue, *J. Photochem. Photobiol., B*, 2010, **98**, 77.
5 R. Sanchez-Martin, M. Muzerelle, N. Chitkul, S. How, S. Mittoo and M. Bradley, *ChemBioChem*, 2005, **6**, 1341.
6 R. Sanchez-Martin, M. Cuttle, S. Mittoo and M. Bradley, *Angew. Chem., Int. Ed.*, 2006, **45**, 5472.
7 L. Alexander, K. Dhaliwal, J. Simpson and M. Bradley, *Chem. Commun.*, 2008, 3507.
8 M. Lopalco, E. M. Koini, J. K. Cho and M. Bradley, *Org. Biomol. Chem.*, 2009, **7**, 856.
9 T. S. Wilkinson, K. Dhaliwal, T. W. Hamilton, A. F. Lipka, L. Farrel, D. J. Davidosn, R. Duffin, A. C. Morris, C. Haslett, J. R. Govan, C. D. Gregory, J. M. Sallenave and A. J. Simspon, *Am. J. Pathol.*, 2009, **174**, 1338.
10 I. Bournazou, J. D. Pound, R. Duffin, S. Bournazos, L. A. Melville, S. B. Brown, A. G. Rossi and C. D. Gregory, *J. Clin. Invest.*, 2009, **119**, 20.
11 J. W. Bulte, *AJR, Am. J. Roentgenol.*, 2009, **193**, 314.

12 M. Srinivas, P. A. Morel, L. A. Ernst, D. H. Laidlaw and E. T. Ahrens, *Magn. Reson. Med.*, 2007, **58**, 725.
13 H. Brooks, B. LeBleu and E. Vives, *Adv. Drug Delivery Rev.*, 2005, **57**, 559.
14 C. Berger, H. U. Gremlich, P. Schmidt, C. Cannet, R. Kneuer, P. Hiestand, M. Rausch and M. Rudin, *J. Immunol. Methods*, 2007, **323**, 65.
15 A. Unciti-Broceta, F. Diezmann, C. Y. Ou-Yang, M. A. Fara and M. Bradley, *Bioorg. Med. Chem.*, 2009, **17**, 959.
16 R. Sanchez-Martin, L. Alexander and M. Bradley, *Ann. N. Y. Acad. Sci.*, 2008, **1130**, 207.
17 Y. I. Jeong, H. S. Na, J. W. Nah and H. C. Lee, *J. Pharm. Sci.*, 2009, **98**, 3659.
18 M. Srinivas, A. Heerschap, E. T. Ahrens, C. G. Figdor and I. J. Vries, *Trends Biotechnol.*, 2010, **28**, 363.
19 L. Thiberville, M. Salaün, S. Lachkar, S. Dominique, S. Moreno-Swirc, C. Vever-Bizet and G. Bourg-Heckly, *Eur. Respir. J.*, 2009, **33**, 974.
20 A. Da Silva, J. M. Dinten, J. L. Coll and P. Rizo, *Conf. Proc. IEEE Eng. Med. Biol. Soc.*, 2007, **2007**, 526.
21 A. Signore, S. J. Mather, G. Piaggio, G. Malviya and R. A. Dierckx, *Chem. Rev.*, 2010, **110**, 3112.
22 H. Liu, G. Ren, Z. Miao, X. Zhang, X. Tang, P. Han, S. S. Gambhir and Z. Cheng, *PLoS One*, 2010, **5**, e9470.

A miniaturised integrated biophotonic point-of-care genotyping system

John M. Girkin,*[a] Mazher-Iqbal Mohammed[b] and Elizabeth. M. Ellis[c]

Received 29th April 2010, Accepted 19th July 2010
DOI: 10.1039/c005271j

This paper reports the development of a novel genotyping device specifically designed for point-of-care applications. As the results of the human genome project are applied to clinical practice there is an increasing requirement for simple to operate high-speed, potentially low-cost genotyping devices for use in the clinic. The aim of such devices is not to specifically detect a full gene sequence but to monitor the presence of specific Single Nucleotide Polymorphisms (SNPs). The instrument is designed to fulfil this specific clinical requirement. Using a FRET-based assay the instrument completes a full PCR process and then performs a melting point test to determine the exact SNPs present in the sample. Results are presented in which the instrument produces results within 18 min based upon saliva samples provided by the patient. The paper also reports successful results both with purified DNA samples and saliva-based samples which were taken from subjects after experiments deliberately aimed at confusing the instrument.

1. Introduction

Since the completion of the Human Genome project the focus in research has moved from decoding the gene to using this information to improve clinical practice. In genetic studies related to applications in clinical practice, this work has concentrated on the investigation of Single Nucleotide Polymorphisms (SNPs) and the role that they play in drug metabolism, in a field now known as pharmocogenetics. Significant advances have been made in this field with a wide number of SNPs being linked to drug metabolism and this is opening up the possibility of tailoring the prescription to a patient's specific genotype both improving the efficacy of treatment and in some cases ensuring the correct class of pharmaceutical products are administered as soon as possible. If this long-term goal of personalised medicine is ever to be achieved, the clinician requires a rapid method of genotyping the patient, not for a full genomic sequence, but to identify the SNPs of interest.

To date many of the current techniques rely on the amplification of template DNA so that there are sufficient copies of the required SNP to perform the genotyping analysis. PCR amplification is now a robust process, not perfect, but with a high level of success, and thus provides an accepted method of producing a relatively high concentration of product (the required DNA) from a small sample of "template DNA" provided by the patient. The process of preparing template DNA for PCR amplification is a several-step process, each of which takes time and operator intervention to complete. Current clinical practice is to typically obtain DNA from blood

[a]*Department of Physics, University of Durham, Durham, DH1 3LE, UK*
[b]*Department of Bioengineering, University of Strathclyde, Wolfson building, Glasgow, G4 0NW, UK*
[c]*Department of Bioscience, Strathclyde Institute of Pharmacy and Biomedical Sciences, University of Strathclyde, Glasgow, G1 1XW, UK*

samples, though this is not the preferred option for the patient (fear of needles and potential pain), nor for the clinician as the sample collection requires trained staff. In 1 ml of whole blood there are approximately 4.5–11 $\times$ 10^5 nucleated cells,[1] which, after preliminary preparation of samples to extract the DNA and subsequently remove contaminants, provides a high yield of template DNA once extracted. The preference is thus for other sources of DNA to be used.

All body fluids contain levels of a patient's DNA but saliva is a preferred option as it is easy to collect and requires minimal preparation. Indeed in previous studies[1,2] it has been demonstrated that storage of saliva is also not critical and samples have been stored for periods of greater than 30 days at room temperature and the delivery of patient saliva samples by post provides an option in some cases. The robustness of saliva as a source of template DNA has thus been established and provides a potentially simple source of patient DNA.

However, the DNA is not found freely within the sample. In a 1 ml saliva sample there are approximately 4.3 $\times$ 10^5 epithelial cells, each of which contains DNA, but this DNA has to be extracted. The DNA is generally extracted before analysis using cell lysis, with the subsequent removal of any biological material which might potentially act to hinder PCR. Several purification kits are available for this purpose.[2–4] Cell lysis can be achieved through a variety of established techniques, including mechanical lysis (grinding of cells with rotating blade/glass beads, shearing cells by the forces imposed when forced through a narrow opening, *etc.*),[5] sonication lysis (cell shearing by an ultrasound wave), freeze/thaw lysis (repeated freezing damaging cells by ice crystal formation),[6] chemical lysis (digestion of the cell membrane, swelling and bursting of cells due to a hypotonic buffer, *etc.*),[1–4,7–10] electroporation lysis (electric-field-induced disruption of the transmembrane potential, creating transient holes in the cell membrane)[11] and thermally-induced lysis (thermally-induced degradation of the cell membrane),[12–16] which was the method selected for this work as the PCR cycling process and subsequent melting point tests all require control heating of the sample and thus the system could be designed to undertake both the DNA extraction, amplification and identification, within one instrument, without user intervention. The previous work on thermal cell lysis was to obtain DNA from mycobacteria and little work has been published on its use in human genotyping.

Having selected the method of DNA acquisition it was then decided that a specific SNP should be chosen to demonstrate that the method can be applied to biologically relevant samples. Two genes encoding cytochrome P-450 enzymes were chosen for amplification containing the SNPs CYP3A4*1B and CYP3A5*3 (corresponding to an A > G substitution in both instances). These genes are also known to be active during drug metabolism and so are considered ideal candidates for pharmacogenetic studies related to the treatment of epilepsy.

Genotyping for the SNPs CYP3A4*15B and CYP3A5*3 was achieved by means of a melting curve analysis utilising a Förster resonance energy transfer (FRET)-based dual hybridisation probe assay. Fluorescent probes for this experiment were designed to be complementary to the nucleotides found at the corresponding SNP sites as found in the human genome database (Entrez). This resulted in the CYP3A4 being complimentary to an A nucleotide and the CYP3A5 probe being complementary to a G nucleotide at the SNP site. CYP3A4 detection was achieved using a fluorescein-labelled probe of 5′-TTC AAG TAT TTT GGA ATG AGG ACA GCC ATA GA-3′–FL and a LC Red 640 labelled probe of LC RED640-5′-AAC AGG GCA AGA GAG AGG CGA T-3′–PH. CYP3A5 detection was achieved using a fluorescein-labelled probe of 5′-CAT TTC TAA CCA TAA TCT CTT TAA AGA GCT CTT-3′–FL and a LC Red 640 labelled probe of LC RED640-5′-TGT CTT TCA GTA TCT CTT CCC TGT-3′–PH.

Within the instrument the detection of the hybridisation melting point was detected by monitoring the change in the fluorescein fluorescence. At various points during the instrument development cycle the fluorescence from the LC Red 640 and the change in fluorescence lifetime of the fluorescein were explored but neither proved to be as robust and as simple as the direct monitoring of the fluorescein emission. It is possible that a slightly more sensitive system could have been constructed using a lifetime-based assay but the high-speed detector and associated electronics would have significantly increased the cost of the instrument. Recently advances in single photon avalanche photodiodes (SPADs) along with integrated processing electronics, have been developed[17] and these would be considered in a future design.

As a control, samples from the volunteers were also processed in a commercial genotyping system available through the Glasgow Western Infirmary. The 12 samples were evaluated using the methodology used by the hospital service and the results compared with those obtained with the new instrument. During the genotyping neither party was aware of the results of the other instrument, thus a basic comparative control was established.

2. Results

2.1 PCR results

Following the optimisation of the amplification conditions on samples of purified DNA, samples of saliva were taken from the volunteers and run through the cycle described above but without the addition of the fluorescent tags. A range of volumes of saliva (from 1 to 4 μl) from three individuals were run to determine the most suitable volume to be used in the instrument. It was found that the optimum volume of saliva was a compromise between having sufficient template DNA, while diluting down any potential PCR inhibitors contained within the reaction mixture. The highest yield of amplified product was obtained for volumes of either 1 μL or 4 μL of saliva, where average product yields of approximately 1–8 μg μL^{-1} were obtained. Fig. 1 illustrates the typical DNA output from a 4 μl sample for three subjects.

We believe that amplification for the 1 μl volume is likely to have produced a comparatively high yield of product due to a lower concentration of potential PCR inhibitors in the reaction mixture. However, a higher yield is observed for 4 μl, compared to 2–3 μl saliva, as the concentration of initial PCR template is higher, overcoming any potential inhibitor effects. Based on these results a 4 μl volume of saliva was used for all subsequent amplification reactions. All runs, once optimised using 4 μl of saliva produced positive results when examined using gel electrophoresis.

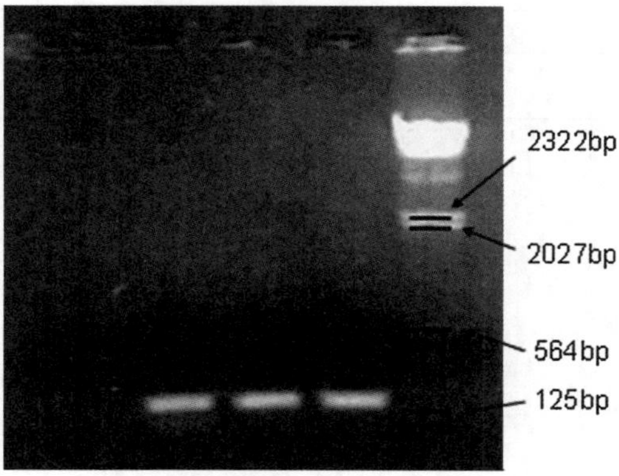

Fig. 1 Photograph of an agarose gel electrophoresis run of amplified DNA of the CYP3A5 gene from three saliva samples.

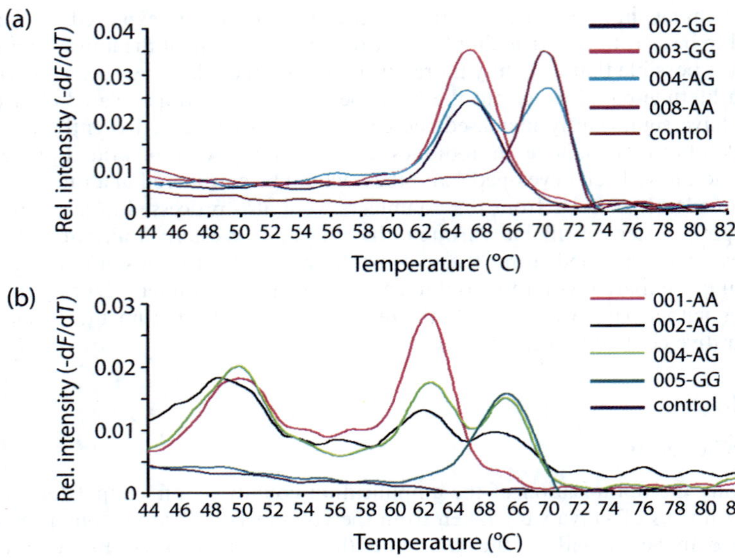

Fig. 2 Melting peak data for the genotyping of (a) CYP3A4*15B and (b) CYP3A5*3 for various DNA amplified saliva samples. In each instance an example of each allele type is illustrated relative to a control sample, which contained D_2O in place of PCR template DNA.

2.2 SNP results

Following the successful demonstration of the PCR process the samples of the 12 volunteers were run through a complete PCR cycle and a subsequent melting point curve assay within the developed instrument. An output of typical melting peaks observed during the are run is shown Fig. 2. It was found that from the 12 samples genotyped, only sample 006 when genotyping the CYP3A5*3 SNP showed any level of ambiguity. This result still, however, provided sufficient fluorescent signal and resolution of the melting peak to allow a reasonably confident genotyping result.

Table 1 Results for genotyping of DNA obtained from the amplification of various participant saliva samples for the SNPs CYP3A4*15B and CYP3A5*3. Ticks indicate agreement with commercial process, crosses indicated disagreement

Genotyping results		
Participant sample	CYP3A5*15B	CYP3A5*3
001	GG ✓	AA ✓
002	GG ✓	AA ✓
003	AA ✓	AG ✓
004	AG ✓	AG ✓
005	AA ✓	GG ✓
006	AA ✓	AG ✗
007	AA ✓	AA ✓
008	AA ✓	GG ✓
009	AA ✓	GG ✓
010	AA ✓	GG ✓
011	AA ✓	GG ✓
012	AA ✓	GG ✓

When inspecting the melting peak data, it was observed that for the genotyping results of the CYP3A5*3 SNP there was an additional peak at approximately 50 °C [Fig. 2(b)]. It is likely that this peak is due to the presence of non-specific amplifications products or primer–dimers resulting from the PCR assay. The presence of this non-specific peak does not affect the ability to successfully genotype a particular SNP, as the peak is observed at a temperature much lower than that of the hybridisation probes. A summary of genotyping results for the two examined SNPs for all twelve participants can be found in Table 1. Using the genotyping undertaken by the hospital genotyping procedure as the control the agreement between the two methods is shown in the table with a tick, a cross indicates that the commercial system provided a different result.

3. Discussion

We have demonstrated that it is possible to perform PCR using unpurified samples of human saliva, under standard assay conditions, by means of a simple thermal lysis stage prior to commencement of PCR. The thermal lysis phase required only six minutes to perform and was carried out within the same reaction vessel in which PCR was performed. This technique removes the need to perform a purification step, which can be time consuming, prone to contamination and at an additional cost. The results were compared with those taken from a commercial system and are in agreement except for one sample giving an overall 96% agreement. The one sample that did not agree did show some ambiguity in the new instrument and it is believed that by developing software to analyse the melting point curves in greater detail a level of confidence could be placed on the results obtained by the new instrument. This will be investigated in the future.

Each saliva sample used was obtained with no requirements on the participant to limit oral consumption. A number of participants were known to have recently consumed food, drink (tea, coffee, wine) and smoked cigarettes, however, no oral preparation and rinsing was found to be required. The successful amplification of these non-ideal samples demonstrates the robust and reproducible nature of using saliva cells, in conjunction with thermal lysis, as template DNA for PCR. A full analysis of such confounding factors is currently underway.

The detection of the SNPs of interest using the FRET-based assay was shown to agree with conventional sequencing of the gene sequence illustrating that in addition to the DNA being correctly amplified the subsequent hybridisation and SNP detection was not affected by the use of unpurified saliva samples. With a low component cost and simple design it has been demonstrated that potential point-of-care genotyping is possible.

If real clinical advances are to be achieved through the sequencing of the human genome several significant further advances have to take place. The identification of specific SNPs with diseases both for diagnosis and disease detection as well as subsequent drug treatment is required. In the case of the pharmacogentic aspects of this work this may well be time-consuming and one could postulate that this might become a requirement in the longer term of the drug certification process. However, even when this significant body of biomolecular work has been undertaken, new and dedicated instruments will be required to identify a patient's SNPs relevant to the disease under examination. This places a task upon instrument developers and companies to come up with novel and innovative ideas to enable rapid, cost effective and clinically relevant genotyping. It is generally accepted that a patient who attends an outpatient clinic will wait for up to an hour for their results and thus any screening, that does not require the recall of the patient, should be undertaken within this waiting time. The instrument developed in this project fulfils all of these requirements and thus potentially points the way forward for changes in clinical practice. Alongside the technological advances, however, a consideration of the funding of such instruments should be considered. In the instrument described above, no novel components are protected and thus it might be hard to see how an instrument company might exploit such advances. Joint ventures between advanced instrumentation companies and pharmaceutical companies may provide the route to exploiting these advantages both for improved financial gain, but more crucially more clinically effective treatments.

4. Method and instrumentation

4.1 Saliva PCR and hybridisation

Saliva samples were collected anonymously from 12 staff and students at the University of Strathclyde following ethical permission provided by the University of Strathclyde Ethics Committee and prior to the procurement of saliva samples all subjects were asked to sign the relevant ethical consent forms. Samples were collected by asking the 12 volunteers to collect saliva already present in their mouth and to spit this directly into a sterilised vial. This avoided the need to utilise additional equipment and solutions (*e.g.* buccal swabs or mouthwash collection). Upon collection, assays were performed immediately and all remaining saliva samples were stored at −20 °C for the duration of the study. In the later tests undertaken in an attempt to confound the instrument samples were taken ten minutes after subjects had drunk tea, red wine, coffee and smoked a cigarette.

The complete reaction process was conducted within a single vessel, containing standard PCR reagents. The conditions for the amplification of the genes of interest were initially optimised for human DNA using a conventional thermocycler (Hybaid PCRSprint, Thermo Scientific, USA). PCR amplification for the CYP3A4 region of interest was achieved using a forward primer 5′-ACA GGC ACA CTC CAG GCA TAG-3′ and a reverse primer 5′-CAG AAA CTC AAG TGG AGC CAT TG-3′ giving an expected product size of 149 bp. Amplification of the CYP3A5 region of interest was achieved using the forward primer 5′-ACC CAG CTT AAC GAA TGC TCT ACT-3′ and a reverse primer 5′-TAC GTT CTG TGT GGG GAC AA-3′ giving an expected product size of 173 bp. All Oligonucleotide primers were designed using Primer Express Software v3.0 (Applied Biosystems, CA, USA) and purchased from Invitrogen (Paisley, UK).

The PCR reaction mixture employed for the amplification of both CYP3A4 and CYP3A5 was contained within PCR microtubes with a final reaction volume of 50 µl. This contained 5 µL of 10 × Taq Buffer (100 mM Tris-HCl (pH 9.0), 15 mM $MgCl_2$, 500 mL KCl, 1% Triton®X-100, 0.1% (w/v) stabilizer), 0.4 mM of dNTP (2 µl), 4 µl of raw saliva, 160 nmol of Forward Primer (8 µl), 160 nmol of Reverse Primer (8 µl), 2.5 unit of Super Taq Polymerase (0.5 µl), 200 nM fluorescein probe, 200 nM LC Red 640 probe and 22.5 µl of D_2O. All stated concentrations are the final concentrations for a given reaction mixture and all reaction mixtures were prepared on ice. Prior to placing into the instrument described below all samples were gently vortexed and centrifuged at 3000 rpm for 30 s to collect the contents at the bottom of the PCR tubes. The PCR instrument was programmed for an initial thermal lysing at 95 °C for 13 min, which also served to perform the initial DNA denaturing step. This was directly followed by 30 cycles of amplification consisting of denaturing at 94 °C for 5 s, annealing at 54 °C for 20 s and polymerisation at 72 °C for 20 s with a final polymerisation step for 30 s.

In the initial experiments, samples of the amplified DNA were extracted (in experiments in which the fluorescent probes were not added) and run in a gel electrophoresis to ensure that the system was producing a high level of the required DNA sequence. The samples were analysed by separation on a 1% agarose gel in 1 × TAE (Tris acetate EDTA) buffer, and visualized following ethidium bromide labelling and transillumination with UV light. Sizes of products were compared against lambda DNA cut with HindIII restriction endonuclease or a standard 1 kb DNA ladder. Yield was quantified by comparison of band intensity with known concentrations of DNA ladder.

4.2 PCR instrument

The full details of the miniature PCR instrument in terms of its thermal design have been published previously[18] the outline of the instrument is presented here and the optical configuration is presented for the first time. The main sample block consists of an aluminium block machined to take the sample of interest (Fig. 3). This block was then mounted onto a SuperCool Peltier device (RS Components LTD, UK) which is specifically designed for use between −40 and 120 °C, with an active area of 30 × 30 mm. Good thermal contact was ensured by the use of thermal transfer compound. The reverse of the Peltier was mounted onto a heat sink block, again using thermal compound material. During the cooling cycle forced cooling was provided by a small DC fan mounted in the light tight box which enclosed the entire thermal and optical system.

The temperature of the system was monitored using a T-type thermocouple (10 kΩ at 20 °C) mounted in very close proximity to the sample tube. In trial experiments the full thermal properties of this system were analysed but are beyond the scope of this paper.[19] The system was controlled using a USB D/A and A/D card (National Instruments) and controlled using custom-written software in LabView which allowed for automated control from the input of desired cycling temperature parameters. It was found that the Peltier heater and fan required higher currents than were attainable from the DAQ card and so separate high-current drive circuits were constructed that could achieve the desired current outputs for both devices. These were then regulated using the voltage outputs from the DAQ card as illustrated in Fig. 3(a). The driver circuit only allowed heating of the Peltier, eliminating the need for a high current, bipolar driver circuit. The low-current DC fan acted as a cooling aid and was only utilised for the transition between the denaturing and the annealing temperatures.

The optical configuration is shown in Fig. 4. Light from a blue LED (controlled *via* the A/D card) with a centre wavelength of 460 nm was directed through a blue glass filter (Comar Instruments, Cambridge UG) to remove the long wavelength "pedestal" of light found from blue GaN LEDs before entering the sample tube at an angle of 40 degrees. The entire sample was illuminated by using the aluminium

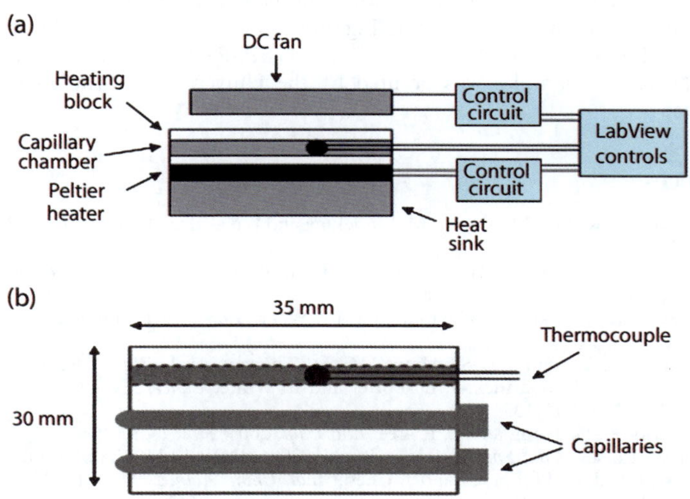

Fig. 3 Diagrams of the experimental apparatus used to create the thermocycler. (a) Side profile illustrating the electrical and control components employed in the thermocycler and (b) overhead view of the heating block illustrating the placement of the temperature sensor and two glass capillaries holding the PCR reaction mixtures.

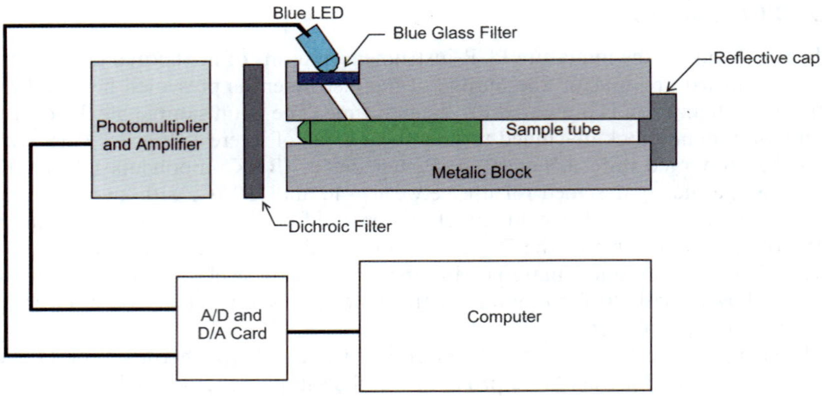

Fig. 4 Optical configuration of PCR and melting point detection instrument.

block to direct the light through the sample to ensure optimal optical coupling and excitation. The fluorescence excitation was also preferentially directed out of the end of the capillary through the channel and the use of a reflective white cap on the capillary. The end of the capillary was rounded which acted as a lens partially collimating the light emerging from the tube. This light then passed through a dichroic filter (Chroma, Vermont US) to remove the excitation light but to pass light in the 500–600 nm region. The remaining fluorescein fluorescence was then directed onto a photomultiplier (H671G, Hamamatsu, Japan) which incorporated an inbuilt amplifier. The output of the photomultiplier was then directed to the A/D card and processed using the LabView software. The LED was run in a pulsed mode at around 760 Hz and the signal detected *via* a lock-in amplifier developed in the LabView software.

Acknowledgements

We would like to thank the staff and students at the University of Strathclyde for the use of their saliva in this study. All funding for this project was provided by the Engineering and Physical Sciences Research Council (EPSRC). Ethics approval over the use of the patient samples was granted by the University of Strathclyde Ethics Committee

References

1 D. Quinque, R. Kittler, M. Keyser, M. Stoneking and I. Nasidze, *Anal. Biochem.*, 2006, **353**, 272–277.
2 T. Rylander-Rudqvist, N. Hakansson, G. Tybring and A. Wolk, *Cancer Epidemiol., Biomarkers Prev.*, 2006, **15**(9), 1742.
3 L. Todesco, M. Török, S. Krahenbuhl and M. Wenk, *Eur. J. Clin. Pharmacol.*, 2003, **59**, 343–346.
4 I. S. About, I. B. King, J. S. About, M. D. Thornquist, J. Bigler, R. E. Patterson, A. R. Kristal, A. L. Shattuck, J. D. Potter and E. White, *Cancer Epidemiol., Biomarkers Prev.*, 2002, **11**, 1130–1133.
5 D. D. Carlo, K. H. Jeong and L. P. Lee, *Lab Chip*, 2003, **3**, 287–291.
6 William P. Tansey, *Cold Spring Harbor Protoc.*, 2006, DOI: 10.1101/pdb.prot4614.
7 X. Chen, D. Cui, C. Liu, H. Li and J. Chen, *Anal. Chim. Acta*, 2007, **584**, 237–243.
8 L. Zhang, F. Dang, N. Kaji and Y. Baba, *J. Chromatogr., A*, 2006, **1106**, 175–180.
9 S. J. Park, J. Y. Kim, Y. G. Yang and S. H. Lee, *J. Forensic Sci.*, 2008, **53**, 335–341.
10 X. Xie, X. Zhang, B. Yu, H. Gao, H. Zhang and W. Fei, *J. Magn. Magn. Mater.*, 2004, **280**, 164–168.
11 H. Lu, M. A. Schmidt and K. F. Jensen, *Lab Chip*, 2005, **5**, 23–29.

12 C. B. Kim, D. K. Yi, P. S. S. Kim, W. Lee and M. J. Kim, *J. Nanosci. Nanotechnol.*, 2009, **9**, 2841–2845.
13 K. Zhu, H. Jin, Y. Ma, Z. Ren, C. Xiao, Z. He, F. Zhang, Q. Zhu and B. Wang, *J. Biotechnol.*, 2005, **118**, 257–264.
14 R. H. Liu, J. Yang, R. Lenigk, J. Bonanno and P. Grodzinski, *Anal. Chem.*, 2004, **76**, 1824–1831.
15 I. U. H. Khan and J. S. Yadav, *J. Clin. Microbiol.*, 2004, **42**, 453–457.
16 C. Y. Lee, G. B. Lee, J. L. Lin, F. C. Huang and C. S. Liao, *J. Micromech. Microeng.*, 2005, **15**, 1215–1223.
17 B. Rae, K. R. Muir, Z. Gong, J. McKendry, J. M. Girkin, E. Gu, D. Renshaw, M. D. Dawson and R. Henderson, *Sensors*, 2009, **9**, 9255–9274.
18 M.-I. Mohammed, G. J. Sills, G. J. Brodie, J. Martin, E. M. Ellis and J. M. Girkin, *Sens. Actuators, B*, 2009, **139**, 83–89.
19 M.-I. Mohammed, PhD thesis, University of Strathclyde, 2009.

Small-molecule diagnostics based on functional DNA nanotechnology: a dipstick test for mercury

Seyed-Fakhreddin Torabi[a] and Yi Lu*[ab]

Received 1st June 2010, Accepted 24th July 2010
DOI: 10.1039/c005404f

Detecting small molecular targets such as metal ions is just as important as detecting large molecules such as DNA, RNA and proteins, but the field of metal ion sensors has not yet been well developed. A good example of a metal ion target is mercury, which is highly toxic, widely distributed in the environment and affects human health. To develop a diagnostic platform for metal ions, we demonstrate that functional DNA-linked gold nanoparticles (AuNPs) can quickly and simply detect and quantify Hg^{2+} ions in aqueous solution, with high sensitivity and selectivity over competing metal ions. A linker DNA molecule containing thymine residues and sequences complementary to the DNA on the AuNPs was designed to aggregate DNA-functionalized AuNPs. When Hg^{2+} ions were introduced into this system, they induced the linker DNA to fold by forming thymine–Hg^{2+}–thymine bonds. The linker DNA's folding caused the AuNPs to rapidly disassemble, which caused a discernable color change in the solution from purple to red. The limit of detection for Hg^{2+} in the present method is 5.4 nM, which is below the 10 nM maximum contaminant level defined by the US Environmental Protection Agency (EPA) for drinking water. Our results show that this Hg^{2+} detection method has excellent selectivity over other divalent metal ions (*e.g.* Pb^{2+}, Cu^{2+}, Mn^{2+}, Co^{2+}, Zn^{2+}, Cd^{2+}, Mg^{2+}, Ca^{2+}, and Ba^{2+}). This system has been converted into a dipstick test using lateral-flow devices, making it even more practical for point-of-care diagnostics.

1. Introduction

Importance of detecting small-molecule targets

Novel analytical techniques improve diagnostics and theranostics for applications in healthcare and related fields.[1–3] In contrast to the tremendous progress in nucleic acid and protein detection, small-molecule analysis of disease markers and inorganic or organic cofactors in biological pathways, has lagged behind.[4–6] These targets are just as important as their larger counterparts, as they play critical roles in biological function and disease development. For example, for many years a protein-based marker called prostate-specific antigen (PSA) has been used as a marker for prostate cancer.[7] However, it has been recognized that PSA is a poor marker for the disease because it gives high rates of false positives and false negatives.[8] On the other hand, a recent report demonstrated that sacrosine, a small organic molecule similar to glycine, may be a better marker for the onset of prostate cancer than PSA.[9] Therefore, detecting small-molecule targets is a major but exciting challenge in bioanalytical chemistry and medical diagnostics.

[a]*Department of Biochemistry, University of Illinois at Urbana-Champaign, Urbana, IL 61801, USA. E-mail: yi-lu@illinois.edu*
[b]*Department of Chemistry, University of Illinois at Urbana-Champaign, Urbana, IL 61801, USA*

Importance of detecting small-molecule targets such as metal ions

Among the small-molecule targets, metal ions such as calcium or iron are known to benefit human health while other metal ions, such as lead and mercury, are deleterious. With the exception of a few metal ions such as calcium, we know very little about the mechanisms by which metal ions are transported into cells, speciated into different forms, and enabled to carry out their beneficial or deleterious functions.[10] Tsien and others' calcium sensors are an excellent example of how metal-ion sensors can revolutionize biochemistry, cellular biology, and related fields.[11]

Challenges in detecting small-molecule targets such as metal ions

More than half of the elements in the periodical table form metal ions that are found in biological systems and the environment, and some metals form many different oxidation states. Unlike nucleic acid and protein detection where nucleic acid or antibody arrays can be used to detect almost any biomolecular target, there are few general sensing strategies that can be used to produce a sensor for any metal ion or a specific oxidation state of a given metal ion.[12] Antibodies that can be raised against large molecular targets such as proteins often fail to recognize metal ions; they are too small to elicit a strong immunological response, and some metal ions are too toxic for animals or cells to produce antibodies against them.[13] Recently, several whole-cell bacterial biosensors were developed to detect metal ions such as mercury. However, most of them were insufficiently selective.[14,15] Also, whole-cell biosensors are limited by the expense, fragility, growth requirements of cells, and the specialized training needed to handle them.[16,17]

Functional DNA or RNA as an excellent alternative in detecting small-molecule targets such as metal ions

Functional DNA and RNA are nucleic acids with either enzymatic or specific binding activities.[18] Examples of functional DNA include ribozymes and deoxyribozymes (also called DNAzymes), which are similar to protein enzymes, and riboswitches and aptamers, which are similar to antibodies. Ribozymes and riboswitches have been isolated from cells while DNAzymes and DNA aptamers are obtained by *in vitro* selection[19,20] or Systematic Evolution of Ligands by Exponential Enrichment (SELEX).[21,22] The selection process can be tailored to obtain functional DNA or RNA to bind almost any target of interest, making it, like antibody selection, a general method for sensing molecules.[23] Unlike antibody selection, however, these functional DNA and RNA molecules have been shown to bind selectively not only to large molecules, but also to small-molecule targets, with affinities down to the picomolar range for the dissociation constant. In particular, functional DNA molecules are much more stable than antibodies and can be denatured and renatured many times without losing their function.[12] Because of these advantages, we and others have developed functional nucleic acid sensors for a wide range of targets including metal ions, small organic molecules, proteins, and cells.[18] By combining the functional DNA with fluorophores/quenchers, gold nanoparticles, carbon nanotubes, quantum dots, and iron oxide nanoparticles, we and others have developed highly sensitive and selective, fluorescent, colorimetric, and electrochemical sensors and MRI contrast agents, with detection limits as low as 45 pM (11 ppt) and over million-fold selectivities.[24–28]

Functional DNA sensors for mercury

Mercury is a highly toxic heavy metal ion that can cause a number of severe adverse health effects such as damage to the nervous, immune, and excretory systems. Sensors that can detect mercury *in situ* and in real time can greatly increase our understanding of its toxicity and prevent its toxic effects. Toward this goal, we

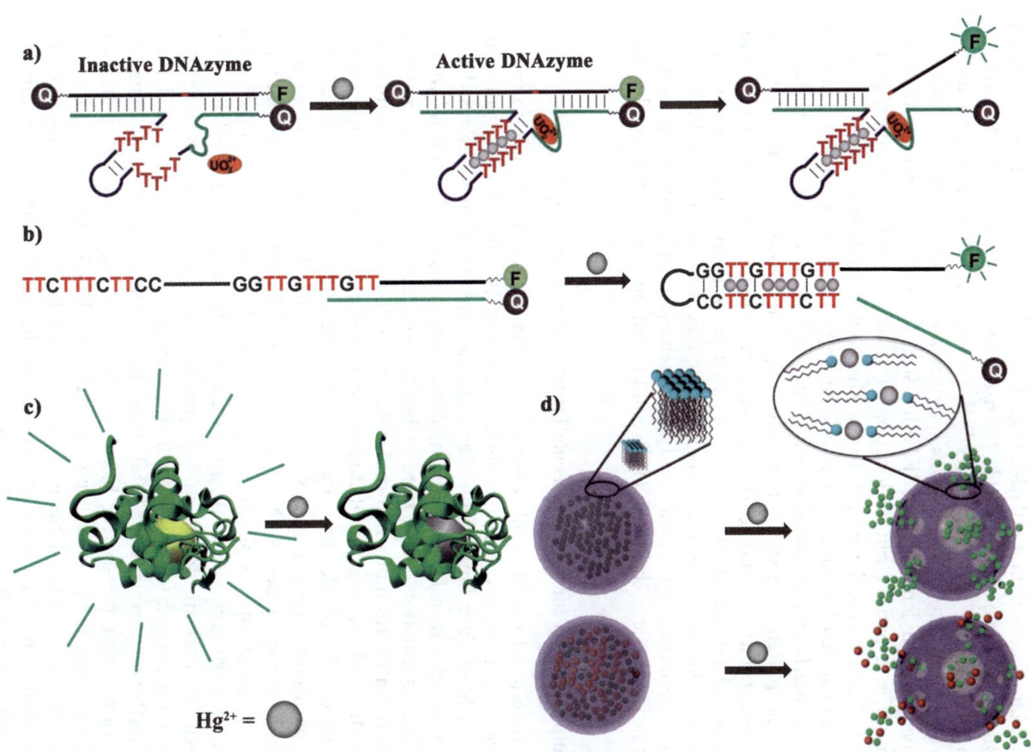

Fig. 1 Schematic illustration of sensors and theranostic agents for mercury. a) A mercury sensor based on allosteric enhancement of uranium-specific DNAzyme activity by Hg^{2+} binding.[29] b) A simplified "turn-on" mercury sensor based on structure-switching DNA that releases quencher-modified DNA in the presence of Hg^{2+}.[27] c) A lysozyme-stabilized gold fluorescent cluster as a Hg^{2+} sensor.[30] d) A theranostic agent for Hg^{2+} based on Hg^{2+}-responsive PEG-liposomes that results in the release of both fluorescent probe and detoxifying agent.[36]

have designed a highly sensitive and selective catalytic beacon sensor for Hg^{2+} using the thymine–mercury–thymine interaction to modulate DNAzyme activities through allosteric interactions (Fig. 1a).[29] Its detection limit of 2.4 nM is lower than the 10 nM maximum contamination level defined by the US EPA for drinking water. We then further simplified the sensor by using the thymine–mercury–thymine interaction without the DNAzyme to create a "turn-on" fluorescent sensor for mercury. This only slightly changed the sensor's properties, giving it a detection limit of 3.2 nM and high selectivity (Fig. 1b).[27] In addition to DNA molecules, we also used proteins in lysozyme-stabilized gold fluorescent clusters Hg^{2+} sensors, which also had a similar detection limit (10 nM) (Fig. 1c).[30] These results complement those of other groups who have developed Hg^{2+} sensors using functional DNA and lysozyme-stabilized gold fluorescent clusters.[31–35] Finally, we also developed a Hg^{2+} theranostic agent that can simultaneously detect and detoxify mercury using an ion-responsive poly(ethylene glycol) (PEG) liposome (Fig. 1d).[36] Since the distribution of mercury in a patient can be location- and time-specific, treating a patient exposed to Hg^{2+} can be challenging. Using a Hg^{2+} detoxification agent at too high a concentration can lead to side-effects, and in any case a single dose of such an agent may be inefficient in dealing with the Hg^{2+} concentration distribution throughout the patient's tissues. The novel liposome-based system shown in Fig. 1d can detect and respond to Hg^{2+} at concentrations as low as 10 nM. This system adjusts the release of Hg^{2+} chelators to the local concentration of Hg^{2+}, releasing more chelators in regions of high Hg^{2+} concentration and no chelators in regions of low concentration. This "budgeted" release profile will be particularly useful in situations where the local Hg^{2+} contamination levels vary, or fluctuate over time.

Developing dipstick tests for small-molecule targets such as metal ions

While these functional nucleic acid sensors are useful as analytical tools to understand biological systems and serve as diagnostics and theranostics in healthcare, they are not without their limitations. For example, the techniques described thus far require the precise transfer of solutions containing reagents,[18] and thus are not suitable to address a major unmet need in healthcare: rapid, on-site and real-time detection and quantification of small-molecule targets. To overcome this limitation, we have transformed the functional nucleic acid sensors into dipstick tests using lateral-flow devices.[37,38] Similar to a pregnancy test that uses antibodies as the diagnostic agent, we used aptamers and DNAzymes that are much more stable than antibodies, and gold nanoparticles (AuNPs) whose extinction coefficient is 10–100 times greater than the most intense organic dyes currently used in commercial dipstick tests.[39] More importantly, we took advantage of the distance-dependent color changes of AuNPs that are due to surface plasmon effects. As a result, when the target is absent, the aggregated AuNPs will appear blue, but when the target is introduced the AuNPs will appear red.[38] While this unique concept made adenosine and cocaine detection in aptamer-AuNP-based lateral flow devices possible, it has not been feasible to detect metal ions using an analogous DNAzyme system. Instead, AuNPs serve only as a color-labeling agent.[37,40] To take full advantage of the distance-dependent optical changes of AuNPs, we herein describe a new dipstick test that changes color in the presence of mercury.

2. Experimental

Materials

All DNA samples were purchased from Integrated DNA Technologies, Inc. (Coralville, IA). The linker DNA was purified by gel electrophoresis, whereas the thiol-modified DNA molecules were purified by HPLC. Streptavidin was purchased from Promega. AuNPs that were 13 nm in diameter were synthesized by the

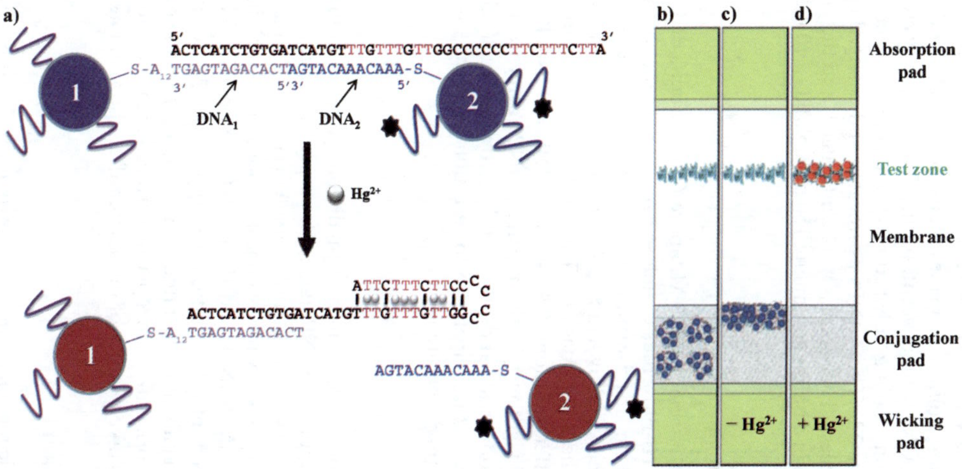

Fig. 2 Design of a colorimetric dipstick sensor for mercury. a) A schematic representation of the colorimetric detection of Hg^{2+}. In the DNA sequences, A_{12} in DNA_1 (purple) denotes a 12-mer polyadenine chain and biotin is denoted as a black star (✱). AuNPs 1 and 2 are functionalized with two different DNA molecules through thiol-gold chemistry. The two kinds of AuNP are linked by 45-mer linker DNA strand (in black) containing several thymine residues (in pink). In the presence of Hg^{2+}, the aggregates disassemble to give dispersed red AuNPs. b) Assembled lateral-flow device loaded with AuNP aggregates (on the conjugation pad) and streptavidin (on the membrane in cyan) before use. c) Negative control. In the absence of Hg^{2+}, unreacted AuNP aggregates stagnate on the boundary between the conjugation pad and the membrane. d) Positive test. In the presence of Hg^{2+}, AuNP aggregates disassemble and the dispersed AuNP products migrate, when they are captured at the test zone through the biotin–streptavidin interaction, a red line forms.

citrate-reduction method.[41] Thiol-modified DNA was activated with two equivalents of tris(2-carboxyethyl) phosphine hydrochloride (TCEP). The TCEP-activated, thiol-modified DNA and AuNPs were mixed at room temperature for at least 16 h. (typically 9 μL of 1 mM DNA was added to 3 mL of AuNPs), and NaCl (100 mM) was added to the solution. The solution was left at room temperature for another two days, and the DNA-functionalized AuNPs were purified by centrifugation and removal of the supernatant before use.

Sensor preparation

Solutions of thiol-modified DNA AuNPs were mixed in the presence of $NaNO_3$ (300 mM), MOPS (3-(N-morpholino) propanesulfonic acid) buffer (10 mM, pH 7.2), and 45-mer linker DNA (varying concentrations). The samples were annealed by heating at 75 °C for 2 min, cooling down to room temperature over ~1 h, and then being stored at 4 °C for 2 h. The AuNPs aggregated and changed color from red to purple during this process. The samples were centrifuged, and the precipitates were collected and dispersed in a buffer containing 150 mM $NaNO_3$; 10 mM MOPS; pH 7.2.

Dipstick test sample preparation

The Millipore Hi-Flow Plus Assembly Kit (Millipore Corporation, Bedford, MA) was used. The kit contains a Hi-Flow Plus Cellulose Ester Membrane with a nominal capillary flow time of 240 s/4 cm, an absorption pad, wicking, and a glass fiber conjugation pad. The device assembly is shown in Fig. 2b. One DNA arm contains a biotin moiety at the 5′-end while the other does not. The purified AuNP aggregates were dispersed in designated buffer solutions (8% sucrose, 150 mM $NaNO_3$, 10 mM MOPS; pH 7.2) and agitated vigorously with a pipette. AuNP aggregates (4 μL) were spotted onto each conjugation pad, and 10 mg mL^{-1} streptavidin (1.5 μL) was applied to the membrane by a 2 μL pipette to form a thin line. The loaded strips were stored in a drawer overnight before use.

Detection

To detect Hg^{2+} using a solution-based sensor, a small volume of concentrated Hg^{2+} solution was added to 100 μL of the AuNP-aggregate suspension. The ensuing color change was monitored with a UV–vis spectrophotometer (Hewlett-Packard 8453) at room temperature. For dipstick tests, varying concentrations Hg^{2+} were dissolved in a flow buffer containing 150 mM $NaNO_3$ and 10 mM MOPS at pH 7.2. The wicking pad of each device was dipped into the solutions for approximately two minutes to fully hydrate the conjugation pad, and the liquid began migrating on the membrane. Then the device was placed horizontally on a plastic surface for the flow to continue. A digital camera was used to take pictures of the devices after approximately 10 min.

3. Results and discussion

We constructed a colorimetric mercury sensor as shown in Fig. 2a. It is based on our recent fluorescent Hg^{2+} sensor that uses structure-switching DNA and on colorimetric sensors for adenosine and cocaine using aptamers and AuNPs.[27,38,42] The sensor system consists of two sets of 13-nm AuNPs, with each set of AuNP functionalized with either DNA_1 (in purple) or DNA_2 (in blue). A 45-mer linker DNA is then used to bring the two sets of DNA-functionalized AuNPs together. This linker contains three segments. The first segment is complementary to DNA_1, the second segment is complementary to DNA_2, and the third segment remains in single-stranded form. Critical to this sensor design is the presence of seven thymine–thymine mismatched pairs (T–T shown in pink) separated by five self-complementary base pairs located within the second and third segments of the linker

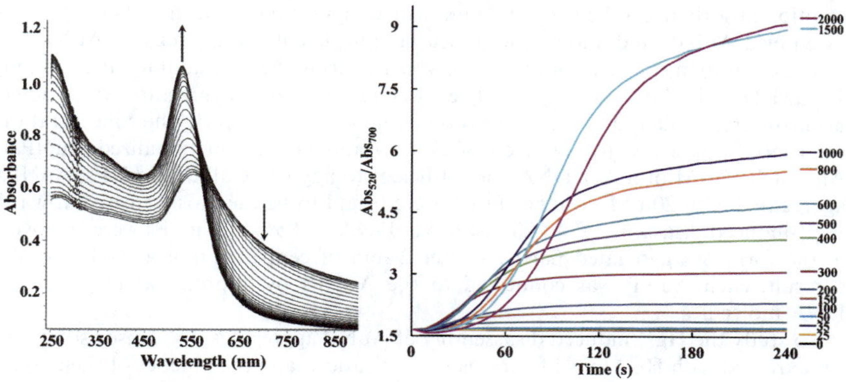

Fig. 3 Kinetics of the color change at various concentrations of Hg^{2+}. a) UV–vis spectra of the colorimetric mercury sensor whilst 1.5 μM of Hg^{2+} is been added over the course of four minutes. A decrease in the absorbance at 700 nm and an increase at 520 nm is shown. b) The kinetics of the absorbance ratio increase as the Hg^{2+} concentration increases.

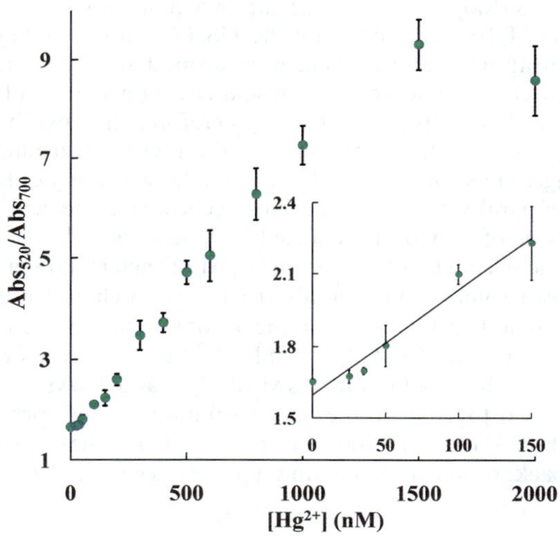

Fig. 4 Quantifying the Hg^{2+} concentration by monitoring the absorbance ratio four minutes after adding mercury. The calibration curve of the colorimetric mercury sensor saturated at concentrations higher than 1500 nM. A linear relationship between the absorbance ratio and Hg^{2+} concentration was observed for a wide concentration range (0–1500 nM). The data were obtained from three independent measurements and the error bars indicate the standard deviation.

DNA. In this way, the addition of the linker DNA should aggregate the AuNPs functionalized with DNA_1 and DNA_2 to form blue- or purple-colored aggregates due to the surface plasmon effect. Since Hg^{2+} is known to form strong thymine–Hg^{2+}–thymine base pairs,[31,43] the presence of Hg^{2+} would induce the folding of the last two segments of the linker DNA into a hairpin structure as shown in Fig. 2a. As a result, only five base pairs would remain between DNA_2 and the linker DNA, which is not enough to keep both strands in a stable double-stranded form at room temperature. Therefore, DNA_2 and the attached AuNPs would be released from the linker DNA, the AuNPs would disassemble, and their color would change from purple to red. This phenomenon can be monitored by measuring changes in the solution absorbance at 520 and 700 nm, with a high ratio indicating red-colored disassembled AuNPs and a low ratio indicating purple-colored aggregated AuNPs.[42]

The colorimetric spectra of the sensor system during the four minutes after adding 1.5 μM Hg^{2+} is shown in Fig. 3a. The assembly of AuNP aggregates was studied using different linker DNA concentrations from 4 nM to 100 nM. The concentration of AuNPs was 8 nM (4 nM each of DNA_1- and DNA_2-functionalized AuNPs). Although 40 nM of linker DNA was sufficient to aggregate all the AuNPs, AuNPs aggregated with 20 nM of linker DNA were found to be the most sensitive toward Hg^{2+}-induced disassembly. At 20 nM linker DNA, ~ 85% of AuNPs were removed in the form of aggregated particles after 2 min of centrifugation at 400 × *g*. As a result, each AuNP was connected to the AuNPs by approximately six linker DNA molecules.

To study the Hg^{2+}-induced disassembly of AuNP aggregates, the sensor solutions were treated with Hg^{2+} ions of various concentrations and the kinetics of their color changes were monitored by UV–vis spectroscopy (Fig. 3b). Higher concentrations of Hg^{2+} ions increased the rate at which the solution changed to red. Quantitative analysis was carried out by monitoring the increase in the absorbance ratio four minutes after adding different Hg^{2+} concentrations. As illustrated in Fig. 4, a linear relationship between the absorbance ratio and Hg^{2+} concentration was observed in the wide range of Hg^{2+} concentrations from 0–1500 nM. The sensor has a detection limit of 5.4 nM based on 3σ/slope, which is about two times lower than the maximum contaminant level of 10 nM defined by the US EPA for drinking water. At very low Hg^{2+} concentrations, because there were around six linkages for each AuNP and each linkage contains seven T–T mismatches, only a small fraction of the AuNPs was released from the aggregates. Therefore, the absorbance ratio could not substantially increase in the presence of the high background absorbance of the remaining aggregates. Despite this limitation, the sensitivity of the current detection system is comparable to our recent fluorescence-based detection method,[27] and is greater than those of previously reported Hg^{2+} sensors.[31,44–46]

To determine the selectivity of the sensor, 1 μM of each metal ion was individually added to the sensor solution and the absorbance ratio change was monitored. As shown in Fig. 5 (blue bars), among the metal ions tested (Zn^{2+}, Pb^{2+}, Cu^{2+}, Co^{2+}, Cd^{2+}, Mn^{2+}, Ba^{2+}, Ca^{2+}, $Mg^{2+}$and Hg^{2+}), only Hg^{2+} resulted in a significant increase in the absorbance ratio. In addition, 1 μM of Hg^{2+} and 1 μM of another divalent metal ion were added together to the sensor solution. The absorbance ratio of the Hg^{2+}–M^{2+} pair (Fig. 5, red bars) indicates that the sensor has a high selectivity for Hg^{2+} against a background of competing analytes. The excellent selectivity of the

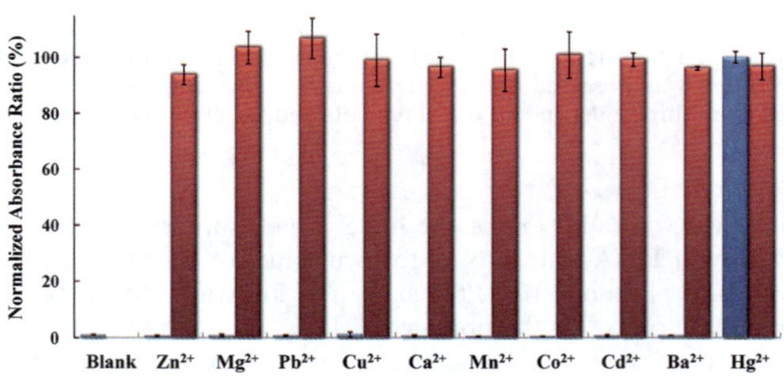

Fig. 5 Selectivity of the Hg^{2+} sensor. Blue bars represent normalized absorbance ratios four minutes after adding 1 μM of the other metal ions (from left to right: blank, Zn^{2+}, Mg^{2+}, Pb^{2+}, Cu^{2+}, Ca^{2+}, Mn^{2+}, Co^{2+}, Cd^{2+}, Ba^{2+} and Hg^{2+}). Red bars represent sensor responses after adding 1 μM of Hg^{2+} and 1 μM of another metal ion. The last red bar represents the sensor response when 1 μM of mercury was added to the pond water; the dilution factor was 1.5.

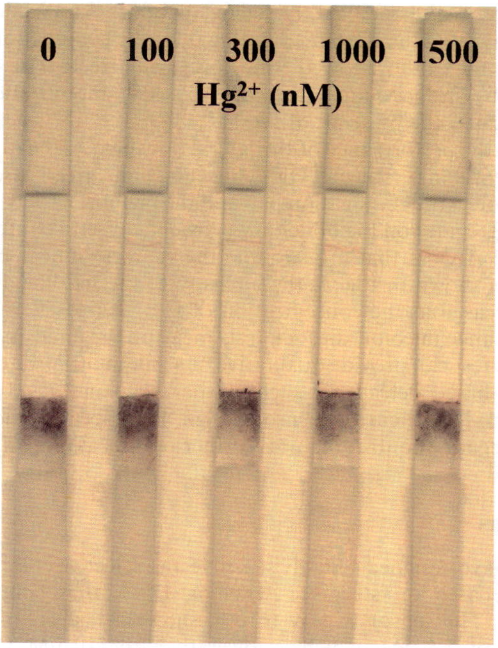

Fig. 6 Lateral-flow-based detection of Hg^{2+}. Test of the mercury-sensing lateral-flow device with varying concentrations of Hg^{2+}.

sensor is likely due to the high specificity of the interaction between T–T mismatches and Hg^{2+} which has been previously demonstrated.[47]

Since it had exhibited exquisite sensitivity and selectivity in buffer solutions, our sensor was further tested with pond water collected on the University of Illinois campus. Using the standard addition method, Hg^{2+} ions were added to the sensor solution in the pond water up to a final concentration of 1 μM. The sensor responded to Hg^{2+} very similarly, whether it was in pure or pond water (Fig. 5). Similar responses were also obtained with 100 and 300 nM mercury in pure and pond water. This is an indicator that our sensor is able to detect Hg^{2+} in pond water with little interference.

Although the success of this Hg^{2+} colorimetric sensor is an important step toward real-time sensing, the next step would be to convert this sensor into an even simpler Hg^{2+} dipstick test that does not require instruments such as a UV–vis spectrophotometer or quantitative solution transfer. As in the colorimetric sensing system, Hg^{2+}-responsive AuNP aggregates containing two kinds of DNA-functionalized AuNPs and a 45-mer linker DNA were prepared. In this case, two kinds of thiol-modified DNA_2 (biotinylated and non-biotinylated) were used to functionalize particle 2. Biotin modifications on the DNA strands (✱) allowed the AuNPs to be captured by streptavidin. The optimal 1 : 1 ratio between two DNA molecules, which has been shown in our previous work,[38] was used in this study.

Four overlapping pads of a lateral flow device were attached to a paper backing as follows (from top to bottom): absorption pad, HiFlow Plus membrane, glass fiber conjugation pad, and wicking pad. Aggregated AuNPs were prepared in a buffer containing 10 mM MOPS (pH 7.2), 150 mM $NaNO_3$, and 8% sucrose, producing a dark-purple solution. Sucrose was added to the buffer to keep the DNA hybridized and facilitate the rehydration of the aggregates. The lateral-flow device was allowed to dry for 8 h. Aggregated AuNPs were spotted onto the conjugation pad, and streptavidin was applied to the HiFlow Plus membrane as a thin line (Fig. 2b). When the bottom of the dipstick was placed into the sample solution, the Hg^{2+} ions migrated to the aggregates and disassembled the aggregated particles. While the large size of the aggregates prevented them from migrating onto the conjugation pad,[42] the small size of the dispersed AuNPs allowed them to migrate freely with the solution flow and be captured by streptavidin. Their capture and immobilization formed a thin red line that indicated the presence of Hg^{2+}.

In order to test the performance of the lateral flow device, the dipstick was placed into flow buffers containing varying Hg^{2+} concentrations. In all cases, a dark band at the boundary between the conjugation pad and the membrane was observed (Fig. 6). The unreacted aggregates on the boundary provided a useful control which was due to the aggregated AuNPs that are so large they can no longer migrate along the membrane.[38,42] As shown in Fig. 6, the intensity of the red line at the streptavidin capture zone increased as the concentration of Hg^{2+} increased. While colorimetric tests monitor a change in color, dipstick tests measure the intensity of a single color. Thus, the dipstick test can provide qualitative or semiquantitative results. In fact, for most lateral-flow-based detections, such as the pregnancy test, only a yes or no answer is required. In our current sensor design, the appearance of a faint red line in the case of a negative control requires further optimization, which is in progress.

4. Conclusions

In summary, we have designed a highly sensitive and selective colorimetric sensor for mercury that is based on structure-switching DNA that contains mismatched thymine residues. The sensor is easy to use, shows a fast color change, and has a detection limit of 5.4 nM, which is lower than the 10 nM US EPA limit for Hg^{2+} in drinking water. Furthermore, we have immobilized AuNP aggregates onto a lateral-flow device, resulting in the first easy-to-use dipstick test for mercury. Their simple design makes the highly sensitive and selective colorimetric sensor and corresponding dipstick tests capable of carrying out real-time mercury detection in environmental and medical applications.

Acknowledgements

We wish to thank Mr Zidong Wang for helpful discussions and Ms. Hannah Ihms for proof-reading. The work described in this work is supported by the US National Institute of Health (ES16865), Department of Energy (DE-FG02-08ER64568), and the National Science Foundation (DMR-0117792, CTS-0120978 and DMI-0328162).

References

1 A. K. H. Cheng, D. Sen and H. Z. Yu, *Bioelectrochemistry*, 2009, **77**, 1–12.
2 J. C. Lindon, E. Holmes and J. K. Nicholson, *Curr. Opin. Mol. Ther.*, 2004, **6**, 265–272.
3 M. Famulok, J. S. Hartig and G. Mayer, *Chem. Rev.*, 2007, **107**, 3715–3743.
4 J. Majidi, J. Barar, B. Baradaran, J. Abdolalizadeh and Y. Omidi, *Hum. Antibodies*, 2009, **18**, 81–100.
5 D. Majumdar, X. H. Peng and D. M. Shin, *Curr. Top. Med. Chem.*, 2010, **10**, 1211–1226.
6 B. Lu, M. R. Smyth and R. O'Kennedy, *Analyst*, 1996, **121**, 29R–32R.
7 G. Wu, R. H. Datar, K. M. Hansen, T. Thundat, R. J. Cote and A. Majumdar, *Nat. Biotechnol.*, 2001, **19**, 856–860.
8 A. Semjonow, F. Oberpenning, C. Weining, M. Schon, B. Brandt, G. De Angelis, A. Heinecke, M. Hamm, P. Stieber, L. Hertle and H. P. Schmid, *Clin. Chem.*, 2001, **47**, 1472–1475.
9 A. Sreekumar, L. M. Poisson, T. M. Rajendiran, A. P. Khan, Q. Cao, J. Yu, B. Laxman, R. Mehra, R. J. Lonigro, Y. Li, M. K. Nyati, A. Ahsan, S. Kalyana-Sundaram, B. Han, X. Cao, J. Byun, G. S. Omenn, D. Ghosh, S. Pennathur, D. C. Alexander, A. Berger, J. R. Shuster, J. T. Wei, S. Varambally, C. Beecher and A. M. Chinnaiyan, *Nature*, 2009, **457**, 910–914.

10 R. A. Yokel, S. M. Lasley and D. C. Dorman, *J. Toxicol. Environ. Health, Part B*, 2006, **9**, 63–85.
11 A. E. Palmer, C. Jin, J. C. Reed and R. Y. Tsien, *Proc. Natl. Acad. Sci. U. S. A.*, 2004, **101**, 17404–17409.
12 C. K. O'Sullivan, *Anal. Bioanal. Chem.*, 2002, **372**, 44–48.
13 D. Hilvert, *Annu. Rev. Biochem.*, 2000, **69**, 751–793.
14 A. Ivask, M. Virta and A. Kahru, *Soil Biol. Biochem.*, 2002, **34**, 1439–1447.
15 E. M. Nolan and S. J. Lippard, *Chem. Rev.*, 2008, **108**, 3443–3480.
16 H. A. Fishman, D. R. Greenwald and R. N. Zare, *Annu. Rev. Biophys. Biomol. Struct.*, 1998, **27**, 165–198.
17 J. Zhao, S. S. Jedlicka, J. D. Lannu, A. K. Bhunia and J. L. Rickus, *Biotechnol. Prog.*, 2006, **22**, 32–37.
18 J. Liu, Z. Cao and Y. Lu, *Chem. Rev.*, 2009, **109**, 1948–1998.
19 R. R. Breaker and G. F. Joyce, *Chem. Biol.*, 1994, **1**, 223–229.
20 D. L. Robertson and G. F. Joyce, *Nature*, 1990, **344**, 467–468.
21 A. D. Ellington and J. W. Szostak, *Nature*, 1990, **346**, 818–822.
22 C. Tuerk and L. Gold, *Science*, 1990, **249**, 505–510.
23 D. H. Bunka and P. G. Stockley, *Nat. Rev. Microbiol.*, 2006, **4**, 588–596.
24 T. Lan, K. Furuya and Y. Lu, *Chem. Commun.*, 2010, **46**, 3896–3898.
25 J. H. Lee, Z. Wang, J. Liu and Y. Lu, *J. Am. Chem. Soc.*, 2008, **130**, 14217–14226.
26 J. H. Lee, M. V. Yigit, D. Mazumdar and Y. Lu, *Adv. Drug Delivery Rev.*, 2010, **62**, 592–602.
27 Z. Wang, J. Heon Lee and Y. Lu, *Chem. Commun.*, 2008, 6005–6007.
28 J. Liu, J. H. Lee and Y. Lu, *Anal. Chem.*, 2007, **79**, 4120–4125.
29 J. Liu and Y. Lu, *Angew. Chem., Int. Ed.*, 2007, **46**, 7587–7590.
30 H. Wei, Z. Wang, L. Yang, S. Tian, C. Hou and Y. Lu, *Analyst*, 2010, **135**, 1406–1410.
31 A. Ono and H. Togashi, *Angew. Chem., Int. Ed.*, 2004, **43**, 4300–4302.
32 J. S. Lee, M. S. Han and C. A. Mirkin, *Angew. Chem., Int. Ed.*, 2007, **46**, 4093–4096.
33 X. Xue, F. Wang and X. Liu, *J. Am. Chem. Soc.*, 2008, **130**, 3244–3245.
34 D. Li, A. Wieckowska and I. Willner, *Angew. Chem., Int. Ed.*, 2008, **47**, 3927–3931.
35 J. P. Xie, Y. G. Zheng and J. Y. Ying, *Chem. Commun.*, 2010, **46**, 961–963.
36 M. V. Yigit, A. Mishra, R. Tong, J. Cheng, G. C. Wong and Y. Lu, *Chem. Biol.*, 2009, **16**, 937–942.
37 D. Mazumdar, J. Liu, G. Lu, J. Zhou and Y. Lu, *Chem. Commun.*, 2010, **46**, 1416–1418.
38 J. Liu, D. Mazumdar and Y. Lu, *Angew. Chem., Int. Ed.*, 2006, **45**, 7955–7959.
39 C. A. Mirkin, *Inorg. Chem.*, 2000, **39**, 2258–2272.
40 W. Zhao, M. M. Ali, S. D. Aguirre, M. A. Brook and Y. Li, *Anal. Chem.*, 2008, **80**, 8431–8437.
41 J. Liu and Y. Lu, *Nat. Protoc.*, 2006, **1**, 246–252.
42 J. Liu and Y. Lu, *Angew. Chem., Int. Ed.*, 2006, **45**, 90–94.
43 Y. Tanaka, S. Oda, H. Yamaguchi, M. Kudo, Y. Kondo, C. Kojima and A. Ono, *Nucleic Acids Symp. Ser.*, 2006, **50**, 47–48.
44 M. Kumar, A. Dhir, V. Bhalla, R. Sharma, R. K. Puri and R. K. Mahajan, *Analyst*, 2010, **135**, 1600–1605.
45 L. Zhang, T. Li, B. Li, J. Li and E. Wang, *Chem. Commun.*, 2010, **46**, 1476–1478.
46 H. Wang, Y. Wang, J. Jin and R. Yang, *Anal. Chem.*, 2008, **80**, 9021–9028.
47 Y. Miyake, H. Togashi, M. Tashiro, H. Yamaguchi, S. Oda, M. Kudo, Y. Tanaka, Y. Kondo, R. Sawa, T. Fujimoto, T. Machinami and A. Ono, *J. Am. Chem. Soc.*, 2006, **128**, 2172–2173.

General discussion

Professor Cass opened the discussion of the paper by Dr Ko Ferrigno: If you take the same scaffold and sample, does the overall performance of the sensor depend more on the base sensor (label *vs.* label-free or resonant *vs.* non-resonant) or more on the biological element?

Dr Ko Ferrigno answered: We have done this experiment in the sense that the same model peptide aptamers and samples have been used in several labs, who were using different detection technologies (microcantilevers;[1] electrical;[2] SPR[3]). Even better, in some cases, especially Estrela *et al.*, 2009, the same systems have been assayed using different techniques (*e.g.* QCM and electrical) in the same lab. In all cases, the overall performance is driven by the base sensor.

1 Shu *et al.*, 2008.
2 Estrela, 2009, 2010; Evans, 2008.
3 Johnson, 2008; Davis *et al.*, 2007, 2009.

Professor Vadgama asked: How is the scaffold protein attached to the aptamer? Is it an engineered fusion construct?

Dr Ko Ferrigno replied: Yes, these are engineered fusions. We have manipulated the gene encoding the scaffold protein so that we can insert oligonucleotides into specific positions. The oligonucleotides can encode either determined or random peptides, which are translated as part of the new fusion protein. The idea is that the three-dimensional structure of the scaffold protein will force the encoded and inserted peptides to adopt a specific conformation, leading to increased binding specificity and affinity with a target protein.[1]

1 For details see: Woodman *et al.*, 2005 and Hoffmann *et al.*, 2010.

Professor Vadgama asked: Where is the 'variable' peptide optimiser – in the tail, orientated away from the surface and freely flexible?

Dr Ko Ferrigno responded: Each peptide aptamer (comprising the scaffold protein and one or more inserted peptides that constitute the recognition/interaction surface – I like the term 'optimiser'!) are fused at their amino-terminus to a 51 amino acid polypeptide that is encoded in large part by the protein expression vector we use, pET30(a)+. This polypeptide includes a hexahistidine tag that we use for purification and immobilisation, as well as other functionalities, some of which we added ourselves and that were first described by Davis *et al.*[1] In the arrays, the whole protein is immobilised using functions in the 51 amino acid amino terminus, ensuring that the recognition moiety is oriented away from the surface and exposed for interaction with molecules in solution.

1 Davis *et al.*, 2007.

Professor Vadgama remarked: How extensive does the peptide sequence need to be in order to be selective?

Dr Ko Ferrigno answered: We don't know. In nature, as you know, selectivity can be achieved by very short peptides – glutathione, for example, is only 3 amino acids long. In our scaffold, we use short peptides (6 residues) in the shorter loop and peptides 12 amino acids long in the second loop. We also find functional peptide aptamers where concatamerisation of oligonucleotides leads to very long peptide inserts of 36 amino acids or more. I think the key is to have access to one or more scaffolds that afford the versatility required for the exploration of biological space.

Dr Brown asked: Do your peptide aptamers show any bias towards different classes of proteins and what is your success rate? Also, what strategy do you employ in cases where you are unable to obtain a peptide aptamer against a specific target?

Dr Ko Ferrigno responded: We have not performed enough screens to be able to say that some classes of targets are better than others, and nor have we screened multiple libraries against a given bait. We do know that some targets are not amenable to yeast two hybrid screening, but this has mostly been described by others. If we cannot identify a binder to a given target, we have several options. For example, the combinatorial nature of our scaffold means that we can screen one loop at a time, or combinations of loops, *etc.* If a screen fails in yeast two hybrid, we can turn to phage display or other technologies. We can also seek to change the bait, rather than adapting the library or the screening platform. Thus, if a screen against a full length protein fails, we can screen against peptides derived from that protein (or vice versa).

Professor Cooks asked: Is the success rate (80%) that you mention for the detection of particular compounds including those in cell lysates where 1 : 10^6 ratios of proteins of interest to others obtain?

Dr Ko Ferrigno responded: The quoted success rate of 80% applies exclusively to yeast two hybrid screens, not to cell lysates. Our 1 : 5 failures are probably more to do with problems expressing the target protein than with the size of our libraries, as we do not routinely codon optimise for expression in yeast, *etc.*

Professor Cooks asked: What is the specificity attributable to the aptamer itself rather than the read-out system? What type of selectivity can be achieved for analytes in complex matrices?

Dr Ko Ferrigno answered: All of the specificity in our detection systems comes from the peptide aptamer. Any labelled molecule recruited to the surface of the optical arrays would be detected, so it is crucial that the peptide aptamers be specific. The same is true for our label-free systems, where any molecule recruited would lead to a signal.

In terms of selectivity in complex matrices, we have shown specific detection of a target protein in lysates of yeast cells,[1] cultured human cells[2] and placenta.[2] In both yeast and human cell lysates, we have been able to show selectivity for the active over the inactive form of CDK2, which differ only in their relative post-translational modification and associated conformational changes. In unpublished work we have been able to show selective detection of a mutant protein kinase in proteins extracted from formalin-fixed, paraffin-embedded cancer biopsies.[3]

1 Davis *et al.*, 2009; Estrela *et al.*, 2009 and 2010; Evans *et al.*, 2008; Johnson *et al.*, 2008; Shu *et al.*, 2008.
2 See our paper: Q. Song, L. K. J. Stadler, J. Peng and P. Ko Ferrigno, *Faraday Discuss.*, 2011, **149**, DOI: 10.1039/c005376g.
3 Stadler, Yeh *et al.*, unpublished.

Professor Tsien asked: Granted that Stefin A is an inhibitor of the cathepsins, has the scaffold been tested for resistance to other types of proteases?

Dr Ko Ferrigno replied: These experiments are in progress. Preliminary data indicate that scannins are stable in cell lysates, and the purified protein shows resistance to trypsin. We are testing a range of other proteases.

Professor Guiseppi-Elie commented: Printing is inherently noisy by its nature; hence printing replicates is a step in the correct direction. However, did you statistically evaluate the need for anything other than duplicates: triplicate or quadruplicates, for example?

Dr Ko Ferrigno answered: We have not yet evaluated this.

Professor Guiseppi-Elie said: There is an implication that the recognition loop has no potential for interaction with the surface and hence its activity is not altered by immobilization. Is this your conclusion?

Dr Ko Ferrigno replied: Yes, so far all data indicate that chemisorption driven by the cysteine-thiol group onto the gold surface is dominant over physisorption, where the recognition loop would be expected to bind the surface at least a proportion of the time. We have explored this in published work.[1]

1 Davis *et al.*, 2007, and Johnson *et al.*, 2008.

Professor Tsien commented: How large and diverse a library is needed to produce hits reliably against arbitrary ligands?

Dr Ko Ferrigno answered: The smallest library we have screened by yeast two hybrid comprised 250 000 naive scannins, from which we obtained 3 hits, one of which was able to compete *in vivo* against natural partners of the target.[1] Using the microarray format described, 10 000 scannins are sufficient to obtain hits. This is consistent with work using arrayed antibodies.[2]

1 Chattopadhyay *et al.*, 2006.
2 For example: H. Mao, J. J. Graziano, T. M. A. Chase, C. A. Bentley, O. A. Bazirgan, N. P. Reddy, B. D. Song and V. V. Smider, *Nat. Biotechnol.*, 2010, **28**, 1195, DOI: 10.1038/nbt.1694.

Professor Cass opened the discussion of the paper by Professor Ikebukuro:† How dependent is your aptamer design on a detailed (atomic level) knowledge of the three-dimensional structure of the aptamer-target complex?

Dr Abe responded: Our design strategy is not dependent on 3D structural information. We have already reported the design of a capturable aptamer using mice prion protein-binding aptamer with predicted 2D structural information. In addition, we did not use 2D and 3D structural information for designing a capturable aptamer for thrombin in our paper in this conference although we have 3D structural information of a thrombin binding aptamer. We only assigned the regions which additional DNA hybridize with and examined some CaDNAs which have different lengths (Fig. 2b of our paper). Although 3D structural information is useful to design a capturable aptamer, we consider the most important information would be the regions which affect binding ability to maintain binding ability of aptamer.

Professor Beynon commented: Is it possible to use two selection criteria in screening for suitable aptamers – in other words, a suitable structure has to (a) bind analyte, and (b) reveal the capture oligonucleotide? Is such screening possible, as that would seem to enhance specificity.

Dr Abe responded: Yes, it is. We have already reported evolution of aptamer function such as inhibitory activity using a genetic algorithm. This method would be

† Professor Ikebukuro's paper was presented by Dr Abe, Tokyo University of Agriculture and Technology, Tokyo, Japan.

applied to improvement of any factor which can be experimentally evaluated. Therefore we can directly evolve a capturable aptamer sequence by evaluating two selection criteria.

Professor Lu commented: Since three-dimensional (3D) structural information is not available for most aptamers and since mfold can predict only double-stranded and single-stranded regions of the aptamer, I am wondering how to design the CaDNA so that it hybridizes only the single-stranded regions that are exposed, instead of single-stranded regions that are critical to forming a 3D binding pocket for the aptamer target. If one can find such a CaDNA to hybridize to the single-stranded regions that are exposed, is it possible for the hybridization to weaken the aptamer binding to the target?

Dr Abe replied: Additional sequences in a capturable aptamer can hybridize internal sequence of the aptamer and the CaDNA immobilized on the solid support. Therefore CaDNA is not complementary to aptamer sequence except additional sequence and it would not hybridize with a single-stranded region that is critical to form a 3D binding pocket for the aptamer target. We designed it in that way.

Decrease of binding ability caused by hybridization might occur. However, the capturable aptamer that we designed in this study kept binding ability comparable to intact thrombin binding aptamer. Since we do not yet know a general rule of how to design capturable aptamer without decrease of binding ability, trial and error is still required.

Professor Tsien asked: Can the aptamers discriminate between thrombin and prothrombin? The two are radically different in biological role, and prothrombin is much more abundant under normal nonactivated conditions.

Dr Abe replied: Although we do not have information about affinity against prothrombin of the 31-mer thrombin aptamer that we used in this study, the 15-mer thrombin aptamer named HD1 recognizes prothrombin with similar affinity to thrombin.[1] The dissociation constant of HD1 against thrombin is 34 nM and that against prothrombin is 86 nM. However, the 29-mer aptamer named HD22 that can recognize different epitopes from HD1 can discriminate thrombin and prothrombin.[1]

In addition, thrombin aptamers can inhibit thrombin activity and they are expected to be anti-coagulant agents and it has been reported that binding of aptamer to prothrombin enhances anti-coagulant activity. Therefore if we use it as an anti-coagulant agent, we do not have to discriminate thrombin and prothrombin.

1 C. A. Kretz, A. R. Stafford, J. C. Fredenburgh and J. I. Weitz, *J. Biol. Chem.*, 2006, **281**, 37477.

Professor Tsien asked: If the current aptamers do not discriminate between thrombin and prothrombin, why not incorporate positive selection for thrombin and negative selection for prothrombin into the SELEX evolution?

Dr Abe replied: The 29-mer thrombin aptamer can discriminate thrombin from prothrombin (see previous question). When we select a specific aptamer against a target molecule, negative selection would be a useful method. However, it sometimes requires skilful examination of conditions.

Professor Beynon opened a general discussion of the papers by Dr Ko Ferrigno and Professor Ikebukuro: It is worth remembering how little of protein 'space' has been explored by living processes. There are roughly 10^{400} different proteins of about 35 kDa (best seen in the context of there being 10^{80} atoms in the universe, and it may be that the whole of life has explored a few million or tens of millions of these possibilities. It does not seem unreasonable to conjecture that there are near-perfect protein or nucleic acid aptamers 'out there' waiting to be discovered. The challenge is being able to explore such large libraries, or direct the search optimally. A supplementary thought for Dr Ko Ferrigno: how many of your aptamer sequences on the stefin backbone have precise equivalents in the known protein world (found by BLASTing them against global databases)?

Similar arguments to Dr Abe: Is your thrombin aptamer a wholly unique DNA sequence?

Dr Ko Ferrigno replied: In my experience of looking at the sequences of peptide inserts, only once have we seen a BLAST match that may be relevant. The question of how proteins have evolved to use the proportion of protein space that they do is deeply fascinating. There are many experiments we would like to do to explore the relationship between our random libraries and 'current biological protein sequence space'. For example, if we spike a library with a known peptide interactor from a natural protein partner of the target, will it be the best hit out of a library of 1000, 10 000, 100 000, 1 000 000, *etc.* random competitors? If we take a natural peptide and make a single change to decrease its effectiveness, then create a library of random secondary mutations, how many ways are there to rescue binding efficiency? If we make a library of loops predicted to be encoded by the human genome, do we see more binders than from a library of random loops when screening against human targets than against, say, bacterial targets? On a more detailed level, to explore the evolution of an individual protein, if we graft a protein interaction loop from a human protein into our scaffold and observe a loss of binding, can the binding be rescued by changing the scaffold, or must the binding be rescued by changing the loop itself? Grant reviewers, take note!

Dr Abe replied: We cannot evaluate whole DNA sequences even if we use a 40-mer DNA random library (it requires about 27 kg of DNA). Therefore we consider affinity maturation following *in vitro* selection is important. We have already reported affinity maturation using genetic algorithms, which is an evolutional technology using a computer.[1] This method consists of two steps: computational sequence generation and evaluation of their sequence by *in vitro* assay. We can expect to expand selection space without *in vitro* experiment biases.

1 K. Ikebukuro, Y. Okumura, K. Sumikura and I. Karube, *Nucleic Acids Res.*, 2005, **33**(12), e108; N. Savory, K. Abe, K. Sode, K. Ikebukuro, *Biosens. Bioelectron.*, 2010, DOI: 10.1016/j.bios.2010.07.057.

Professor Vadgama said: We certainly need to have much wider screening of probes in order to move to new targets such as peptides and hormones. However, we don't have to go to exceptionally high numbers, after all the body does not make an open-ended number of antibodies; it relies on a finite repertoire.

Professor Cooks commented: Yesterday the Chair told us how impressed he was at the rate of development of genomics-based methodology and how this contrasted with the slow development of spectroscopy, specifically mass spectrometry, which he observed was a much slower method. This morning he has noted the limitations in specificity of genomics-based methods. Perhaps in his earlier comments he conflated chromatography with mass spectrometry: by itself mass spectrometry is extremely fast and separations are not necessary. It is not necessary to separate sheep from goats to be able to tell them from each other in a mixed herd.

Professor Beynon answered: Indeed, I was specifically comparing 'genomics' with 'proteomics' (for most practitioners, this largely means LC-MS/MS at present), and there is no doubt that the speed of a high resolution chromatographic separation is so low as to impair throughput. A typical MS-based proteomics workflow might do well to identify and crudely quantify 500 proteins or peptides per hour. The second factor is that of sheer sample complexity that is often streamed into the mass spectrometer. It would be rather difficult to separately quantify sheep and goats if 10 000 different types of mammals were streaming through a gate every minute! To continue the analogy, in many analytical challenges we are required to discriminate the equivalent of Merinos from Texels. The more subtle the difference, the longer we have to spend on discrimination. Genome and transcriptome profiling methods have one huge advantage, namely massive parallelisation, that an essential serial LC delivery cannot hope to match, even if we can achieve modest parallel MS methods (hundreds rather than tens of millions).

Professor Lu said: Regarding selectivity for the target among hundreds of competing molecules, it is extremely difficult, if not impossible, to achieve 100% selectivity and no binding of any competing molecules at all. Instead of perfect selectivity by a single sensor for its target, we should try to mimic how our nose or tongue work by assembling a collection of many sensors, each of which has good selectivity (but not necessarily perfect selectivity) toward its target and use the combined signals for more accurate detection.

Dr Davies responded: You seem to be describing what used to be called “artificial noses”. They had arrays of sensors which produced a 3D response pattern for the given volatile in a mixture of volatiles. A training set of vapours and some powerful software was used to initially set up the sensor. They do not appear to have survived the real world. Aromascan, based in Crewe, UK, in the 1990s, was one of the largest companies and sold to the perfume and motor industry – control over the smell of the leather was seen as important in selling cars. But even with good funding it was unable to succeed. One problem, I believe, was that when the sensor head failed the new head had to be trained as the old training set was not transferable.

Professor Cooks replied: One might argue that the nose and tongue work rather poorly even in the minimally demanding tasks they are set. If we are to use animal analogs then the gypsy moth would be a better guide. Unfortunately for the argument, these moths detect a single compound (7*R*,8*S*)-7,8-epoxy-2-methyloctadecane, with high molecular specificity.

Professor Lu replied: You are certainly correct that many “artificial noses or tongues” have not survived the real world because they are not robust enough, like our own nose and tongue, that can be used numerous times. Other improvements are also needed, such as improving selectivity of each receptor/sensor toward its targets and increasing the number of such receptor/sensors. These technical issues do not necessarily invalidate the approach; it is just not ready for real world applications.

Professor Beynon remarked: Following on from my previous comment about combinatorial space, it is clear that we cannot explore all of it, so we need to be content with a ‘good enough’ exit solution. What are the criteria for a ‘good enough’ solution, and how do we know when we reach it?

Dr Ko Ferrigno answered: In the context of healthcare, “good enough” means “better than” the existing test, if there is one. “Better” can mean faster, cheaper, more sensitive and/or more specific. In the context of combinatorial space, then, we need to identify reagents that enable one or more of these improvements, which will usually mean identifying a recombinant alternative to a biological agent (*e.g.* antibody) with binding affinities and specificities that match or surpass those of existing probes. Our experience is that you do not need huge libraries of combinatorial recombinant probes for this to be achievable if the screening of the library is performed in an environment that matches the end application. In this regard, the introduction of yeast two hybrid screening of peptide aptamer libraries by Colas *et al.*[1] was inspired, as it enabled the identification of probes to a fully folded target protein expressed at low levels in the context of a eukaryotic cell, against a background of thousands or even millions of copies of thousands of competing biological molecules. We have performed successful screens of libraries comprising as few as 250 000 combinatorial probes[2] and generally aim to screen libraries of complexity between 1 and 10 million.

1 Colas *et al.*, *Nature*, 1996.
2 Wickramasinghe, *et al.*, *BMC Cell Biol.*, 2010.

Professor Vadgama said: You need to define what final probe sensitivity/selectivity you want during the search of libraries. There is no need for perfection for clinical assays. Current *in vivo* monitoring has a finite capability and is characteristically not as reliable as *in vitro* assay, yet clinical utility is unquestionable.

Dr Ko Ferrigno answered: This is unquestionably true. Our goal will be to enable clinical assays that are needed but for which there is no solution, and to improve assays that are available but may be too slow, too costly, insufficiently specific or insufficiently robust. I also believe that the clinical paradigm is shifting. It will soon no longer be acceptable to perform one or a few tests, wait hours to days for a negative result, then need to perform a series of back-up tests. Multiplexed assays that address multiple clinical possibilities will rapidly become the diagnostic and prognostic tool of choice. This is particularly obvious with molecularly targeted therapies in cancer, where prescription of, for example, a tyrosine kinase inhibitor can only be justified if the protein kinase is expressed, has not been mutated during carcinogenesis, and is not rendered irrelevant by mutations in other genes in the same or in redundant pathways. Patients therefore need to be tested for multiple genetic events before a single treatment can be prescribed, saving patients from unnecessary or ineffective treatments and the heath service from unnecessary calls against their drugs budgets.

Professor Cass mentioned: I'd like to sound a counter-note to the view that multiplex measurements are a better approach than defined single analyte measurements. The use of 'electronic noses' and infra-red spectroscopy have shown the limitations of this approach in complex and variable backgrounds.

Dr Ko Ferrigno answered: My view is that multiplex measurements are fast becoming a clinical necessity, and that we need technology to deliver them. The limitations of electronic noses and infra-red spectroscopy do not appear to be directly relevant to other technologies, and the data delivered by our collaborators (notably Walti and Davies in Leeds, Estrela and Migliorato in Cambridge and Bath, and Davis in Oxford) certainly appear to promise to deliver the capability needed. It is very clear that the solution cannot be delivered by one lab alone, but requires a combination of excellence in surface chemistry, (bio-)sensing, signal processing and probe technology.

Professor Beynon said: I want to open the discussion a little to address detection technology. Dr Ko Ferrigno has intimated that the optimal detection modality might be electrochemical, rather than optical. I thought this would be a good point to open out the discussion about the best detection modalities.

Dr Campbell replied: I think I take the opposite stance on this matter (of electrochemistry *vs.* optics).

Where you have a large number of probe molecules the challenges of depositing them are too great using multiplexed electrodes (and too expensive). I think that for large numbers a disposable and cheap substrate, without any built-in intelligence (like electrodes), is preferable. However, when you have a small number of probe molecules in a point of care test, then electrochemistry becomes more competitive.

Dr Ko Ferrigno responded: The problems of multiplexing are probably the greatest hurdle to the implementation of our technology. The good news is that, generally speaking, clinicians are happy with a low level of multiplexing (10s) and you are quite right that electrochemistry is probably the best solution for clinical applications on this scale. However, highly multiplexed and high density solutions exist, whether mechanical (dip-pen nanolithography, ink jet printing) or electrochemical, and I do not think that this will be a real problem even in the short to medium term.

Professor Schuhmann remarked: There is an ongoing discussion if optical or electrochemical detection schemes are to be preferred for point of care (POC) diagnostics. There is no doubt that for large libraries of compounds which have to be screened optical (mainly fluorescence-based) detection schemes have advantages due to their parallel readout. However, in POC devices electrochemistry may compete successfully with optical techniques due to the low device cost and the possibility to design label-free detection schemes.

Dr Ko Ferrigno answered: I agree with this statement and have nothing to add.

Professor Graham asked: Is there any advantage to miniaturising the peptide arrays in terms of enhancing detection?

Dr Ko Ferrigno responded: I am not sure whether miniaturising would enhance detection. From a healthcare perspective, it would enable analysis of samples that are limiting, such as blood from a neonate, brain tissue at the margins of a tumour, *etc.* From a technical perspective, smaller arrays may be easier to subject to highly efficient mixing that could be used to reduce non-specific binding. One might also consider that smaller arrays, using less probe, may lead to lower limits of detection, but would have a decreased dynamic range at higher analyte concentrations. Finally, from a commercial perspective, smaller arrays means less consumables, *i.e.* cheaper manufacturing costs.

Dr Davies commented: Your slide showing an array of *ca.* 17 000 optically addressable aptamers tends to argue against your choice of an electrochemical array platform as the most likely to succeed. Such optical arrays have been around for years, and usually require a large piece of equipment to interrogate them – there may be an inverse relationship between the spot and equipment sizes. However, I would, like you, place my money on electrochemistry: it gets better as the electrodes get smaller and the reading equipment does not need to get larger, making for a lower cost, and more portable system. Furthermore, this system does not suffer when the sample is not transparent or fluoresces.

Dr Ko Ferrigno replied: I should clarify that my response (electrochemistry is better) was in the specific context of healthcare applications, where, as you say, the devices can be smaller and cheaper. There is a place for optical detection, such as that shown here in research. Given that our optical detection platform uses legacy equipment (printers and scanners) from DNA microarrays, we actually don't see the need for large pieces of equipment to be a problem, assuming you were referring to these scanners.

Professor Gooding said: In response to the suggestion that electrochemistry may be best for high density arrays, I would like to add to the statements made previously as to why this may not be the case. Additional to the points raised, the fact that each electrode in an array must be independently addressable means that leads and bonding pads may take up too much valuable real estate on the chip. Furthermore, in relation to affinity assays, we currently do not have good surface chemistry to resist fouling of electrodes by proteins that does not, itself, passivate the electrode.

Dr Ko Ferrigno responded: The semi-conductor industry has essentially solved the real estate issue. Mobile phone screens are one example of a device comprising hundreds of thousands of individually addressed electrodes. Work in Christoph Walti's laboratory at the University of Leeds is showing that devices suitable for biological detection can be made.[1] With regard to surface chemistry, we have co-published several papers with collaborators who have used self-assembling monolayers to resist non-specific binding of proteins to electrode surfaces. In 2008 we used dual polarisation interferometry to characterise a simple maleimide layer on gold,[2] similar to the surfaces we use for electro-chemical assays, and found that the layer formed was susceptible to non-specific binding of proteins from cell lysates, in that 0.3 ng of mass was loaded for each square millimetre of surface. In contrast, the binding of a specific protein present in a matched cell lysate led to an increase in mass of 1.2 ng mm^{-2}, which indicates that our detection layers bind very little bulk protein from cell lysates. We have also used PEG to protect electrodes.[1] Careful surface chemistry has also been described by Davis *et al.*[3] Electronic measurements using peptide aptamers on protected gold surfaces have been described by Estrela *et al.*[4] In sum, our collaborators have shown that electrical/electrochemical measurements can be used to detect specific proteins even in complex biological mixtures, which in our experiments has up until now usually involved lysates from yeast cells expressing the human protein to be detected. These lysates are complex (the complexity of the yeast genome is roughly 1/5 the complexity of the human genome) and competing yeast proteins are present at concentrations several orders of magnitude greater than the target human protein.

1 For example: D. Evans, S. Johnson, S. Laurenson, A. G. Davies, P. Ko Ferrigno and C. Wälti, *J. Biol.*, 2008, **7**, 3.
2 Johnson *et al.*, *Anal. Chem.*, 2008
3 Davis *et al.*, *Anal. Chem.*, 2007, and *Anal. Chem.*, 2009.
4 Estrela *et al.*, 2009 and 2010.

Professor Guiseppi-Elie said: Thanks to Michael Faraday, electrochemistry does allow us to count single electrons.

Professor Girkin said: It should be remembered that the final device format may be set by the production methods available. The large and expensive infrastructure in the electronics industry for making Si chips may direct which method actually reaches the market.

Dr Ko Ferrigno replied: This is true, and the good news is that there are many fabrication facilities whose technology is outdated by the standards of the semi-conductor industry that would be able to produce chips with dimensions and pitches appropriate for clinical devices. Thus, the high costs of establishing a fabrication infrastructure will not be a limitation to the clinical implementation of electrochemical detection.

Professor Schuhmann commented: For optical detection schemes relatively large molecules (fluorophores or even green fluorescent protein) have to be attached to the compound under investigation. This may dramatically change the properties of the compound and even completely mask events, for example binding events. Hence, the presence of the rather large label has to be considered. In contrast, modification of surfaces can be detected with electrochemical impedance spectroscopy without the need for attaching large molecules.

Dr Ko Ferrigno responded: This is the driving force for our work on label-free detection. If we are to understand life at the molecular level, we need to be able to explore the contribution of protein conformation to complexity, and the conformational dynamics of protein surfaces in cells and organisms. In my view, post-translational modifications serve to stabilise or alter protein interaction surfaces (including the active sites of enzymes). The addition of a large hydrophobic dye molecule when studying a protein will almost inevitably alter its conformational dynamics, creating or destroying the very information we need to access.

Dr Kendall remarked: Don't rule out optical technologies, although fluorescence usually requires the use of dyes, markers or probes, other optical technologies such as vibrational spectroscopy can provide multiplexed biochemical information without staining or the addition of any markers.

Professor Cass said: Comparing different measurement methods (electrochemical, optical, mass based) on the basis of what can be done in the laboratory, rather than what provides a suitable route to scale-up and manufacture, could be misleading.

Professor Beynon commented: Nobody has mentioned, for example, hologram-based detectors. Could these be superior, inasmuch as they require no detection instrumentation sitting behind the test?

Dr Davies responded: They exist, and are cheap to make and can display the "result" either as a colour or even as a symbol that appears above a threshold. However, as far as I know, they have not had success in making good quality measurements in biological matrices, but can measure ethanol in breath, water in organic solvents and there are some interesting ideas on glucose measurements in contact lenses bearing a soft hologram.

Professor Beynon commented: For the final part of the discussion, I wanted to create an opportunity for some 'crystal ball gazing', and incite speculation of the nature of tests and devices that would find their way into routine point of care use. How will the terrific ideas we're hearing here be translated into cheap, disposable, non-invasive, and useable tests? When will there be an iPhone app for this?

Dr Ko Ferrigno answered: Give us 10 years.

Professor Girkin said: Large sensor multinational companies (who are not traditionally in the medical field except, perhaps, for pressure sensors for gas handling, for example) are looking at the home monitoring of certain key parameters of body function that relate to ill health. Long term changes could then be seen leading to pro-active early treatment. Other companies (*e.g.* Unilever) are looking at "function foods" which could be tailored to individuals for significantly improved quality of life. This could mean again home monitoring of key parameters.

Dr Davies responded: I don't share the "neutraceutical" dream of the multinational food companies. I doubt that Mum, who still does most of the food preparation, will have space in her kitchen for a variety of person-tailored foodstuffs. When she makes the sandwiches, she just wants one margarine, not one per person, one type of ham, *etc.* I think the same goes for clinical tests, they will not succeed unless the user has a real or strongly perceived need for it. Such a person may have suffered a stroke, heart attack, or have an implant, or is on therapy that would benefit from regular monitoring. An at home measurement, provided it is easy to do, and which has proven clinical benefit and sanctioned by their clinician would be convenient. However, for both home use and POC tests there are other barriers to widespread adoption, among these are cost (who pays?) and quality control. As for science fiction – we've already seen that, Dr McCoy's medical tricorder and detachable scanner which worked on the fully clothed body. Our "wet tests" are way behind that, and even the non-invasive near-infrared and body gas measurements have a way to go to achieve credibility.

Professor Cooks replied: The word sensor usually means a single analyte but the type of device that one can imagine if spectroscopy is included can be simple and yet multiplexed *via* the spectroscopic measurement. One can for example rapidly measure a large number of components in whole blood using a simple "sensor" system with a disposable front-end.

Professor Guiseppi-Elie commented: There is every indication that science fiction has inspired science fact, an example is the development of fully integrated implantable biochips that sense and report telemetrically. There are challenges, of course: power sources and power management among them.

Dr Ko Ferrigno answered: In my opinion, all of the components of a Dr McCoy device are in place. The hand-held scanner he uses could be both an RFID-type reader and a source of inductive power for an implanted chip that could be measuring real-time or device-lifetime levels of analytes of interest. We are now limited not by the technology, but by our understanding of biology and molecular pathology and by the availability of probes to measure the biomarkers of interest.

Dr Campbell asked: Since it seems unlikely that a mass spectrometer (as we are used to seeing them in chemistry labs) would be suitable for point of care diagnostics, how much scope is there for miniaturisation?

Dr Kendall mentioned: When considering point of care testing we need to keep the wider adoption issues in mind, which relate to who it is designed for, the setting and how they will access the technology. Development of good technology is not sufficient if the target group do not access or use it; perhaps, an issue with testing for sexually transmitted disease.

Professor Gooding commented: Another factor that seems to play a big role in adoption of technology is how the information that the device gives the end user is packaged. For example, the Roche xCELLigence system is an impedance method of determining the response of mammalian cells to certain stimuli such as drugs and has been well adopted by biologists. Part of the reason seems to be that, not only is the instrument compatible with normal cell biology, the electrodes being housed in microtitre plates for example, but the output is packaged in such a way that the information the instrument provides appears to be easy to understand. So there is no mention of impedance, resistance, capacitance, *etc.*, but these quantities are packaged as a single number the instrument provides called the cell index.

Professor Cooks said: I am probably the only person here who has actually been involved in attempting to incorporate analytical instrumentation into a cell phone. This was a mass spectrometer/Blackberry experiment of some 7 years ago which involved working with Sandia in developing an array of a million micron-sized ion traps. Simulations show that each trap will coulombic empty to the point of "an ion trap for every ion". The experiment ultimately failed because one needs physical size to perform some of the operations of mass spectrometry. A compromise 3 kg tandem mass spectrometer (total system) was subsequently developed, which does function as a point of care device.

Professor Vadgama mentioned: We need to have continuous monitoring to help predict clinically significant or dangerous events. We really need biomolecule binding reversibility to do such continuous monitoring; hormones and other complex targets have dynamic profiles *in vivo*, yet we can only attempt these as single discrete assays.

Dr Mount asked: How suitable is this technology for developing aptamers not for the highest sensitivity but for thermodynamically reversible binding with insensitivity to clinical sample variation to enable continuous monitoring applications?

Dr Ko Ferrigno answered: In principle the technology is eminently suitable and we are actively exploring this potential.

Dr Davies said: For continuous *in vivo* or *ex vivo* monitoring, Professor Vadgama would like a sensor surface whose occupancy with analyte quickly responds to changing analyte levels. To do this we may have to cheat, rather than beat, the thermodynamics that controls the interactions between antigen and antibody (or apatamer). The slow off-rates, characteristic of high affinity interactions and needed for sensitive assays are the problem. So why not add an external push, in the form of a small amount of electrochemically generated acid to take the pH to around 3 to 4. The BiaCore instrument uses an acid regeneration step, and this works well. Professor Vadgama may be horrified at this, but before the acid leaves the measurement cell to re-enter the body we would neutralise it. There may be other methods to encourage the complex to dissociate, *e.g.* strong and divergent electric fields (dielectrophoretic force) or localised heating, that may distort the binding site. Forced dissociation may be an area of research worth study if Professor Vadgama's well-known goal of continuous *in vivo* monitoring is itself realised to be advantageous, which seems to be so for human blood glucose.

Professor Vadgama responded: This is an interesting suggestion, *i.e.* to change the microenvironment that led to the high-affinity binding in the first place, and to so alter it that the binding is reversed. I fully agree that this has been realised for antibodies, which can be stable to an acidified environment. It is worth considering such an approach to achieve quasi-continuous tracking of dynamic change of a target analyte. Two key issues need to be taken on board, however: the *in vivo* safety and compatibility of any reversing step, *e.g.* a high electric field or an extreme electrochemical polarizing voltage, and also that whilst affinity may be high with an antibody, there can be a need for a prolonged incubation – too slow for real-time monitoring. However, it may be possible to circumvent the latter by tracking only initial binding and so obviating the need for equilibrium binding. To attempt microenvironmental manipulation for binding reversal, however, is a worthwhile ambition. Also, it would be rewarding to study the fundamentals of cell membrane receptor binding to targets; these, admittedly labile, proteins remarkably combine high affinity with binding reversibility.

Dr Mount commented: This approach does not cheat thermodynamics in my opinion, but harnesses changes in the thermodynamics as well as the kinetics with external stimulus to achieve this dissociation. One issue is that relatively rapid binding reversal is typically not part of antibody or other binding partner selection strategies at present, which typically identify the partner with the highest selectivity and affinity with little regard to kinetics. I agree that cell membrane receptors may give insights into how nature achieves this without apparently relatively harsh external perturbation.

Professor Cass remarked: Lowering or raising the pH would induce dissociation although the buffer capacity of physiological fluids is high, precisely to avoid this, and so may be difficult to achieve in practice.

Professor Vadgama said: Though high affinity and sensitivity have a price, *i.e.* low target dissociation rates, nature uses cell membrane receptors that combine high affinity and ready binding reversibility, so we should model such systems and design similar structures.

Professor Graham opened the discussion of the paper by Dr Dhaliwal: Were there any issues with dilution of the microspheres when using the macrophages as a delivery mechanism?

Dr Dhaliwal answered: No real issues were encountered. The primary macrophages we used ate the microspheres avidly.

Professor Graham asked: Have you tried anything other than the microspheres, and what imaging systems did you use? Also, were there an issues with depth penetration?

Dr Dhaliwal responded: Our paper describes alternative macrophage sensors, namely the cell delivery peptoid and MRI contrast reagents. The imaging systems we used were a 7T Varian MRI, Xenogen Spectrum, Kodak FX PRO and we have also used the Visen FMT 2500.

Dr Ko Ferrigno enquired: How, and how quickly, do the polystyrene microspheres clear?

Dr Dhaliwal replied: They are likely to be present for weeks to months. The fate and clearance remains to be fully investigated. There have been no toxicity issues in the preclinical models tested.

Dr Mount commented:Protracted imaging of fluorescently labelled microspheres *in vivo* is said to be limited by the contribution of cell free microspheres to the signal. We have established that time-resolved fluorescence anisotropy and fluorescence lifetime imaging can be used to distinguish between populations of fluorophores in different environments for clinically relevant measurements.[1] Could such techniques be used to distinguish between cell-free and cell-bound microspheres and enable protracted imaging?

1 For example: G. Giraud, H. Schulze, T. T. Bachmann, C. J. Campbell, A. R. Mount, P. Ghazal, M. R. Khondoker, S. W. J. Ember, I. Ciani, C. Tlili, A. J. Walton, J. G. Terry and J. Crain, *Chem. Phys. Lett.*, 2010, **484**(4–6), 309–314; G. Giraud, H. Schulze, T. T. Bachmann, C. J. Campbell, A. R. Mount, P. Ghazal, M. R. Khondoker, A. J. Ross, S. W. Ember, I. Ciani, C. Tlili, A. J. Walton, J. G. Terry and J. Crain, *Int. J. Mol. Sci.*, 2009, **10**(4), 1930–1941.

Dr Dhaliwal replied: Yes, this would be a very interesting way to potentially delineate if the microspheres are still within the cell or have been released.

Professor Bradley responded: Practically, the microspheres are typically delivered into the cells, and excess microspheres removed prior to application. This means that the microspheres should remain inside cells until the cells die. The timeframe will, of course, depend on the cell type. Thus if the cells are those that will liberate their cargos and rapidly die (such as monocytes) then indeed time-resolved fluorescence anisotropy and fluorescence lifetime imaging could be useful in distinguishing between "free" and internalised microspheres. In other cases (cancer cells) then the particles might be expected to remain for some considerable period of time – although one could also imagine a therapeutic agent that kills cancer cells specifically could be imaged using the fluorescent techniques you describe.

Professor Vadgama remarked: There would be problems with regulation (FDA) if polystyrene is used as it is not an appropriate material for implantation. It would be better to invest in an established degradable polymer, *e.g.* poly-L-lactic acid. However, perfluoro compound emulsions with short *in vivo* life-times would be even better; they are used already and can be readily eliminated from the body.

Dr Dhaliwal replied: Yes, I agree. The ultimate degradation fate and persistence of the polystyrene micropheres in the host is not known so may have to be designated as a device to pass through FDA. The perfluorocarbons are a much faster way into the clinic.

Dr Brown commented: Could you comment in detail about the optimum sizes of the microspheres you use for tracking different cell types?

Dr Dhaliwal replied: For macrophages detailed in this report, we used 500 micron microspheres. These were considered optimal. For other cells, we have found it is necessary to optimise microsphere size for each individual cell type to prevent under or overloading.

Professor Bradley opened the discussion of the paper by Professor Girkin: There are issues with regard to contamination between samples. Is there any risk of sample-to-sample transfer that could then lead to false positives following PCR.

Professor Girkin responded: A new sample tube is used for each sample taken along with standard cross-contamination precautions (*i.e.* single use pipette tips, *etc.*). There is a risk that the glass tube could be broken when inserted into the sample holder (as with any instrument using such tubes). In this case the electronics and optics are all shielded and any spilled sample could be removed using conventional cleansing agents. Even then it would be hard for any contamination with another sample as it will be housed in its own tube. Thus, in summary, as long as the standard precautions are taken the risk is minimal.

Professor Cass asked: Companies such as DNAelectronics and IonTorrent are using CMOS to build the sensing elements as well as the instruments. How would these compare in a point of care device with the photonic devices?

Professor Girkin answered: I am not sure I could comment directly in terms of sensitivity which is very often the critical aspect in all such systems (the specificity normally being excellent). One advantage that we have with the photonic device is that the only disposable, apart from the chemicals, is the low-cost glass micro-test-tube, which also limits any risk of cross-contamination (see previous question). I believe that with these CMOS devices the entire CMOS chip must be removed for each test.

Dr Ko Ferrigno said: How many SNPs can be analysed at one time?

Professor Girkin replied: The system was specifically optimised for a single SNP, at low cost. By adding extra detectors up to four could be detected without significant changes to the methodology. However, the preferred option is to consider several such system running in parallel.

Dr Davies asked: You placed the saliva and PCR reaction mixture, after centrifugation – which I presume removes air bubbles – into the capillary in your PCR instrument. Did you encounter any problems with the saliva sample diluted 12.5-fold in the PCR reagent?

You had a final concentration of 0.1% Triton X-100 in the reaction mixture. I would expect this to have lysed the cells, and so save 13 min of high power use, which is a large part of your amplification energy budget. This might be important if you were to make, for example, a battery-powered instrument.

Did you attempt to dry the PCR mixture, perhaps from a dilute sugar solution, inside the capillary and add neat saliva to it? This would simplify the process and would be an obvious next step. If not, do you propose to simplify the test procedure?

Professor Girkin replied: We did have reduced success with the PCR process when the saliva was diluted greater than about 15-fold (as stated in the paper) but the 12.5-fold used did not create any problems.

We had not looked at power consumption directly in the design as the aim was that the system would always work on a desktop (*i.e.* with power around) but your thoughts about further reduction of power consumption would make sense and we will look at this if we take the project further. Currently this was a PhD student's work and we are speaking to two bodies about a continuation, but at present there is no further work taking place.

Professor Cass asked: Our experience of saliva in microfluidic channels it that it behaves much more like water with respect to back pressures in hydrostatically driven flow than it does in bulk. This may be due to a combination of the low channel widths and flow thinning.

Professor Girkin answered: I would echo these comments, especially when the saliva has been diluted slightly. I believe that saliva is a very complex fluid (without doubt non-Newtonian) and thus the local conditions may also play an important role. Probably an area of further research.

Professor Nakano opened the discussion of the paper by Professor Lu: Under particular biological environments, most metal ions, including biologically active small molecules, exist as certain types of chemically stabilized species such as metal chelates. So any functional DNA molecules are necessary required to first decompose them then bind to form the chemical species intended. In this regard, the molecular design of DNAzyme needs to offer sufficient stabilizing energy to every target ion or molecule to compensate for the decomposition energy of their naturally occurring forms. So, do you think this is always possible or not?

Professor Lu answered: This is an excellent point that all sensor designs need to consider, as most targets are not alone or "bare"; even metal ions in the absence of biological fluid but in solution alone have water around them, *i.e.*, the so-called metal aqua species. Assuming kinetic inertness is not a major issue, the sensor (including functional DNA sensors) is in equilibrium with other biomolecules to compete with the targets such as metal ions. As long as the affinity of the functional nucleic acids for the target metal ion is stronger than those of other molecules in the biological environment, there is a good chance that the functional DNA sensors can compete and sense the metal ion. That's why we have emphasized the importance of tunable affinity or dynamic range for the sensors; if the sensor has a weaker affinity, it will not be able to compete and sense metal ions, while if the sensor has too strong an affinity it may perturb the equilibrium drastically. Therefore, we have developed a general methodology to fine-tune the affinity or dynamic range of the sensor.[1]

1 See: J. Liu and Y. Lu, A Colorimetric Lead Biosensor Using DNAzyme-Directed Assembly of Gold Nanoparticles, *J. Am. Chem. Soc.*, 2003, **125**, 6642–6643; Y. Xiang, A. Tong and Y. Lu, Abasic Site-Containing DNAzyme and Aptamer for Label-Free Fluorescent Detection of Pb^{2+} and Adenosine with High Sensitivity, Selectivity, and Tunable Dynamic Range, *J. Am. Chem. Soc.*, 2009, **131**, 15352–15357.

Dr Mount asked: What limitations are there in the ability of this approach to detect selectively a wide range of ions and multiplexed detection in clinically relevant samples?

Professor Lu responded: We are taking two approaches to multiplexed detection of a wide range of ions. First, because the catalytic beacon method is very versatile in that it can use any fluorophore/quencher pairs, we are developing "one-pot" multiplex sensing by labeling different DNAzymes in the same solution with difference fluorophore/quencher pairs that fluoresce at difference wavelengths, and the ions can be identified and quantified by each unique wavelength of emission. A fundamental limitation of this approach is that most organic fluorophores have broad absorption and emission wavelengths that could overlap with each other and in practice the number of distinct fluorophore/quencher pairs that do not interference with each other is very limited, thus restricting the number of ions we can detect simultaneously. We have overcome this limitation somewhat by using quantum dots that have much narrower emission.[1] To allow detection of an even wider range of ions, we are developing a second approach of using functional microarray, similar to the DNA microarray for gene detection but with the goal of detecting metal ions or other types of molecules. I cannot think of any fundamental limitations of this approach; the limitations are more technical, such as how to keep the functional DNA active on the microarray surface. We have made some progress in overcoming this technical limitation.[2]

1 J. Liu, J. H. Lee and Y. Lu, Quantum Dot Encoding of Aptamer-Linked Nanostructures for One Pot Simultaneous Detection of Multiple Analytes, *Anal. Chem.*, 2007, **79**, 4120–4125.
2 D. P. Wernette, C. Mead, P. W. Bohn and Y. Lu, Surface Immobilization of Catalytic Beacons Based on Ratiometric Fluorescent DNAzyme Sensors – A Systematic Study, *Langmuir*, 2007, **23**, 9513–9521.

Dr Mount commented: The extension of this technology to multiple ion detection requires not only specific binding for each detector but also the ability to detect ions in different environments and at clinically relevant concentrations. How is sensor selectivity affected by differences in ionic speciation and relative ion concentrations in clinical samples?

Professor Lu replied: We are just at the beginning of answering this question in clinical samples, as most of our work so far has focused on detection in environmental samples, which share similar issues of different ion speciation and concentration ranges. Our experience told us that the ion speciation and concentration range in environmental samples is even broader than those of clinical samples, as the environmental sample can come from very strange places, such as different waste waters, and under extreme non-physiological conditions. Our strategy in addressing this issue is to customize the *in vitro* selection conditions toward the particular applications. In another words, we are developing multiplex sensing of not only different metal ions, but also different speciation states of the same metal ions. In addition, we have demonstrated a general methodology of fine-tuning the dynamic range of the DNAzyme sensors.[1] Therefore we can add another dimension in the multiplex sensing microarray responsive to different concentration ranges.

1 See: J. Liu and Y. Lu, A Colorimetric Lead Biosensor Using DNAzyme-Directed Assembly of Gold Nanoparticles, *J. Am. Chem. Soc.*, 2003, **125**, 6642–6643; Y. Xiang, A. Tong and Y. Lu, Abasic Site-Containing DNAzyme and Aptamer for Label-Free Fluorescent Detection of Pb^{2+} and Adenosine with High Sensitivity, Selectivity, and Tunable Dynamic Range, *J. Am. Chem. Soc.*, 2009, **131**, 15352–15357.

Dr Campbell said: It seems that, compared to proteomics, there are fewer potential molecules (*i.e.* other metal ions) that your aptamers could react with. How many distinct metal ion species are there in the human body and how many do you negatively select against?

Professor Lu replied: With 4–5 natural nucleotides instead ~20 natural amino acids as building blocks, functional nucleic acids certainly have fewer functional groups than proteins and thus in principle there are fewer potential molecules that they can react with. However, the *in vitro* selection is a powerful method that allows us to obtain functional nucleic acids that can bind a wide variety of molecules that we have set out to do. Experimentally we are only at the beginning of exploring the potential functional nucleic acids in binding other molecules and I do not think the technology has reached its full potential. Furthermore, some work has already been published to introduce new functional groups, including those found in proteins such as imidazole or thiol groups, into functional nucleic acids. These hybrid molecules can certainly match proteins in their potentials to bind other molecules.

There are numerous metal ion species in human bodies, including alkaline earth metal ions such as magnesium and calcium and first row transition metal ions such as iron and copper. Instead of doing negative selection against all metal ions species found in human body, we have normally carried out a few rounds of positive selections without negative selection, and then paused to perform a survey of metal ion selectivity against such metal ions as those mentioned above, focusing on those metal ions that most likely resemble the target metal ions. If we found competitive binding from those other metal ions, we would apply negative selection using those metal ions to get rid of DNA populations that bind those competing metal ions. One benefit of *in vitro* selection is that the selection condition can be customized to the desired conditions. For example, if the above method of negative selection fails, we could use human serum as a negative selection mixture.

Professor Vadgama said: For us in medicine a high priority would be Cu, Zn and Al ions as they are difficult to assay, yet needed routinely, so use up a lot of skilled analytical resource as trace metals. The ions are just various size spheres with hydration shells, so how do you achieve analytical differentiation?

Professor Lu responded: Thanks for your comment and suggestions of important metal ions to pursue. We have already reported DNAzyme sensors for Cu[1] and we are working on developing a sensor for Zn. We will look into Al sensors in the future. You are right in that, unlike organic molecules with many different sizes and shapes, ions are mostly spherical with hydration shells that differ slightly in diameter, and therefore it has been much more challenging to use a rational design method to differentiate one ion with one size sphere from another one with a slightly different size sphere. In principle, one can design sensors with size selection, *i.e.*, smaller ions fit perfectly in the sensor while larger ions do not. However, the size difference is often too subtle to allow such a strategy to work well. In some cases, two metal ions, such as Zn^{2+} and Co^{2+}, have almost the same ionic radius, and it is even harder to differentiate the two. Fortunately, inorganic coordination chemistry has accumulated a vast amount of knowledge about other properties that may be different even though the sizes of the spheres are similar. These differences include slightly different preferences for coordinating ligands or geometry of the metal-binding site. Even armed with those knowledge, it is still difficult to rationally design metal ion sensors. That's why I think *in vitro* selection is a powerful approach in that we can obtain DNAzyme from a large DNA library with many sequences that can interact with specific metal ions and, through negative selection, do not interact other competing metal ions. Previous work has shown this strategy works much better than rational design in giving DNAzymes that can interact with a much broader variety of metal ions. Now we have obtained many of these DNAzymes, we are in the process of using biophysical tools to understand why certain *in vitro* selection DNAzymes binds its target metal ion selectivity. We hope we can provide concrete structural information to answer your question in the near future.

1 See: J. Liu and Y. Lu, A DNAzyme Catalytic Beacon Sensor for Paramagnetic Cu^{2+} Ions in Aqueous Solution with High Sensitivity and Selectivity, *J. Am. Chem. Soc.*, 2007, **129**, 9838–9839.

Professor Bradley asked: The DNA is attached by a thioether bond. These are known to react with Hg(II). Would not Hg(II) therefore react with the thioether and cleave off the DNA, distorting the function of the sensor?

Professor Lu responded: The DNA is attached to gold nanoparticles through one or more thiol-gold bonds. You are right that this kind of thiol-gold bond is known to react with other thiols in biological fluids or in environmental samples. Fortunately, it has been found by Mirkin's group that even at elevated levels of biological thiols such as glutathione (10 mM), >60% of the oligonucleotides remain bound to the gold particle after 48 h.[1] Because the sensor itself is stored in the absence of those other thiols and the sensing operations such as dipstick tests can be accomplished in minutes upon exposure to the test samples, such a complication has not been shown to be an issue.

1 See: N. L. Rosi, D. A. Giljohann, C. S. Thaxton, A. K. R. Lytton-Jean, M. S. Han and C. A. Mirkin, Oligonucleotide-modified gold nanoparticles for intracellular gene regulation, *Science*, 2006, **312**, 1027.

Professor Gooding asked: In the examples you have shown from your work the metals these DNAzymes bind are all soft metals which is consistent with nitrogen atoms on the DNA bases binding to the metals. Are there examples where this technology could be used for selective binding of hard metals?

Professor Lu answered: Yes, there are examples where this technology could be used for selective binding of hard metal ions. A prominent example is the 10–23 DNAzyme selected by Gerald Joyce's group that binds Mg^{2+}.

Professor Cass asked: A traditional method of performing multiplex metal ion analysis is to use Anodic Stripping Voltammetry. How do the limits of detection compare with your aptamer-based approach?

Professor Lu responded: Anodic stripping voltammetry (ASV) has limits of detection (LODs) that are among the lowest, typically in 10^{-9}–10^{-10} M range. Some of the functional DNA sensors have comparable LODs (*e.g.*, a uranyl DNAzyme sensor has a LOD of 45 pM[1]), while many others have LODs typically in nanomolar range. Despite the differences, the LODs of functional DNA sensors (*e.g.*, Pb, U, Cu and Hg) are all below the maximum contamination levels of respective metal ions in drinking water defined by the US Environmental Protection Agency, and much lower than the safe levels in blood defined by the US Center for Disease Control. In addition, the functional DNA sensors have the added advantage of easier and faster operation and no need to use toxic metal ions (such as mercury in ASV) in the detection process. Finally, since ASV covers typically only 12–15 metal ions and mostly is not oxidation state or speciation state specific, the functional DNA sensors can potentially sense many more metal ions and can differentiate different oxidation or speciation states of the same metal ion.

1 J. Liu, A. K. Brown, X. Meng, D. M. Cropek, J. D. Istok, D. B. Watson and Y. Lu, *Proc. Natl. Acad. Sci. U. S. A.*, 2007, **104**, 2056.

Dr Zhou asked: Your functional DNA-based small molecule sensors are very interesting. One question regarding the mercury sensor, you have shown that it is very sensitive for free Hg^{2+} ions, can your sensor also work for organometallic mercury, such as dimethyl mercury?

Professor Lu responded: Our Hg^{2+} sensor can detect only Hg^{2+}, not organometallic mercury, which we know there is a need for detection. The functional DNA sensor is very selective toward different speciation states of the same metal ion. To obtain functional DNA sensors toward organometallic mercury, we need to carry out *in vitro* selection using organometallic mercury as the target.

Dr Brown asked: What counterions have you used in your selection procedures and do you think that altering the counterions you currently use could improve the affinity of your sensors?

Professor Lu answered: We normally use monovalent metal ions, such as sodium, that are commonly found in physiological buffers and in the environment as counterions. These monovalent counterions do play a role in the binding and function of the DNAzymes. We have just surveyed their effects in a recent publication.[1]

1 D. Mazumdar, N. Nagraj, H.-K. Kim, X. Meng, A. K. Brown, Q. Sun, W. Li and Y. Lu, Activity, folding and Z-DNA formation of the 8-17 DNAzyme in the presence of monovalent ions, *J. Am. Chem. Soc.*, 2009, **131**, 5506–5515.

Dr Krause said: Are the mercury ions released after DNA cleavage and does this lead to amplification of the signal?

Professor Lu replied: In the design described in this paper, we have not found any evidence that mercury ions are released that lead to amplification of the signal. However, in the catalytic beacon sensors for lead and uranyl reported earlier in my group, we have found evidence that the metal ions are released after DNA cleavage, resulting multiple turnovers and amplification of the signal.[1]

1 A. K. Brown, C. M.-B. Pavot, J. Li and Y. Lu, A Lead-dependent DNAzyme with a Two-Step Mechanism, *Biochemistry* 2003, **42**, 7152–7161; A. K. Brown, J. Liu, Y. He and Y. Lu, Biochemical Characterization of a Uranyl Ion-Specific DNAzyme, *ChemBioChem*, 2009, **10**, 486–492.

Dr Brown opened a general discussion of the papers by Dr Dhaliwal, Professor Girkin and Professor Lu: Diagnostics aimed at detecting infectious diseases cannot necessarily be generalized into a "universal" assay system. Development of good diagnostics targeted to infectious disease detection clearly demands "systems" knowledge (*e.g.*, molecular, cellular, physiological) of the infection process and for many pathogens such knowledge is often lacking. In addition, for these assays to genuinely be utilised for diagnosis of an infectious disease "in the field" one also needs to be fully informed of the types of samples and platforms/methods which can be used in the context of the clinical setting where the organism/disease is to be detected. Particularly in developing countries, one must be sensitive to technical and financial limitations and cultural differences that can constrain the use of certain samples and diagnostic formats.

Professor Vadgama commented: We need to be careful about how, if at all, insurance companies are to be involved in the process of disease prediction, *e.g.* through access to genetic data, since such knowledge could well affect individual insurance premiums. Resulting ethical issues are going to be quite substantial if we provide diagnostics with strong disease predictive capability, as is the case with some current molecular biology tests.

Professor Lu said: In addition to monitoring of a known target, such as glucose, for a long time, there is a space for rapid diagnosis in point of care (POC) at home, including detection of infectious diseases such as HIV, as people care about it. There are differences between diagnosis by POC at home and by professional tests in clinical labs. Diagnosis by POC at home needs only to be accurate enough to tell the users that they may have a problem or not, so that they can decide whether or not to go to see a doctor for more accurate tests in clinical labs. The success of POC tests will depend on how good the tests are in terms of accuracy. The more accurate tests with low rates of false positive or false negative results can find success.

Dr Davies replied: Getting tests suitable for use in the home is the goal of many companies. The big numbers tests, glucose and pregnancy, have been done, although innovation continues in these areas. Undoubtedly there are areas where home testing could be of value, despite the numbers being smaller – warfarin therapy monitoring is one recent development. However, the important factors deciding which tests will be successful will depend upon the user's need to do them, the ease with which they can be done (this is very important), the cost and who pays (user, insurer or state), the test accuracy – FDA tend to insist upon matching accuracy to hospital lab tests – and how to inform the user of the meaning, or potential meaning, of the result. The latter two are interesting. Test accuracy is really how it performs in users' hands and not in the lab in the hands of skilled personnel. Users vary. What does the test mean? – we really cannot scare people but do want them to do the right thing as a result of the test. This is a difficult issue, particularly for a cancer test. We all know that early detection of cancer improves outcome, but if you were told you had cancer would you act rationally? Would a leaflet in the box help you do the right thing? We could lower the threshold for, say, a PSA test so that, as you say, it may indicate a problem at 2 ng ml^{-1} (but you might also say it probably doesn't so that you do not worry the user) and suggest a visit to the doctor. The doctor refers you for tests and after doing this for several patients may conclude that the test is of no use if all his referrals are reported as fine. The setting of a threshold is difficult. The whole area is difficult.

Dr Ko Ferrigno said: New and rapid technologies are needed in the home, in the GPs office, in the clinic and in the hospital, and in off-site diagnostic labs. Some tests will be suited to the home and can be used by the individual to help with wellbeing (what is my cholesterol today? should I cut back on snacks?). Some tests will be suited to the GP's office (should this patient be referred for follow-up? is this pain caused by infection, inflammation or musculo-skeletal disease?). In the clinic, there is an unmet need for tests that allow rapid decision-making in, for example, the emergency or intensive care setting, or in the operating theatre. In out-patient clinics, faster diagnostic tests could mean that patients would not need multiple return visits, which would have benefits to them, to the healthcare provider and to the transport infrastructure. Finally, diagnostic laboratories would benefit from faster, higher-throughput and less expensive tests.

Dr Dhaliwal responded: I agree with all of this as long as the tests unequivocally guide therapy, if not they are unlikely to penetrate the market. We are data rich with these tests but the lack of validated interventions may limit their use by clinicians and cause uncertainty in patients with home testing with the primary care physician not able to respond either with enough speed (appointments may take 48 hours to 5 days to see a GP) or with effective intervention.

Professor Cass replied: I think that is true if we look at the position from the point of view of a technology developer, however from that of the GP, clinic or hospital there would need to be suitable decision support infrastructure to help in the interpretation and the clinicians (and medical laboratory scientists) would have to adapt to this.

Professor Girkin said: The crucial aspect here is that the information provided by the instrument should provide the clinician with more information to aid his clinical judgment. The instruments should help with detection, diagnosis and long term monitoring and not provide a red/green light treat/don't treat. This is true for both home and GP tests, where in the former case the clinician may be provided with the results, when, added to the clinical judgement, a treatment protocol is decided for that patient.

Very quick reverse transcription polymerase chain reaction for detecting 2009 H1N1 influenza A using wire-guide droplet manipulations†

David J. You,[a] Phat L. Tran,[b] Hyuck-Jin Kwon,[a] Deepa Patel[b] and Jeong-Yeol Yoon*[ab]

Received 11th May 2010, Accepted 22nd July 2010
DOI: 10.1039/c005326k

Reverse transcription polymerase chain reaction (RT-PCR) is currently a gold standard in identifying influenza A virus, especially H1N1 flu. Typical RT-PCR assays take about 1–2 h for thermocycling, and there is a growing need to further speed up the thermocycling to less than 30 min. Additionally, the PCR assay system should be made portable as a point-of-care detection tool. There have been attempts to further speed up the PCR assays by reducing its volume. There have also been attempts to use droplet microfluidics technology to PCR, primarily to automate the PCR enrichment processes and take advantage of its small volume. In all these attempts, heating and cooling is made by conduction heat transfer. Rapid movements of droplets (immersed in oil) over three different temperature zones make very quick PCR possible, as heating/cooling will be made by convection heat transfer, whose heat transfer coefficients are much higher than that of conduction. We used our newly-invented method of wire-guide droplet manipulations towards very quick RT-PCR. Computational fluid dynamics (CFD) simulation of our system revealed that heating/cooling for each temperature change takes 1–4 s for a 10 μL droplet, as compared to >30 s in the other quick PCRs. Theoretically a 30-cycle process can take as short as 13 s × 30 cycles = 6 min 30 s. The entire system was made as a single instrument, with the components made by a milling machine and a rapid prototyping device. No additional equipment and external computers are required. With this newly developed system, 160 bp gene sequence was amplified from 2009 H1N1 influenza A (human origin). The 30-cycle process took as short as 6 min 50 s for a 10 μL droplet (with additional 4 min for reverse transcription). Its product was confirmed by traditional gel electrophoresis, subsequent imaging as well as gene sequencing, which has been very difficult with the other stationary droplet/nanodrop approaches. The proposed system has a potential to become an extremely rapid, portable, point-of-care tool for detecting influenza A.

[a]*Department of Agricultural and Biosystems Engineering, The University of Arizona, Tucson, Arizona, 85721-0038, United States*
[b]*Biomedical Engineering Graduate Interdisciplinary Program, The University of Arizona, Tucson, Arizona, 85721-0038, United States*

† Electronic supplementary information (ESI) available: A full movie showing a single wire-guide PCR cycle. See DOI: 10.1039/c005326k

Introduction

Inluenza A H1N1 2009 is a highly infectious disease that initiated a global pandemic in April 2009.[1] As of 25th of April 2010, the WHO has reported at least 17 919 deaths of H1N1 2009 worldwide and reported confirmed laboratory cases in more than 214 countries and overseas territories and communities.[2] H1N1 2009 is defined as an acute febrile respiratory illness, in which the most common symptoms were fever, coughing, and sore throats.[3] In June 2009, the WHO raised the level of influenza alert from phase 5 to phase 6, indicating that a global pandemic was underway.[4] Currently, polymerase chain reaction (PCR) technology and viral cultures are the only CDC (US Centers for Disease Control and Prevention) recognized and FDA (US Food and Drug Administration) approved methods for the detection of influenza A subtype (H1N1) virus.[5]

PCR is a well-established method for the detection and amplification of DNA and RNA with an unlimited sensitivity and unparalleled specificity. It is an essential tool in medical research and clinical medicine, used extensively for the detection of infectious disease organisms as well as detecting mutations in genes. Reverse transcription PCR (RT-PCR) is a variant of PCR in which an RNA strand is first transcribed to complementary DNA (cDNA) using the enzyme reverse transcriptase, which is then amplified using traditional PCR.

PCR works by copying a specific DNA sequence of interest. The process of PCR amplification requires three important steps: denaturing, annealing, and extension. Each of these steps is performed at three specific temperatures. Commercial PCR systems generally incorporate a high powered Peltier module in order to heat and cool the chamber in which the samples reside. This is the major disadvantage of current systems: conduction is the least efficient method for heating and cooling when comparing the three modes of heat transfer (conduction, convection and radiation). Additionally, most commercial systems utilize a single heating block. The time required to reach the desired temperatures usually comprises a significant portion of the total PCR cycling time.

There has been significant interest towards the reduction of PCR cycling times, primarily through the implementation of microfluidic devices, in which small volumes of fluid travel through microchannels. Conventional methods of conductive heat transfer for heat cycling mean that it often takes a couple of minutes to complete a single cycle (and thus over an hour to finish typical 25–30 cycles of PCR). However, if this liquid is made to move over three different temperature areas within the microfluidic channel, the time required for heating and cooling may be significantly reduced, thus leading to a faster PCR assay. Additionally, the liquid volume that is needed to be heated and cooled is so small that the required heat transfer can be completed in a much shorter amount of time. The first such demonstration was made by Kopp *et al.*,[6] where a single serpentine microfluidic channel travels through three different temperature zones to achieve 20-cycle PCR. In this manner, less than 20 min or even less than 10 min 20-to-30-cycle PCR has become a possibility. This concept has been combined with microfluidic cell lysis,[7] capillary electrophoresis to confirm PCR products,[8,9] or fluorescence microscopy for real-time monitoring of PCR products.[10] Liquid flows through a microchannel continuously, or as discrete liquid plugs within a microchannel. Although this microchannel PCR is faster than conventional PCR assays, it has several limitations in terms of reconfigurability or adaptability; for example, the number of cycles is set by the microchannel layout (if microvalves are not employed); sample-dependent pre-processing steps such as cell lysis, DNA/RNA extraction and/or reverse transcription (RT) are rather difficult to be implemented into microchannel layout, as it is not reconfigurable. Additionally, these are closed systems, which mean that once liquid is inserted into the microchannel inlets, the process cannot be interrupted or augmented until it has completed its course or has been expelled from the system.

Droplet microfluidics (sometimes referred as "digital" microfluidics) are another area of significant interest for speeding up the PCR cycling process. These include electrowetting-based microfluidics,[11] magnetofluidics,[12] and wire-guide microfluidics.[13] The reconfigurability and reprogrammability feature of droplet microfluidics allows for the strong adaptability and expandability over the microfluidic microchannels. Specifically, these systems can be easily modified to be combined with cell lysis, DNA/RNA extraction and subsequent detection and sequencing of PCR products. Therefore, droplet microfluidic platforms are extremely adaptable to the variety of available and newly emerging PCR protocols. Unfortunately, attempts to perform PCR in droplet microfluidics format have been relatively rare, and focused primarily on electrowetting-on-dielectrics (EWOD), which is comparatively more difficult to fabricate and operate.[14] Magnetofluidics does not involve complicated fabrication, as all you need is a superhydrophobic surface. However, proper heat transfer *via* a superhydrophobic surface can be inefficient, and the effect of magnetic fields and the presence of magnetic particles on PCR thermocycling and subsequent identification of its products remain to be investigated and resolved.[15] Wire-guide microfluidics is the simplest method for droplet manipulation and does not require the addition of particles that may interfere with the PCR thermocycling. Wire-guide microfluidics can be implemented on a variety of droplet environments, including superhydrophobic surfaces[13] and oil submersion, of which the latter is introduced in this paper.

This paper presents a novel method to significantly reduce the time for PCR thermocycling by employing wire-guide droplet microfluidics. A syringe needle tip is used to control the movement of a droplet submerged in silicone oil over three temperature regions. The 3-axis computer-numerically-controlled (CNC) system offers the benefit of reconfigurability and reprogrammability of droplet (digital) microfluidics. The submersed droplet continuously moves in silicone oil, which allows for convective heat transfer from the surrounding oil to the droplet. This method of very quick PCR will allow for the rapid analysis of viral RNA for the purpose of detecting and preventing the spread of infectious diseases before they become pandemic.

This work should not be confused with droplet or nanodrop PCR,[16] where the PCR assay is performed in a very small droplet to speed up the conduction-based heat transfer, which is not moving. This "stationary" droplet PCR has succeeded in speeding up the thermocycling significantly faster than other conventional methods. However, with this tiny volume (typically in nanolitre range), it becomes very difficult to confirm its products by traditional means such as gel electrophoresis, subsequent imaging and/or gene sequencing. Additionally, there is a concern in assay reproducibility and reliability, as aliquoting the low-concentration sample into nanolitre volume may occasionally result in no target molecule in the droplet. In wire-guide microfluidics, however, the volume of the droplet can be made large enough (in this case, 10 μL) to enable confirmation of its products (gel imaging and gene sequencing) as well as ensuring the assay reliability.

Materials and methods

Temperature controller and PCB heater

Temperature monitoring and heating of the silicone oil in the chambers were conducted on a single-sided printed circuit board (PCB) with exposed copper traces. Positive photoresist-coated copper-clad boards, resist etching solution, and ferric chloride solution were obtained from Marlin P. Jones & Assoc. (MPJA; Lake Park, FL, USA). Photoresist template was created using Adobe Photoshop CS4 (San Jose, CA, USA). Trace width (w, in mm) for the heating element was calculated using equations from IPC-2221, generic standard on printed circuit boards:[17]

$$w = \frac{\left[I/k(\Delta T)^b\right]^{1/c}}{h \times 1.378}$$

where I is current in amperes (A), ΔT is temperature rise in °C, h is thickness as represented by how many ounces (oz) of copper to cover one square foot of circuit board, the constant 1.378 has the unit of mils oz^{-1}, where mils is 1/1000 of an inch and k, b and c are constants resulting from curve fitting to the IPC-2221 curves (for IPC-2221 external layers: $k = 0.048$, $b = 0.44$ and $c = 0.725$).

The photoresist template design was printed onto a clear transparency using a 2400 dpi laser printer. The coated copper-clad board was then exposed to an incandescent 100 W bulb and dissolved in photoresist developer (MPJA) at 46 °C for 6 min. The board was then transferred to ferric chloride (0.240 g mL^{-1} bath (MPJA) at 55 °C for 10 min. Remaining photoresist was removed using pure anhydrous acetone and drilled using a 0.025 inch (0.635 mm) drill bit.

Temperature monitoring for the PCR chambers was conducted using DS18B20 1-Wire temperature sensors, an Arduino Duemilanove microcontroller, and a serial-enabled liquid crystal display (LCD) obtained from Sparkfun Electronics (Boulder, CO, USA). The PCR chamber was created through rapid prototyping from a Dimension 1200ES (Stratasys; Eden Prairie, MN, USA) using a high temperature acrylonitrile butadiene styrene (ABS) polymer. Silicone oil (catalog number S159-500, Fisher Scientific; Pittsburgh, PA, USA) was used to fill the PCR chambers, which were heated with 2 A benchtop power supplies (MPJA). Solid-state relays, also acquired from MPJA, were used to switch power to the heating elements and were controlled by the Arduino through a transistor circuit using common NPN bipolar junction transistors (BJT; 2N2222), diodes (1N4148), and 1 kΩ resistors. A proportional-integrative-derivative (PID) algorithm was implemented to control the heating of the silicone oil. This setup allowed for precise temperature control to within ±0.25 °C of the target temperatures.

Computer-controlled wire-guided manipulator

The computer-controlled wire-guided apparatus was fabricated and assembled for manipulating the PCR solution droplets in the silicone oil (Fig. 1). The base was

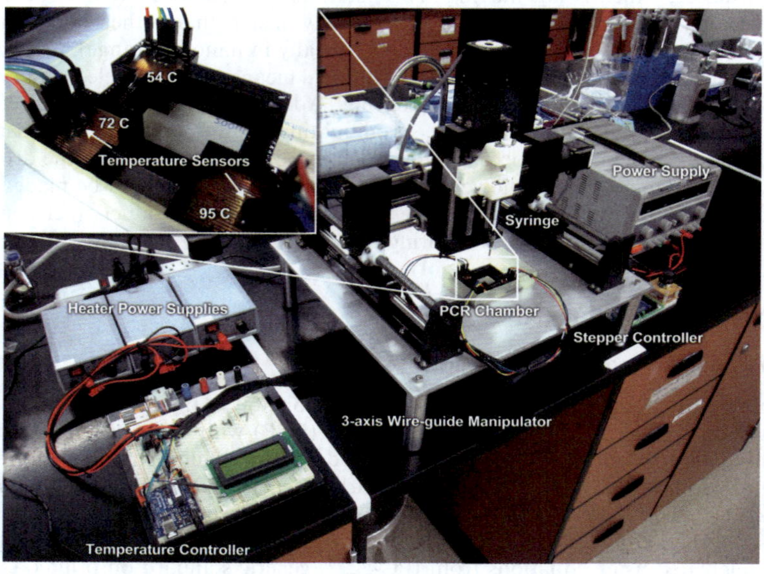

Fig. 1 Droplet PCR Apparatus.

constructed of aluminium onto which Thomson 1/2 inch (12.7 mm) linear rails and twin ball bearing pillow blocks (Thomson; Radford, VA, USA) were mounted, allowing precise movement along the *X*, *Y*, and *Z* axes. Computer-controlled linear movement of each axis was performed with Nema 17 stepper motors obtained from Automation Direct (Cumming, GA, USA), connected to 1/2–10 inch (12.7–10 mm) Delrin® lead screw assemblies obtained from Precision-CNC-Router (Melbourne, FL, USA). Stepping of each motor was performed with an Arduino Mega microcontroller and Easydriver stepper motor controllers utilizing A3967 chips (Sparkfun Electronics). The stepper motors were powered by a 0–30 V, 3 A bench-top power supply (MPJA). All of the mechanical components of the system were connected by ABS structures formed from rapid prototyping (Dimension 1200ES).

For droplet insertion and extraction, disposable syringes were used with Luer-Lok blunt-ended needles (OK International; Garden Grove, CA, USA). 1 mL disposable plastic syringes with Luer-Lok tips were obtained from Fischer Scientific, and fitted into a quick-release plastic holder attached to the *Z*-axis of the apparatus. Precise droplet insertion/extraction from a 14 gauge (1.628 mm inner diameter) blunt-ended needle was accomplished with a 5 V mini linear step motor (MPJA) attached to the syringe plunger.

Primer design

RNA polymerase subunit (PA gene) nucleotide sequences of H1N1 influenza A virus with human origin (GB CY061288.1 and CY061152.1), swine origin (GB AB514935.1), and avian origin (GB GU168243.1) were pooled from GenBank (http://www.ncbi.nlm.nih.gov/genbank/). These RNA sequences were then subjected to multiple alignment analysis using ClustalX[18] to search for conserved regions among 3 different species such that consensus primers could be designed. Selected primers were then analyzed for appropriate melting temperature as well as any possible hairpin or self-dimerization by using OligoAnalyzer 3.1 (IDT Corporation, Coralville, IA, USA). These primers are PAh1n1_583F (5′-CAGTCCGAAAGAGGCGAAG-3′) and PAh1n1_745R (5′-GAAAGCTTGCCCTCAATGCAG-3′); resulting in a product of about 160 base pairs (bp).

RNA extraction

NATtrol influenza A H1N1 virus external run control (catalog number NAT-FLUAH1N1-ERCM) from ZeptoMetrix Corporation (Buffalo, NY, USA) in medium was used as a target. This virus sample is 2009 H1N1 flu with human origin, purified/modified by the manufacturer. Its RNA was extracted using the E.Z.N.A. viral RNA kit (catalog number R6874-01, Promega Bio-tek; Madison, WI, USA) through the spin column protocol. Briefly, 500 μL of QVL lysis buffer was mixed with 150 μL cell media, then precipitated with absolute ethanol. The mixture was then passed through a spin column, and then cleaned with provided RWB buffer from the kit. Finally, RNA was eluted with elution buffer. RNA was then subjected to one-step RT-PCR.

RT-PCR

Reverse transcription PCR (RT-PCR) was first performed under a conventional PCR machine to ensure proper design of primers and proper extraction of RNA. RT-PCR was run according to what was recommended by the AccessQuick™ RT-PCR system (catalog number A1701, Promega Bio-Tek, Madison, WI, USA). Once positive results were verified, we employed our wire-guide droplet PCR system to detect the presence of influenza A H1N1 virus.

AccessQuick™ RT-PCR system kit was used for the wire-guide droplet PCR reaction. 10 μL droplets were used in 30 PCR cycles. After initial extraction of the PCR solution into the syringe, the system inserted the droplet into the 37 °C chamber for 4 min for reverse transcription. After formation of the cDNA template, the chamber would heat to 54 °C degrees during an initial 30 s denaturing step in the 95 °C chamber. The droplet began the cycles in the 95 °C chamber for denaturing, 54 °C chamber for annealing of primers, and then 72 °C chamber for extension of the products. The final step consisted of final annealing for 20 s.

During each of these stages, the system would move the inserted droplet in a circular motion to aid in convective heat transfer. The chambers were connected with narrow bridges to allow the syringe needle to stay submerged in the silicone oil, preventing the possibility of evaporation or droplet loss due to the effects of surface tension from inserting and removing the needle from the oil.

PCR analysis

PCR product was determined through gel electrophoresis and fluorescent imaging of gel. PCR products were applied into 2% low melting agarose gel (catalog number E-3126-25, ISC BioExpress; Kaysville, UT, USA), using a Fisher Scientific (catalog number FB200) power supply at 80 V and 1 A for 60 min in a 1× tris-acetate-EDTA (TAE) buffer solution (catalog number BP13324, Fischer Scientific). The gel was soaked in about 1 μg mL^{-1} ethidium bromide for *ca.* 6 min and imaged using a Gel Doc 1000 imaging system (Bio-Rad Laboratories, Hercules, CA, USA).

DNA preparation and sequencing

PCR products from conventional and wire-guide droplet system were cleaned with QIA quick PCR purification kit (catalog number 28106, Qiagen, Valencia, CA, USA) for sequencing to ensure proper detection of influenza A H1N1 virus. Amplified PCR products were sequenced using the Applied Biosystems 3730 DNA analyzer (Carlsbad, CA, USA). Sequences are then viewed and corrected using Chromos Lite V 2.01 (Technelysium Pty Ltd, Southport, Queensland, Australia) and finally blasted against GenBank database.

CFD modeling

The commercial software (FLUENT®6.3 and GAMBIT®1.3; FLUENT Inc., Lebanon, NH, USA) was used to build mathematical grids for our PCR system using the finite volume method (FVM).[19] A 3-D model was used in this study rather than a 2-D model as our system was not symmetric in shape.

The grid size used in the simulation consisted of 285 376 cells, 876 984 faces and 305 725 nodes. The shape of all meshes was hexagonal to decrease the number of meshes. First-order schemes were used as it is known to provide better convergence of calculations than the second-order, although it provides less accurate results due to the increased error in numerical discretization.[20] A steady-state solution was used because both the outdoor temperature and the amount of supplied heat to each chamber were fixed. To simulate heat transfer in the system, the energy equation option was chosen in FLUENT.[19] To consider the convectional heat flow inside the chamber by heating, the viscous heating and buoyancy heating options were chosen. Since the Reynolds number of the flow around the μL-sized droplet was much smaller than 5000, the turbulence model was not used in this simulation. Since the droplet is relatively small and do not mix with oil, and its movement is forced by external wire-guide manipulation not by surface tension, it can be treated as a solid sphere and thus the volume of the fluid model was not used. Outdoor temperature was set to the measured room temperature, 25 °C.

Results and discussion

Multi-chamber design

Several different designs of multi-chambers were considered, initially with three chambers connected in-line. This in-line chamber system worked fine for PCR thermocycling, and corresponding CFD simulation revealed that each chamber could maintain its designated temperature even though they are connected through "tunnels". However, this design was inefficient because the droplet travelled through the middle chamber twice in one cycle, and was dismissed.

The next design was a rectangular/circular one as shown in Fig. 1. Corresponding CFD simulation (Fig. 2) shows that steady state for the three necessary PCR cycling temperatures, 95 °C, 54 °C and 72 °C are achievable in a connected design, when the actual temperature of the oil is set to 105 °C (10 °C higher), 55 °C (1 °C higher) and 80 °C (8 °C higher), respectively. Temperatures near the connecting tunnels are not very different from the other parts of the chamber, indicating the effect of heat transfer through the tunnels can be disregarded.

As the temperature sensors were placed near the outer wall (as shown in Fig. 1 and 2), significant heat loss to the ambient air (25 °C) is expected through the walls of the chambers, especially at high temperatures (105 °C and 80 °C), which led to a lower temperature reading by about 10 °C. Therefore, the set temperatures of a PID controller were set to 95 °C, 54 °C and 72 °C (the actual temperatures within droplets), which would heat up the nearby oil to 105 °C, 55 °C and 80 °C, respectively.

While the temperature variation parallel to the heating element did not vary more than ±1.0 °C, except near the walls, the depth of the thermocouple resulted in a more significant temperature variation of ±2.0 °C. Thus, the depth of the thermocouple placement for the temperature controller in relation to the depth of the droplet was considered to ensure proper heating of the droplet PCR solution. It is important to note that the mixing motion during the heat cycling process was not included in the CFD simulation of the silicone oil. Most likely this mixing assisted in the temperature distribution.

Convective heat transfer to a droplet

In addition to the chamber design, CFD modeling was also performed on the droplet that travels through three chambers (Fig. 3). The parameters of the simulation consisted of a 10 μL droplet traveling at a rate of 2 cm s^{-1} through uniformly

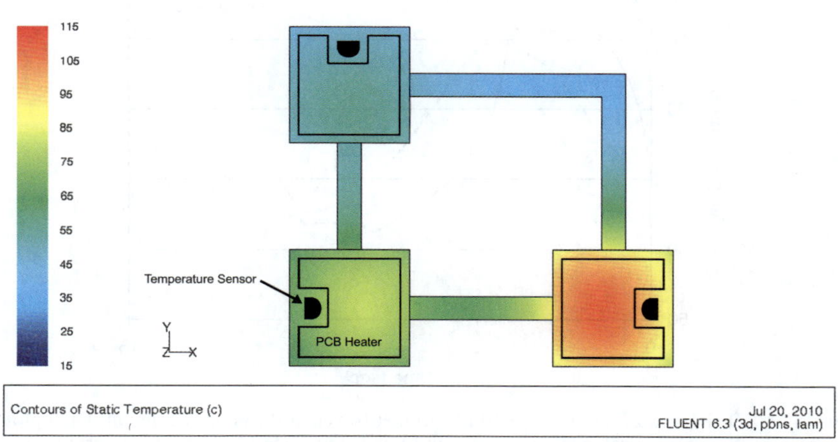

Fig. 2 CFD simulation of three-chambered PCR cycler. FLUENT simulation showing steady state temperatures of silicone oil heated by PCB modules in ABS polymer chamber.

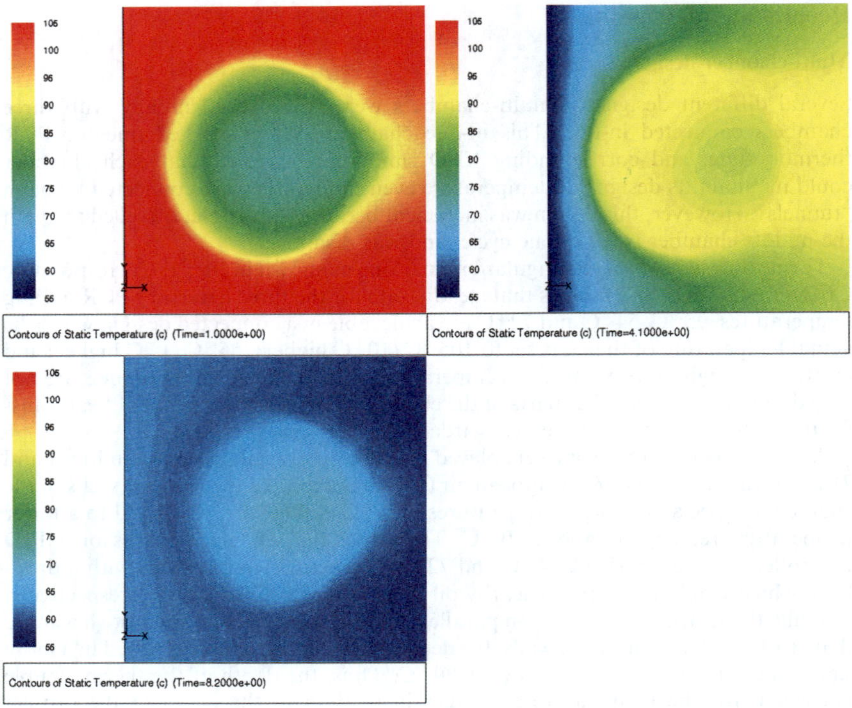

Fig. 3 CFD simulation of a 10 μL droplet in silicone oil. Snapshot of FLUENT simulation showing the temperature gradient of droplet traveling through silicone oil. Left top: 72 °C droplet entered into 95 °C chamber (actual $T = 105$ °C). Right top: The fully heated droplet at 92 °C is now entering into 54 °C chamber (actual $T = 55$ °C). Bottom: The droplet is fully cooled down at 62 °C.

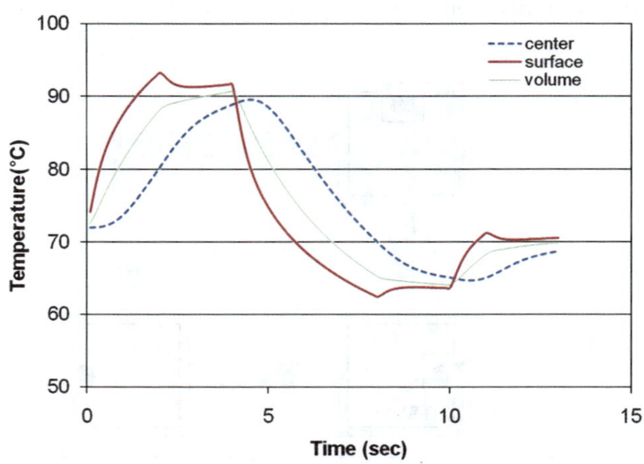

Fig. 4 CFD simulation results for a 10 μL droplet traveling through silicone oil. Each plots represents the surface-averaged ("surface") and volume-average ("volume") temperature s as well as at the droplet center. Graph indicates theoretical PCR cycle can be conducted in 13 s, and 30 cycles in 6 min 30 s.

heated silicone oil. The simulation reveals the effect of convective heat transfer on the PCR droplet, at the surface and throughout its volume.

The simulation from the convective heat transfer to a droplet were datalogged and used to construct a temperature *versus* time graph (Fig. 4). This graph shows the temperature of the droplet at the center, surface, and overall volume. Heating of the droplet from 72 °C to 95 °C takes approximately 2 s (in the 95 °C chamber; actual temperature is 105 °C), cooling from 95 °C to 62 °C takes approximately 4 s (in the 54 °C chamber; actual temperature is 55 °C), and heating from 62 °C to 72 °C takes approximately 1 s (in the 74 °C chamber; actual temperature is 80 °C). Therefore, the overall time spent heating and cooling the droplet for one PCR cycle is theoretically 7 s. However, because particles will primarily accumulate at the oil to water interface, these times may be significantly reduced. If the droplet is additionally maintained at the correct temperatures for 2 s per chamber, then the overall PCR cycle takes only 13 s. Thus, to perform 30 cycles would only require 6 min 30 s with a 10 μL droplet. The droplet size may be decreased to further speed up the heat cycling process. There has been much focus to reduce the size of the droplet, even to the nanolitre scale.[16] However, 10 μL droplets were chosen for these experiments to closely match the volumes used in typical PCR assays, as well as to allow further analysis to be conducted, such as gel electrophoresis, and DNA sequencing. Without these post-analysis procedures, the product cannot be definitively determined, possibly resulting in misdiagnosis. The CDC requires that positive results for H1N1 2009 using their real-time RT-PCR protocol must be sent to them for further analysis.[5]

Droplet insertion and retraction

With this new design, the droplet stayed "hanging" (*i.e.* pendant drop) throughout the entire thermocycling. This configuration did not create any problems within each chamber, as the droplet only needed to stay in the chamber for 4–5 s, thus did not require fast transportation. Within each chamber, the speed of the wire-guide movements were set to 2 cm s^{-1}, the same as the CFD modeling. When the droplet moves from one chamber to another, however, the transport should be made as quick as possible to speed up the entire process. (This transportation time was not taken into consideration in the above CFD simulation.) This high speed transition from chamber to chamber, however, resulted in the droplet falling off the needle.

To prevent the droplet from falling off the needle during this transport, the system would retract the droplet from the oil inside the needle prior to movement, and then re-insert after the chamber-to-chamber transition completed. This protocol significantly reduced the time needed for thermocycling, which will be reported later in this section. Extracting additional oil after the droplet to provide a barrier from the air was considered and tested during initial chamber designs, however this also created intermittent droplet loss and was dismissed.

The linear step motor for the syringe plunger was calibrated to insert and extract precisely 10 μL of PCR solution into each chamber. The droplet was completely evacuated from the syringe needle while thermocycling in each chamber. Conductive heat loss or gain from the stainless steel syringe needle towards the droplet was considered, but because the droplet was a complete pendant drop, the contact area of the droplet with the syringe needle in relation to the contact area of the droplet to the surrounding silicone oil was significantly smaller. Additionally, because the droplet and syringe needle moved together and were in direct contact, the method of heat transfer between the two was conduction, whereas the heat transfer between the droplet and the silicone oil was primarily convection, which provides a much higher rate of heat transfer. Thus the effects of the syringe needle on the thermocycling temperatures were dismissed in the CFD analysis and experimental protocol.

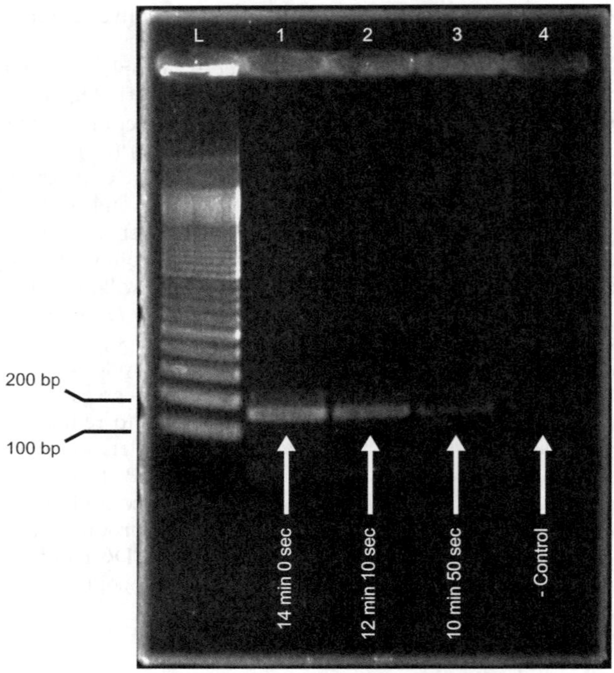

Fig. 5 Wire-guide droplet RT-PCR results. The shortest 30 cycles were completed in 6 min 50 s with 4 min of reverse transcription. L is the 1-kb plus ladder, lanes 1–3 are the 2009 influenza A H1N1 viruses, and lane 4 is the negative control.

Table 1 DNA sequencing results. DNA sequencing confirms that the product from wire-guide droplet RT-PCR is 99% to 100% homologous to influenza A H1N1 virus

Accession	Influenza A virus description	Max score	Total score	Query coverage	*E*-value	Max ident
CY056121.1	A/District of Columbia/ INS31/2009(H1N1)	219	219	95%	3.00×10^{-54}	100%
CY053282.1	A/Brownsville/ 39H/2009(H1N1)	219	219	95%	3.00×10^{-54}	100%
CY044906.1	A/New York/3611/ 2009(H1N1)	219	219	95%	3.00×10^{-54}	100%
CY050026.1	A/Mexico City/ WR1307N/2009(H1N1)	215	215	96%	4.00×10^{-53}	99%
CY061804.1	A/swine/Hong Kong/ 189/2010(H1N1)	213	213	95%	2.00×10^{-52}	99%
CY062323.1	A/swine/Thailand/ CU-RA9/2009(H1N1)	213	213	95%	2.00×10^{-52}	99%
CY062255.1	A/California/VRDL86/ 2009(H1N1)	213	213	95%	2.00×10^{-52}	99%
CY062191.1	A/New York/3251/ 2010(H1N1)	213	213	95%	2.00×10^{-52}	99%
CY061591.1	A/Texas/JMS403/ 2009(H1N1)	213	213	95%	2.00×10^{-52}	99%

Gel electrophoresis and DNA sequencing

Primers were specifically designed to work across human, swine, and avian strains of influenza A H1N1. Fig. 5 shows a gel image of about 160 bp products of influenza A H1N1 (human origin), in which 30 cycles were successfully performed in 6 min 50 s (Fig. 5, lane 3), with an additional 4 min reverse transcription step. This minimal reverse transcription time was determined through series of PCR experiments. Additionally, 30 cycles in 8 min 10 s (Fig. 5, lane 2) and in 10 min (Fig. 5, lane 1) are also shown as well as a negative control experiment (Fig. 5, lane 4) in which the PCR cycling was performed under the same conditions. Results show a decrease in PCR amplification from the 8 min 10 s heat cycling to 6 min 50 s. However, the signal is still relatively strong and can be considered a positive result from the negative control. Obviously the longer thermocycling correlates to the brighter band suggesting that time is essential for Taq polymerase to amplify DNA. If there were more DNA in the target sample or simply prolonging the reverse transcription time, 20 PCR cycles would be sufficient, which would require only 13 s $\times$ 20 cycles = 4 min 20 s. In addition, if a smaller droplet is used, *e.g.* 1 μL, the time for each cycle can be further reduced down a few second scale, possibly leading to <3 min 20 PCR cycles.

Although the PCR products are at the right band size (Fig. 5), further verification is required to validate the content of the band. PCR products were cleaned and submitted for sequencing. The genomic content of the band is then blasted against the nucleotide collection database on The National Center for Biotechnology Information. The query sequence is 99% to 100% homologous to various samples of 2009 H1N1 influenza A virus with an *E*-value of 3×10^{-50} (Table 1) suggesting that our wire-guided droplet manipulation PCR system is working and able to detect virus in 10 min and maybe less.

Conclusion

This paper has successfully demonstrated the wire-guide droplet RT-PCR cycling of 2009 H1N1 influenza A in 10 min 50 s. Specifically, 30 cycles of heat cycling were performed in 6 min 50 s, demonstrating the effectiveness of utilizing wire-guide droplet manipulation in conjunction with convective heat transfer. With further optimization of the system, this heat cycling time is expected to be reduced to less than 3–4 min. Simulation studies were employed to develop a multi-chambered heat cycler and to establish the effectiveness of wire-guide droplet PCR with standard PCR assay solution volumes. DNA sequencing confirmed the product created in the wire-guide droplet PCR system. The computer controlled wire-guided manipulator was constructed for general use in a variety of droplet microfluidics applications. However, when considering commercialization potential for the methods introduced in this paper, the system can be easily miniaturized and simplified. For example, the entire 3-chamber design with embedded temperature sensors can be constructed on a single circuit board incorporating a simple 2-dimensional table for moving the syringe. The entire system can be made relatively small and has the potential for portability due to the low power consumption of the PCB heaters. Alternatively, for high-throughput applications, an array of syringe needles can be used to run multiple PCR assays. Implementing a rotating circular chamber design with multiple heating chambers would allow for the continuous and uninterrupted heat cycling of many solution samples with different start and end times simultaneously. With further optimization, the wire-guide droplet PCR will prove to be an effective method for significantly accelerating PCR assays with the advantages of reconfigurability and reprogrammability necessary for the variety of infectious disease protocols, while still allowing for post analysis procedures to be performed.

References

1 *Influenza-like illness in the United States and Mexico: global alert and response (GAR)*, World Health Organization, Geneva, 20 July 2009, http://www.who.int/csr/don/2009_04_24/en/index.html.

2 *Pandemic (H1N1) 2009 – update 98*, World Health Organization, Geneva, 30 April 2010, http://www.who.int/csr/don/2010_04_30a/en/index.html.

3 S. Fatimah, M. D. Dawood, M. D. Seema Jain, L. Finelli, W. S. Michael, S. Lindstrom, R. J. Garten, L. V. Gubareva, X. Xu, C. B. Bridges, and T. M. Uyeki., Emergence of a novel swine-origin influenza A (H1N1) virus in humans, *N. Engl. J. Med.*, 2009, **360**, 2605–2615.

4 M. Chan, *World now at the start of 2009 influenza pandemic*, World Health Organization, Geneva, 20 June 2009, http://www.who.int/mediacentre/news/statements/2009/h1n1_pandemic_phase6_20090611/en/.

5 *CDC protocol of real-time RT-PCR for influenza A (H1N1)*, World Health Organization, Geneva, 28 April 2009, http://www.who.int/csr/resources/publications/swineflu/CDCRealtimeRTPCR_SwineH1Assay-2009_20090430.pdf.

6 M. U. Kopp, A. J. de Mello and A. Manz, Chemical amplification: continuous-flow PCR on a chip, *Science*, 1998, **280**, 1046–1048.

7 K. Y. Lien, W. C. Lee, H. Y. Lei and G. B. Lee, Integrated reverse transcription polymerase chain reaction systems for virus detection, *Biosens. Bioelectron.*, 2007, **22**, 1739–1748.

8 F. C. Huang, C. S. Liao and G. B. Lee, An integrated microfluidic chip for DNA/RNA amplification, electrophoresis separation and on-line optical detection, *Electrophoresis*, 2006, **27**, 3297–3305.

9 R. Pal, M. Yang, R. Lin, B. N. Johnson, N. Srivastava, S. Z. Razzacki, K. J. Chomistek, D. C. Heldsinger, R. M. Haque, V. M. Ugaz, P. K. Thwar, Z. Chen, K. Alfano, M. B. Yim, M. Krishnan, A. O. Fuller, R. G. Larson, D. T. Burke and M. A. Burns, An integrated microfluidic device for influenza and other genetic analyses, *Lab Chip*, 2005, **5**, 1024–1032.

10 S. Mohr, Y.-H. Zhang, A. Macaskill, P. J. R. Day, R. W. Barber, N. J. Goddard, D. R. Emerson and P. R. Fielden, Numerical and experimental study of a droplet-based PCR chip, *Microfluid. Nanofluid.*, 2007, **3**, 611–621.

11 S. K. Cho, H. Moon and C.-J. Kim, Creating, transporting, cutting, and merging liquid droplets by electrowetting-based actuation for digital microfluidic citcuits, *J. Microelectromech. Syst.*, 2003, **12**, 70–80.

12 A. Egatz-Gómez, S. Melle, A. A. García, S. A. Lindsay, M. Márquez, P. Domínguez-García, M. A. Rubio, S. T. Picraux, J. L. Taraci, T. Clement, D. Yang, M. A. Hayes and D. Gust, Discrete magnetic microfluidics, *Appl. Phys. Lett.*, 2006, **89**, 034106.

13 J.-Y. Yoon and D. J. You, Backscattering particle immunoassays in wire-guide droplet manipulations, *J. Biol. Eng.*, 2008, **2**, 15.

14 Y.-H. Chang, G.-B. Lee, F.-C. Huang, Y.-Y. Chen and J.-L. Lin, Integrated polymerase chain reaction chips utilizing digital microfluidics, *Biomed. Microdevices*, 2006, **8**, 215–225.

15 T. Ohashi, H. Kuyama, N. Hanafusa and Y. Togawa, A simple device using magnetic transportation for droplet-based PCR, *Biomed. Microdevices*, 2007, **9**, 695–702.

16 H. Kim, S. Dixit, C. J. Green and G. W. Faris, Nanodroplet real-time PCR system with laser assisted heating, *Opt. Express*, 2009, **17**, 218–227.

17 *IPC-2221: genericstandard on printed board design*, The Institute for Interconnecting and Packaging Electronic Circuits, Bannockburn, IL, February 1998.

18 M. A. Larkin, G. Blackshields, N. P. Brown, R. Chenna, P. A. McGettigan, H. McWilliam, F. Valentin, I. M. Wallace, A. Wilm, R. Lopez, J. D. Thompson, T. J. Gibson and D. G. Higgins, ClustalW and ClustalX version 2.0, *Bioinformatics*, 2007, **23**, 2947–2948.

19 *FLUENT 5 user's guide*, version 5, Fluent, Inc., Lebanon, NH, 1998.

20 P. S. Mirade and J. D. Daudin, Computational fluid dynamics prediction and validation of gas circulation in a cheese-ripening room, *Int. Dairy J.*, 2006, **16**, 920–930.

Multiphoton microscopy of transdermal quantum dot delivery using two photon polymerization-fabricated polymer microneedles

Shaun D. Gittard,[a] Philip R. Miller,[a] Ryan D. Boehm,[a] Aleksandr Ovsianikov,[b] Boris N. Chichkov,[b] Jeremy Heiser,[cd] John Gordon,[cd] Nancy A. Monteiro-Riviere[ae] and Roger J. Narayan*[a]

Received 18th May 2010, Accepted 22nd June 2010
DOI: 10.1039/c005374k

Due to their ability to serve as fluorophores and drug delivery vehicles, quantum dots are a powerful tool for theranostics-based clinical applications. In this study, microneedle devices for transdermal drug delivery were fabricated by means of two-photon polymerization of an acrylate-based polymer. We examined proliferation of cells on this polymer using neonatal human epidermal keratinocytes and human dermal fibroblasts. The microneedle device was used to inject quantum dots into porcine skin; imaging of the quantum dots was performed using multiphoton microscopy.

Introduction

Theranostics, the combination of therapeutic and diagnostic technologies into a single platform, is a rapidly developing technology for treating a variety of medical conditions. For example, theranostics technologies may provide disease imaging and chemotherapy delivery for treatment of cancer, a group of conditions in which cells exhibit uncontrolled growth and invasion.[1–5] Other applications of theranostics that are being investigated include use in diabetes, ocular disease, obesity, cardiovascular diseases, and infectious diseases.[1,3,6]

Quantum dots (QD) are 2–10 nm diameter fluorescent semiconductor nanostructured materials that are being considered for use in several theranostics applications. Due to the fact that the exciton Bohr radius (average electron-hole distance) is larger than the nanoparticle radius, these materials undergo quantum confinement. As a result, quantum dots demonstrate characteristic excitation states as well as larger bandgap values than bulk material. In addition, emission wavelength in these materials is size dependent; bandgap values for quantum dots increase as the nanoparticle radius decreases.[7–10] Core-shell quantum dots are composed of a core material with a smaller bandgap than the surrounding shell material; zinc sulfide shells surrounding cadmium telluride cores, zinc sulfide shells surrounding cadmium

[a] *Joint Department of Biomedical Engineering, University of North Carolina State University, Campus Box 7115, Raleigh, NC, 27695-7115, USA. E-mail: roger_narayan@ncsu.edu; Fax: +1 509 696 8481*
[b] *Department of Nanotechnology, Laser Zentrum Hannover, Hollerithalle 8, 30419 Hannover, Germany*
[c] *Ceramatec Inc., 2425 South 900 West, Salt Lake City, UT, 84119, USA*
[d] *MicroLin LLC & Technology Holding LLC, 350 West 800 North, Suite 250, Salt Lake City, UT, 84103, USA*
[e] *Center for Chemical Toxicology Research & Pharmacokinetics, North Carolina State University, Raleigh, NC, 27606, USA*

selenide cores, and cadmium sulfide shells surrounding cadmium selenide cores are some examples of this type of quantum dots.[7,11–13]

Quantum dots exhibit unique photoluminescence behavior; some advantages of quantum dots over conventional fluorophores include greater resistance to light-dependent oxidation (photobleaching) and higher brightness, which is attributed to high quantum yield values and large molar extinction coefficient values.[14] Passive mechanisms (*e.g.*, preferential retention of quantum dots within tumor cells) may be utilized. Quantum dots may also be conjugated with peptides, antibodies, aptamers, pharmacologic agents, and other tumor-specific molecules that allow for specific tumor-cell interactions.[15–20] For example, Derfus *et al.* conjugated tumor-homing peptides and small interfering RNA molecules to PEGylated quantum dots. They demonstrated that the F3 tumor-homing peptide enabled tumor cell internalization.[16] When F3 tumor-homing peptide/small interfering RNA-quantum dots were delivered to enhanced green fluorescent protein-transfected HeLa cells, knockdown of the enhanced green fluorescent protein signal was observed. Bagalkot *et al.* demonstrated processing of conjugates containing A10 aptamer (an RNA molecule that recognizes the extracellular domain of prostate-specific membrane antigen), doxorubicin (an anthracycline antineoplastic agent), and fluorescent quantum dots; *in vitro* studies demonstrated that these mutifunctional nanoparticles may be used for simultaneous targeted imaging, therapy, and sensing of prostate cancer cells.[19] Tada *et al.* conjugated trastuzumab to quantum dots coated with polyethylene glycol amine; these structures were subsequently labeled with monoclonal anti-HER2 antibody. The antibody-conjugated therapeutic nanoparticles were successfully demonstrated using a dorsal skinfold chamber in a murine model.[15] Cai *et al.* conjugated a thiolated arginine-glycine-aspartic acid (RGD) peptide with polyethylene glycol-coated quantum dots; these structures were successfully utilized for *in vivo* tumor imaging in a murine model.[20]

It should be noted that quantum dots and other nanoscale pharmacologic agents cannot be administered in enteral form since they may be sequestered within the intestine, kidneys, liver, or lungs prior to entering systemic circulation.[21,22] Intravenous delivery is the most common mechanism for quantum dot administration at this time because it enables instantaneous delivery of the quantum dots into the bloodstream. However, use of hypodermic needles is associated with several shortcomings, including injection site trauma, patient pain, and difficulty in providing sustained administration over an extended period of time. One option for quantum dot delivery involves the use of microneedles, which are needle-, lancet-, or thorn-shaped structures in which one dimension is less than 500 μm. These devices penetrate the stratum corneum layer of the epidermis; this 15 μm layer contains keratinized dead cells and serves as a barrier against transdermal movement of pharmacologic agents through the skin. Quantum dots may subsequently enter dermal blood vessels by passing through the lower part of the epidermis. Hollow microneedles enable pressure-driven or diffusion-driven delivery of quantum dots through the skin to be modulated over an extended period of time. In addition, microneedles may be used both for delivery of quantum dots as well as for withdrawal of blood and/or interstitial fluid. Microneedle devices commonly contain several microneedles; use of microneedle arrays is associated with injection at higher rates as well as injection over wider areas than use of solitary microneedles.[23] Furthermore, redundant microneedles within the microneedle arrays may be utilized if some of the microneedles within the array are obstructed or damaged during use. Use of microneedles is associated with minimal tissue compression, nerve ending interaction, pain sensation, injection site trauma, and injection site inflammation.[24–27]

Recent studies have demonstrated delivery of nanomaterials by means of microneedles. For example, Coulman *et al.* used microneedles prepared by means of wet-etch techniques for delivery of fluorescent nanospheres across human epidermal membranes.[28] McAllister *et al.* demonstrated that pores in the stratum corneum layer of human cadaver abdomen or back skin created by a solid microneedle array enabled transport of polystyrene latex nanoparticles[29] Bal *et al.* recently utilized laser scanning microscopy to visualize movement of sodium fluorescein dye thorugh conduits in human skin, which were created by 300 μm long DermaStamp® microneedles.[30] They proposed that laser scanning microscopy may be used to observe the dimensions of microneedle-fabricated conduits in viable human skin over time.

Microneedles with various geometries as well as microneedle arrays may be fabricated using a laser-based rapid prototyping process known as two photon polymerization.[31–37] Two-photon polymerization involves temporal and spatial overlap of photons, which enables highly localized polymerization of a photosensitive resin. Nearly simultaneous absorption of two photons creates a so-called virtual state; this electronic excitation is analogous to excitation by a single photon with a higher energy. A diagram of the two-photon polymerization process is provided in Fig. 1a. The minimum feature size obtained using two-photon polymerization is can be altered by changing the optics (*e.g.*, the objective used to focus the laser beam). Two photon polymerization provides several advantages over conventional microelectronics-based techniques for fabrication of microneedles and other small-scale medical devices. The photosensitive materials used in two photon polymerization are widely available and are obtained at low cost. In contrast with cleanroom-based microfabrication techniques, photon polymerization can be performed in conventional environments. As opposed to conventional multiple-step microfabrication methods, two photon polymerization can create microneedles and other complex small-scale structures in a single-step process. Two photon polymerization has been used to fabricate a variety of small-scale medical devices, including scaffolds for regenerative medicine, ossicular replacement prostheses, and microneedles.[31–38] For example, Doraiswamy *et al.* showed that organically-modified ceramic (Ormocer®) microneedles did not exhibit fracture during compression load testing with porcine skin.[31] Ovsianikov *et al.* fabricated off-center microneedles using two photon polymerization; these off-center microneedles were created by altering the position of the bore relative to the central axis.[32] The off-center microneedle arrays successfully penetrated porcine adipose tissue; no fracture was observed. Doraiswamy *et al.* recently demonstrated delivery of a quantum dot solution using microneedles fabricated using two photon polymerization in a porcine skin model.[25] In this study, Ormocer® microneedles were produced on perforated glass substrates by two photon polymerization. Quantum dot dispersion was observed by sectioning, differential interference contrast (Nomarski) microscopy, and widefield fluorescence microscopy over a period of one hour. The two photon polymerization-fabricated organically-modified ceramic microneedles enabled more rapid distribution of

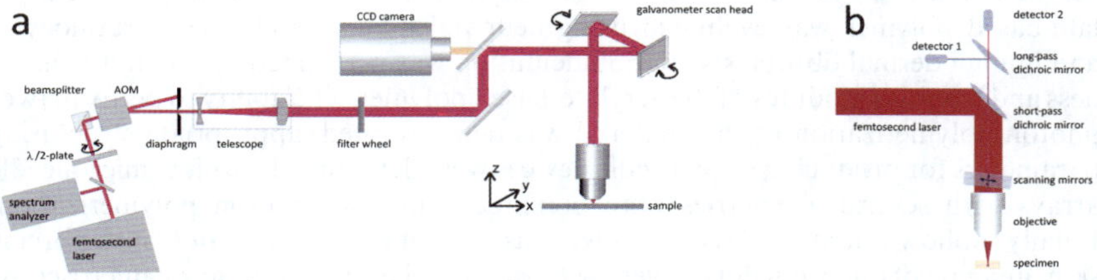

Fig. 1 (a) Schematic of the two photon polymerization system. Femtosecond laser pulses pass through a beam expander and a filter wheel before being focused on the objective. Scanning mirrors control the location of the laser spot in the focal plane. Polymerization of the resin occurs at the focal point. A charge-coupled device camera behind a dichroic mirror is used for visualizing the two photon polymerization process. (b) Schematic of multiphoton microscopy system. Femtosecond laser pulses pass through the objective before being focused on the specimen, where excitation occurs at the focal point. Scanning mirrors control the location of the laser spot in the focal plane. Dichroic mirrors are used to separate excitation and emission light and to separate emitted light into two channels.

polyethylene glycol-amine quantum dot solution to the deep epidermis and the dermis of porcine skin than topical administration.

Multi-photon fluorescence microscopy is a microscopy technique that may be used for imaging quantum dots as well as other fluorophores.[39] Multi-photon fluorescence microscopy involves nearly simultaneous absorption of two long-wavelength photons from a femtosecond laser by a fluorophore; unlike traditional fluorescence microscopy, the excitation wavelength is longer than the emission wavelength. A diagram of the multi-photon microscopy technique is provided in Fig. 1b. This technique enables imaging with depth resolutions of 800 μm, three-dimensional data acquisition from differentially labelled fluorescent structures, as well as imaging of living structures for extended periods of time at high spatial resolutions. For example, Wang *et al.* used multi-photon confocal microscopy to observe microscopic holes created by a microneedles in skin biopsy samples.[40] Kim *et al.* used multiphoton microscopy to observe surfactant (*N*-lauroyl sarcosine)-mediated enhancement of fluorescein penetration into human cadaver skin.[41] Choi *et al.* observed delivery of calcein into DU145 human prostate cancer cells by means of a microneedle array with electroporation functionality.[42] The non-invasive nature, absence of significant phototoxicity, and tissue penetration depth provided by multi-photon microscopy make it an appealing tool for theranostics applications. For example, Maffia *et al.* have used multiphoton microscopy to examine immune cell behavior in an artery disease model.[43] In their study, multiphoton microscopy was used to image fluorescently tagged lymphocyte migration within the adventitia of intact carotid arteries in apolipoprotein-E–deficient mice. Stroh *et al.* used quantum dots to label bone marrow-derived precursor cells; multiphoton intravital microscopy was used to observe movement of bone marrow-derived precursor cells to tumor vessels.[44]

Quantum dots are appealing fluorophores for use in conjunction with multi-photon microscopy in theranostic applications. According to Larson *et al.*, quantum dots exhibit multi-photon excitation cross-sections of up to 47 000 Goeppert-Mayer units.[45] They examined cadmium selenide–zinc sulfide quantum dots, which were located inside capillaries hundreds of micrometres deep within the skin of living mice; no fluorescence intermittency ("blinking") over nanosecond-millisecond time scales was observed. Lee *et al.* examined infusion of COOH-coated QD and QD-PEG in isolated perfused skin by means of inductively coupled plasma emission spectrometry as well as fluorescence spectroscopy.[46] Unlike arterial extraction of conventional pharmacologic agents (*e.g.*, carboplatin, cisplatin, lidocaine, and testosterone), arterial extraction of quantum dots showed unique periodic behavior.

In this study, hollow microneedle devices for transdermal delivery of quantum dots and other nanoscale pharmacologic agents were fabricated out of a commercial acrylic polymer by means of two-photon polymerization. Cell viability of the acrylate-based polymer was evaluated using neonatal human epidermal keratinocytes and human dermal fibroblasts. Nanoindentation was performed to obtain the hardness and Young's modulus of the acrylate-based polymer. The ability to perform two-photon polymerization of this material was demonstrated; appropriate structuring parameters for producing microneedle devices were determined. Hollow microneedle arrays with several geometries were prepared using two-photon polymerization. Finally, hollow microneedles devices were used to inject quantum dots into porcine skin; imaging of quantum dot delivery was performed using multiphoton microscopy. This study suggests that two photon polymerization may provide a unique approach for fabricating polymeric microneedle arrays, which may be used for transdermal delivery of quantum dots for theranostics and other medical applications.

Methods

An ultraviolet light-sensitive acrylate-based polymer, e-shell 300 (EnvisionTEC, Gladbeck, Germany), was examined in this study. This rigid, tough, perspiration-resistant material is processed using single-photon polymerization for use in a variety of medical applications, including hearing aid shells and other Class IIa medical devices. Class IIa medical devices include non-invasive active medical devices for channeling and/or storing liquids for administration to the human body.[47] Information provided by the supplier indicates that e-shell is comprised of 10–25 wt% urethane dimethacrylate (CAS 72869-86-4) and 10–20% tetrahydrofurfuryl-2-methacrylate (CAS 2455-24-5).

The two photon polymerization process was used to produce several types of microneedle structures from the acrylate-based polymer. Femtosecond pulses (60 fs, $\lambda = 780$ nm) from a Chameleon titanium:sapphire laser (Coherent, Santa Clara, CA) were used to polymerize the liquid acrylate-based polymer resin. The resin was polymerized along the trace of the laser focus, which was moved in three dimensions. A hurrySCAN® galvano scan head (Scanlabs, Puchheim, Germany) was used to control laser writing in lateral (X- and Y-) dimensions. Movement in the height (Z-) dimension and movement from one structure to another structure were achieved using three C-843 linear translation stages (Physik Instrumente, Karlsruhe, Germany). Fabrication of the microneedle structures was guided by input stereroli-thography (STL) files, which were prepared using Solidworks Education Edition 2009 commercial software (Dassault Systemes SA, Velizy, France). The unpolymer-ized acrylate-based polymer resin was placed in between two glass cover slips; the resin was enclosed using a 1 mm thick polydimethylsiloxane ring. The structures were washed using isopropanol in order to remove the unirradiated resin. The structures were subsequently exposed to an ELC-410 ultraviolet curing lamp (Electro-Lite, Bethel, CT).

Solid microneedles were initially fabricated using several processing parameters in order to obtain appropriate conditions for subsequent fabrication of hollow micro-needles. Above an upper threshold of laser intensity, burning of the resin will occur. Below a lower threshold of laser intensity, polymerization of the resin will not occur. Processing parameters also alter voxel height and voxel width; these parameters affect shape and integrity of two photon polymerization-fabricated structures. An array of solid microneedles was initially fabricated using two photon polymeriza-tion; structures were fabricated with an input file height of 750 μm and base diameter of 125 μm. The objective, laser energy, laser wavelength, layer spacing, raster spacing, and mark speed were maintained at 10×, 320 mW, 780 nm, 10 μm, 1.5 μm, and 50–250 (arbitrary units) respectively. Hollow microneedles were subsequently fabricated on glass cover slips by means of two photon polymerization. One set of microneedles was produced using an input file height of 375 μm, base diameter of 250 μm, and channel diameter of 30 μm. Another set of microneedles was produced using an input file height of 500 μm, base diameter of 250 μm, and channel diameter of 30 μm. The objective, laser energy, laser wavelength, layer spacing, raster spacing, and mark speed were maintained at 10×, 320 mW, 780 nm, 10 μm, 1.5 μm, and 100 respectively. Freestanding microneedle arrays, containing both the substrate and the microneedles, were then fabricated using two photon polymerization. The 5× objective was used instead of the 10× objective due to limitations associated with radial laser intensity degradation in the focal plane. Changing the objective resulted in a reduction in two photon processing resolution; however, changing the objective was necessary in order to fabricate the large freestanding microneedle array structure. Freestanding microneedle arrays were fabricated in three steps due to radial laser intensity degradation. The border was initially fabricated; this structure enables handling of the freestanding microneedle array without damage to the microneedles. The objective, laser energy, laser wavelength, layer spacing, raster spacing, and mark speed were maintained at 5×, 570 mW, 780 nm, 25 μm, 2 μm, and 40 respectively. The substrate was then fabricated; it exhibited a three-by-three array structure with 500 μm spacing. The objective, laser energy, laser wavelength, layer spacing, raster spacing, and mark speed were maintained at 5×, 370 mW, 780 nm, 25 μm, 1.4 μm, and 60 respectively. A 100 μm

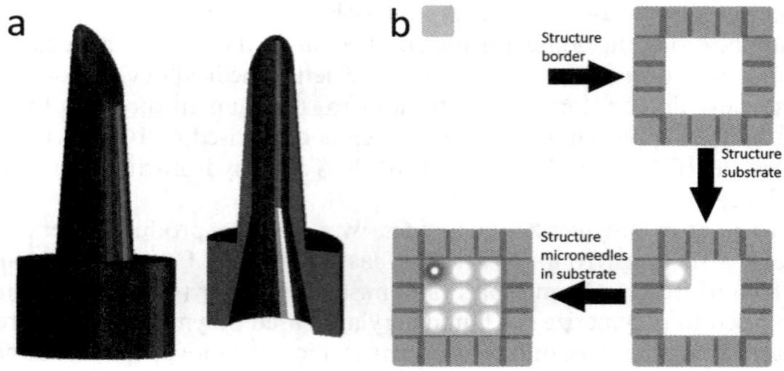

Fig. 2 (a) Input computer-aided design drawing of hollow microneedles processed *via* two photon polymerization. (b) Diagram of hollow microneedle device fabrication by means of two photon polymerization.

overlap between the border and the substrate was used to ensure good bonding between structures. An array of microneedles was subsequently fabricated on the substrate. A computer-aided design drawing of the microneedle structure is shown in Fig. 2a. Microneedles were produced with a conical opening at the base; structures were fabricated with an input file cylindrical base height of 250 μm, cylindrical base diameter of 400 μm, needle height of 608 μm, needle base diameter of 200 μm, and channel diameter of 110 μm. The objective, laser energy, laser wavelength, layer spacing, raster spacing, and mark speed were maintained at 5×, 370 mW, 780 nm, 25 μm, 1.4 μm, and 60 respectively. A 60 μm overlap between the substrate and the microneedles was used to ensure a strong bonding between structures. The structures were subsequently exposed to an ultraviolet curing lamp. The structures were subsequently removed from the glass cover slips. The structures were washed using isopropanol in order to remove the unirradiated resin. Details on fabrication of free-standing microneedle arrays are provided in Fig. 2b. To confirm the unobstructed nature of the channels, 200 mbar vacuum suction was applied; isopropanol was placed on the array and pulled through the microneedles.

Scanning electron microscopy imaging of the microneedle arrays was performed using an S-3200 variable pressure instrument (Hitachi, Tokyo, Japan) with a Robinson™ backscattered electron detector. Energy dispersive X-ray spectroscopy was used to obtain the elemental composition of the acrylate-based polymer microneedle arrays. Fourier transform infrared spectroscopy (FTIR) was performed using a Nexus 470 system with a continuum microscope and an OMNI sampler (Thermo Fisher, Waltham, MA); spectral analysis was performed with OMNIC™ software (Thermo Fisher, Waltham, MA). Material for Fourier transform infrared spectroscopy was prepared by sandwiching resin between two glass cover slips with a polydimethylsiloxane spacer (thickness = 1 mm, diameter= 5 mm); the resin was polymerized using an ELC-410 ultraviolet curing lamp.

Elastic modulus (relative stiffness of a material against elastic deformation) and hardness (resistance of a material to penetration and/or plastic deformation) are critical parameters for materials used in fabrication of microneedle devices. Material for nanoindentation testing was prepared by sandwiching resin between two glass cover slips with a polydimethylsiloxane spacer (thickness = 1 mm, diameter= 5 mm); the resin was polymerized using an ELC-410 ultraviolet curing lamp. An Ultra Nano Hardness Tester (CSM Instruments, Needham, MA) containing an indenter tip with Berkovitch geometry was used to obtain elastic modulus and hardness data for the acrylate-based polymer. Ten indentations were performed; the indenter tip was driven into the surface of the material with up to 2 mN normal load, 4 mN min^{-1} loading rate, and 4 mN min^{-1} unloading rate. Partial or complete relaxation was achieved by reducing the load after the maximum load had been achieved.

Indentation hardness and elastic modulus were obtained from the load-displacement data using the Oliver and Pharr method.[48]

Proliferation of cells on the acrylate-based polymer was assessed by the MTT (3-(4,5-dimethylthiazol-2-yl)2,5-diphenyl tetrazolium bromide) assay; this assay is based on reduction of a yellow tetrazolium salt (MTT) to a purple formazan dye by the succinic dehydrogenase enzyme within mitochondria.[49] Glass cover slips (diameter = 12.5 mm) (n = 4), empty wells within γ-irradiated sterile polystyrene plates (n = 4) (Sigma-Aldrich, St. Louis, MO), and e-shell 300 acrylate-based polymer wafers (n = 4) were evaluated. Cylindrical wafers (thickness = 1.5 mm, diameter = 14.0 mm) of the acrylate-based polymer were prepared using a Perfactory® Standard SXGA+ UV (EnvisionTEC, Gladbeck, Germany) stereolithography system. This system uses stereolithography (STL) files to guide an ultraviolet light beam over the acrylate-based polymer resin. Illumination in the X- and Y-dimensions was regulated by directed light projection (DLP) optics (Texas Instruments, Dallas, TX); the chip used for wafer fabrication exhibits 1280 × 1024 pixel resolution. Acrylate-based polymer samples were subsequently washed in isopropanol and post-cured using the ELC-410 UV curing lamp. The acrylate-based polymer wafers and glass cover slips were rinsed twice in 70% ethanol for thirty minutes and then exposed to ultraviolet B light for two hours. The acrylate-based polymer wafers and glass cover slips were subsequently rinsed twice in Hanks' Balanced Salt Solution (HBSS) and once in culture medium. The acrylate-based polymer wafers and glass cover slips were then placed in 2 mL of medium and stored in an incubator until seeding. Cryopreserved neonatal human epidermal keratinocytes (HEK) and human dermal fibroblasts (HDF) were purchased from a commercial source (Lonza, Walkersville, MD). Cells were propagated in 75 cm^2 flasks, grown to 75% confluency, harvested, and seeded (40 000 cells per well) on acrylate-based polymer wafers, on glass cover slips, and in polystyrene well plates. Culturing was performed with keratinocyte growth media (KGM-2) (Lonza, Walkersville, MD) and fibroblast growth media (FGM-2) (Lonza, Walkersville, MD). The acrylate-based polymer wafers were compared to both empty polystyrene well plates and glass cover slips. Since aspiration of solutions was associated with detachment of cells from the test materials, all media changes and material rinsing were accomplished by moving the test materials from one solution to the other with forceps. Aspiration was performed for evaluation of cell proliferation within the empty wells. Test materials were placed in fresh medium after forty-eight hours, which correlated with 80% confluency of both neonatal human epidermal keratinocytes and human dermal fibroblasts; cell proliferation on test materials was assessed twenty-four hours later. The test materials were rinsed with Hanks' Balanced Salt Solution, desorbed with isopropyl alcohol, and agitated. 100 µl of isopropyl alcohol was transferred to a new twenty-four well plate; absorbance was spectrophotometrically evaluated (λ = 550 nm) using a Multiskan RC plate reader (Labsystems Inc, Franklin, MA).

Since human skin is similar to porcine skin, an *ex vivo* porcine skin model was used to evaluate quantum dot delivery.[50] Full thickness skin was obtained from euthanized female weanling Yorkshire pigs. The back area of the pig was clipped; two days later, full thickness skin was surgically removed. The skin was refrigerated until the quantum dot injection, which occurred one day after surgical removal. Animal care and experimental use were performed according to approved guidelines by the local Animal Care and Use Committee (North Carolina State University, Raleigh, NC).

A LSM-710 multiphoton microscope (Carl Zeiss AG, Oberkochen, Germany) was used for imaging delivery of quantum dots into porcine skin. Qdot® 565 ITK™ carboxyl quantum dots (amount = 250 µL, concentration = 8 µM, pH = 9.0) were obtained from a commercial source (Invitrogen, Carlsbad, CA). Information provided by the supplier indicates that the quantum dots contained a CdSe core and a crystalline ZnS shell; an amphophilic carboxyl coating enables dispersion of quantum dots in aqueous solutions and may serve to minimize generation of free cadmium.

The freestanding microneedle array was initially pressed into the skin. A piece of polydimethylsiloxane containing a hole that was larger than the microneedle array but smaller than the substrate was placed on the backside of the microneedle array in order to obtain a seal. A syringe was filled with the quantum dot solution. A piece of polydimethylsiloxane with a Luer-lock fitting was attached in order to completely seal the connection between the array and the syringe. 250 μL of 8 μM quantum dot solution was diluted in deionized water to a total volume of 1 mL prior to injection; 250 μl of this diluted solution was injected *via* the microneedle array prior to imaging. The microneedle array was kept in the skin during imaging. Quantum dot solution was topically applied to the surface of a separate porcine skin for comparison purposes.

The microneedles and the quantum dots were simultaneously imaged since multiphoton excitation occurs for both the e-shell 300 acrylate based polymer and the Qdot® 565 ITK™ carboxyl quantum dots at $\lambda = 800$ nm. Multiphoton microscopy was performed with a femtosecond laser ($\lambda = 800$ nm) (Chameleon, Coherent, Santa Clara, CA); a pixel dwell time of 1.58 μs were used for imaging. A 20× plan-apochromat water immersion objective was utilized in this study; a droplet of water was placed on the backside of the microneedle array. Water was the only medium between the objective and the sample. Z-stack images were obtained of (a) a porcine skin section containing topically applied quantum dot solution, (b) a porcine skin section containing a freestanding microneedle array without injection of quantum dots, and (c) a porcine skin section containing a freestanding microneedle array immediately after quantum dot injection; image acquisition was completed less than fifteen minutes after quantum dot injection. Imaris 7.0 image analysis software (Bitplane AG, Zurich, Switzerland) was used to produce surface rendering images of the microneedles and maximum projection images of the quantum dots from the Z-stack data; these images provide enhanced visualization of quantum dot delivery.

Results and discussion

The Fourier transform infrared spectrum of e-shell 300 acrylate based polymer is shown in Fig. 3b. Sharp peaks located at 2960.8 cm^{-1}, 1721.5 cm^{-1}, 1509.9 cm^{-1}, 1454.6 cm^{-1}, 1296.6 cm^{-1}, 1248.8 cm^{-1}, 1160.8 cm^{-1}, 1135.8 cm^{-1}, 1065.3 cm^{-1}, and 831.7 cm^{-1} were observed. The peak at ~2961 cm^{-1} was attributed to aliphatic hydrocarbon bond asymmetric stretching. The peak at ~1721 cm^{-1} was attributed to C=O bond stretching. The broad peak at ~3373 cm^{-1} was attributed to N–H bond stretching.[51] The energy dispersive X-ray spectrum of the e-shell 300 acrylate based polymer indicates that it contains carbon and oxygen; other elements, including those with known toxicity, were not observed. Nanoindentation testing of e-shell 300 acrylate based polymer provided hardness and Young's modulus values of 162.9 ± 2.3 MPa and 2.8 GPa ± 0.0, respectively. Park *et al.* indicated that microneedles produced from materials with a Young's modulus values greater than ~1 GPa possess fracture forces that surpass skin insertion forces.[52] Park *et al.* recently demonstrated fabrication of microneedles out of poly-lactic-*co*-glycolic acid, which exhibits a Young's modulus of 3 GPa; they described the relationship between Young's modulus, aspect ratio, and failure force.[53]

The twenty-four hour MTT assay for proliferation of neonatal epidermal keratinocytes and human dermal fibroblasts on e-shell 300 acrylate based polymer are shown in Fig. 4. Statistical analysis of the cell proliferation data was performed using the Student's t-test ($p < 0.05$). The test materials remained stable and intact in the cell culture media. MTT viability of neonatal human keratinocytes was significantly higher for e-shell 300 acrylate-based polymer than for glass. MTT viability of neonatal human keratinocytes was significantly lower for e-shell 300 acrylate-based polymer than for polystyrene well plates. MTT viability of human dermal fibroblasts on e-shell 300 was significantly lower than on glass or on polystyrene well plates.

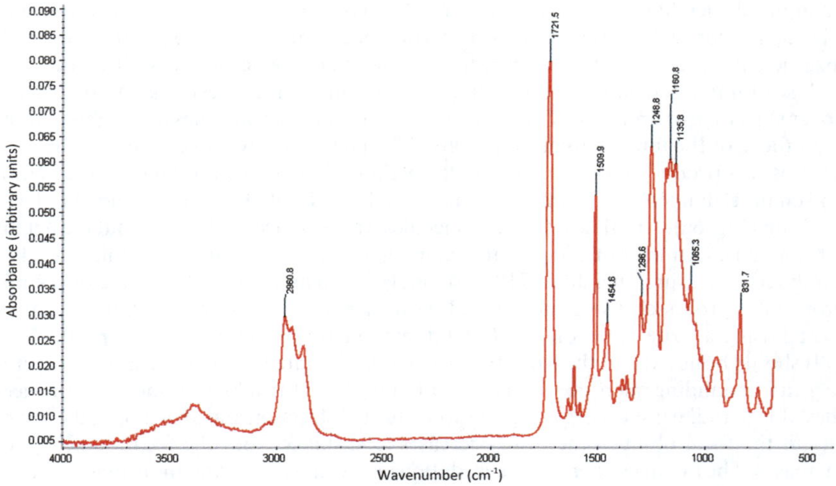

Fig. 3 Fourier transform infrared spectrum of e-shell 300 acrylate-based polymer.

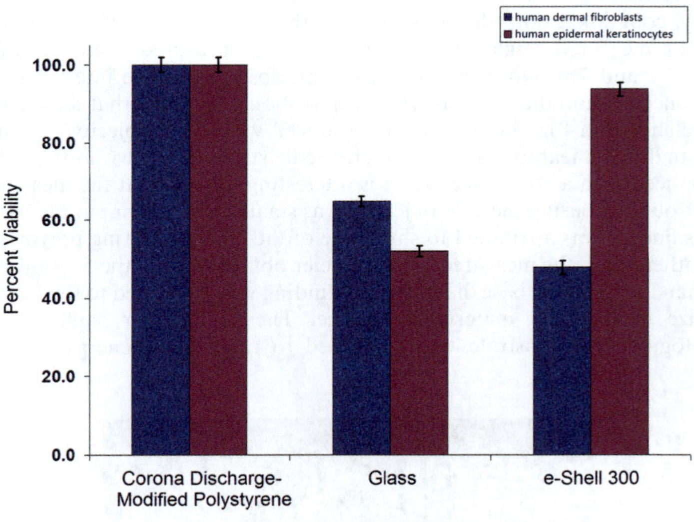

Fig. 4 MTT viability of neonatal human epidermal keratinocytes and human dermal fibroblasts on e-shell 300 acrylate-based polymer, polystyrene well plate, and glass is shown. Error bars indicate standard error of the mean.

Information provided by the supplier indicates that the surfaces of the polystyrene wells were modified with corona discharge in order to incorporate oxygen-containing chemical groups within surface polystyrene chains; increase surface hydrophilicity; and enhance cell spreading and attachment. The glass cover slips were not modified to enhance cell spreading and attachment. The acrylate-based polymer did not exhibit cytotoxicity in a manner that would raise concerns regarding potential *in vivo* use in a microneedle device.

A limitation on the writing speed of the two photon polymerization process for e-shell 300 acrylate-based polymer microneedle fabrication was obtained from scanning electron microscopy images of solid microneedles, which were prepared using an input file height of 750 μm and an input file base diameter of 125 μm. As shown in Fig. 5, microneedles processed using write speeds above 150 exhibited tips with insufficient structural integrity; bending of the microneedle tips was observed.

A scanning electron microscopy image of a hollow microneedle array on a glass cover slip is shown in Fig. 6. The measurements of the microneedles are slightly larger than those of the input stereolithography (STL) files. For example, one set of microneedles was produced using an input file height of 375 μm, base diameter of 250 μm, and channel diameter of 30 μm; these microneedles exhibited a measured height of 614 ± 12 μm (Fig. 6a). Another set of microneedles was produced using an input file height 500 μm, base diameter of 250 μm, and channel diameter of 30 μm; these microneedles exhibited a measured height of 710 ± 10 μm (Fig. 6b and 6c). Unlike some of the microneedles processed using more rapid write speeds (Fig. 5), the microneedles processed using a write speed of 100 (Fig. 6d) exhibited tips that remained upright.

It should be noted that the 10× objective is not well suited for making large structures (*e.g.*, a freestanding microneedle array, containing both the substrate and the microneedles) due to radial laser energy degradation. Due to refraction and aberration, the intensity at the focal point in the resin decreases as distance from the focal plane center increases. The radial laser energy degradation between the edge and the center of a large structure (*e.g.*, a freestanding microneedle array) was so great that the structure could not be fabricated using a 10× objective. In moving from the 10× objective to the 5× objective, the radial laser energy degradation was reduced by approximately half.

Fig. 7 contains scanning electron microscopy images of a freestanding microneedle array, containing both the substrate and the microneedles. These microneedles exhibited a measured height of 508 ± 33 μm and a measured diameter of 212 ± 3 μm (Fig. 7a and 7b). No discontinuities were observed at the interfaces between the microneedles and the substrate (Fig. 7c). It should be noted that the microneedle structure shown in Fig. 7a, which was produced with a 5× objective, exhibits less precise small-scale features than the microneedle structure shown in Fig. 6a, which was produced with a 10× objective. It is interesting to note that the measured base diameter obtained using the 10× objective was smaller than the input file base diameter; this finding was attributed to shrinkage of the material during polymerization. On the other hand, the measured base diameter obtained using the 5× objective was larger than the input file base diameter; this finding was attributed to the fact that the voxel size exceeds the material shrinkage. The nonuniform "pillow top"-like morphology of the substrate was attributed to radial laser energy degradation.

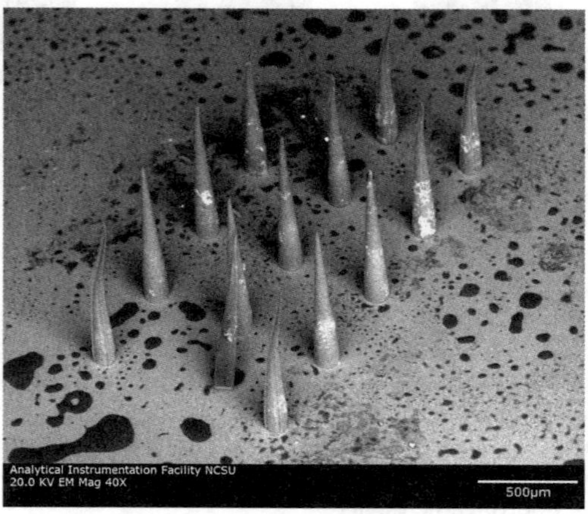

Fig. 5 Scanning electron microscopy image obtained at 45° tilt of solid e-shell 300 microneedles, which were created by means of two photon polymerization. Mark speeds (from front to back) of 250, 50, 100, 150, and 200 are shown.

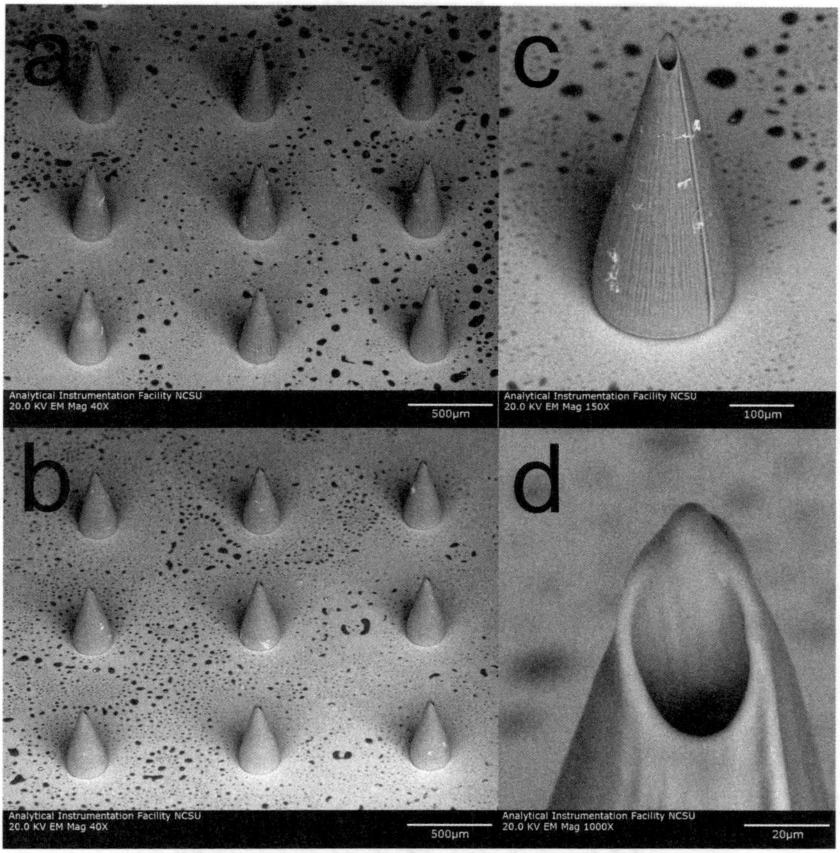

Fig. 6 Scanning electron microscopy images obtained at 45° tilt of e-shell 300 hollow microneedles on glass substrates, which were produced using two photon polymerization. (a) Image of 614 ± 12 μm long microneedle array. (b) Image of 710 ± 10 μm long microneedle array. (c) Image of 710 ± 10 μm long individual microneedle. (d) Image of 710 ± 10 μm long individual microneedle. The base diameter of these microneedles is 226 ± 5 μm. Dimensions are shown as average ± standard deviation.

Minimized processing time and minimized processing cost are obtained by performing two photon polymerization using the most rapid possible write speed. Objective election and mark speed are two of several laser-material interaction parameters that can be readily altered to produce structures with optimized geometries using two photon polymerization.

Multiphoton microscopy was used to observe quantum dot delivery into porcine skin *via* a two photon polymerization-fabricated e-shell 300 microneedle array as well as *via* topical application. Z-stack images of a microneedle in porcine skin prior to quantum dot injection (Fig. 8a), a microneedle in porcine skin after quantum dot injection (Fig. 8b), and quantum dots topically applied to porcine skin (Fig. 8c) are shown. The microneedles are presented as surface renderings (in gray) and the quantum dots are presented as maximum projections (in red). Thresholding values for surface renderings were 62.800 μm^2 for the Z-stack images of the microneedle in porcine skin prior to quantum dot injection and 35.040 μm^2 for the Z-stack images of the microneedle in porcine skin after quantum dot injection.

The e-shell 300 acrylate-based polymer was noted to fluoresce under multiphoton excitation; the use of e-shell 300 or similar materials may be beneficial for

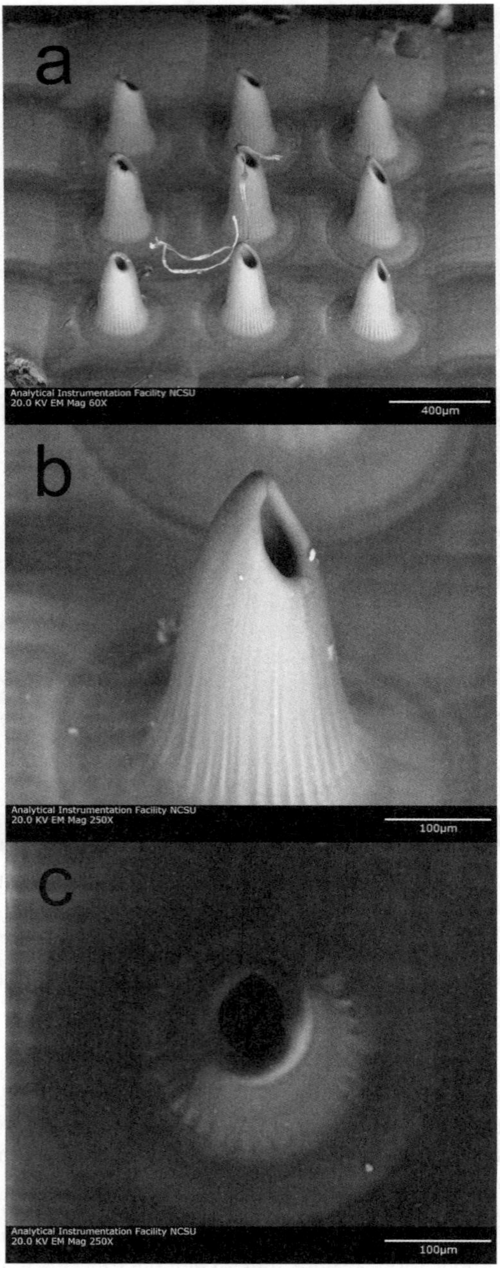

Fig. 7 Scanning electron microscopy images of e-shell 300 hollow microneedle array, which was produced using two photon polymerization. (a) Image of microneedle array obtained at 45° tilt. (b) Image of individual microneedle obtained at 45° tilt. (c) Image of individual microneedle obtained at 0° tilt. The length and base diameters of these microneedles are 508 ± 33 μm and 212 ± 3 μm, respectively. Dimensions are shown as average ± standard deviation.

transdermal theranostics activities involving multiphoton microscopy. Since time is needed to acquire Z-stack image data and adjust image acquisition settings, quantitative time-dependent data on quantum dot movement over time was unable to be

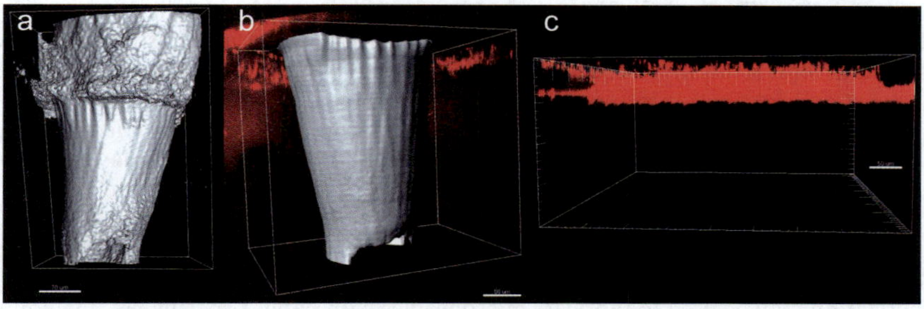

Fig. 8 Multiphoton microscopy images of quantum dot injection *via* two photon polymerization-fabricated e-shell 300 microneedle array as well as *via* topical application. The microneedles are presented as surface renderings (in gray) and the quantum dots are presented as maximum projections (in red). (a) Microneedle in porcine skin prior to quantum dot injection. (b) Microneedle in porcine skin after quantum dot injection. A broad distribution of the quantum dots in the deep epidermis and dermis was observed. (c) Quantum dots topically applied to porcine skin. The topically applied quantum dots exhibited poor penetration and remained in the topmost 50 μm region of the epidermis.

obtained. Another drawback of the multiphoton microscopy setup involved the use of a water immersion lens, which required a water column between the objective and the sample. Water was placed on the backside of the microneedle array. It is possible that capillary forces in the channels pulled some of the water through the microneedles and even into the skin, resulting in dilution of quantum dot solution. Refraction from the skin resulted in reduced image quality from deep tissue regions; neither the e-shell 300 acrylate-based microneedles nor the quantum dots could be imaged beyond a depth of ~500 μm from the skin surface. Levene *et al.* recently demonstrated the use of gradient index lenses with needle-like dimensions for multiphoton microscopy with subcellular resolution; *in vivo* images of quantum dots within tissues that were several millimetres within an anesthetized, intact murine model (*e.g.*, hippocampal tissue and cortical layer tissue) were obtained.[55]

The topically applied quantum dots exhibited poor penetration; they remained in the topmost 50 μm region of the epidermis. Ryman-Rasmussen *et al.* investigated topical administration of Qdot® 565 ITK™ carboxyl quantum dots; the carboxyl quantum dots were shown to penetrate the stratum corneum and primarily localize in the epidermis by eight hours; fluorescence intensity maps suggested that penetration occurred by means of a passive diffusion mechanism.[54] They suggested that quantum dot penetration by carboxylic acid-coated quantum dots (hydrodynamic size = 14 nm) occurred by means of diffusion within the intracellular space. After being pressed against the skin, the microneedles penetrated to nearly their full length. After quantum dot injection *via* the microneedle array, broad distribution of the quantum dots in the deep epidermis and dermis was observed. Quantum dots administered *via* microneedle injection were well distributed within fifteen minutes. Quantum dots were also observed in between the microneedles and the skin; this finding was attributed to migration of the quantum dot solution along the microneedle-created pores. In previous work, Gittard *et al.* used solid e-Shell 200 acrylate-based polymer microneedle arrays to create pores in human stratum corneum and epidermis. Optical imaging of the microneedle-created pores revealed that they were irregular in shape and were smaller than the microneedle dimensions; this finding was attributed to anisotropic tensile forces associated with collagen and other skin tissue components.[33] The small microneedle channel dimensions may enable precise delivery of a theranostics agent.

Conclusions

In this study, we demonstrated that two-photon polymerization can be used to produce a freestanding microneedle array, containing both the substrate and the microneedles. The e-shell 300 acrylate-based polymer microneedles successfully created pores in the stratum corneum layer, which enabled administration of the quantum dot solution to the deep epidermis and dermis. Microneedles enabled more rapid distribution of quantum dots to deep epidermal and dermal layers of porcine skin than topical administration. Multiphoton microscopy enabled imaging of quantum dots as well as microneedles within the skin; this approach may be useful for theranostics applications involving a variety of tissues. The ability to transdermally deliver quantum dots by means of microneedles is advantageous for theranostics applications due to the fact that microneedles are associated with minimal pain sensation, injection site trauma, and injection site inflammation. Two photon polymerization may be used to fabricate microneedles with patient-specific use, including (a) use for depth-dependent transdermal delivery and (b) use for delivery of quantum dots as well as for withdrawal of blood and/or interstitial fluid. It is anticipated that microneedles fabricated by means of two photon polymerization may enable precise transport of theranostics agents into epidermal, dermal, or subdermal tissues.

References

1 F. J. Picard and M. G. Bergeron, *Drug Discovery Today*, 2002, **7**, 1092–1101, DOI: 10.1016/S1359-6446(02)02497-2.
2 S. Warner, *Scientist*, 2004, **18**, 38–39.
3 L. Bissonnette and M. G. Bergeron, *Expert Rev. Mol. Diagn.*, 2006, **6**, 433–450, DOI: 10.1586/14737159.6.3.433.
4 P. Debbage and W. Jaschke, *Histochem. Cell Biol.*, 2008, **130**, 845–875, DOI: 10.1007/s00418-008-0511-y.
5 B. Sumer and J. Gao, *Nanomedicine*, 2008, **3**, 137–140, DOI: 10.2217/17435889.3.2.137.
6 D. Roland, *Pers. Med.*, 2007, **4**, 439–444, DOI: 10.2217/17410541.4.4.439.
7 M. Bruchez, M. Moronne, P. Gin, S. Weiss and A. P. Alivisatos, *Science*, 1998, **281**, 2013–2016.
8 C. M. Niemeyer, *Angew. Chem., Int. Ed.*, 2001, **40**, 4128–4158.
9 E. Klarreich, *Nature*, 2001, **413**, 450–452.
10 C. Seydel, *Science*, 2003, **300**, 80–81.
11 M. E. Åkerman, W. C. W. Chan, P. Laakkonen, S. N. Bhatia and E. Ruoslahti, *Proc. Natl. Acad. Sci. U. S. A.*, 2002, **99**, 12617–12621, DOI: 10.1073/pnas.152463399.
12 Z. A. Peng and X. G. Peng, *J. Am. Chem. Soc.*, 2001, **123**, 183–184.
13 X. Michalet, F. F. Pinaud, L. A. Bentolila, J. M. Tsay, S. Doose, J. J. Li, G. Sundaresan, A. M. Wu, S. S. Gambhir and S. Weiss, *Science*, 2005, **307**, 538–544, DOI: 10.1126/science.1104274.
14 Y. Ho and K. W. Leong, *Nanoscale*, 2010, **2**, 60–68, DOI: 10.1039/b9nr00178f.
15 H. Tada, H. Higuchi, T. M. Wanatabe and N. Ohuchi, *Cancer Res.*, 2007, **67**, 1138–1144, DOI: 10.1158/0008-5472.CAN-06-1185.
16 A. M. Derfus, A. A. Chen, D. Min, E. Ruoslahti and S. N. Bhatia, *Bioconjugate Chem.*, 2007, **18**, 1391–1396.
17 S. Dwarakanath, J. G. Bruno, A. Shastry, T. Phillips, A. John, A. Kumar and L. D. Stephenson, *Biochem. Biophys. Res. Commun.*, 2004, **325**, 739–743.
18 Y. Li, Z. He, P. Zhang, J. Gao, C. Cheng and H. Zhang, *J. Nanosci. Nanotechnol.*, 2010, **10**, 520–524.
19 V. Bagalkot, L. Zhang, E. Levy-Nissenbaum, S. Jon, P. W. Kantoff, R. Langer and O. C. Farokhzad, *Nano Lett.*, 2007, **7**, 3065–3070.
20 W. Cai and X. Chen, *Nat. Protoc.*, 2008, **3**, 89–96.
21 W. I. Hagens, A. G. Oomen, W. H. de Jong, F. R. Cassee and A. J. A. M. Sips, *Regul. Toxicol. Pharmacol.*, 2007, **49**, 217–229, DOI: 10.1016/j.yrtph.2007.07.006.
22 K. R. Vega-Villa, J. K. Takemoto, J. A. Yanez, C. M. Remsberg, M. L. Forrest and N. M. Davies, *Adv. Drug Delivery Rev.*, 2008, **60**, 929–938, DOI: 10.1016/j.addr.2007.11.007.
23 H. J. G. E. Gardeniers, R. Luttge, E. J. W. Berenschot, M. J. de Boer, S. Y. Yeshurun, M. Hefetz, R. van't Oever and A. vanden Berg, *J. Microelectromech. Syst.*, 2003, **12**, 855–862, DOI: 10.1109/JMEMS.2003.820293.
24 H. S. Gill, D. D. Denson, B. A. Burris and M. R. Prausnitz, *Clin. J. Pain*, 2008, **24**, 585–594.
25 S. D. Gittard and R. J. Narayan, in *Toxicology of the Skin*, ed. N. A. Monteiro-Riviere, Informa Healthcare, New York, NY, 2010, pp. 301–316.

26 L. Nordquist, N. Roxhed, P. Griss and G. Stemme, *Pharm. Res.*, 2007, **24**, 1381–1388, DOI: 10.1007/s11095-007-9256-x.
27 S. D. Gittard, A. Ovsianikov, B. N. Chichkov, A. Doraiswamy and R. J. Narayan, *Expert Opin. Drug Delivery*, 2010, **7**, 513–533, DOI: 10.1517/17425241003628171.
28 S. A. Coulman, A. Anstey, C. Gateley, A. Morrissey, P. McLoughlin, C. Allender and J. C. Birchall, *Int. J. Pharm.*, 2009, **366**, 190–200, DOI: 10.1016/j.ijpharm.2008.08.040.
29 D. V. McAllister, P. M. Wang, S. P. Davis, J. H. Park, P. J. Canatella, M. G. Allen and M. R. Prausnitz, *Proc. Natl. Acad. Sci. U. S. A.*, 2003, **100**, 13755–13760, DOI: 10.1073/pnas.2331316100.
30 S. Bal, A. C. Kruithof, H. Liebl, M. Tomerius, J. Bouwstra, J. Lademann and M. Meinke, *Laser Phys. Lett.*, 2010, **7**, 242–246, DOI: 10.1002/lapl.200910134.
31 A. Doraiswamy, C. Jin, R. J. Narayan, P. Mageswaran, P. Mente, R. Modi, R. Auyeung, D. B. Chrisey, A. Ovsianikov and B. Chichkov, *Acta Biomater.*, 2006, **2**, 267–275, DOI: 10.1016/j.actbio.2006.01.004.
32 A. Ovsianikov, B. Chichkov, P. Mente, N. A. Monteiro-Riviere, A. Doraiswamy and R. J. Narayan, *Int. J. Appl. Ceram. Technol.*, 2007, **4**, 22–29.
33 S. D. Gittard, A. Ovsianikov, N. A. Monteiro-Riviere, J. Lusk, P. Morel, P. Minghetti, C. Lenardi, B. N. Chichkov and R. J. Narayan, *J. Diabetes Sci. Technol.*, 2009, **3**, 304–311.
34 S. D. Gittard, R. J. Narayan, C. Jin, A. Ovsianikov, B. N. Chichkov, N. A. Monteiro-Riviere, S. Stafslien and B. Chisholm, *Biofabrication*, 2009, **1**, 041001, DOI: 10.1088/1758-5082/1/4/041001.
35 S. D. Gittard, A. Ovsianikov, H. Akar, B. Chichkov, N. A. Monteiro-Riviere, S. Stafslien, B. Chisholm, C. Shin, C. Shih, S. Lin, Y. Su and R. J. Narayan, *Adv. Eng. Mater.*, 2010, **12**, B77–B82, DOI: 10.1002/adem.200980012.
36 A. Doraiswamy, A. Ovsianikov, S. D. Gittard, N. A. Monteiro-Riviere, R. Crombez, E. Montalvo, W. Shen, B. N. Chichkov and R. J. Narayan, *J. Nanosci. Nanotechnol.*, 2010, **10**, 6305–6312.
37 S. D. Gittard and R. J. Narayan, *Expert Rev. Med. Devices*, 2010, **7**, 343–356.
38 A. Ovsianikov, B. Chichkov, O. Adunka, H. Pillsbury, A. Doraiswamy and R. J. Narayan, *Appl. Surf. Sci.*, 2007, **253**, 6603–6607, DOI: 10.1016/j.apsusc.2007.01.062.
39 R. Weissleder and M. J. Pittet, *Nature*, 2008, **452**, 580–589, DOI: 10.1038/nature06917.
40 P. M. Wang, M. Cornwell and M. R. Prausnitz, *Diabetes Technol. Ther.*, 2005, **7**, 131–141, DOI: 10.1089/dia.2005.7.131.
41 Y. Kim, P. J. Ludovice and M. R. Prausnitz, *J. Controlled Release*, 2007, **122**, 375–383, DOI: 10.1016/j.jconrel.2007.05.031.
42 S. Choi, Y. Kim, J. Park, J. Hutcheson, H. Gill, Y. Yoon, M. Prausnitz and M. Allen, *Biomed. Microdevices*, 2010, **12**, 263–273.
43 P. Maffia, B. H. Zinselmeyer, A. Ialenti, S. Kennedy, A. H. Baker, I. B. McInnes, J. M. Brewer and P. Garside, *Circulation*, 2007, **115**, e326–e328, DOI: 10.1161/CIRCULATIONAHA.106.658492.
44 M. Stroh, J. P. Zimmer, D. G. Duda, T. S. Levchenko, K. S. Cohen, E. B. Brown, D. T. Scadden, V. P. Torchilin, M. G. Bawendi, D. Fukumura and R. K. Jain, *Nat. Med.*, 2005, **11**, 678–682, DOI: 10.1038/nm1247.
45 D. R. Larson, W. R. Zipfel, R. M. Williams, S. W. Clark, M. P. Bruchez, F. W. Wise and W. W. Webb, *Science*, 2003, **300**, 1434–1436.
46 H. A. Lee, M. Imran, N. A. Monteiro-Riviere, V. L. Colvin, W. W. Yu and J. E. Riviere, *Nano Lett.*, 2007, **7**, 2865–2870, DOI: 10.1021/nl071563c.
47 S. Frank, in *A New Model for European Medical Device Regulation: A Comparative Legal Analysis in the EU and the USA*, Europa Law Publishing, Groningen, 2003, pp. 15–69.
48 W. C. Oliver and G. M. Pharr, *J. Mater. Res.*, 1992, **7**, 1564–1583.
49 T. Mosmann, *J. Immunol. Methods*, 1983, **65**, 55–63.
50 C. G. Grant, R. Warren and E. Salter, *Burns*, 1980, **7**, 57–69.
51 C. A. Khatri, J. W. Stansbury, C. R. Schultheisz and J. M. Antonucci, *Dent. Mater.*, 2003, **19**, 584–588, DOI: 10.1016/S0109-5641(02)00108-2.
52 H. Park, M. G. Allen and M. R. Prausnitz, *J. Controlled Release*, 2005, **104**, 51–66, DOI: 10.1016/j.jconrel.2005.02.002.
53 J. H. Park and M. R. Prausnitz, *J. Korean Phys. Soc.*, 2010, **56**, 1223–1227, DOI: 10.3938/jkps.56.1223.
54 J. P. Ryman-Rasmussen, J. E. Riviere and N. A. Monteiro-Riviere, *Toxicol. Sci.*, 2006, **91**, 159–165, DOI: 10.1093/toxsci/kfj122.
55 M. J. Levene, D. A. Dombeck, K. A. Kasischke, R. P. Molloy and W. W. Webb, *J. Neurophysiol.*, 2004, **91**, 1908–1912, DOI: 10.1152/jn.01007.2003.

Single-cell analysis of phosphoinositide 3-kinase and phosphatase and tensin homolog activation

Dechen Jiang,[a] Christopher Eldridge Sims[a] and Nancy Lynn Allbritton*[ab]

Received 17th May 2010, Accepted 21st July 2010
DOI: 10.1039/c005362g

A single-cell assay was developed to measure the activation of phosphoinositide 3-kinase (PI3K) using microanalytical chemical separations and a fluorescently labeled lipid substrate. Phosphatidyl-inositol 4,5 bisphosphate labeled on its acyl chain with Bodipy fluorescein (Bodipy Fl PIP_2) was utilized as a substrate for both *in vitro* and cell-based assays. Detection limits for the substrate and product of the PI3K reaction were 10 to 20 zeptomol. *In vitro* assays with PI3K with and without pharmacologic inhibitors demonstrated that Bodipy Fl PIP_2 was converted to phosphatidyl-inositol 3,4,5 trisphosphate (Bodipy Fl PIP_3). Bodipy Fl PIP_3 could be back converted to Bodipy Fl PIP_2 by the phosphatase PTEN. When Bodipy Fl PIP_2 was added to a cell lysate, 1.4 fmol of the Bodipy Fl PIP_3 were produced per ng of protein in the cytoplasmic extract in 10 min. Addition of Bodipy Fl PIP_3 to a cell lysate yielded 3 fmol of Bodipy Fl PIP_2 per ng of protein in 8 min. Both Bodipy Fl PIP_2 and Bodipy Fl PIP_3 were measureable in single cells and the two species could be inter-converted. Under the appropriate conditions, a fluorescent diacylglycerol was also detected in single cells. When the FcεR1 receptor on the cells loaded with the fluorescent lipid was cross-linked, the amount of Bodipy Fl PIP_3 generated per cell increased 4-fold over that of unstimulated cells. This production of Bodipy Fl PIP_3 was blocked by wortmannin. Chemical cytometry utilizing the fluorescent lipids will be of value in understanding lipid metabolism at the single-cell level.

Introduction

Lipid signaling is now recognized to have essential roles in health and disease, particularly cancer and inflammatory diseases.[1–8] The PI3K pathway is especially important as the products of numerous oncogenes provide constitutive input signals to PI3K.[9,10] PI3K itself has been shown to be mutated in a variety of cancers as has the tumor suppressor phosphatase and tensin homolog (PTEN), a 3-lipid phosphatase, which normally down-regulates the PI3K pathway by dephosphorylating the PI3K product phosphatidyl-inositol 3,4,5-trisphosphate (PIP_3) to generate phosphatidyl-inositol 4,5-bisphosphate (PIP_2).[11] Immediately downstream of PI3K lies protein kinase B (PKB, *aka* Akt) which functions as an important signaling node for numerous cancer-promoting activities including cell cycle entry, resistance to apoptosis, and enhanced cell migration.[12,13] Not surprisingly, the PI3K pathway is an active therapeutic target and novel PI3K inhibitors, both broad spectrum and isoform specific, have entered clinical trials.[14–16]

[a]*Department of Chemistry, University of North Carolina, Chapel Hill, NC 27599, USA*
[b]*Department of Biomedical Engineering, University of North Carolina, Chapel Hill, NC 27599, North Carolina State University, Raleigh, NC 27695, USA. E-mail: nlallbri@unc.edu; Fax: +919-843-7825*

Despite its importance, direct analysis of PI3K signaling in living cells is a difficult task, particularly in small samples such as those obtained from patients.[17–19] Radioactivity-based thin-layer chromatography or mass spectrometry are commonly used analytical techniques for the analysis of lipids; however, they are technically demanding, of limited sensitivity and specificity, or require sophisticated equipment.[20–26] PI3K signaling has also been analyzed by fluorescence-based methods. For example, high-throughput assays are based on phosphoinositide-binding pleckstrin homology (PH) domains as detectors in measuring the production or localization of PIP_3, but such assays require large numbers of cells.[27,28] GFP-tagged PH domains have been used in microscopy as an indirect assay of the enzymatic activities of PI3K and PTEN, but these molecularly engineered cell-based assays cannot be used in clinical samples.[29–31] These various limitations necessitate development of new technologies for analysis of lipid signaling in primary cells and in small samples such as from needle biopsies or fine needle aspirates.

Microanalytical chemical separation applied to high-sensitivity analyses of single cells, known as chemical cytometry, has such potential.[32–35] Chemical cytometry is one of the fastest growing fields within bioanalytical chemistry as it holds great promise for understanding cell metabolism and signaling.[36–40] Our group has been active in this area having developed single-cell assays for protein kinases[41] and sphingosine kinase.[38] In these assays, cells are loaded with fluorescent substrates (reporters) followed by laser-based lysis of a single cell and loading of its contents into an overlying capillary where chemical separation is accomplished *via* capillary electrophoresis (CE). In CE, substrate and product forms of the reporter are readily separated, detected with high-sensitivity using laser-induced-fluorescence (LIF), and identified by their characteristic migration times. The ratio of the peak areas of the substrate and product are then used as a measure of the enzyme's activation. Routine detection limits correspond to an intracellular concentration of $\leq$10 nM.[41] The ability to detect the reporter at nanomolar concentrations means that concentrations can be used at or below those of endogenous substrates so that competition between the endogenous substrate and exogenous reporter is minimal.[41] Phosphorylated products can also be dephosphorylated by phosphatases; therefore, the measurement reflects the dynamic equilibrium of the kinase and phosphatase activity in the cell.[38,41] Furthermore, the ability to analyze multiple analytes in individual cells enables the identification of cell-to-cell heterogeneity with respect to biochemical behavior.

Understanding this heterogeneity is particularly important for lipid signal transduction since lipid levels are dynamic with biosynthesis and metabolism being robust in most cells.[17–19] Recently chemical reagents have become available to enable biochemical studies of these rich metabolic networks, particularly in cell-based assays.[17,42] Fluorescently labeled lipids have been shown to localize and translocate in the same manner as the endogenous substrates, and have been useful for monitoring dynamic changes in cells.[42–44] Lipids tagged with fluorophores have also been developed as substrates for a variety of enzymes, often with similar kinetics to the endogenous substrate.[38,45–59] When coupled with chemical cytometry, these new chemical tools may make it possible to perform direct, quantitative measurements of the activation of PI3K and other lipid metabolizing enzymes in single cells. In the current work, a single-cell biochemical assay for PI3K was developed and validated using a tumor cell line. A fluorescently labeled PIP_2 was used as a substrate of PI3K. An *in vitro* reaction with purified PI3K using CE-LIF to monitor the phosphorylation of the substrate demonstrated the feasibility of the assay format. The fluorescent PIP_2 was then tested in a cell lysate to confirm that its phosphorylation occurred and a separation could be performed in the presence of cellular constituents. Microscopy was used to monitor and optimize the loading of the exogenous substrate into cells. Lastly, single-cell assays were undertaken to measure the phosphorylation and dephosphorylation of substrate by PI3K and PTEN, respectively. Specific inhibition of the enzymes provided additional validation for the single-cell measurements.

Results and discussion

Separation and detection limits of Bodipy Fl PIP_2 and Bodipy Fl PIP_3 by CE

Separation of Bodipy Fl PIP_2 and Bodipy Fl PIP_3 was accomplished as previously reported.[60,61] The current work sought to apply this separation scheme to single-cell analyses of PIP_2 and PIP_3 metabolism; therefore, it was necessary to extend the earlier work to establish that adequate sensitivity could be attained to accomplish this goal. Due to the biochemical nature of the enzyme measurements, it was desired to introduce the fluorescent substrate at the lowest possible cellular concentration; otherwise, excessive amounts of the exogenous lipid could serve to overwhelm the cellular signaling networks with a substrate that might perturb cellular physiology. In mammalian cells the estimated concentration of PIP_2 (averaged across the cell volume) is 10 μM, or 10^{-18} mol in a typical mammalian cell of 1 pL.[62]

A series of separations using standard solutions of Bodipy Fl PIP_2 and Bodipy Fl PIP_3 were undertaken to establish the sensitivity and reproducibility in the separation of these analytes. Free fluorescein was employed as an internal standard in the electrophoretic traces. Fluorescein (1 nM), Bodipy Fl PIP_2 (1 nM) and Bodipy Fl PIP_3 (1nM) and were loaded simultaneously into a capillary and separated. The peak area of Bodipy Fl PIP_2 and Bodipy Fl PIP_3 was divided by that for the fluorescein internal standard. The migration times of Bodipy Fl PIP_2 and Bodipy Fl PIP_3 were also normalized by division of the migration time of the fluorescein peak.[63] The relative standard deviation (RSD) for the lipids was calculated from this normalized data. The migration times for Bodipy Fl PIP_2 and Bodipy Fl PIP_3 were highly reproducible with respect to that of fluorescein with RSDs of 0.6% and 0.7%, respectively ($n = 10$). The reproducibility of the peak areas for Bodipy Fl PIP_2 and Bodipy Fl PIP_3 was less consistent with RSDs of 10.0% and 14.2%, respectively ($n = 10$). This relatively large RSD for the peak area was likely due to variability in the injected volume of sample since analyte loading into the capillary was performed manually; however, the ratio of the normalized peak areas (1.0 ± 0.1) for Bodipy Fl PIP_2 and Bodipy Fl PIP_3 was excellent. The method can accurately measure the ratio of these two lipid species. To determine the detection limits of the lipids in the capillary, the lipid/fluorescein solution was progressively diluted and then separated. When 10^{-20} mol of either lipid was loaded into the capillary, the noise height of the baseline was one third of the peak height of the lipids. Under the conditions used, the limit of detection for both species was 1×10^{-20} mol, which was 100-fold less than the expected amount of endogenous PIP_2 in a single cell (see above). The detection limits were approximately 10-fold higher than other small molecule analytes such as fluorescein-labeled peptides.[41] This may in part be due to losses on vessel walls during dilution of the lipid prior to electrophoresis.

Phosphorylation of Bodipy Fl PIP_2 by PI3K

A fluorescein-labeled PIP_2 was previously shown to function as an *in vitro* substrate of PI3K.[64] Thus, it was likely that the Bodipy Fl PIP_2 could also serve as a substrate for PI3K. PI3K was incubated with Bodipy Fl PIP_2 with or without ATP (1 mM), and the reaction mixture was analyzed by CE after varying incubation times. In the presence of ATP, the electropherogram revealed a peak possessing a longer migration time than that of Bodipy Fl PIP_2 (Fig. 1A). The area of this peak increased as the incubation time increased (Fig. 1B). This peak co-migrated with a standard of Bodipy Fl PIP_3. When the assay was performed in the absence of ATP, this additional peak was not observed. In the presence of wortmannin (100 nM), an irreversible inhibitor of PI3K[65] and ATP (1 mM), the latter peak was not observed over 180 min of incubation (Table 1). Based on these data, this latter peak was identified as Bodipy Fl PIP_3. In the presence of ATP and without wortmannin, $30 \pm 4\%$ ($n = 3$) of Bodipy Fl PIP_2 was phosphorylated over 180 min under these conditions (Fig. 1B, Table 1). In addition to the expected lipid and internal standard peaks, an additional

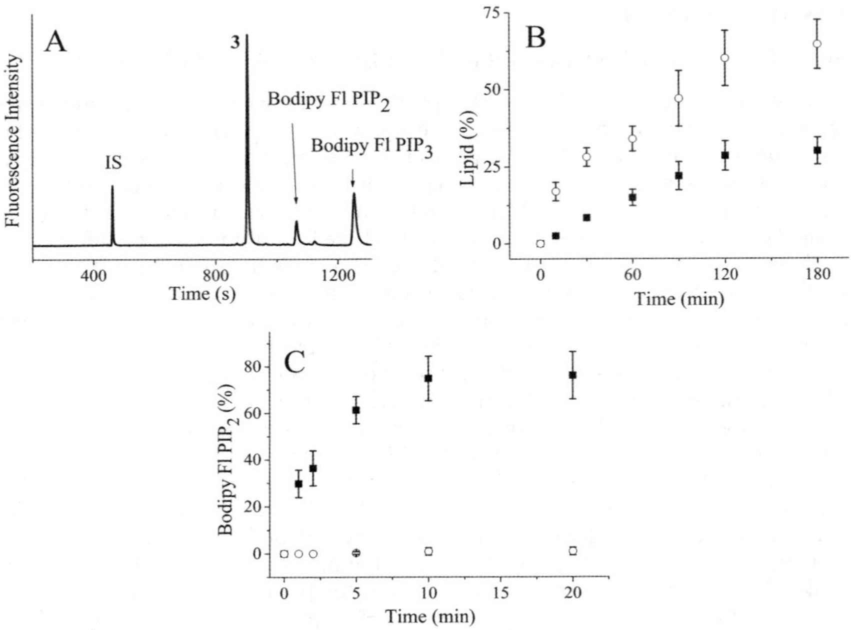

Fig. 1 Incubation of PI3K and PTEN with fluorescent lipids *in vitro*. (A) Electropherogram of the reaction mixture of Bodipy Fl PIP_2 and PI3K after 120 min. The internal standard fluorescein is marked as "IS" while the peak co-migrating with Bodipy Fl PIP_1 is marked with a "3". (B) Time course of the reaction of Bodipy Fl PIP_2 with PI3K. ■ represents the percentage of Bodipy Fl PIP_3 while ○ reflects the percentage of the lipid in the peak co-migrating with Bodipy Fl PIP_1 (peak 3 above), respectively. (C) Time course of the reaction of Bodipy Fl PIP_3 with PTEN. ■ and ○ are the percentage of Bodipy Fl PIP_3 in the absence and presence of 500 nM bpv (pic), respectively. The data in panels B and C represent the average of three experiments and the error bars depict the standard deviation of the data points.

peak (peak 3) was observed that co-migrated with a standard of Bodipy-fluorescein labeled phosphatidyl-inositol 5-phosphate (Bodipy Fl PIP_1). However, the addition of a phosphatase inhibitor cocktail containing orthovanadate, β-glycerophosphate and the PTEN inhibitor bpv(pic) was ineffective in preventing the formation of the analyte co-migrating with Bodipy Fl PIP_1 (data not shown). While it was likely that peak 3 was a Bodipy Fl PIP_1, its identity must be confirmed by additional means. The presence of a 3rd peak during *in vitro* reactions with purified PI3K has been reported previously for fluorescein-labeled PIP_2.[64] Similarly, PI3K assays using thin-layer chromatography or HPLC have shown additional species consistent with the current findings.[66,67]

Dephosphorylation of Bodipy Fl PIP_3 by PTEN

In cells, dephosphorylation of PIP_3 by the phosphatases PTEN or Src homology 2 domain-containing inositol 5′ phosphatase (SHIP2) acts to convert PIP_3 to PIP_2. To determine if Bodipy Fl PIP_3 might be dephosphorylated, this lipid was incubated with PTEN phosphatase for varying times and the reaction mixture was separated by CE. Under the conditions used, $75 \pm 10\%$ ($n = 3$) of Bodipy Fl PIP_3 was dephosphorylated to form Bodipy Fl PIP_2 over 10 min (Fig. 1C, Table 1). In contrast to the PI3K assay, only peaks co-migrating with Bodipy Fl PIP_2, Bodipy Fl PIP_3, or the fluorescein standard were present on the electropherogram. The addition of fresh PTEN into the reaction mixture at 10 min intervals resulted in additional conversion of Bodipy Fl PIP_3 to Bodipy Fl PIP_2 suggesting that PTEN was not stable under

Table 1 Metabolism of lipid substrates

Substrate	Bodipy Fl PIP_2		Bodipy Fl PIP_3	
Products formed	Bodipy Fl PIP	Bodipy Fl PIP_3	Bodipy Fl PIP	Bodipy Fl PIP_2
Commercial PI3K	64 ± 8%	30 ± 4%		
Commercial PTEN			0	76 ± 10%
Cell lysate	82 ± 5%	11 ± 4%	17 ± 4%	59 ± 4%
+bpv(pic)	58 ± 2%	28 ± 2%	0	0
+Wortmannin & bpv(pic)	92 ± 5%	0		
Single cell (100 nM substrate)	0	2 ± 1%	0	9 ± 3%
+Wortmannin	0	0		
+bpv(pic)			0	0
+Antigen/IgE	2.3 ± 0.8%	9 ± 4%		
Single cell (2 μM substrate)	28 ± 5%	15 ± 6%		
+Wortmannin	4.4 ± 1.5%	0		

these conditions. The assay was then performed in the presence of bpv(pic) (500 nM), a PTEN inhibitor. No peak co-migrating with a Bodipy Fl PIP_2 standard was observed over 1 h even with addition of fresh PTEN at 10 min intervals. These data suggest that the Bodipy Fl PIP_2 and Bodipy Fl PIP_3 might serve as enzyme substrates within cells.

Separation and detection of fluorescent lipids loaded into cell lysates

A challenge that exists in the use of these fluorescent lipid probes for cell-based assays by CE is whether these probes can be successfully extracted from cellular membranes, since the lipids can be expected to partition into the lipid bilayer of the plasma membrane and organelles. Numerous intracellular proteins are also known to associate with these lipids.[9] For successful CE-based analysis, these reporters must be extracted from membrane and cytosol, separated and detected. To evaluate the potential to use these compounds for performing assays in the presence of cellular constituents, Bodipy Fl PIP_2 (5 μM) or Bodipy Fl PIP_3 (5 μM) was incubated with a cell lysate followed by CE analysis using a protocol similar to that used for the assays with purified proteins. The RSDs for the normalized peak areas for Bodipy Fl PIP_2 and Bodipy Fl PIP_3 in the lysate were 9.0% and 11.5%, respectively ($n = 3$). These results were similar to those achieved in the assay of the lipid using purified proteins. Based on the peak areas of each analyte compared with that of standards, the percentage of each analyte recovered in the sample was 73 ± 7% for Bodipy Fl PIP_2 and 64 ± 5% for Bodipy Fl PIP_3. Thus, recovery of each of the lipids from the cellular milieu is similar. When separated from the cell lysates, the limit of detection for both species was 2×10^{-20} mol (S/N = 3). These data support the feasibility of retrieving and accurately measuring the substrate and product lipid forms in the presence of cellular components.

Phosphorylation and dephosphorylation of fluorescent lipids in cell lysates

To determine whether PI3K in cell lysates might metabolize Bodipy Fl PIP_2, this fluorescent lipid (5 μM) was added to cell lysates preincubated with bvp(pic) (10 μM, 30 min) to inhibit PTEN activity and the assay mix was incubated for varying times at room temperature followed by CE-based separation. A peak co-migrating with Bodipy Fl PIP_3 was observed (Fig. 2A) suggesting that the conversion of Bodipy Fl PIP_2 to Bodipy Fl PIP_3 by PI3K in the cell lysates was occurring. Of note, a peak co-migrating with peak 3 (or Bodipy Fl PIP_1) was also present on the

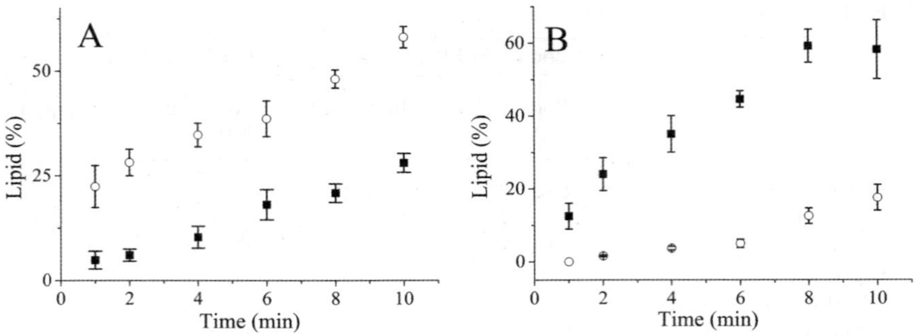

Fig. 2 Incubation of the fluorescent lipids with a cell lysate. (A) Time course of cell lysate incubated with Bodipy Fl PIP_2. ■ and ○ are the percentages of Bodipy Fl PIP_3 and peak 3 (identified in Fig. 1A), respectively. (B) Time course of cell lysate incubated with Bodipy Fl PIP_3. ■ and ○ are the percentage of Bodipy Fl PIP_2 and peak 3, respectively. The data represent the average of three experiments and the error bars depict the standard deviation of the data points.

electropherogram and the area of this peak was greater than that of Bodipy Fl PIP_3. After 10 min of incubation with the cell lysate, 14 ± 3% of the Bodipy Fl PIP_2 remained while 28 ± 2% had been converted to Bodipy Fl PIP_3 and 58 ± 2.5% to the peak-3 analyte (Table 1). At this time point, 1.4 fmol of the Bodipy Fl PIP_3 were produced per ng of protein in the cell lysate. In the presence of the PI3K inhibitor wortmannin (100 nM) and bpV(pic), no Bodipy Fl PIP_3 was present at the 10-min time point while 92 ± 5% of the fluorescent lipid was present in peak 3. In the absence of wortmannin and bpV(pic), 11 ± 4% was present in the Bodipy Fl PIP_3 peak and 82 ± 5% in the peak co-migrating with Bodipy Fl PIP_1 at 10 min. As with the purified PI3K, at least two enzymes metabolized the Bodipy Fl PIP_2 in the cell lysate.

To determine whether Bodipy Fl PIP_3 might also be metabolized in the lysate, Bodipy Fl PIP_3 (5 μM) was incubated with the cell lysate and at varying times aliquots were analyzed by CE. A peak with a migration time corresponding to that of Bodipy Fl PIP_2 was observed after 1 min (12 ± 3%), which suggested that Bodipy Fl PIP_3 was serving as a phosphatase substrate in the lysate (Fig. 2B, Table 1). The peak area of Bodipy Fl PIP_2 continued to increase with time, but after 2 min a peak corresponding to the analyte co-migrating with Bodipy Fl PIP_1 peak (peak 3) was also present. After 8 min of incubation, the appearance of Bodipy Fl PIP_2 reached a maximum at 59 ± 4% and then began to decline as the peak area of the peak-3 species increased proportionately. At this 8-min time point, 3 fmol of the Bodipy Fl PIP_2 were produced per ng of protein in the cell lysate. When the lysate was incubated with bpV(pic) (500 nM, 30 min) and then Bodipy Fl PIP_3 added, 19 ± 2% was present as Bodipy Fl PIP_2 and analyte corresponding to peak 3 was observed. When the inhibitor concentration was increased to 10 μM, no peaks co-migrating with either Bodipy Fl PIP_2 or peak 3 were observed. These results suggest that the peak-3 analyte is derived from PIP_2 and is very likely a form of Bodipy Fl PIP_1.

Loading cells with fluorescent lipids

To detect the activation of PI3K and lipid phosphatases in intact cells, the fluorescent lipids must be translocated across the cell membrane so that the lipids are accessible to cytosolic enzymes. In the current experiments, the cells were loaded with the lipid by first forming a 1 : 1 complex between the lipid and histone.[59,68,69] To examine the rate at which Bodipy Fl-PIP_2 loaded into cells, the fluorescence of the cells was measured after addition of Bodipy Fl-PIP_2:histone (2 μM) to the cells. Cellular fluorescence increased within 30 s of addition of the Bodipy Fl-PIP_2:histone to the cell

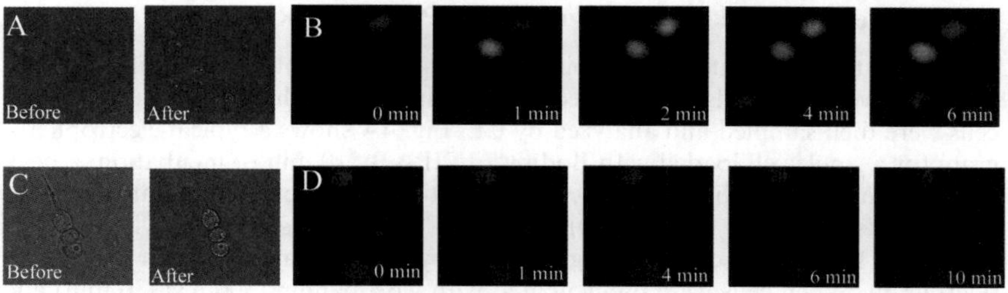

Fig. 3 Increase in $[Ca^{2+}]_i$ in cells incubated with PIP_2:histone. (A) Bright-field images of RBLs before and after incubation with 2 μM PIP_2:histone for 10 min. (B) Fluorescence images of Fluo-3-loaded RBL cells incubated with 2 μM PIP_2:histone for varying times. (C) Bright field images of RBL cells before and after incubation with 0.5 μM PIP_2:histone for 10 min. (D) Fluorescence images of Fluo-3-loaded RBL cells incubated with 0.5 μM PIP_2:histone for varying times.

chamber suggesting that the fluorescent lipid was rapidly incorporated into the cells. The fluorescence of the cells reached a plateau within 5 min indicating that no more lipid was loaded after that time (data not shown). When cells were examined by bright-field microscopy following addition of Bodipy-Fl-PIP_2:histone, the cells were noted to become more refractile with small granular objects surrounding the cell (Fig. 3A). Cells incubated under identical conditions without Bodipy-Fl-PIP_2:histone did not undergo this morphologic change. Since RBL cells contain secretory granules, it was possible that the observed material surrounding the lipid-loaded cells was secreted in response to the incorporation of Bodipy Fl-PIP_2 into the cells.

Prior investigators have observed increases in $[Ca^{2+}]_i$ when cells were loaded with a high concentration of PIP_2-histone complex (40 μM of a 1 : 1 complex).[59] The elevation in $[Ca^{2+}]_i$ was thought to be the result of conversion of the PIP_2 to inositol 1,4,5-trisphosphate (IP_3) by phospholipase C (PLC). This hypothesis was supported by data showing that pre-exposure of cells to the PLC inhibitor U73122 abrogated the fluorescence increase.[59] Conversion of PIP_2 into other active signaling species by the process of cell loading could yield substantial artifacts in a cell-based assay designed to measure the physiologic conversion of PIP_2 into its metabolites.[1,70] In the current experiments, the concentration of Bodipy Fl-PIP_2 incubated with the cells was 1/20 that loaded into cells by Ozaki *et al.* and 1/5 that the averaged PIP_2 cellular concentration.[62] However, since Bodipy Fl-PIP_2 is likely more concentrated in the membrane relative to that in the surrounding medium, the uptake into the cells may have been sufficient to increase flux through the PLC pathway, generating IP_3 and increasing $[Ca^{2+}]_i$.

To determine whether loading cells using the PIP_2:histone procedure increased $[Ca^{2+}]_i$, RBL cells were first loaded with the calcium indicator Fluo-3 and then incubated with nonfluorescent PIP_2 in an equimolar complex with 2 μM histone. The fluorescence intensity of the cells increased within 30 s after addition of the complex and persisted beyond 10 min suggesting that incubation of cells in as little as 2 μM PIP_2:histone resulted in an increase in $[Ca^{2+}]_i$ (Fig. 3B). Notably addition of histone alone (2 μM) led to a transient increase in Fluo-3 fluorescence which returned to baseline within 3 min. These data suggested that the rise in $[Ca^{2+}]_i$ was the result of an initial perturbation of the plasma membrane by histone with a more prolonged effect due to the PIP_2, perhaps as a result of its conversion to IP_3.

To determine whether cells could be loaded with PIP_2 without an increase in $[Ca^{2+}]_i$, Fluo-3-loaded cells incubated with PIP_2:histone (0.5 μM) while monitoring their fluorescence intensity. No change in the fluorescence intensity of the cells was observed over 10 min (Fig. 3D). Concomitant observation by brightfield demonstrated that the morphology of the cells was also not altered by the loading procedure (Fig. 3C). Loading cells with a low concentration of the PIP_2:histone complex did not lead to a detectable rise in $[Ca^{2+}]_i$ and pathways downstream of PIP_2 were likely not activated.

Formation of Bodipy Fl PIP_3 and Bodipy Fl PIP_2 in single RBL cells

To demonstrate that fluorescent lipids from single cells could be quantified by CE, serum-starved RBL cells were loaded with Bodipy Fl PIP_2:histone (0.5 μM). Single cells were then sampled and analyzed by CE. Fig. 4A shows a typical electropherogram for a single cell loaded with Bodipy Fl PIP_2. By 10 min of incubation, a peak co-migrating with Bodipy Fl PIP_3 was present on the electropherogram. In these experiments, only $2 \pm 1\%$ ($n = 10$) of Bodipy Fl PIP_2 was phosphorylated in the cell (Table 1). No peak co-migrating with Bodipy Fl PIP_1 was evident on the electropherogram. When cells were preincubated with wortmannin (100 nM, 30 min) and then incubated with Bodipy Fl PIP_2 for 10 min, no peak corresponding to Bodipy Fl PIP_3 was observed.

To investigate lipid phosphatase activity, serum-starved RBL cells were incubated with Bodipy Fl PIP_3:histone (0.5 μM). Individual cells were then analyzed by CE (Fig. 4B). In these experiments, $9 \pm 3\%$ ($n = 5$) of the Bodipy Fl PIP_3 was converted to Bodipy Fl PIP_2 by 10 min. No peak corresponding to Bodipy Fl PIP_2 was present when cells were preincubated with the inhibitor bpV(pic) (500 nM, 30 min), and then incubated with Bodipy Fl PIP_3:histone (0.5 μM). The ability of lipid phosphatases to act on Bodipy Fl PIP_3 suggests that the single-cell measurements may reflect the combined actions of PI3K and phosphatase in the cells.

Formation of Bodipy Fl PIP_2 metabolites in single RBL cells loaded with high concentrations of Bodipy Fl PIP_2

To investigate the fate of Bodipy Fl PIP_2 in activated cells, RBL cells were loaded with a high concentration Bodipy Fl PIP_2:histone complex (2 μM). After incubation for 10 min, single cells were analyzed by CE (Fig. 4C). A peak co-migrating with

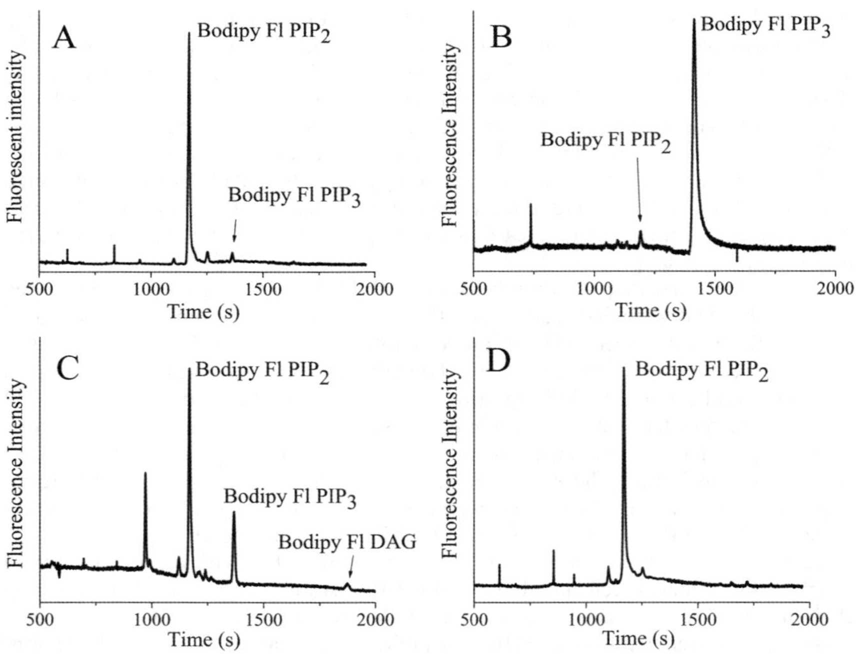

Fig. 4 Electropherograms obtained from single RBL cells loaded with fluorescent lipids. (A) An RBL cell loaded with 0.5 μM Bodipy-Fl PIP_2:histone. B) An RBL cell loaded with 0.5 μM Bodipy-Fl PIP_3:histone. (C) and (D) RBL cells were loaded with 2 μM Bodipy-Fl PIP_2. The cells were incubated with (C) 0 or (D) 100 nM wortmannin prior to loading the fluorescent lipid.

Bodipy Fl PIP_3 was present on the electropherogram and comprised $15 \pm 6\%$ ($n = 5$) of the fluorescent lipid (Table 1). Multiple additional peaks were observed on the electropherogram which did not co-migrate with either Bodipy Fl PIP_2 or peak 3. The final peak on the electropherogram co-migrated with a standard of Bodipy Fl DAG generated by reacting Bodipy Fl PIP_2 with purified PLC. These data suggest that in the intact cells, Bodipy Fl PIP_2 could be acted upon by both PI3K and PLC. The formation of Bodipy Fl DAG by PLC implies the formation of IP_3 which is consistent with the observed rise in $[Ca^{2+}]_i$ and degranulation of RBL cells loaded with 2 μM Bodipy Fl PIP_2:histone.

To determine whether inhibition of PI3K by wortmannin could be measured in cells, RBL cells were preincubated with wortmannin (100 nM, 30 min), and then incubated with 2 μM Bodipy Fl PIP_2:histone for 10 min. No peak corresponding to Bodipy Fl PIP_3 or Bodipy Fl DAG was observed in the single cells (Fig. 4D). These results are consistent with wortmannin's reported inhibition of both PI3K and PLC-mediated signaling in RBL cells.[71,72]

Formation of Bodipy Fl PIP_3 in single cells after Fcε receptor activation

To determine if metabolism of the fluorescent lipids was increased in single cells following physiologic activation, RBL cells were loaded with Bodipy Fl PIP_2:histone (0.5 μM) followed by cross-linking of the Fcε receptor by antigen. Single cells were then analyzed by CE (Fig. 5A). As a percentage of the total fluorescent lipid in the cell, 4-fold more Bodipy Fl PIP_3 was produced in cells stimulated by cross-linking their Fcε receptor relative to those without stimulation (Fig. 5B, Table 1). To inhibit the activity of PI3K, cells were preincubated with 100 nM wortmannin and then stimulated with antigen. No peak corresponding to Bodipy Fl PIP_3 was observed suggesting that the fluorescent lipids in combination with CE can be used to assess the activation of PI3K in single cells.

Experimental

Materials

Sodium deoxycholate (SDC), 1-propanol, adenosine triphosphate (ATP), protease inhibitor (P8340) and 4-(2-hydroxyethyl) piperazine-1-ethanesulfonic acid (HEPES) were purchased from Sigma-Aldrich Inc. (St. Louis, MO, USA). The fluorescent lipid reagents, Bodipy fluorescein phosphatidyl-inositol 4,5-bisphosphate (Bodipy Fl PIP_2, C6) and Bodipy fluorescein phosphatidyl-inositol 3,4,5-trisphosphate

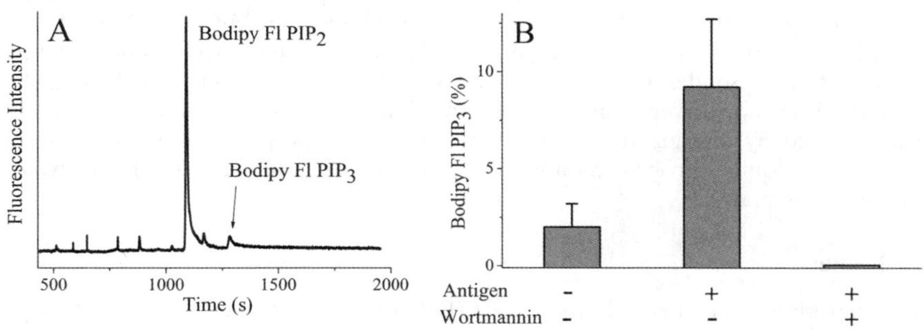

Fig. 5 Physiologic activation of RBL cells loaded with Bodipy Fl PIP_2. (A) Electropherogram obtained from an RBL cell loaded with 0.5 μM Bodipy Fl PIP_2:histone and then stimulated with antigen. (B) The percentage of fluorescent lipid present as Bodipy Fl PIP_3 in serum-starved cells (control), antigen-activated cells, and antigen-activated cells in the presence of wortmannin. The data represent the average of five experiments and the error bars depict the standard deviation of the data points.

(Bodipy Fl PIP_3, C6), PIP_2 and PIP_3 were obtained from Echelon Biosciences Inc. (Salt Lake City, UT, USA). PI3K (p110α/p85α), and PTEN were procured from Millipore Corp. (Billerica, MA, USA). Fluo-3 AM (fluo-3 acetoxymethyl ester), fluorescein (Fl), bovine serum albumin labeled with 2,4-dinitrophenyl (DNP-BSA), *anti*-DNP rat IgE, DMEM media and all other tissue culture reagents were purchased from Invitrogen (Carlsbad, CA, USA). EOTrol LR (low reverse) polymer solution was acquired from Target Discovery (Palo Alto, CA, USA). Wortmannin and dipotassium bisperoxo (picolinato) oxovanadate (V) (bpV(pic)) were obtained from Enzo Life Sciences International Inc. (Plymouth Meeting, PA, USA). All other reagents were procured from Fisher Scientific (Pittsburgh, PA, USA).

PI3K assay

In vitro assays were performed at a PI3K concentration of 10 nM and a Bodipy Fl PIP_2 concentration of 5 μM with a total reaction volume of 40 μL at room temperature. The assay was conducted in the presence or absence of 1 mM ATP. In some experiments, 100 nM wortmannin was added to the reaction mixture before addition of the lipid. Aliquots (5 μL) were removed from the reaction mixture at 0, 10, 30, 60, 90, 120 and 180 min time points. The reaction was stopped by diluting the aliquot in propanol (5 μL) followed by flash freezing in liquid nitrogen. Samples were thawed and diluted 200-fold in water just prior to CE analysis. The percentage of Bodipy Fl PIP_3 present in the mixture was determined by dividing the peak area of the Bodipy Fl PIP_3 by the area of all of the peaks on the electropherogram.

PTEN assay

The assay was performed with a PTEN concentration of 200 nM and a Bodipy Fl PIP_3 concentration of 5 μM with a total reaction volume of 40 μL (5 mM HEPES, pH 8) at room temperature. The assays were performed without and with 500 nM bpV(pic), a pharmacologic inhibitor of PTEN. Aliquots (5 μL) were removed from the reaction mixture at 0, 1, 2, 5, 10, and 20 min time points. The reaction was stopped by diluting the aliquot with 5 μL propanol followed by flash freezing in liquid nitrogen. Samples were thawed and diluted 200-fold in water just prior to analysis. The percentage of Bodipy Fl PIP_2 present in the mixture was determined by dividing the peak area of the Bodipy Fl PIP_2 peak area by the area of all of the peaks on the electropherogram.

Preparation of Bodipy Fl diacylglycerol (DAG) standard

Bodipy Fl PIP_2 (5 μM) was incubated with phospholipase C (PLC, 1 ng) in a total reaction volume of 40 μL (50 mM HEPES, 35 mM KCl, 3 mM $CaCl_2$, 2 mM DTT, 3 mM EGTA, pH 7.2) at room temperature. The PLC protein used was provided by Professor John Sondek (Department of Biochemistry and Biophysics, UNC). An aliquot (5 μL) was taken from the reaction mixture after 120 min and the reaction was stopped by diluting the sample volume in 5 μL propanol followed by flash freezing in liquid nitrogen. Samples were thawed and diluted 200-fold in water just prior to analysis.

Cell culture

Rat basophilic leukemia 2H3 mast cells (RBL) were used since they express high levels of the p110α isoform.[73] RBL cells were cultured in DMEM supplemented with 10% fetal bovine serum (FBS), 584 mg L^{-1} L-glutamine, 100 units/mL penicillin and 100 μg mL^{-1} streptomycin. Cells were maintained at 37 °C in a humidified 5% CO_2 atmosphere. Cells for use in single-cell, CE experiments were plated the day before the experiment in custom-made cell chambers created from a #1 glass cover-slip to which a silicon "O" ring was attached using poly(dimethyl siloxane) (Sylgard

184).[74] For experiments using serum-starved cells, the media was exchanged with DMEM lacking FBS 8–12 h before the start of the experiment.

Preparation of RBL cell lysates

A suspension of 400 000 RBL cells in 10 mL media was centrifuged, and the pellet was washed in deionized water (1 mL) × 3 at 4 °C. The washed cell pellet was then suspended in 200 μL of deionized water and monitored under the microscope for complete lysis before use.[75] Cell debris and nuclei were removed by centrifugation at 16 000 × g for 10 min at 4 °C. The protein concentration in the lysate was determined by reaction with 3 mg mL^{-1} fluorescamine for 5 min followed by measurement of fluorescence and comparison to a standard curve. For assay of PI3K or phosphatase activity, the lysate was diluted in water to a final protein concentration of 1 mg mL^{-1} and incubated with either Bodipy Fl PIP_2 or Bodipy Fl PIP_3. For PI3K assays, ATP (1 mM) was added to the reaction mix.

Cell loading with fluorescent lipid reagents

Bodipy Fl PIP_2 or Bodipy Fl PIP_3 was loaded into RBL cells as described previously.[59,68,69] Briefly, the lipid was mixed with histone in a 1 : 1 molar ratio and then incubated for 10 min at room temperature. The mixture was then diluted in a physiological extracellular buffer (ECB: 135 mM NaCl, 5 mM KCl, 1 mM $MgCl_2$, 1 mM $CaCl_2$, 10 mM HEPES, 10 mM glucose, pH 7.4) to the desired final concentration. The lipid:histone complex was incubated with cells at room temperature for 5 min. After loading, the cells were washed five times with ECB and allowed to recover for 5 min at 37 °C.

Fluorescence microscopy and image analysis

Fluorescence imaging was performed using a 20× 0.50 N.A., Fluor, air objective on an inverted fluorescence microscope (Nikon Ti, Melville, NY) and a standard fluorescein filter set (ex/em 488 nm/535 nm). Transmitted light and fluorescence images were obtained with a cooled CCD camera (Photometrix, Phoenix, AZ) using Metamorph image acquisition software (Molecular Devices, Sunnyvale, CA).

Detection of increases in intracellular free calcium ($[Ca^{2+}]_i$) in RBL cells

The calcium indicator dye Fluo-3 ($k_D = 400$ nm for Ca^{2+}) was used to measure relative increases in $[Ca^{2+}]_i$ during the loading of lipids. Cells were cultured on a glass coverslip in DMEM without serum overnight. Prior to the experiment, the cells were incubated in ECB with 1 μM Fluo-3 AM at 22 °C for 15 min. Cells were then washed with ECB and allowed to recover in ECB with 10 mM glucose at 37 °C for 30 min. Fluorescence imaging of the cells was performed at room temperature. Images were collected every 30 s for 10 min to confirm that the fluorescence intensity was stable, then the cells were incubated with the PIP_n-histone complex at various concentrations and fluorescence images were again collected every 30 s for 10 min.

Activation of RBL cells through the FcεR1 receptor

RBL cells possess the tyrosine kinase linked FcεR1 receptor that when bound to IgE and cross-linked with antigen leads to activation of PI3K.[73] The cells were stimulated as previously described.[76] Briefly, RBL cells were preincubated overnight with immunoglobulin E (IgE) against dinitrophenyl (DNP, 1 mg mL^{-1}). The cells were washed to remove excess IgE. At the desired time, cells were activated by the addition of polyvalent antigen (0.1 mg mL^{-1} DNP-BSA) for 10 min at 37 °C.

Capillary electrophoresis

CE analysis of lipid analytes was performed using a custom-built CE system with laser-induced fluorescence detection scheme as previously described.[60] Fused-silica capillaries (40 cm length, 50 μm i.d., 360 μm o.d., Polymicro Technologies, Phoenix, AZ) were used for the analyte separations. A voltage of 15 kV was applied across the capillary during electrophoresis. For CE analysis of the *in vitro* reaction mixtures, sample volumes (1 nL) were loaded by hydrodynamic injection. The number of mol of analyte present in each sample was determined by comparing the peak areas on the electrophoretic trace with standards as described previously.[77] Separation of Bodipy Fl PIP_2 and Bodipy Fl PIP_3 was performed in 100 mM Tris, 10 mM SDC, 1 mM $MgCl_2$, 30% 1-propanol, and 5% EOTrol LR, at pH 8.5. Fluorescein (1 nM mixed with the analyte) was used as an internal standard. Prior to each run, the capillary was flushed with 1 M NaOH for 3 min, deionized H_2O for 3 min, and the separation buffer for 3 min using a pressurized washing system at 20 psi.

Single-cell CE analysis

A custom-built CE-laser-induced fluorescence (LIF) system mated with a microscope was used for the single-cell analyses.[77] A custom-made chamber in which cells were plated was placed on an inverted microscope (Nikon TE2000, Melville, NY), and the inlet of the capillary was positioned 50 μm above the cell to be analyzed. A pulsed Nd:YAG laser (5 ns, 532 nm, Minilase, New Wave Research, Fremont, CA) integrated with the microscope was used to lyse the cell. Simultaneously with the firing of the laser, an electric field of 25 V cm^{-1} was applied across the capillary for 10 s to load the cell contents into the capillary. The capillary was then moved to a reservoir containing the electrophoresis buffer and electrophoresis initiated (300 V cm^{-1}, inlet at ground). Peaks of fluorescent cellular analytes were detected by LIF and recorded as electropherograms. In some experiments, the sample was spiked with a standard of Bodipy Fl PIP_2 or Bodipy Fl PIP_3 (1 nM) immediately prior to cell lysis to confirm the identity of the peak in the electropherogram.

Conclusion

An assay has been validated for PI3K and related phosphatases, including PTEN, with a number of advantages over conventional methods, particularly for cell-based assays. This method is enabled by combining microanalytical chemical separations with fluorescently labeled lipids to provide quantitative information on the conversion of a lipid substrate to metabolic end products. The format possesses the flexibility to study purified enzymes, cell lysates, or individual cells. Importantly, the chemical separation step enables the quantitative determination of both substrate and product(s) as well as side products. The resolving power of CE permits quantitation of not only the reversible conversion of a fluorescently labeled PIP_2 to the PIP_3 product, but also the measurement of multiple metabolic products simultaneously in the same sample. Moreover, with the sensitivity limits of LIF detection in a capillary, analysis of these metabolites in single cells has been achieved. Such information will be critical to enhance understanding of important signal transduction and homeostatic networks at the single-cell level. With the increasing interest in biochemical measurements of the PI3K pathway for cellular signal transduction, this assay can be expected to find broad application in biological and biomedical investigations. Information rich cell-based assays of pharmaceutical compounds targeting this pathway can now be readily performed. Furthermore, the small sample size needed in this method opens up the possibility of performing drug assays directly on samples of patient cells.

Acknowledgements

This research was supported by the NIH (CA139599 and CA140173). We express our appreciation for many helpful discussions with Qisheng Zhang.

References

1 M. P. Wymann and R. Schneiter, *Nat. Rev. Mol. Cell Biol.*, 2008, **9**, 162–176.
2 I. Ader, B. Malavaud and O. Cuvillier, *Cancer Res.*, 2009, **69**, 3723–3726.
3 H. Sheng, B. Niu and H. Sun, *Curr. Med. Chem.*, 2009, **16**, 1561–1587.
4 P. Dent, D. T. Curiel, P. B. Fisher and S. Grant, *Drug Resist. Updates*, 2009, **12**, 65–73.
5 T. Carlo and B. D. Levy, *Allergology International*, 2008, **57**, 299–305.
6 N. P. Duroudier, A. S. Tulah and I. Sayers, *Allergy*, 2009, **64**, 823–839.
7 J. Rivera, N. A. Fierro, A. Olivera and R. Suzuki, *Adv. Immunol.*, 2008, **98**, 85–120.
8 K. Ito, G. Caramori and I. M. Adcock, *J. Pharmacol. Exp. Ther.*, 2007, **321**, 1–8.
9 J. A. Engelman, J. Luo and L. C. Cantley, *Nat. Rev. Genet.*, 2006, **7**, 606–619.
10 M. P. Wymann and R. Marone, *Curr. Opin. Cell Biol.*, 2005, **17**, 141–149.
11 R. C. Castellino and D. L. Durden, *Nat. Clin. Pract. Neurol.*, 2007, **3**, 682–693.
12 D. A. Guertin and D. M. Sabatini, *Cancer Cell*, 2007, **12**, 9–22.
13 G. Fritz and B. Kaina, *Curr. Cancer Drug Targets*, 2006, **6**, 1–4.
14 Q. W. Fan and W. A. Weiss, *Cell Cycle*, 2006, **5**, 2301–2305.
15 T. A. Yap, M. D. Garrett, M. I. Walton, F. Raynaud, J. S. de Bono and P. Workman, *Curr. Opin. Pharmacol.*, 2008, **8**, 393–412.
16 P. Workman, P. A. Clarke, S. Guillard and F. I. Raynaud, *Nat. Biotechnol.*, 2006, **24**, 794–796.
17 T. Balla, Z. Szentpetery and Y. J. Kim, *Physiology (Bethesda)*, 2009, **24**, 231–244.
18 G. Di Paolo and P. De Camilli, *Nature*, 2006, **443**, 651–657.
19 R. S. Rapaka, D. Piomelli, S. Spiegel, N. Bazan and E. A. Dennis, *Prostaglandins Other Lipid Mediators*, 2005, **77**, 223–234.
20 S. K. Dove, F. T. Cooke, M. R. Douglas, L. G. Sayers, P. J. Parker and R. H. Michell, *Nature*, 1997, **390**, 187–192.
21 L. E. Hokin and M. R. Hokin, *J. Biol. Chem*, .1958, **233**, 805–810.
22 M. Whitman, C. P. Downes, M. Keeler, T. Keller and L. Cantley, *Nature*, 1988, **332**, 644–646.
23 T. E. Rusten and H. Stenmark, *Nat. Methods*, 2006, **3**, 251–258.
24 F. F. Hsu and J. Turk, *J. Am. Soc. Mass Spectrom.*, 2000, **11**, 986–999.
25 Y. Ishida, O. Nakanishi, S. Hirao, S. Tsuge, J. Urabe, T. Sekino, M. Nakanishi, T. Kimoto and H. Ohtani, *Anal. Chem.*, 2003, **75**, 4514–4518.
26 M. R. Wenk, L. Lucast, G. Di Paolo, A. J. Romanelli, S. F. Suchy, R. L. Nussbaum, G. W. Cline, G. I. Shulman, W. McMurray and P. De Camilli, *Nat. Biotechnol.*, 2003, **21**, 813–817.
27 A. Gray, H. Olsson, I. H. Batty, L. Priganica and C. Peter Downes, *Anal. Biochem.*, 2003, **313**, 234–245.
28 B. E. Drees, A. Weipert, H. Hudson, C. G. Ferguson, L. Chakravarty and G. D. Prestwich, *Comb. Chem. High Throughput Screening*, 2003, **6**, 321–330.
29 M. Sato, Y. Ueda, T. Takagi and Y. Umezawa, *Nat. Cell Biol.*, 2003, **5**, 1016–1022.
30 T. Balla and P. Varnai, *Science's STKE*, 2002, **2002**, 3pl.
31 M. A. Lemmon, *Traffic*, 2003, **4**, 201–213.
32 L. Borland, S. Kottegoda, K. S. Phillips and N. L. Allbritton, *Annu. Rev. Anal. Chem.*, 2008, **1**, 191–227.
33 E. H. Turner, D. Cohen, H. R. Pugsley, D. Gonzalez Gómez, C. D. Whitmore, C. Zhu and N. J. Dovichi, *Anal. Bioanal. Chem.*, 2008, **390**, 223–226.
34 J. Kraly, M. A. Fazal, R. M. Schoenherr, R. Bonn, M. M. Harwood, E. Turner, M. Jones and N. J. Dovichi, *Anal. Chem.*, 2006, **78**, 4097–4110.
35 N. J. Dovichi and S. Hu, *Curr. Opin. Chem. Biol.*, 2003, **7**, 603–608.
36 C. D. Whitmore, O. Hindsgaul, M. M. Palcic, R. L. Schnaar and N. J. Dovichi, *Anal. Chem.*, 2007, **79**, 5139–5142.
37 K. M. Fuller, C. F. Duffy and E. A. Arriaga, *Electrophoresis*, 2002, **23**, 1571–1576.
38 K. J. Lee, S. M. Mwongela, S. Kottegoda, L. Borland, A. R. Nelson, C. E. Sims and N. L. Allbritton, *Anal. Chem.*, 2008, **80**, 1620–1627.
39 A. Malek and M. G. Khaledi, *Anal. Biochem.*, 1999, **270**, 50–58.
40 L. Zhang, S. Hu, L. Cook and N. J. Dovichi, *Electrophoresis*, 2002, **23**, 3071–3077.
41 G. D. Meredith, C. E. Sims, J. S. Soughayer and N. L. Allbritton, *Nat. Biotechnol.*, 2000, **18**, 309–312.

42 G. D. Prestwich, *Chem. Biol.*, 2004, **11**, 619–637.
43 G. D. Prestwich, R. Chen, L. Feng, S. Ozaki, C. G. Ferguson, B. E. Drees, D. A. Neklason, M. J. Mostert, P. A. Porter-Gill, V. H. Kang, J. C. Shope, P. O. Neilsen and D. B. Dewald, *Adv. Enzyme Regul.*, 2002, **42**, 19–38.
44 G. D. Prestwich, *Prostaglandins Other Lipid Mediators*, 2005, **77**, 168–178.
45 G. S. Taylor and J. E. D. Dixon, *Anal. Biochem.*, 2001, **295**, 122–126.
46 C. E. Senkal, S. Ponnusamy, M. J. Rossi, J. Bialewski, D. Sinha, J. C. Jiang, S. M. Jazwinski, Y. A. Hannun and B. Ogretmen, *Mol. Cancer Ther.*, 2007, **6**, 712–722.
47 N. Lipsky and R. Pagano, *Proc. Natl.Acad. Sci. U. S. A.*, 1983, **80**, 2608–2612.
48 J. P. Vos, M. L. Giudici, P. van der Bijl, P. Magni, S. Marchesini, L. M. van Golde and M. Lopes-Cardozo, *FEBS Lett.*, 1995, **368**, 393–396.
49 P. Paul, Y. Kamisaka, D. L. Marks and R. E. Pagano, *J. Biol. Chem.*, 1996, **271**, 13088–2293.
50 Y. Hayashi, Y. Horibata, K. Sakaguchi, N. Okino and M. Ito, *Anal. Biochem.*, 2005, **345**, 181–186.
51 Y. Horibata, H. Higashi and M. Ito, *J. Biochem.*, 2001, **130**, 263–268.
52 J. M. Figueiredo, W. B. Dias, L. Mendonça-Previato, J. O. Previato and N. Norton Heise, *Biochem. J.*, 2005, **387**, 519–529.
53 W. Zhong, D. J. Murphy and N. H. Georgopapadakou, *FEBS Lett.*, 1999, **463**, 241–244.
54 M. Tani, N. Okino, M. Mutsutake and M. Ito, *J. Biochem.*, 1999, **125**, 746–749.
55 A. Abe, J. A. Shayman and N. S. Radin, *Lipids*, 1992, **27**, 1052–1054.
56 C. Mao, M. Wadleigh, G. M. Jenkins, Y. A. Hannun and L. M. Obeid, *J. Biol. Chem.*, 1997, **272**, 28690–28694.
57 A. Loidl, R. Claus, H. P. Deigner and A. Hermetter, *J. Lipid Res.*, 2002, **43**, 815–823.
58 D. B. Dewald, S. Ozaki, S. Malaviya, J. C. Shope, K. Manabe, L. Crosby, P. Neilsen, D. Johnston, S. Harihar and G. D. Prestwich, *Cell Calcium*, 2005, **38**, 59–72.
59 S. Ozaki, D. B. DeWald, J. C. Shope, J. Chen and G. D. Prestwich, *Proc. Natl. Acad. Sci. U. S. A.*, 2000, **97**, 11286–11291.
60 S. M. Mwongela, K. Lee, C. E. Sims and N. L. Allbritton, *Electrophoresis*, 2007, **28**, 1235–1242.
61 A. C. Otieno, E. W. Quainoo and S. M. Mwongela, *J. Pep. Sci.*, 2008, **31**, 3894–3901.
62 S. McLaughlin, J. Wang, A. Gambhir and D. Murray, *Annu. Rev. Biophys. Biomol. Struct.*, 2002, **31**, 151–175.
63 C. P. Palmer and B. G. M. Vandeginste, *J. Chromatogr., A*, 1995, **718**, 153–165.
64 Caliper LifeSciences, http://www.caliperls.com/assets/013/6998.pdf, 2008.
65 Z. A. Knight and K. M. Shokat, *Biochem. Soc. Trans.*, 2007, **35**, 245–249.
66 M. J. Fry, in *Lipid Signaling Protocols*, ed. B. Larijani, R. Woscholski and C. A. Rosser, Human Press, Totowa, 2009, pp. 345–362.
67 L. A. Serunian, K. R. Auger and L. C. Cantley, *Methods Enzymol.*, 1991, **198**, 78–87.
68 F. Wang, P. Herzmark, O. D. Weiner, S. Srinivasan, G. Servant and H. R. Bourne, *Nat. Cell Biol.*, 2002, **4**, 513–518.
69 O. D. Weiner, P. O. Neilsen, G. D. Prestwich, M. W. Kirschner, L. C. Cantley and H. R. Bourne, *Nat. Cell Biol.*, 2002, **4**, 509–513.
70 C. Walker, M. Thomas and M. J. Edwards, *Drug Discovery Today Dis. Mech.*, 2006, **3**, 63–69.
71 S. A. Barker, K. K. Caldwell, J. R. Pfeiffer and B. S. Wilson, *Mol. Biol. Cell*, 1998, **9**, 483–496.
72 J. I. Wilde and S. P. Watson, *Cell. Signalling*, 2001, **13**, 691–701.
73 A. J. Smith, Z. Surviladze, E. A. Gaudet, J. M. Backer, C. A. Mitchell and B. S. Wilson, *J. Biol. Chem.*, 2001, **276**, 17213–17220.
74 A. R. Nelson, N. L. Allbritton and C. E. Sims, *Methods Cell Biol.*, 2007, **82**, 709–722.
75 *Short Protocols in Molecular Biology*, ed. F. M. Ausubel, R. Brent, R. E. Kingston, D. D. Moore, J. G. Seidman, J. A. Smith and K. Struhl, Wiley, New York, 2002.
76 N. L. Allbritton, E. Oancea, M. Kuhn and T. Meyer, *Proc. Natl. Acad. Sci. U. S. A.*, 1994, **91**, 12458–12462.
77 C. E. Sims, G. D. Meredith, T. B. Krasieva, M. W. Berns, B. J. Tromberg and N. L. Allbritton, *Anal. Chem.*, 1998, **70**, 4570–4577.

Peptide-tags for enhanced DNA microarray performance†

Holger Schulze,[a] Alan J. Ross,[a] Stuart W. J Ember,[a] Julie Luby,[a] Mizanur Khondoker,[a] Gerard Giraud,[b] Ilenia Ciani,[c] Chaker Tlili,[c] Davide Papale,[c] Jonathan G. Terry,[d] Andrew R. Mount,[c] Anthony J Walton,[d] Jason Crain,[be] Peter Ghazal,[a] Till T. Bachmann[a] and Colin J. Campbell*[ac]

Received 13th July 2010, Accepted 23rd July 2010
DOI: 10.1039/c005491g

DNA microarrays are powerful tools for gene expression analysis and genotyping studies in research and diagnostic applications. A high sensitivity and short time-to-result are prerequisites for their practical application in the clinic. The hybridization efficiency of DNA microarrays depends on the probe density and the probe orientation and thus their accessibility for target molecules. In order to find an optimal probe immobilization procedure a set of different oligonucleotide modifications was tested on epoxy silane functionalized glass slides. It was found that histidine-tagged oligonucleotides resulted in the highest amount of bound probe and by far the best hybridization efficiencies. The detection limit obtained with histidine-tagged probes was up to two orders of magnitude lower compared to commonly used probe modifications. In order to further investigate the binding mechanism of histidine-tags towards functionalized glass substrates a set of different peptide-tags with and without free terminal amino-groups and with different amino acid compositions was tested. The results indicate an impact of the terminal amino group on the covalent surface binding and of aromatic amino acid residues on the enhanced hybridisation efficiency.

Introduction

DNA and protein microarrays are powerful tools for clinical diagnostics as they enable the detection of a high number of biomarkers in parallel with the possibility

[a]*Division of Pathway Medicine, College of Medicine and Veterinary Medicine, The University of Edinburgh, Chancellor's Building, 49 Little France Crescent, Edinburgh, EH16 4SB, Scotland, UK. E-mail: holger.schulze@ed.ac.uk; alan.j.ross@ed.ac.uk; stuart.ember@yahoo.co.uk; colin.campbell@ed.ac.uk; mizanur.khondoker@kcl.ac.uk; pghazal@uun.ed.ac.uk; till.bachmann@ed.ac.uk; Fax: +44 131 242 6244; Tel: +44 131 242 6280*

[b]*School of Physics and Astronomy, The University of Edinburgh, The King's Buildings, West Mains Road, Edinburgh, EH9 3JZ, Scotland, UK. E-mail: gerard.giraud@ed.ac.uk; jason.crain@npl.co.uk*

[c]*School of Chemistry, The University of Edinburgh, Joseph Black Building, West Mains Road, Edinburgh, EH9 3JJ, Scotland, UK. E-mail: iciani@staffmail.ed.ac.uk; a.mount@ed.ac.uk; chaker_tlili@yahoo.fr*

[d]*Institute for Integrated Systems, School of Engineering, The University of Edinburgh, The King's Buildings, West Mains Road, Edinburgh, EH9 3JF, Scotland, UK. E-mail: anthony.walton@ed.ac.uk; Jon.Terry@ee.ed.ac.uk*

[e]*National Physical Laboratory, Hampton Road, Teddington, Middlesex, TW11 0LW, UK*

† Electronic supplementary information (ESI) available: Calibration plots obtained with differentially functionalized HCMV and HCV probes hybridized with Cy3-labelled complementary targets. See DOI: 10.1039/c005491g

to multiplex relevant information like cell surface antigens,[1] pathogen detection,[2] virulence factors, antibiotic resistance or gene expression on a single test.[3–10] The production methods for DNA microarrays are well established and mainly use either *in situ* synthesis or spotting of oligonucleotide detection probes.[11–14] Several microarrays have been developed for the identification of pathogens, such as bacteria, fungi and viruses.[15–20] A prerequisite for their practical application is a high sensitivity and short time-to-result. There are different approaches to increase the sensitivity of DNA microarrays, *e.g.* by decreasing the reaction volume in microfluidic systems and integrated lab-on-a-chip devices, as summarized in recent reviews.[21–24] The efficiency and speed of DNA target binding to probe molecules immobilized on a solid support can also be improved in the standard DNA microarray slide format by optimizing the mixing conditions during hybridization, *e.g.* by the means of automated hybridization stations.[25] Another parameter which has a major impact on the hybridization efficiency in solid phase systems is the surface chemistry and the type of probe binding. Non-covalent approaches, such as the use of polylysine coated slides, which rely on electrostatic interaction of the negatively charged DNA backbone with the positively charged slide surface[26] are relatively simple to use but covalent, terminal attachment of the probe molecules to the slide surface is preferable in terms of probe layer stability. The most common slide types used for covalent attachment of oligonucleotide probes on DNA microarrays are glass substrates functionalized either with aldehyde or epoxy silane functional groups.[27–29] We focused our studies on epoxy silane modified glass slides as they have been shown to be an excellent matrix for various microarray applications, *e.g.* for SNP detection and are commercially available with a high quality standard.[3,7] For epoxy silane modified surfaces, the most commonly used probe modifications are amino functional groups,[29] whereas there are also examples of thiol modified oligonucleotides used on epoxy silane coated slides.[30–32]

In this study, we tested peptide tags as new probe modifications for covalent attachment of oligonucleotide probes onto epoxy silane coated substrates. We compared the amount of immobilized peptide tagged probes with that of other modifications using fluorescence-labelled probes and measured the influence of the different probe modifications on the hybridization efficiency using an HCV viral load assay.

Results and discussion

Comparative probe binding

Commercial epoxy silane coated glass slides were chosen as microarray substrate for their competitive advantages over other slide types. The electrophilic epoxy silane group forms stable covalent bonds with DNA probe molecules which are modified with a terminal nucleophilic functional group without the need of any additional chemicals. Aldehyde coated slides form less stable bonds or require an additional chemical reduction step with sodium borohydride to form stable bonds.[29,33] In this study we examine the influence of different probe modifications on probe binding to the slide and on assay performance. The relative amount of immobilized probe molecules with different functional groups was determined with a set of 3′ fluorescently labelled probes. These probe modification included amino, thiol, biotin and modification with an oligopeptide containing six histidines at their 5′ end. Histidine and biotin-tags were included in the study because we initially tested Nickel-NTA and streptavidin coated glass slides but found them to give non satisfactory results (data not shown). We chose a probe sequence specific for HCMV, which has been previously used in a gene expression array.[15] Fig. 1 shows the relative fluorescence intensities measured at spots of Cy3-labelled oligonucleotides with different modifications. The highest fluorescence intensities and thus the highest amount of bound probe were obtained with histidine-tagged probes, followed by thiol modified, amino modified and biotinylated probes.

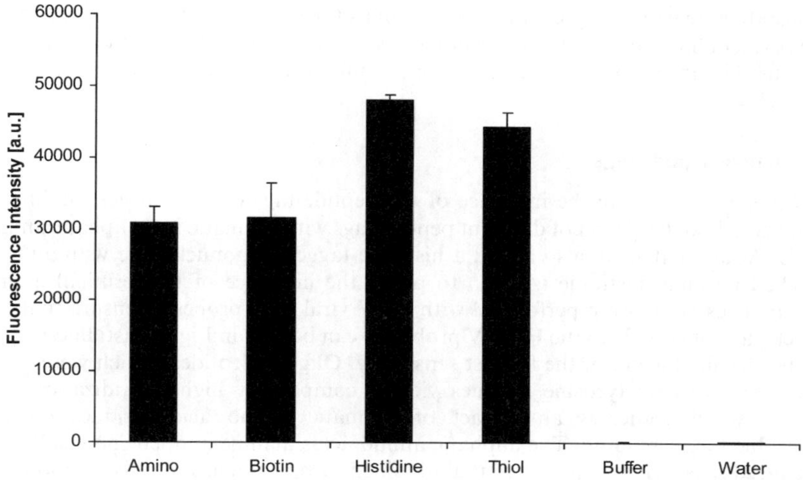

Fig. 1 Mean fluorescence intensities of differentially functionalized, Cy3-labelled HCMV oligonucleotides (20 μM) spotted on epoxy silane slides. Histidine = 6× histidine; error bars represent standard deviations of 30 replicates.

Hybridization efficiency

The impact of different probe modifications on the hybridization efficiency was tested with unlabelled, but differently modified 75mer probes, which (after immobilization) were hybridized with Cy3-labelled 75mer target molecules complementary to the probe molecules. The type of probe immobilization had a dramatic impact on the hybridization efficiency as can be seen in Table 1 and Fig. S1 and S2 (ESI†). By far the highest sensitivity with a detection limit of 1 pM was obtained with histidine-tagged probes. Unmodified probes were least sensitive towards the target with a detection limit of 100 pM. This low sensitivity can be explained with a probe immobilization *via* the DNA backbone and not *via* terminal reactive groups causing the probe to be less accessible for the target. The detection limit obtained with amino modified probes is consistent with that obtained by Zammatteo *et al.* for HCMV. However, they used more than three times longer amino functionalized probes (255 bp) on aldehyde coated glass slides.[29]

In contrast, a peptide tag consisting of six histidine residues with a free terminal amino group improved the hybridization efficiency compared to commonly used amino modification. This was a rather unexpected result as this is, to the best of our knowledge, the first example of the use of a histidine-tag for covalent attachment of oligonucleotides on a solid support. Up to now histidine-tags have predominantly been used for affinity purification or affinity binding of peptides to microarray slides.[34;35] It has been reported previously that imidazole residues, the building block

Table 1 Detection limit obtained with different probe modifications against fluorescence labelled HCMV target

Modification	Detection limit [pM]
None	100
Amino	10
Thiol	10
Biotin	10
Histidine-tag	1

of histidine, react with epoxy functional groups forming a covalent linkage.[36] Thus, the covalent linkage between the histidine-tag and the epoxy silane surface can either be caused by the terminal amino group or the imidazole residues or a combination of both effects.

Alternative peptide tags

In order to investigate the influence of the peptide-tags on the surface binding in more detail, we used a set of different peptide tags with aromatic and aliphatic amino acids. Additionally, we also added a histidine-tagged oligonucleotide with an acyl blocked terminal histidine (AcHis) to probe the influence of the terminal amino group. These tests were performed with HCV viral load probes to ensure that the effect was not specific to the HCMV probe. As can be seen in Fig. 2, histidine-tagged probes resulted again in the highest sensitivity. Oligonucleotides, which were functionalized with six tyrosine residues yielded comparably high hybridization efficiency, which indicates an impact of aromatic amino acid residues on the favorable target binding. The aliphatic amino acids alanine and leucine resulted in a reduced sensitivity compared to that obtained with histidine and tyrosine modified

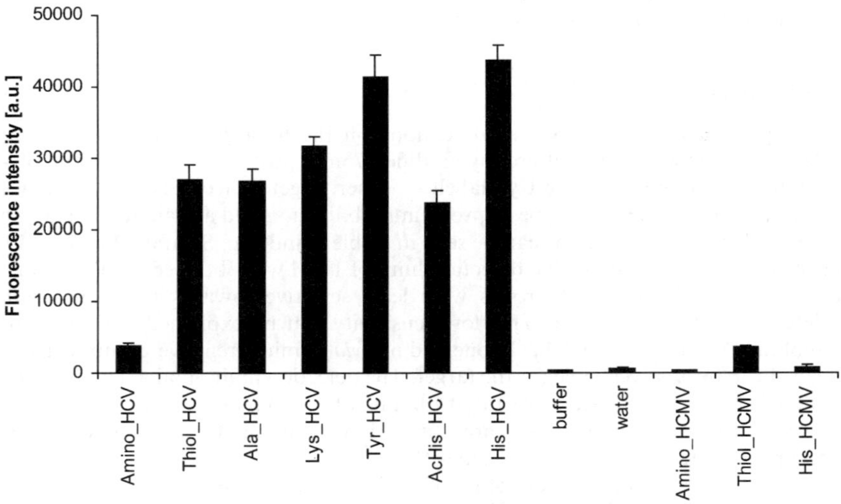

Fig. 2 Mean fluorescence intensities of differentially functionalized probes (20 μM) on epoxy silane slides hybridized with 10 pM Cy3-labelled HCV target; Ala = 6× alanine; Lys = 6× lysine; Tyr = 6× tyrosine; AcHis = 5 histine + terminal histidine with acetylated terminal amino group; His = 6× histidine; error bars represent standard deviations of 5 replicates.

Table 2 Detection limit obtained with different probe modifications against fluorescence labelled HCV target

Modification	Detection limit [pM]
Amino	10
Thiol	1
Alanine/Leucine-tag	1
Lysine-tag	0.1
Tyrosine-tag	0.1
Acetylated-Histidine-tag	1
Histidine-tag	0.1

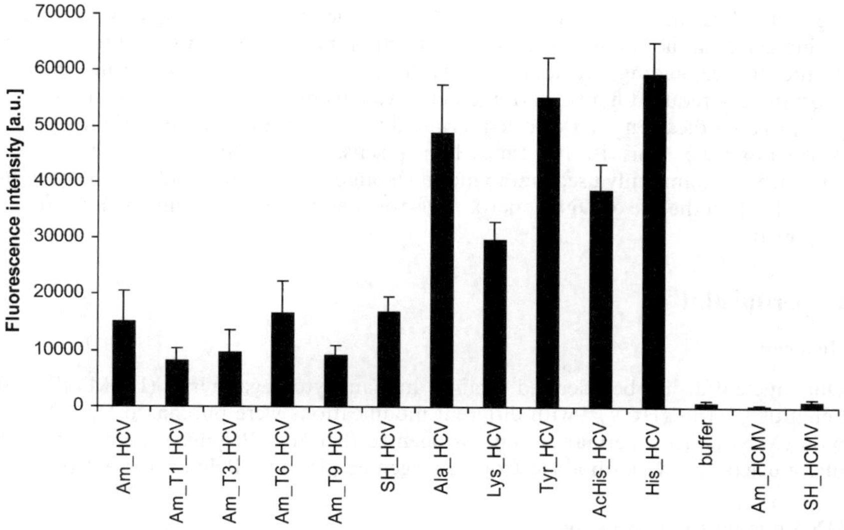

Fig. 3 Mean fluorescence intensities after hybridization with 100 pM Cy3-labelled HCV target on an array of differentially functionalized probes (20 μM) on epoxy silane slides. Am = amino; T1 = 1 thymine spacer; T3 = 3 thymine spacer; T6 = 6 thymine spacer; T9 = 9 thymine spacer; SH = thiol; Ala = 6× alanine; Lys = 6× lysine; Tyr = 6× tyrosine; AcHis = 5 histine + terminal histidine with acetylated terminal amino group; His = 6× histidine. Error bars represent standard deviations of 5 replicates.

probes. The reduced sensitivity of histidine-tagged probes without terminal amino group clearly indicate the impact of this terminal amino group to the covalent attachment of the peptide tags to the epoxy silane coated microarray slides. Tests with different amounts of 40mer fluorescence labelled HCV targets revealed a detection limit two order of magnitude lower when using a histidine-tagged compared to amino modified probes (see Table 2 and ESI Fig. S3 and S4†). Fig. S4 shows that histidine-tagged probes enabled a clear detection of 10 pM target with a five times higher signal compared to that obtained with amino modified probes.

Spacer effect

We also tested a range of poly thymine spacers from one to nine additional thymine residues at the 5′ end of the specific probe sequence in order to find out if the superior performance of the peptide tags is only related to a spacer effect. The size of a peptide consisting of six amino acids is in the range between 1 and 2 nm. A DNA double helix of 10 nucleotide bases is in the order of 3.4 nm long.[37] This means that a peptide tag of six amino acids should have a comparable spacer length to a polyT spacer of three to six thymine residues. Fig. 3 shows that the superior effect of the histidine and tyrosine tag can not be explained only by a spacer effect. The peptide tags seem to result in an optimized orientation of the surface bound probe molecules making them more accessible for the target molecules in solution and thus increasing the sensitivity of the DNA microarray.

Conclusion

We have shown in this study that peptide tags, especially histidine-tags, used as a new type of oligonucleotide probe modification can substantially increase the sensitivity of DNA microarrays by up to two orders of magnitude. Similar results were obtained with HCMV and HCV probes. Tests of different type of peptide tags with aliphatic and aromatic amino acid residues revealed an impact of aromatic amino acids on the enhanced sensitivity and an impact of the terminal amino group on the surface binding. Histidine-tags without a free terminal amino group showed substantially reduced hybridization efficiency. This influence of the terminal amino group is an indication for a directed terminal immobilization of the histidine-tagged probe molecule. This 10–100 times higher sensitivity of histidine-tagged probes compared to commonly used amino modified oligonucleotides could have a substantial impact on the use of DNA microarrays for practical applications such as clinical diagnostics.

Experimental

Reagents

Oligonucleotide probes derived against human cytomegalovirus (HCMV)[15] and hepatitis C virus (HCV)[38] with different modifications were purchased from Metabion (Martinsried, Germany) and Eurogentec (Seraing, Belgium). Peptide tagged oligonucleotides were obtained from Eurogentec (Seraing, Belgium) (see Table 3).

DNA microarray fabrication

Modified oligonucelotides were spotted in 1× Schott Nexterion spot buffer (20 μM) on Schott Nexterion Slides E (epoxy silane modified surface) with a Microgrid II spotter (Digilab, Huntingdon, UK) using 200 μm solid pins. Thiol-modified oligonucleotides were spotted in 1× Schott Nexterion spot buffer containing 5 mM tris(2-carboxy-ethyl)phosphine hydrochloride (TCEP) to cleave the mercapto-ethyl protection group. TCEP containing spotting solutions were incubated for 30 min at RT before printing. The oligonucleotide probe molecules were immobilized by incubating the slides in a humidity chamber for 1 h followed by storage over night at room temperature (RT) under dry conditions. The slides were then washed with 0.1% TritonX-100 solution under constant mixing for 5 min at RT, with 1 mM HCl solution for 4 min, with 100 mM KCl solution for 10 min, and with deionized water for 1 min. The slides were blocked with 50 mM ethanolamine + 0.1% sodium dodecyl sulfate (SDS) in 0.1 M Tris buffer (pH 9) for 15 min at 50 °C. After blocking the slides were washed in deionized water for 1 min and then dried by centrifugation (2 min at 800 g).

Arrays of Cy3-labelled probes were spotted in 3× sodium chloride, sodium citrate (SSC) solution (450 mM NaCl + 45 mM Na-citrate) on Erie Scientific Superchip epoxy silane slides (Thermo Fisher Scientific, Waltham, USA) with a Microgrid II spotter using 200 μm solid pins. Probes were immobilized 60 min after printing by UV crosslinking at 350 mJ with a UV crosslinker device (UVP, Upland, USA). Slides were blocked with 1% (w/v) bovine serum albumin in 3× SSC solution for 10 min, followed by 15 min wash in water and four additional rinsing steps in water. Slides were then dried by centrifugation (2 min at 800 *g*) with an Eppendorf centrifuge 5810 R (Eppendorf, Hamburg, Germany).

Hybridization and washing

Arrays were hybridized with 50 μL Cy3-labelled 40mer target solution in 4× SSC buffer + 0.01% SDS with an Agilent 8 well gasket slide in an Agilent hybridization oven at 55 °C under agitation (rotation speed 4). After hybridization the arrays were washed with 2× SSC + 0.2% SDS solution for 10 min at RT under constant mixing, with 2× SSC solution for 10 min at RT, and with 0.2× SSC for 10 min. After dipping into water the slides were dried by centrifugation (2 min at 800 *g*).

Hybridization with HCMV target was performed in a Tecan HS400Pro automated hybridization station (Tecan, Maennedorf, Switzerland). Slides were washed for 2 min with 0.1% Triton-X-100 solution, 2 min with 10 pM HCl solution, 5 min

Table 3 List of oligonucleotides used in this study

Oligo type	Sequence (5′–3′)	5′ modification	3′
HCMV	CAAATACCGTGGG ACGACACGCACCGGC AGTGCGCAGGCAGC GTCGGACACAACA CGCTTACGGCC CTCAACACT	biotin	—
HCMV	CAAATACCGTGGGA CGACACGCACCGG CAGTGCGCAGGCAG CGTCGGACACAACA CGCTTACGGCC CTCAACACT	NH_2 (C6)	—
HCMV	CAAATACCGTGGGA CGACACGCA CCGGCAGT GCGCAGGCAGCGTCG GACACAACACGCTTAC GGCCCTCAACACT	SH (C6)	—
HCMV	CAAATACCGTGGGA CGACACGCACCGGCAG TGCGCAGGCAGCGTCG GACACAACACGCTT ACGGCCCTCAACACT	NH_2-His-His-His-His-His-His-Cys	—
HCMV	CAAATACCGTGGGAC GACACGCACCGGCAG TGCGCAGGCAGCGTC GGACACAACACGCTT ACGGCCCTCAACACT	—	—
HCMV	CAAATACCGTGGGACG ACACGCACCGGCAGT GCGCAGGCAGCGTC GGACACAACACGCTTA CGGCCCTCAACACT	biotin	Cy3
HCMV	CAAATACCGTGGGA CGACACGCACCGGC AGTGCGCAGGCAGCG TCGGACACAACACG CTTACGGCC CTCAACACT	NH_2 (C6)	Cy3
HCMV	CAAATACCGTGGGA CGACACGCACCGGC AGTGCGCAGGCAGC GTCGGACACAACA CGCTTACGGCC CTCAACACT	SH (C6)	Cy3
HCMV	CAAATACCGTGGGA CGACACGCACCGG CAGTGCGCAGGCAGCGT CGGACACAACACGC TTACGGCCCTCAACACT	NH_2-His-His-His-His-His-His-Cys	Cy3
HCMV target	AGTGTTGAGGGCC GTAAGCGTGTTGTGT CCGACGCTGCCTGCG CACTGCCGGTGCGTG TCGTCCCACGGTATTTG	Cy3	—

Table 3 (*Contd.*)

Oligo type	Sequence (5′–3′)	5′ modification	3′
HCV	TCGCAAGCACCCT ATCAGGCAGTACCACA AGGCCTTTCGC	NH_2-Ala-Ala-Leu-Ala-Leu-Ala-Cys	—
HCV	TCGCAAGCACCC TATCAGGCAGTACC ACAAGGCCTTTCGC	NH_2 (C6)	—
HCV	TCGCAAGCACCCT ATCAGGCAGTACCAC AAGGCCTTTCGC	SH (C6)	—
HCV	TTCGCAAGCACCCT ATCAGGCAGTACCAC AAGGCCTTTCGC	NH_2 (C6)	—
HCV	TTTTCGCAAGCAC CCTATCAGGCAGTA CCACAAGGCCTTTCGC	NH_2 (C6)	—
HCV	TTTTTTTCGCAAGC ACCCTATCAGGCAGTA CCACAAGGCCTTTCGC	NH_2 (C6)	—
HCV	TTTTTTTTTTCGCA AGCACCCTATCAGG CAGTACCACAAGG CCTTTCGC	NH_2 (C6)	—
HCV target	GCGAAAGGCCT TGTGGTACTGCCTGAT AGGGTGCTTGCGA	Cy3	—

with 100 mM KCl solution, and 1 min with water, respectively. Blocking with 50 mM ethanolamine solution was performed for 10 min at 50 °C, followed by 1 min washing with water. Hybridization was performed at 55 °C for 2 h. Slides were then washed with 2× SSC + 0.1% SDS for two times 2 min, with 0.2× SSC for 2 min and with water for 30 s. Slides were automatically dried by nitrogen flow for 2 min.

Image acquisition and data analysis

Fluorescence images were generated with a Tecan LS Reloaded fluorescence scanner (Tecan, Maennedorf, Switzerland) with excitation at 532 nm and emission at 575 nm at PMT 200 (chosen to maximise signal without pixel saturation). Quantification of fluorescence signal intensities was performed with the Quantarray software (Perkin Elmer, Waltham, MA) using the Histogram quantification method. For further analysis, the mean signal intensity minus local background intensity was processed with Excel (Microsoft Corp., Redmond, USA) and the mean and standard deviation of all replicates were calculated. The detection limit was determined using the mean of the fluorescence intensity of the negative control probe plus three times the standard deviation.

Acknowledgements

This work was supported by ITI Techmedia (Glasgow, Scotland, UK) under the BioSensing Platform program. We would like to thank Mark Bradley and Juan Jose Diaz-Mochon for technical advice. CC is supported by an EaStCHEM fellowship.

References

1 C. J. Campbell, N. O'Looney, K. M. Chong, J. S. Robb, A. J. Ross, J. S. Beattie, J. Petrik and P. Ghazal, *Anal. Chem.*, 2006, **78**, 1930–38.

2 S. T. Burgess, F. Kenyon, N. O'Looney, A. J. Ross, K. M. Chong, J. S. Beattie, J. Petrik, P. Ghazal and C. J. Campbell, *Anal. Biochem.*, 2008, **382**, 9–15.

3 T. Barl, U. Dobrindt, X. Yu, D. J. Katcoff, D. Sompolinsky, S. Bonacorsi, J. Hacker and T. T. Bachmann, *Int. J. Antimicrob. Agents*, 2008, **32**, 272–77.

4 V. Grimm, S. Ezaki, M. Susa, C. Knabbe, R. D. Schmid and T. T. Bachmann, *J. Clin. Microbiol.*, 2004, **42**, 3766–74.

5 B. Strommenger, C. Schmidt, G. Werner, B. Roessle-Lorch, T. T. Bachmann and W. Witte, *Mol. Cell. Probes*, 2007, **21**, 161–70.

6 X. Yu, M. Susa, C. Knabbe, R. D. Schmid and T. T. Bachmann, *J. Clin. Microbiol.*, 2004, **42**, 4083–91.

7 X. Yu, M. Susa, J. Weile, C. Knabbe, R. D. Schmid and T. T. Bachmann, *Int. J. Med. Microbiol.*, 2007, **297**, 417–29.

8 J. D. Hoheisel, *Nat. Rev. Microbiol.*, 2006, **7**, 200–10.

9 C. J. Campbell and P. Ghazal, *J. Appl. Microbiol.*, 2004, **96**, 18–23.

10 G. Giraud, H. Schulze, T. T. Bachmann, C. J. Campbell, A. R. Mount, P. Ghazal, M. R. Khondoker, A. J. Ross, S. W. J. Ember, I. Ciani, C. Tlili, A. J. Walton, J. G. Terry and J. Crain, *Int. J. Mol. Sci.*, 2009, **10**, 1930–41.

11 G. Palacios, P. L. Quan, O. J. Jabado, S. Conlan, D. L. Hirschberg, Y. Liu, J. Zhai, N. Renwick, J. Hui, H. Hegyi, A. Grolla, J. E. Strong, J. S. Towner, T. W. Geisbert, P. B. Jahrling, C. Buechen-Osmond, H. Ellerbrok, M. P. Sanchez-Seco, Y. Lussier, P. Formenty, S. T. Nichol, H. Feldmann, T. Briese and W. I. Lipkin, *Emerging Infect. Dis.*, 2007, **13**, 73–81.

12 B. E. Cleven, M. Palka-Santini, J. Gielen, S. Meembor, M. Kronke and O. Krut, *J. Clin. Microbiol.*, 2006, **44**, 2389–97.

13 G. Y. C. Cheung, P. Dickinson, G. Sing, M. Craigon, P. Ghazal, R. Parton and J. G. Coote, *Microb. Pathog.*, 2008, **44**, 61–70.

14 S. Burteau, P. Bogaerts, R. de Mendonca, L. Irenge, C. Berhin, J. Hiffe, N. de San, P. Beyne, S. Hamels, Y. Glupczynski, M. Struelens, J. L. Gala and J. Remacle, *Eur. J. Clin. Microbiol. Infect. Dis.*, 2008, **27**, 17–27.

15 J. Chambers, A. Angulo, D. Amaratunga, H. Q. Guo, Y. Jiang, J. S. Wan, A. Bittner, K. Frueh, M. R. Jackson, P. A. Peterson, M. G. Erlander and P. Ghazal, *J. Virol.*, 1999, **73**, 5757–66.

16 C. C. Hsia, V. E. Chizhikov, A. X. Yang, A. Selvapandiyan, I. Hewlett, R. Duncan, R. K. Puri, H. L. Nakhasi and G. G. Kaplan, *Biochem. Biophys. Res. Commun.*, 2007, **356**, 1017–23.

17 D. M. Leinberger, U. Schumacher, I. B. Autenrieth and T. T. Bachmann, *J. Clin. Microbiol.*, 2005, **43**, 4943–53.

18 D. M. Leinberger, V. Grimm, M. Rubtsova, A. Henn, K. Schroppel, T. A. Wichelhaus, C. Knabbe, J. Weile, R. D. Schmid and T. T. Bachmann, *Int. J. Med. Microbiol.*, 2007, **297**, 17.

19 H. Wiesinger-Mayr, K. Vierlinger, R. Pichler, A. Kriegner, A. M. Hirschl, E. Presterl, L. Bodrossy and C. Noehammer, *BMC Microbiol.*, 2007, **7**, 78.

20 S. Burteau, P. Bogaerts, M. R. de, L. Irenge, C. Berhin, J. Hiffe, S. N. de, P. Beyne, S. Hamels, Y. Glupczynski, M. Struelens, J. L. Gala and Remacle, *Eur. J. Clin. Microbiol. Infect. Dis.*, 2008, **27**, 17–27.

21 H. Schulze, G. Giraud, J. Crain and T. T. Bachmann, *J. Biophotonics*, 2009, **2**, 199–211.

22 C. J. Easley, J. M. Karlinsey, J. M. Bienvenue, L. A. Legendre, M. G. Roper, S. H. Feldman, M. A. Hughes, E. L. Hewlett, T. J. Merkel, J. P. Ferrance and J. P. Landers, *Proc. Natl. Acad. Sci. U. S. A.*, 2006, **103**, 19272–77.

23 J. Vanderhoeven, K. Pappaert, B. Dutta, P. Vanhummelen, G. V. Baron and G. Desmet, *Electrophoresis*, 2004, **25**, 3677–86.

24 F. S. Ligler, *Anal. Chem.*, 2009, **81**, 519–26.

25 J. Vanderhoeven, K. Pappaert, B. Dutta, P. Van Hummelen and G. Desmet, *Anal. Chem.*, 2005, **77**, 4474–80.

26 S. Taylor, S. Smith, B. Windle and A. Guiseppi-Elie, *Nucleic Acids Research*, 2003, **31**, e87.

27 R. P. Auburn, D. P. Kreil, L. A. Meadows, B. Fischer, S. S. Matilla and S. Russell, *Trends Biotechnol.*, 2005, **23**, 374–79.

28 E. D. Dawson, A. E. Reppert, K. L. Rowlen and L. R. Kuck, *Anal. Biochem.*, 2005, **341**, 352–60.

29 N. Zammatteo, L. Jeanmart, S. Hamels, S. Courtois, P. Louette, L. Hevesi and J. Remacle, *Anal. Biochem.*, 2000, **280**, 143–50.

30 D. Sethi, A. Kumar, K. C. Gupta and P. Kumar, *Bioconjugate Chem.*, 2008, **19**, 2136–43.
31 S. Mahajan, D. Sethi, S. Seth, A. Kumar, P. Kumar and K. C. Gupta, *Bioconjugate Chem.*, 2009, **20**, 1703–10.
32 C. Preininger, U. Sauer, J. Dayteg and R. Pichler, *Bioelectrochemistry*, 2005, **67**, 155–62.
33 J. Sobek, C. Aquino and R. Schlapbach, *Methods Mol. Biol.*, 2007, **382**, 17–31.
34 F. Khan, M. Y. He and M. J. Taussig, *Anal. Chem.*, 2006, **78**, 3072–79.
35 C. Wingren, C. Steinhauer, J. Ingvarsson, E. Persson, K. Larsson and C. A. K. Borrebaeck, *Proteomics*, 2005, **5**, 1281–91.
36 S. H. Shi, T. Yamashita and C. P. Wong, *Proceedings - International Symposium on Advanced Packaging Materials: Processes, Properties and Interfaces, Braselton, Ga., Mar.14–17, 1999*, 1999, pp. 317–24.
37 M. Mandelkern, J. G. Elias, D. Eden and D. M. Crothers, *J. Mol. Biol.*, 1981, **152**, 153–61.
38 S. W. J. Ember, H. Schulze, A. J. Ross, J. Luby, M. Khondoker, G. Giraud, J. G. Terry, I. Ciani, C. Tlili, J. Crain, A. J. Walton, A. R. Mount, P. Ghazal, T. T. Bachmann and C. J. Campbell, Development of fast DNA and protein microarray tests for the diagnosis of hepatitis C virus infection, submitted.

Multimodal superparamagnetic nanoplatform for clinical applications: immunoassays, imaging & therapy†

Laurence Motte,*[a] Farah Benyettou,[a] Caroline de Beaucorps,[ab] Marc Lecouvey,[a] Irena Milesovic[a] and Yoann Lalatonne[ac]

Received 5th May 2010, Accepted 21st July 2010
DOI: 10.1039/c005286h

A new type of multifunctional magnetic nano-platform for bioassays, diagnosis and therapy applications was designed using superparamagnetic γFe_2O_3 nanoparticles conjugated to hydroxylmethylene bisphosphonate (HMBPs) molecules. The nanoparticle has two essential roles: to act as a probe owing to its specific magnetic properties and to carry on its surface antitumoral molecules (therapeutic applications) or precursor groups for the covalent coupling of biological recognition molecules, such as antibodies, nucleic acids (*in vitro* & *in vivo* diagnosis).

1. Introduction

Nanotechnology-based devices and nanomaterials have been demonstrated to be greatly useful for improving the efficiency and efficacy of bioassays and medical diagnosis. Among different types of nanomaterials, magnetic iron oxide nanoparticles have set the most successful example for medical applications of inorganic nanoparticles. Because of their ultrafine size, biocompatibility, and superparamagnetic properties, iron oxide nanoparticles are already approved for various applications such as cellular labeling or cell separation[1,2] and enhanced resolution magnetic resonance imaging[3,4] or undergoing preclinical and clinical evaluation for applications such as drug delivery,[5] tissue repair,[6] cell and tissue targeting,[7] and transfection.[8]

This paper deals with new results concerning the elaboration of multifunctional magnetic nanoplatform that can be used for bioassays as well as for diagnosis and therapy applications. The nanoplatform is iron oxide nanoparticles coated with hydroxylmethylene bisphosphonate (HMBP) molecules: γFe_2O_3@HMBP.

For the last 30 years, there has also been growing interest in the use of superparamagnetic particles for application in quantitative and highly-sensitive biosensors. The use of them as labels of the biological recognition events and their detection by means of some magnetic methods is one way to realize quantitative high-sensitive multianalyte immunoassays.[9–11] Detection of magnetically labeled biomolecules is a promising method for fast and inexpensive medical diagnosis. However, detection of small numbers of nanometre-scale magnetic particles by magnetoresistive devices is extremely challenging due to the intrinsic noise and design limitations of giant magnetoresistance (GMR) and Hall-type of biosensors.[12–16] So far, scientific efforts have been focused on the magnetic detection itself and have totally left aside the

[a]*Laboratoire CSPBAT, FRE 3043 CNRS Université Paris 13, 93017 Bobigny Cedex, France*
[b]*Magnisense Compagny, 75008 Paris, France*
[c]*Service de Biophysique et Médecine Nucléaire, Hôpital Avicenne, 93009 Bobigny Cedex, France*

† Electronic supplementary information (ESI) available: Fig. S1 and S2. See DOI: 10.1039/c005286h

optimization of intrinsic material magnetic features. To date biological assays are currently performed using commercially micrometre beads. Nanobeads are polymer beads with an even dispersion of superparamagnetic iron oxide nanoparticles. The beads are subsequently coated with a chemical or biological species that selectively binds to the target analyte, these types of particles have been used primarily to separate and concentrate analytes for off-line detection.

On the other hand, superparamagnetic iron oxide nanoparticles have already demonstrated their utility as an important tool for enhancing magnetic resonance contrast allowing monitoring of not only anatomical changes, but physiological and molecular changes as well. A great amount of research has been focused on their use for the delivery of anticancer drugs and noninvasive tumor imaging.[17–19] In this context, two main strategies are explored: either nanocrystal surface functionalization with biomolecules of interest or covalent coupling of the molecules on nanoplatform. In the latter case, it is necessary to introduce surface reactive moieties on nanoparticle surfaces such as carboxylic and amine groups. One of the most widely used conjugation reactions[20] is l-ethyl-3-(3-dimethylaminopropyl)carbodiimide (EDC)- or EDC/*N*-hydroxy-succinimide (EDC-NHS)-mediated amidation reaction for chemically conjugating carboxylic group- or amine group-bearing particles to proteins, peptides, *etc.*

Bisphosphonates (BPs) are currently the major class of drugs for the treatment of osteoporosis and other disease characterized by increasing bone resorption. The

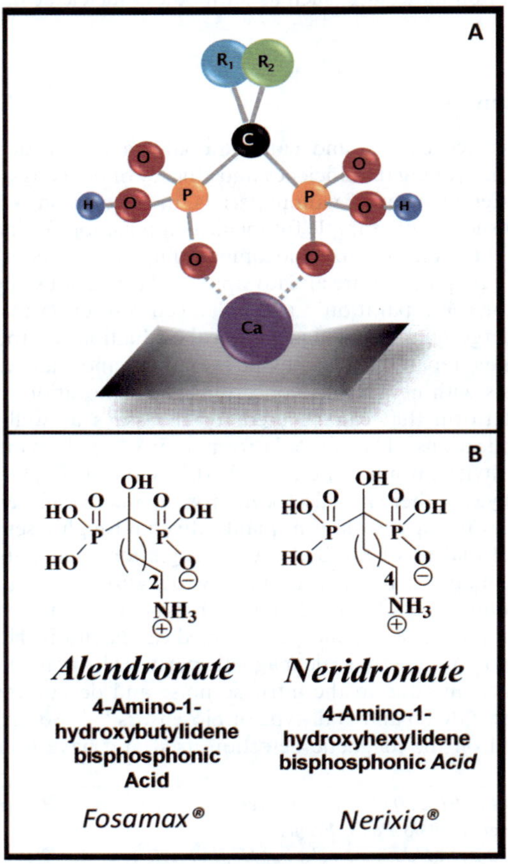

Fig. 1 (A) Schematic representation of HMBP molecule chelating a calcium ion; (B) HMBP molecules used in this work to coated γ-Fe_2O_3 nanoparticles.

classical pharmacological effects of bisphosphonates appear to result from two key properties: their affinity for bone mineral as a consequence of their capacity to chelate calcium ions and their inhibitory effects on osteoclasts.[21] Most BPs contain an hydroxyl group on the P–C–P structure (HMBP hydroxyl methylene bisphosphonate) that confers high affinity binding to calcium phosphate (hydroxyapatite ($Ca_{10}(PO_4)_6(OH)_2$) and is the basis for the bone targeting properties of bisphosphonates[22,23] (Fig. 1.A). The HMBPs described in Fig. 1.B have been investigated in humans with respect to their effects on bone. After the promise shown in the early clinical use of etidronate ($R_1 = OH$, $R_2 = CH_3$) and clodronate ($R_1 = R_2 = Cl$), second generation of bisphosphonates were synthesized containing a primary amino group in an alkyl chain (alendronate, neridronate). This increased the antiresorptive potency by up to one hundred times. Later modifications of the R_2 side chain to produce compounds containing nitrogen heterocycles, such as risedronate and zoledronate, further increased potency. They are up to 100 times more potent than alendronate in some experimental systems. In addition, HMBPs also exhibit direct and indirect antitumour effect against a broad variety of tumour cell lines, such as melanoma, mesothelioma, prostate, breast, lung and myeloma cancer cells *in vitro*.[24] The direct growth inhibitory effects have been attributed to cell cycle arrest and/or induction of apoptosis. It has been shown that nitrogen-containing bisphosphonates such as alendronate, risedronate and zoledronate (Fig. 1.B), act by inhibiting the mevalonate pathway.[25,26] Numerous studies have also described the ability of HMBPs to inhibit tumour cell adhesion and invasion but the mechanism is still unclear.[27] However, the major drawback of these drugs resides in their poor bioavailability, only 7% of the administrated dose reaches the target. This is mainly due to their rapid elimination in the blood circulation due to their large negative charge and their preferential accumulation in bone. To improve the bioavailability of these antitumoral molecules, one strategy is to use the affinity of HMBP functions for cations such as iron ions. In previous work,[28] γFe_2O_3 nanocrystals were surface functionalized with 5-hydroxy-5,5-bis(phosphono)pentanoic acid[29] (called HMBP-COOH in the text, Fig. 1B). We demonstrated the potentiality of this hybrid nanomaterial with no cytotoxicity as MRI contrast agents. In another study, the nanoparticle was surface passivated with 1-hydroxy-2-(imidazol-1-yl)ethylidene-1,1-bisphosphonic acid or zoledronate (Fig. 1B). The biological *in vitro* tests performed on breast cancer cell line showed that the hybrid nanomaterial present antiproliferative activity.[30]

In this study, five HMBP molecules are used to coat the nanoparticles (Fig. 1). First, we demonstrated the used of the as-synthesized nanoplatforms for mono and multiparametric immunoassays. We show that due to their small size, magnetic nanoplatforms bearing carboxylic end functional groups (γFe_2O_3@HMBP-COOH) offer larger surface area-to volume ratios and COOH functionalities than currently used nanobeads, resulting in good biolabelling and fast reaction kinetics for magnetic immunoassays. In the second part, we designed a nitrogen-containing bisphosphonate nanoplatform as nanotool for cancer therapy and diagnostic. The influence of the coating on the *in vitro* cancer theranostic potentialities is evaluated.

2. Immunoassays

The development of reliable monitoring methods is essential for diagnosis of various diseases and accurate evaluation of environmental problems. Immunoassay is one of the most important methods in the field of clinical and environmental analysis. Immunoassay is based on the exceptional specificity that an antibody has for its own target antigen. The binding constants for binding of antibody and antigen can reach up to 10^{11} M and result in an extremely selective and sensitive detection. The highly specific affinity coupled with a very low limit of detection has made immunoassay a widely accepted analytical method in various fields. Immunoassays using a radioisotopic technique have been heavily utilized since the 1960s[31] because

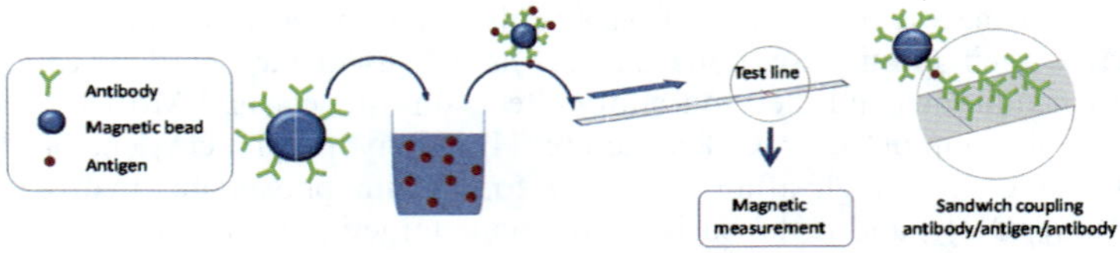

Fig. 2 Schematic principle for a sandwich immunoassay based on magnetic detection.

radioisotopic labels have shown superior performance in both high selectivity and low detection limits. The use of radioisotopes, however, has many problems related to handling and disposal; therefore, immunoassays using a nonisotopic technique have increased dramatically and have gained a certain status as a common analytical technique in recent years. The use of magnetic particles in immunological assays has grown considerably, as the particles' magnetic properties permit their easy separation and/or concentration in large volumes, allowing for faster assays and in some cases improved sensitivity over currently available commercial methods. A typical microtiter plate sandwich immunoassay is schematically presented Fig. 2. One antibody is attached to a superparamagnetic surface (usually microsphere), and the other antibody is adsorbed onto a strip (test line). The antibody-modified magnetic nanobeads are incubated with the sample. The antibody on the beads will bind the target antigen from the sample. After magnetic separation, the antigen-antibody bead is deposited on the strip and migrates by capillarity. The antigen-antibody beads will bind the antigen on the test line, and retain it in the plate, thereby forming an antigen 'sandwich' between the two different antibodies. This binding reaction is measured by detection of magnetic signal on test line.

To detect this magnetic signal, an enormous effort has been spent to develop new detection devices such as giant magneto-resistant (GMR),[32] spinvalve,[33] magnetoelastic sensors[34] or various types of magnetic sensors.[12,35–38] Recently, a new detection and characterization method was developed to take advantage of the non-linearities of superparamagnetic or ferromagnetic materials for Magnetic ImunoAssays (MIA). The Magnisense company developed two different room temperature sensors, called MIAtek® and MIAplex® (respectively Magnetic ImunoAssays technology and multiplex). The MIAtek® sensor measures a signal proportional to the third derivative of the magnetization at zero magnetic field.[39] Hence, the MIAtek® signal is very sensitive and is proportional to the amount of magnetic particles. This allows detecting nanograms of superparamagnetic materials. This technology is used for very sensitive monoparametric immunoassays. The MIAplex® sensor measures a signal which is proportional to the second derivative of a magnetization curve.[12] Fig. 3 B and D show the superposition of MIAplex® signature of two kinds of superparamagnetic nanoparticles with the conventional magnetization curve measured at room temperature with SQUID device. It can be observed that the magnetic detection with MIAplex® is measured at the origin of the magnetization where the magnetization slope is the most important. Hence, comparing to the usual $B(H)$ signature, the measurement of $d^2B(H)/dH^2$ is expected to be more sensitive to slightly changes of the magnetic behaviour. Since the latter usually arises from a change in the nanoparticle size, shape and/or composition, one can imagine detecting simultaneously different types of biological entities by attaching them to nanoparticles.

So far, scientific efforts have been focused on the magnetic detection itself and have totally left aside the optimization of intrinsic material magnetic features. Hence, magnetic immunoassays are currently performed using commercially micrometre beads.

The aim of this study is to emphasise the efficiency of individual nanometre superparamagnetic nanoparticle for magnetic immunoassays. Such applications of these

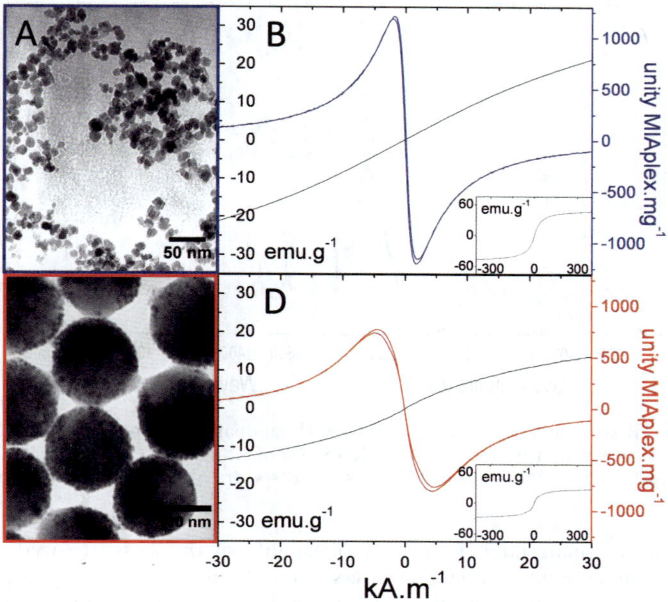

Fig. 3 TEM pictures of γFe_2O_3@HMBP-COOH nanoparticles (A) and Estapor@COOH beads(C) and enlarged view of room temperature SQUID magnetization (black curve) and MI-Aplex® signature of γFe_2O_3@HMBP-COOH nanoparticles (blue curve) (B) and Estapor@-COOH beads (red curve) (D). Inserts: SQUID magnetization.

single nanoparticles, is unexplored and is preliminary demonstrated through the MIAtek® and Miaplex® measurements for monoparametric and multiparamagnetic immunoassays. Two kinds of particles are used: maghemite γ-Fe_2O_3 nanoparticles synthesized in the CSPBAT laboratory (γ-Fe_2O_3@HMBP-COOH) and commercial beads Estapor@COOH (Merck M1-020/50). The synthesis and characterization of maghemite γ-Fe_2O_3@HMBP-COOH nanoparticles were described in previous work.[28] Briefly, the average size of the nanoparticle is 10 nm in diameter, with a size polydispersity of 20%, Fig. 3A. The nanocrystal surface is functionalized with 5-hydroxy-5,5-bis(phosphono)pentanoic acid (call HMBP-COOH in the text, Fig. 1.B). The HMBP function complexes the nanocrystal surface and carboxylate groups at the outer surface and induce electrostatic repulsion between nanocrystals. A stable ferrofluid is obtained on large concentration and pH range. The commercial beads Estapor@COOH are magnetite Fe_3O_4 nanoparticles of 10 nm (polydispersity of 40%) embedded in polystyrene (50% ferrite). The average size of the beads is 165 nm with a size polydispersity of 30%, Fig. 3C. These nanobeads are coated with carboxylate groups. Independently of the manufacturing, squid hysteresis curves are characteristic of superparamagnetic behaviour at room temperature with no remanence and coercivity. The magnetization saturation corresponds to a value of 55 emu g^{-1} and 27 emu g^{-1} (iron oxide weight) for γ-Fe_2O_3@HMBP-COOH nanoparticles (insert Fig. 3B) and Estapor@COOH beads (insert Fig. 3D) respectively.

For magnetic immunoassay, the superparamagnetic nanoparticles have to be coated with recognition biomolecules, such as antibodies. This could be performed taking advantage of carboxylate functions at the outer particle surface either for γ-Fe_2O_3@HMBP-COOH nanoparticles and Estapor@COOH beads. In previous work, the feasibility of such process was demonstrated for γ-Fe_2O_3@HMBP-COOH nanoparticles by the coupling of amino fluorescein to

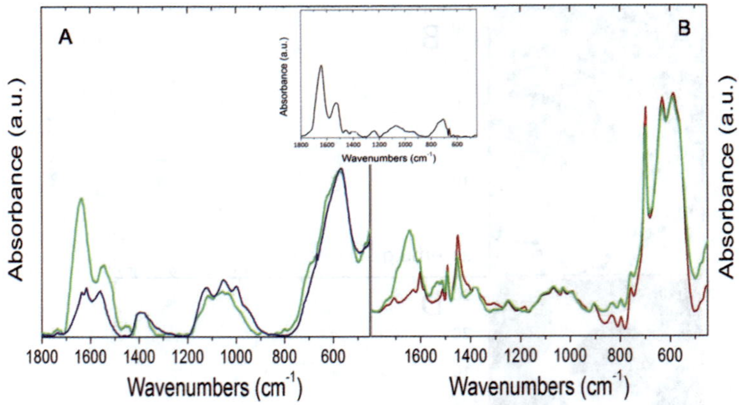

Fig. 4 (A) IR spectra of γFe_2O_3@HMBP-COOH nanoparticles unmodified (blue curve) and coupled to streptavidin (green curve). (B) IR spectra of Estapor@COOH particles (red curve) and coupled to streptavidin (green curve). Insert: IR spectra of free streptavidin (black curve).

the hybrid nanomaterial.[28] To test the feasibility of the as-synthesized nanoplatforms for magnetic immunoassays, a common streptavidin–biotin binding procedure has been chosen. Indeed, a similar carbodiimide coupling procedure was performed using streptavidin as biomolecule on the two kinds of particles and the ability to target biotin is evaluated after migration on the strip. After chemical coupling, the solution is dialysed in order to remove all chemical moieties in solution. The streptavidin coupled on the surface of the nanoparticles is characterized using infra-red spectroscopy (Fig. 4).

Fig. 4 A and B show the IR spectra of free streptavidin (insert, black curve), unmodified particles as well as streptavidin coupled particles (A and B, green curve) using γFe_2O_3@HMBP-COOH nanoparticles (A, blue curve) or Estapor@COOH (B, red curve) respectively. The IR spectra in Fig. 4 show that the characteristic peaks of the functionalized nanoparticles, as well as streptavidin, appeared on the spectrum of the two sample conjugates, indicating successful tailoring of the nanoparticle surface with streptavidin. The presence of streptavidin after dialysis suggests a chemical bonding between the carboxylic group and the terminal amino groups of the streptavidin.

The ability of the two different streptavidin hybrid nanoparticles to target biotin was tested by deposition and migration on a conventional strip. Biotin is sprayed onto the strip (test line). The streptavidin nanoparticles are deposited on the strip at the point 0 mm (Fig. 5). Then the particles migrate by capillarity. The streptavidin particles will bind the biotin on the test line, and retain it in the plate. The detection of magnetic particles is performed over the entire strip measuring the MIAtek® signal (Fig. 5). Such a measurement is directly proportional to the magnetic material quantity.[39] The lab-synthesized nanoparticles (blue curve) present a much better targeting on the biotin test line than the commercial nanoparticles (Fig. 5). The specific targeting of the bisphosphonate coated nanoparticles indicates that the streptavidin is well coupled to the nanoparticles. This better targeting of the as-synthesized nanoparticles compared to commercial beads is attributed to the largest amount of carboxylic functions at the outside surface. Such large number of functionalities (around 1800 molecules per particle) allows both a better efficiency for coupling of streptavidin and a better stability at physiological pH. The particles' stability is confirmed by the lower hydrodynamic diameter for the as-synthesized nanoparticles compared to commercial beads (20 nm compared to 185 nm). This lowest hydrodynamic diameter may also induce a better migration by capillarity. Hence such a magnetic nanoplatform may act as a biolabel for immunoassays and in first

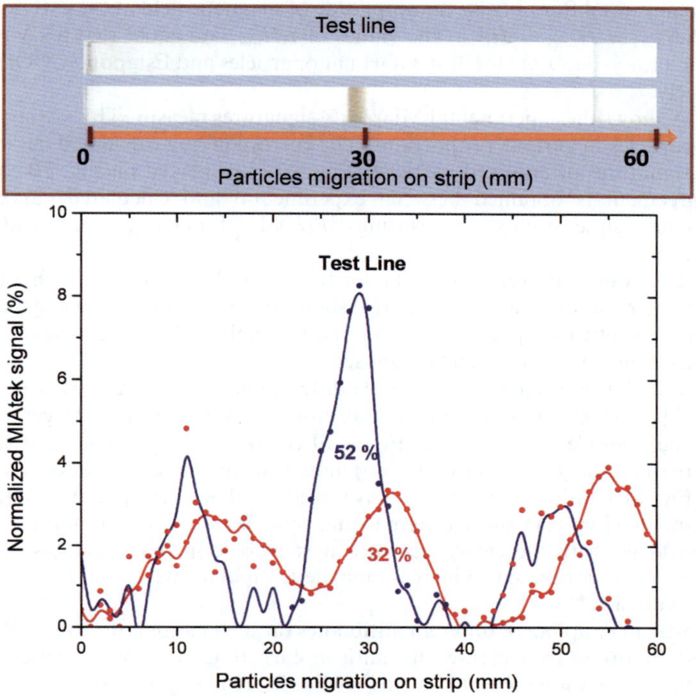

Fig. 5 MIAtek® signal measured over the entire strip after migration of 10nm γFe_2O_3@HMBP-COOH nanoparticles (blue curve) and Estapor@COOH beads (red curve) and optical images of strips before (up) and after (below) migrations of streptavidin coupled to γFe_2O_3@HMBP-COOH nanoparticles.

assumption appears more efficient for antigen targeting than the micrometre-scale superparamagnetic beads conventionally used.

In an attempt for multiparametric bioassays, the two biolabel nano (micro) platforms, γ-Fe_2O_3@HMBP-COOH and Estapor@COOH beads, are mixed at different ratios. The iron concentration of the two ferrofluids was adjusted in order to obtain Miaplex® signature at isointensity. Fig. 6 shows the Miaplex® signatures of the two kinds of particles.

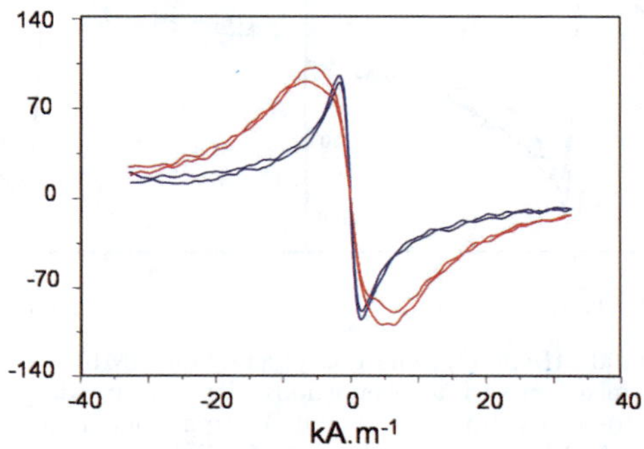

Fig. 6 MIAplex® signature of γFe_2O_3@HMBP-COOH particles (blue curve) and Estapor@-COOH beads (red curve), adjusted at isointensity.

The signatures differ only by the amplitude of magnetic field at the extrema of the signal: $\Delta H_{max} = H_{max} - H_{min}$. The values of ΔH_{max} are equal to 3.6 kA m^{-1} and 8.9 kA m^{-1} for γ-Fe_2O_3@HMBP-COOH nanoparticles and Estapor@COOH beads, respectively.

Fig. 7 A shows the experimental Miaplex® signatures measured by mixing the two kind of particles in various proportions. The corresponding calculated curves using the two signatures of unmixed particles (Fig. 6) are displayed on Fig. 7B. A rather good agreement is obtained between experimental and calculated curves. This confirms the high sensitivity of measuring $d^2B(H)/dH^2$ for characterization and separation purpose.

These first results are very promising for the use of the MIAplex® technology for multiparametric immunoassays. This also shows the opportunity of using individual superparamagnetic nanoparticles instead of an assembly of nanoparticles embedded in a polymer matrix as a magnetic nanosensor.

Today, work is in progress to elaborate different nanoplatforms@HMBP-COOH differing by their size, shape and/or composition and thus presenting larger non-linearities. One example is presented in Fig. 8. Slight changes in the chemical synthesis procedure (increasing iron concentration) induce the formation of nanocrystals with 20 nm (Fig. 8B) and 40 nm (Fig. 8C) average size. The corresponding MIAplex® measurements (Fig. 8A) show different magnetic signature depending on the size of the particles, with the emergence of a hysteresis loop that increases with size. Such a change is correlated to increase magnetic dipolar interaction between nanoparticles with size.[40]

The covalent coupling of different antibodies on such nanoplatform has also to be improved in order to increase the antigen targeting. The optimization of the coupling protocol on nanoplatform@HMBP-COOH is in progress.

We can also modify the terminal end group on the nanoplatform@HMBR-R. This opens the opportunity for covalent coupling of a larger panel of biomolecules. Furthermore, as shown in the second part of this manuscript, this new nanoplatform is an innovative multimodal system for cancer theranostic.

3. Imaging and therapy

Superparamagnetic iron oxide nanoparticles with tailored surface chemistry have emerged as one of the primary nanomaterials for biomedical applications such as

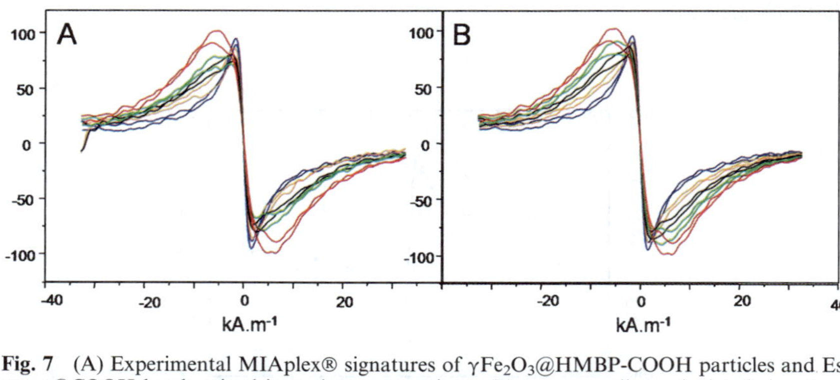

Fig. 7 (A) Experimental MIAplex® signatures of γFe_2O_3@HMBP-COOH particles and Estapor@COOH beads mixed in various proportions; (B) corresponding calculated signatures from unmixed magnetosensors. For both A and B graphs: blue and red curves correspond to 100% of γFe_2O_3@HMBP-COOH and Estapor@COOH, respectively. The proportions of γFe_2O_3@HMBP-COOH/Estapor@COOH are: 75%–25% orange curve, 50%–50% black curve, 25%–75% green curve.

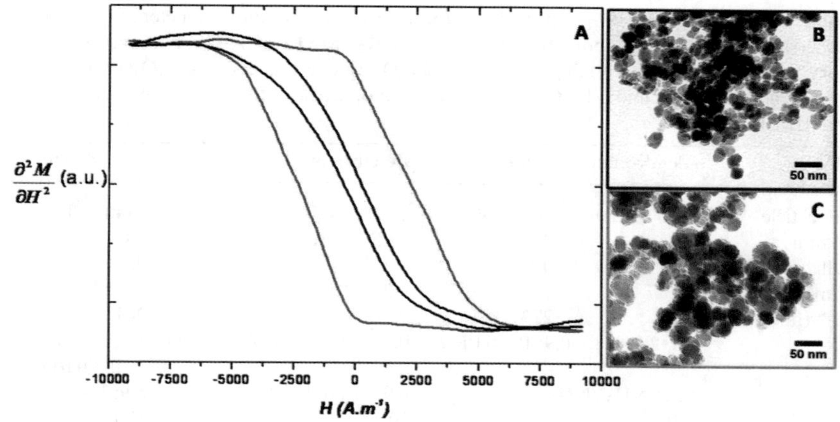

Fig. 8 TEM images (B, C) and MIAplex® signatures (A) of 20 nm (black curve) and 40 nm (green curve) nanocrystals.

detection and cancer therapy.[41–44] The small size of the particles (<50 nm) gives rise to single domain particles with superparamagnetic behaviour. This property is used for magnetic resonance imaging contrast enhancement especially with T_2 relaxation time alteration.[45] On the other hand, drug carriers with superparamagnetic nanoparticles respond to an external static magnetic field and can be used as a magnetic targeting agent for drug delivery.[46,47] In previous work, we demonstrated the potentiality of γFe_2O_3@HMBP–COOH as MRI contrast agent.[28] In another study, the nanoparticle was surface passivated with zoledronate, an anticancer interest molecule of the HMBPs family, Fig. 1. The biological *in vitro* tests performed on breast cancer cell line showed that this hybrid nanomaterial presents antiproliferative activity.[30]

In the present work, γFe_2O_3 nanocrystals are surface passivated with four nitrogen containing bisphosphosphonate drugs (γFe_2O_3@HMBP-$(CH_2)_n$-R) differing either by the terminal end group (R): imidazol (zoledronate),[30] pyridyl (risedronate), amine (alendronate and neridronate) or by the length (n) of the intermediate aliphatic group between HMBP and amine NH_3^+ terminal end group (R): alendronate and neridronate. The chemical formula of the various bisphophonate drugs are displayed in Fig. 1. In this study, the influence of the coating on the *in vitro* cancer theranostic potentialities of these hybrid nanomaterials is evaluated.

The maghemite γFe_2O_3 nanocrystals were synthesized and surface functionalized as described previously.[28,30,48] For all the γFe_2O_3@HMBP-$(CH_2)_n$-R nanocrystals, large changes in IR spectra are observed within the P–O stretching region (1200–900 cm^{-1}), whereas the amine bending vibration band (1600 cm^{-1}) for alendronate and neridronate coated nanocrystals or the imine (1650 cm^{-1}) and pyridyl (1630 cm^{-1}) vibration bands for zoledronate and risedronate, respectively, are unchanged (ESI†). These results indicate that HMBP molecules are grafted onto the nanocrystal surface through the phosphonate groups. ^{31}P NMR titration was used in order to quantify the average number of molecules per nanocrystal. After chemical decomposition of the magnetic γFe_2O_3 core and precipitation of ferrous ions, the supernatant containing free HMBPs is analyzed with ^{31}P NMR. All nanoparticles are stable at physiological pH. The size and the zeta potential of the nanocomplex was determined by dynamic laser light scattering (DLS). This measurement is an indication of surface charge on a particulate species, which plays an important role in determining solution stability, susceptibility to aggregation and precipitation problems, as well as protein and cellular surface binding *in vivo*. Table 1 summarizes the hybrid nanoparticles characterization.

Table 1 Amount of HMBP per particles (N/particle), hydrodynamic diameter (D_h) and polydispersity index (PdI), zeta potential (Z), transverse R_2* MRI relaxivities, half maximal inhibitory concentration (IC_{50}) for γFe_2O_3@alendronate, γFe_2O_3@neridronate, γFe_2O_3@zoledronate and γFe_2O_3@risedronate nanoparticles and pK_a values and IC_{50} for free HMBPs

HMBP	Alendronate	Neridronate	Zoledronate	Risedronate
N/particle	1200 ± 150	600 ± 90	1320 ± 120	1440 ± 120
D_h/nm	32.5	38.1	37.5	28.8
Pdl	0.11	0.17	0.58	0.18
Z/mV	−37.7	−14.2	−21.6	−42.6
R_2* ($mM^{-1}s^{-1}$)	287	277	254	267
pK_a	0.8, 2.2, 6.3, 10.9: P–OH/PO$^-$		0.8, 2.9, 6.6, 11: P–OH/PO$^-$	1.6, 2.2, 6.3, 11: P–OH/PO$^-$
	12.2: NH_3^+/NH_2		6.9: H-imidazole	5.9: H-pyridyl
IC_{50} Free HMBP (μM)	120	—	10	120
IC_{50} γFe_2O_3@ HMBP (μM)	35	19	100	60

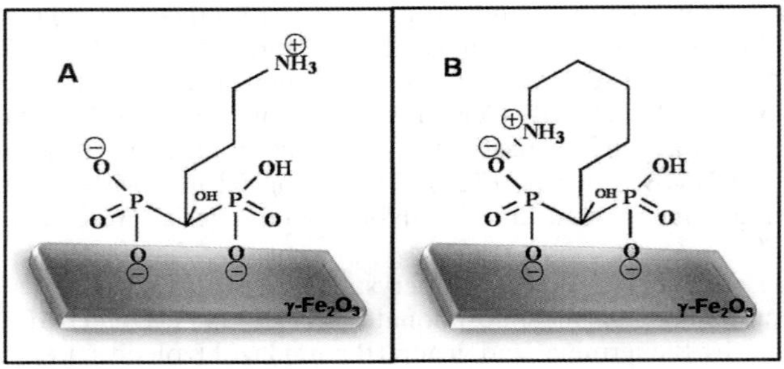

Fig. 9 Proposed structure of alendronate (A) and neridronate (B) interacting on γFe_2O_3 nanocrystals surface at physiological pH.

The HMBP coverage density is very similar, independent of the molecular structure of HMBP, except for neridronate molecules as coating agents. The decrease in coverage density with neridronate is attributed to steric hindrance and relatively disordered structure at the nanocrystal surface when the alkyl chain length between HMBP and amine groups increase (comparing alendronate to neridronate). The long amino side chain of neridronate may fold back on itself with the formation of hydrogen bond between the amino group and one phosphonate group (Fig. 9).[49]

This behaviour is also supported taking into account of the higher hydrodynamic diameter and low electrical surface charge for γFe_2O_3@neridronate compared to γFe_2O_3@alendronate inducing a lower colloidal stability.[50] The value of this charge seems to be correlated with the number of molecules grafted and so to the coverage density. Besides comparing alendronate to neridronate, the neridronate surface charge is twice less important and twice less covered. Similarly, the hydrodynamic diameter increases and the surface charge decreases for γFe_2O_3@zoledronate compared to γFe_2O_3@risedronate nanoparticles. This indicates higher aggregation when the nanoparticles are surface passivated with zoledronate. These differences are attributed to the π–π stacking between imidazol groups less effective than between pyridyl groups, inducing lower stability and more aggregated state. At physiological pH, all γFe_2O_3@HMBPs nanoparticles exhibit a negative zeta potential whereas uncoated nanocrystals has a point of zero charge at pH 7.[51] This negative charge surface is correlated to the pK_a values of HMBPs (Table 1). Considering, that HMBP coordinates iron surface *via* two Fe–O–P bonds (Fig. 1A), at pH 7.4 the negative charge is induced by the third deprotonated hydroxyl groups. Hence, HMBPs functionalized nanoparticles are resulting to stable anionic superparamagnetic particles at physiological pH.

The superparamagnetic behaviour of 10 nm γFe_2O_3 allows use of these particles as contrast agents for magnetic resonance imaging (MRI). This property is currently used for MRI contrast enhancement especially with T_2 relaxation time alteration.[45] To investigate the MR signal enhancement effects, the aqueous as-prepared nanoparticles were measured on a 4.7 T MRI scanner. The measurements were performed at room temperature for various iron concentrations between 0.031 to 1 mM. For T_2* relaxation times, axial T_2*-weighted SE images were obtained with a TR of 800 ms and TE of 8.6 ms at 4.7 T. As shown in Fig. 10A with γFe_2O_3@zoledronate nanoparticles, T_2* weighted images change drastically in signal intensity with an increasing amount of nanoparticles, indicating that the as synthesized nanoparticles generated MR contrast on transverse (T_2*) proton relaxation times weighted sequences.

Fig. 10B shows the relaxation rates $1/T_2$* as a function of the iron concentration. The relaxation rate varied linearly with the iron concentration, as expected. The transverse R_2* relaxivities (corresponding to the slopes of the lines) for the various γFe_2O_3@HMBP nanoplatforms are reported in Table 1. The R_2* values are very similar for all the nanoplatforms. Hence, these hybrid nanoparticles are potential candidates for MRI contrast agents.

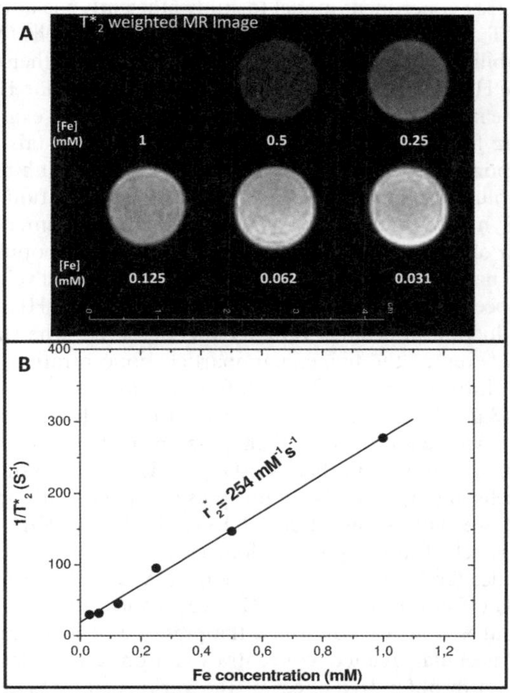

Fig. 10 (A) T_2* weight MR images of aqueous solutions of as-synthesized of γFe_2O_3@zoledronate nanoparticles at different Fe concentrations. (B) T_2* relaxation rates ($1/T_2$*) plotted against the Fe concentration for the various aqueous solutions.

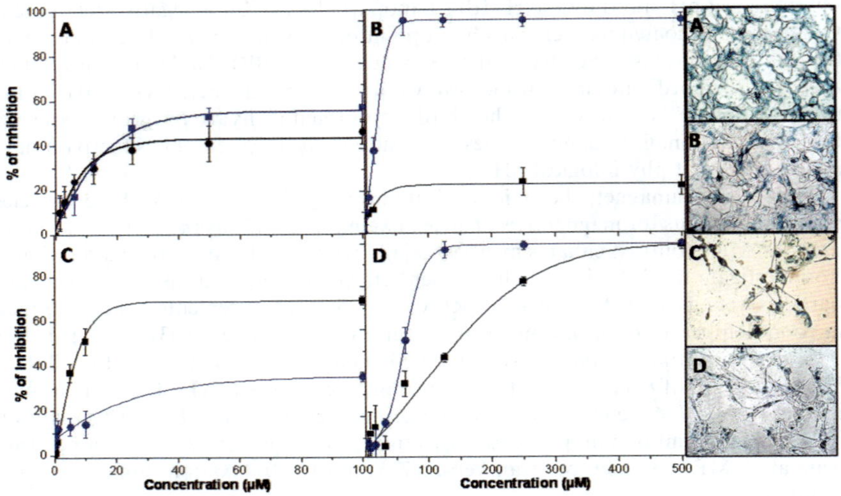

Fig. 11 Comparative effects on MDA-MB-231 cell proliferation after incubation during 48 h of free HMBP (black curve) and γFe_2O_3@HMBP nanocrystals (blue curve). Optical images of mammal cell (MDA-MB-231) stained by Prussian blue and incubated with γFe_2O_3@HMBP for 48 h and corresponding to 0.1 mM in HMBP, showing nanocrystals incorporation into cells. (A) Alendronate, (B) neridronate, (C) zoledronate and (D) risedronate.

In order to assess the as-synthesized nanoparticles as a biocompatible material suitable for antitumoral applications, viability tests were performed on breast cancer cells (MDA-MB231), a pertinent model to study efficiently the antitumoral effect of HMBPs.[52,53] MTT assays were performed on breast cancer cells (MDA-MB231) to evaluate cells viability. These adherent cell lines was treated either with free HMBPs or with γFe_2O_3@HMBP nanocrystals, for various extracellular HMBP concentrations for 48 h. The intracellular uptake of the nanocrystals was examined using Prussian blue staining for the various γFe_2O_3@HMBP nanocrystals. Fig. 11 presents typical dose–response curves of MDA-MB231 cell growth inhibition and typical cells stained by blue Prussian blue images after 48 h incubation with the various γFe_2O_3@HMBP nanoparticles. Prussian blue staining is intended to result in a blue coloration of any iron-containing compounds in cell populations, including γFe_2O_3@HMBP nanocrystals and endogenous iron. Umbilical vein endothelial cells (HUVEC) have been chosen as normal cell line model. With HUVEC cells (ESI†), no significant reduction in cellular viability was observed. The survival rate being higher than 90%, even at the higher nanoparticle concentrations and in presence of magnetic field. Little or practically no difference was shown between the cytotoxicities of γFe_2O_3@alendronate and γFe_2O_3@neridronate. In the bloodstream, nanoparticles interact with endothelium, then pass through the vessel wall to reach tissues. Therefore, it is important that γFe_2O_3@HMBP nanocrystals do not damage the endothelial cells and apparently do not present cytotoxicity on normal cells. In previous work,[28] we have shown that γFe_2O_3@HMBP-COOH nanocrystals are internalized inside cells following the endocytosis pathway and concentrate within intracellular vesicles (endosomes). The γFe_2O_3@HMBP-COOH nanocrystals presents no inhibition effect on MDA-MB 231 breast cancer lines.

The optical images, Fig. 11, confirmed the presence of nanoparticles inside cells and showed intracellular related concentration iron accumulation in the cells. Considering the antiproliferative effect of the various γFe_2O_3@HMBP nanocrystals, Fig. 11 shows that each hybrid nanomaterial presents an antitumoral activity. A significant enhancement of HMBP cytotoxicity results from the anchoring to the nanoparticle surface of a pro-drug to cancer cells, except when zoledronate

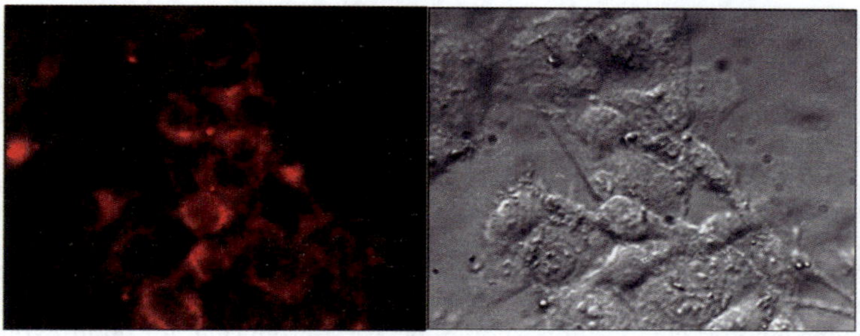

Fig. 12 Magnetic fluorescent nanoparticle cell labelling by microscopy of MDA-MB231 human cancer cells: (A) fluorescence and (B) transmission images.

molecules are used as coating agent. The lower activity observed for γFe_2O_3@zoledronate nanoparticles is correlated to the low electrical surface charge of nanocrystals as well as the aggregated state inducing less cell internalization compared to the other nanoplatforms, (Table 1). This behaviour is clearly shown in Table 1 comparing HMBP and γFe_2O_3@HMBP IC_{50}, half maximal inhibitory concentration. This value is a good indicator of the effectiveness of a compound in inhibiting biological or biochemical function. The antitumoral activity of alendronate (Fig. 11 A) and risedronate (Fig. 11 D) are greatly improved by the delivery system. More spectacular is the neridronate cell penetration induced by nanoparticle surface functionalization. Indeed, free neridronate exhibits no inhibitory effect within the concentration range for 48 h treatment (Fig. 11 B). However, in the presence of γFe_2O_3@neridronate, cell viability is considerably reduced above 19 μM. This clearly indicates that neridronate cell penetration is induced by nanocrystal vectorization through γFe_2O_3 surface functionalization. The mechanism of drug action or drug release is not demonstrated here. However it can be expected that the anchoring of HMBPs to the nanoparticle surface allowed increasing their hydrophobicity and improving their accumulation in cells, especially for γFe_2O_3@neridronate. γFe_2O_3@HMBP-COOH nanocrystals presents no inhibition effect on MDA-MB 231 breast cancer lines.[28] Hence we can attribute this antitumoral effect to the HMBP functionalization with specific activity on cancer cells.[52,53]

Moreover as discussed in the first part, the terminal NH_3^+ surface pendant groups can be used as precursors for the covalent coupling of biomolecules. The feasibility of such process is demonstrated by the coupling of rhodamine B to γFe_2O_3@alendronate nanocrystals. Briefly, the dye is activated and then directly grafting on the nanocrystals. The remaining non-grafted dyes in solution are isolated with dialysis. The fluorescent pattern shown in Fig. 12 demonstrated the capacity of combined fluorescent and magnetic labelling of living cells of the rhodamine coupled nanocrystals.

Hence, we designed new antitumour magnetic nanoparticles by coating iron oxide nanoparticles with clinical antitumour molecules. We showed that these new nanoparticles both have MRI contrast agent and anticancer properties. Besides, the nanoparticle surface functionalization spectacularly improves the neridronate cell penetration and their antitumour effect. These results allow us to consider these nanohybrid particles as a drug delivery system. Moreover, the terminal amine functionalities of alendronate molecules can be used as precursor groups for the covalent coupling of fluorescence markers. Hence, in light of these characteristics, the magneto-fluorescent nanocrystals are potential candidates for medical diagnostics, smart drug delivery system as well as for biosensing.

4. Conclusion

In conclusion, a new type of multifunctional magnetic nano-platform for bioassays, diagnosis and therapy applications was designed using superparamagnetic γFe_2O_3

nanoparticles conjugated to hydroxylmethylene bisphosphonate (HMBPs) molecules. The nanoparticle has two essential roles: to act as a probe owing to its specific magnetic properties (MRI or biosensor) and to carry on its surface antitumoral molecules (therapeutic applications) or precursor groups for the covalent coupling of biological recognition molecules, such as antibodies, nucleic acids (*in vitro* & *in vivo* diagnosis). We show that due to their small size, magnetic nanoplatform bearing carboxylic end functional groups (γFe_2O_3@HMBP–COOH) offer larger surface area-to-volume ratio and COOH functionalities than currently used nanobeads. These characteristics result in good biolabelling and fast reaction kinetics for magnetic immunoassays. New nanoplatforms@HMBP–COOH differing by their size, shape and/or composition should be very promising for multiparametric magnetic immunoassays. We also demonstrated that, depending on the HMBP molecules, these new nanoparticles have both MRI contrast agent and anticancer properties. These results allow us to consider these nanohybrid particles as a drug delivery system. Indeed, the antiproliferative effect on tumor cells is due to the specific coating of the nanoparticle surface. In addition, we have shown that the terminal amine functionalities of alendronate molecules can be used as precursor groups for the covalent coupling of fluorescence markers. Hence, in light of these characteristics, the γFe_2O_3@HMBP–$(CH_2)_3$–NH_3^+ nanoplatform will be used in future work as a biosensor.

Acknowledgements

The immunoassay part was supported by ANR Biotecs and Magnisense Company. The authors are grateful to J. M. Serfatty (Bichat Hospital) for MRI measurements, Nicole Lièvre (UP13) for MET images, Odile Sainte Catherine (UP13) for MTT assays, and Hicham Jouni (UP13) for help in nanocrystals synthesis.

References

1 O. Olsvik, T. Popovic, E. Skjerve, K. I. S. I. Cudjoe, E. I. Hornes, J. Ugelstad and M. Uhlen, *Clin. Microbiol. Rev.*, 1994, **7**, 43–54.
2 A. K.Gupta and A. S. G. Curtis, *Biomaterials*, 2004, **25**, 3029–3040.
3 E. Amstad, S. Zurcher, A. Mashaghi, J. Y. Wong, M. Textor and E. Reimhult, *Small*, 2009, **5**, 1334–1342.
4 R. Qiao, C. Yang and M. Gao, *J. Mater. Chem.*, 2009, **19**, 6274–6293.
5 S. Gai, P. Yang, C. Li, W. Wang, Y. Dai, N. Niu and J. Lin, *Adv. Funct. Mater.*, 2010, **20**, 1166–1172.
6 A. K. Gupta and M. Gupta, *Biomaterials*, 2005, **26**, 3995–4021.
7 V. Tchikov, S. Winoto-Morbach, M. Kronke, D. Kabelitz and S. Schutze, *J. Magn. Magn. Mater.*, 2001, **225**, 285–293.
8 F. Krotz, C. de Wit, H. Y. Sohn, S. Zahler, T. Gloe, U. Pohl and C. Plank, *Mol. Ther.*, 2003, **7**, 700–710.
9 M. Nichkova, et al, *Anal. Biochem.*, 2007, **369**, 34–40.
10 E. Baldrich and F. X. Muñoz, *Analyst*, 2008, **133**, 1009–1012.
11 R. L. Millen, J. Nordling, H. A. Bullen, M. D. Porter, M. Tondra and M. C. Granger, *Anal. Chem.*, 2008, **80**, 7940–7946.
12 L. Lenglet, *J. Magn. Magn. Mater.*, 2009, **321**, 1639.
13 J. C. Rifea, M. M. Millera, P. E. Sheehana, C. R. Tamanahaa, M. Tondrab and L. J. Whitmana, *Sens. Actuators, A*, 2003, **107**, 209–218.
14 M. PA Laitinen, J. Salmela, L. Gilbert, R. Kaivola, T. Tikkala, C. Oker-Blom, J. Pekola and M. Vuento, *Nanotechnol., Sci. Appl.*, 2009, **2**, 13–20.
15 X. J. A. Janssen, A. J. Schllekens, K. Van Ommering, L. J. Van Ijzendoorn and M. W. J. Prins, *Biosens. Bioelectron.*, 2009, **24**, 1937–1941.
16 Y. Kumagai, Y. Imai, M. Abe, S. Sakamoto, H. Handa and A. Sandhu, *J. Appl. Phys.*, 2008, **103**, 07A309.
17 M. Mahmoudi, A. Simchi, M. Imani and Urs O. Häfeli, *J. Phys. Chem. C*, 2009, **113**, 8124–8131.

18 A. Z. Wang, V. Bagalkot, C. C. Vasilliou, F. Gu, F. Alexis, L. Zhang, M. Shaikh, K. Yuet, M. J. Cima, R. Langer, P. W. Kantoff, N. H. Bander, S. Jon and O. C. Farokhzad, *ChemMedChem*, 2008, **3**, 1311–1315.
19 X. Wang, L. y. Yang, Z. Chen and D. M. Shin, *Ca–Cancer J. Clin.*, 2008, **58**, 97–110.
20 G. T. Hermanson, *Bioconjugate Techniques*, Elsevier, Amsterdam, 2nd edn, 2008.
21 R. G. Russell, N. B. Watts, F. H. Ebetino and M. J. Rogers, *Osteoporosis Int.*, 2008, **19**, 733–59.
22 G. H. Nancollasa, R. Tanga, R. J. Phippsb, Z. Hennemana, S. Guldea, W. Wua, A. Mangooda, R. G. G. Russellc and F. H. Ebetinob, *Bone*, 2006, **38**, 617–627.
23 R. Graham and G. Russell, *Bone*, 2007, **40**, S21–S25.
24 V. Stresing, F. Daubiné, I. Benzaid, H. Monkkonen and P. Clézardin, *Cancer Lett.*, 2007, **257**, 16–35.
25 M. A. Merell, S. Wakchoure, P. P. nlehenkari, K. W. Harris and K. S. Selander, *Eur. J. Pharmacol.*, 2007, **570**, 27–37.
26 D. Ribatti, B. Nico, D. Mangieri, N. Maruotti, V. Longo, A. Vacca and F. P. Cantatore, *Clin. Rheumatol.*, 2007, **26**, 1094–1098.
27 K. Hashimoto, K. I. Morishige, K. Sawada, M. Tahara, S. Shimizu, S. I. Ogata, M. Sakata, K. Tasaka and T. Kimura, *Biochem. Biophys. Res. Commun.*, 2007, **354**, 478–484.
28 Y. Lalatonne, C. Paris, J. M. Serfaty, P. Weinmann, M. Lecouvey and L. Motte, *Chem. Commun.*, 2008, 2553.
29 E. Guenin, E. Degache, J. Liquier and M. Lecouvey, *Eur. J. Org. Chem.*, 2004, 2983–2987.
30 F. Benyettou, Y. Lalatonne, O. Sainte-Catherine, M. Monteil and L. Motte, *Int. J. Pharm.*, 2009, **379**, 324.
31 R. S. Yalow and S. A. Berson, *J. Clin. Invest.*, 1960, **39**, 1157.
32 X. J. A. Janssen, L. J. Van Ijzendoorn and M. W. J. Prins, *Biosens. Bioelectron.*, 2008, **23**, 833.
33 D. L. Graham, H. A. Ferreira, P. P. Freitas and J. M. S. Cabral, *Biosens. Bioelectron.*, 2003, **18**, 483–488.
34 S. Huang, H. Yang, R. S. Lakahmanan, M. L. Johnson, J. Wan, I. H. Chen, H. C. Wikle, V. A. Petrenko, J. M. Barbaree and B. A. Chin, *Biosens. Bioelectron.*, 2009, **24**, 1730.
35 P. I. Nikitin, P. M. Vetoshko and T. I. Ksenevich, *J. Magn. Magn. Mater.*, 2007, **311**, 445.
36 M. M. Miller, P. E. Sheehan, R. L. Edelstein, C. R. Tamanaha, L. S. Zhong, L. Bounnak, L. J. Whitman and R. J. Cotton, *J. Magn. Magn. Mater.*, 2001, **225**, 138–144.
37 Y. R. Chelma, H. L. Grossman, Y. Poon, R. McDermott, R. Stevens, R. Alper and J. Clarke, *Proc. Natl. Acad. Sci. U. S. A.*, 2000, **97**, 14268.
38 H. A. Ferreira, D. L. Graham, P. P. Freitas and J. M. S. Cabral, *J. Appl. Phys.*, 2003, **93**, 7281.
39 L. Lenglet, P. Nikitin and C. Péquignot, IVD Technology, July/August , 2008, 43.
40 Y. Lalatonne, Farah Benyettou, Dominique Bonnin, Nicole Lièvre, Philippe Monod, Marc Lecouvey; Pierre Weimann, Laurence Motte, *J. Magn. Magn. Mater.*, 2009, **321**, 1653–1657.
41 K. K. Jain, *Technol. Cancer Reas. Treat*, 2005, **4**, 645.
42 S. Sengupta and R. Sasisekharan, *Br. J. Cancer*, 2007, **96**, 1315.
43 M. D. Wang, D. M. Shin, J. W. Simons and S. Nie, *Expert Rev. Anticancer Ther.*, 2007, **7**, 833–7.
44 H. H. P. Yiu, H. Niu, E. Biermans, G. V. Tandeloo and M. J. Rosseinsky, *Adv. Funct. Mater.*, 2010, **20**, 1599–1609.
45 R. Lawaczeck, M. Menzel and H. Pietsch, *Appl. Organomet. Chem.*, 2004, **18**, 506–513.
46 J. Zhang and R. D. K. Misra, *Acta Biomater.*, 2007, **3**, 838–850.
47 Y. Mi Kyung, Y. J. Yong, P. Jinho, P. Sangjin, W. K. Jin, J. M. Jung, K. Kyuwon and J. Sangyong, *Angew. Chem.*, 2008, **120**, 5442.
48 Y. Lalatonne, M. Monteil, H. Jouni, J. M. Serfaty, O. Sainte-Catherine, N. Lièvre, S. Kusmia, P. Weinmann, M. Lecouvey and L. Motte, *J. Osteoporosis*, 2010, 747852, DOI: 10.4061/2010/747852.
49 J. Robinson, I. Cukrowski and H. M. Marques, *J. Mol. Struct.*, 2006, **825**, 134–142.
50 K. A. Karimi, B. Denizot, F. Hindre, R. Filmon, J. M. Greneche, S. Laurent, T. J. Daou, S. Begin-Colin and J. J. Le Jeune, *J. Nanopart. Res.*, 2010, **12**, 1239.
51 S. A. Gomez-Lopera, R. C. Plaza and A. V. Delgado, *J. Colloid Interface Sci.*, 2001, **240**, 40–7.
52 R. Verdijk, H. R. Franke, F. Wolbers and I. Vermes, *Cancer Lett.*, 2007, **246**, 308–312.
53 S. S. Padalecki and T. A. Guise, *Breast Cancer Res.*, 2002, **4**, 035.

General discussion

Dr Ko Ferrigno opened the discussion of the paper by Dr Yoon:† What are the advantages of the wire guide technology over other techniques for moving droplets of similar volumes?

Mr You replied: Other methods such as magneto-fluidics require the use of magnetic particles in the solution sample. The effect of magnetic fields and the presence of magnetic particles on PCR thermocycling and subsequent identification of its products remain to be investigated and resolved. Our system does not introduce any additional components into the solution other than those required for PCR. Additionally, proper heat transfer *via* a superhydrophobic surface can be inefficient, and the maximum velocity that the droplet can travel in the oil may have limitations due to minimal surface interactions of the magnetic particles with the droplet. Our system is able to tightly hold on to the droplet, which aids in increasing the velocity of the droplet and further speeding up the heating and cooling of the droplet from forced convective heat transfer. Another popular method for moving droplets is electrowetting-on-dielectrics (EWOD). This method is comparatively more difficult to fabricate and operate, and thus expensive. Additionally, the effects of forced convective heat transfer is minimized because of the "wetting" of the droplet to the surface of the chip. If the surface is used as the heating element as opposed to the surrounding environment, then the method of heat transfer remains as conductive, and is less efficient, and thus slower, than forced convection employed in our system.

Professor Gooding asked: Is loss of material from the droplet, say into the silicone oil, a problem?

Mr You answered: We originally used three glass beakers filled with oil on top of heating elements, and programmed the syringe needle to extract the droplet, raise out of the oil bath, move to the next beaker, insert the needle, and reinsert the droplet. During extraction of the solution sample, we would also extract an additional 5 microlitres of oil to act as a vapor barrier to prevent evaporation. However, during experimental testing, we noticed that when the two oil interfaces separated during the removal of the syringe tip from the oil, the sudden separation of the oil to air interface sometimes caused the solution sample to separate into the oil bath, resulting in a very small loss of material into the beaker. This is why we adopted a connected multi-chamber design, in which the syringe needle still extracts and inserts the droplet at each chamber, but it never leaves the oil. There has not been problems with loss of material from the droplet under these conditions.

Professor Vadgama said: Is the current system ready for use in diagnostic pathology?

Mr You responded: While the methodology has been proven to be highly effective at reducing the PCR thermocycling times, the current device is a general system used for all our wire-guided droplet microfluidic experiments. In order for the system to be ready for use in diagnostic pathology, it would need to be scaled up in such a way as to provide high-throughput and real-time detection. In other words, instead of just a single syringe needle, an array of needles in a grid or circular pattern on

† Dr Yoon's paper was presented by Mr You, The University of Arizona, Tucson, Arizona, USA.

conjunction with multiple heating chambers would provide high-throughput. Integrating LEDs and photodetectors directly onto the printed circuit boards would allow for real-time detection.

Professor Vadgama commented: Does the silicone oil interface lead to the degradation or adsorption of the DNA/reactants? Is there a problem with heat loss or gain through the metal of the metal syringe needle to distort the heat cycling profile of the sample drop?

Mr You replied: While it is possible for degradation and adsorption of proteins at the silicone oil interface because of hydrophobic and hydrophilic regions, it is not an issue with DNA because of the highly charged backbone. However, there may be some degradation or adsorption of the reactants, and while it would be detrimental to DNA at its initial low concentration, the extremely high concentration of the reactants would make this loss negligible. In regards to the heat loss or gain through the metal syringe needle: although stainless steel has a high thermal conductivity and therefore would act as a very good heat sink or heating element to the droplet, the large thermal mass of the syringe needle combined with the fact that less than 10 percent is submerged in the oil means that during the short durations in each chamber, the needle would most likely have a low variation in temperature. Additionally, because the syringe needle and droplet are in direct contact all the time, the method of heat transfer between them is conduction, whereas the droplet is undergoing forced convective heat transfer from the oil.

Dr Davies asked: I don't think you answered Professor Vadgama's question about loss of material to the oil/water interface. You said you had initially focused work on the interface but did not say what the losses were, and, as Professor Vadgama says, the lower the drop volume the higher the area/volume and so adsorptive losses to the interface are more important. Are losses to the oil/water interface significant in your method?

One common way to reduce such losses is to use a surfactant to fill the surface. Have you needed to use this approach?

How often do your drops fall off? If they do, does the frequency of drop fall off depend on drop volume? Can you ameliorate this by reducing transport speed, or vibration?

If you use surfactant or some other method to reduce O/W interface losses does this cause a problem with drop fall off?

Mr You replied: We haven't observed any significant losses to the oil/water interface. The solubility of DNA is profoundly greater in water than in oil due to its highly charged backbone. In some of our previous gel electrophoresis images the positive control used was a MiniCycler benchtop thermocycler, and when compared to our system for the same number of cycles our device actually produced a slightly brighter band. Therefore, we do not believe that there are significant losses to the oil/water interface. We have not tried surfactant. We also have not had problems with the droplet falling off. Initially we did on occasion lose the droplet because of a combination of too high a velocity and too small a syringe tip. This was resolved by using a larger syringe needle and a slower mixing velocity; the optimal combination was determined through trial and error.

Dr Mount commented: For a viable fast PCR system for virus detection the issue of the amplification of foreign nucleic acids through cross contamination must be avoided. Can the possibility of all cross contamination, *e.g.* through silicone oil contamination and droplet fragmentation, be excluded in the final device?

Mr You answered: In our experiments, the issue of contamination was avoided simply by using new silicone oil for each trial. However, as a final device, using new oil for every assay would be expensive and wasteful. Since DNA molecules will tend to move towards the water droplet, there is a possibility of cross contamination and this cannot be excluded. We believe this can be resolved by quickly heating all the chambers between assays, using UV irradiation, or adding some anti-viral or anti-bacterial agents into the silicone oil.

Professor Schuhmann commented: Are there any possibilities to read out the amplification by PCR in the described system.

Mr You replied: Yes, it is possible to do real-time detection with our system. All that would be needed in addition to the current system is a light emitting diode (LED) light source and a photodetector, such as an avalanche photodiode. These components would be integrated onto the printed circuit board in the same way as the temperature sensors. When the droplet passes between the light source and photodetector, the fluorescence signal is measured. This measurement can occur at every cycle, therefore allowing a read out of the amplification by PCR in real time.

Professor Graham asked: What are the main unique selling points of your approach?

Mr You replied: The main unique selling point of our approach is the ability to perform a 30 cycle PCR assay in less than 6 min 50 s with a 10 microlitre sample. Using a large sample volume standard with conventional thermocyclers allows for post-analysis procedures such as DNA sequencing and gel electrophoresis, which is extremely difficult with other methods such as nanodroplet PCR. Additionally, our method allows for reconfigurability and can be easily scaled-up for high-throughput systems. Because a simple printed circuit board trace heater is used instead of high powered Peltier modules, the power consumption and overall cost is significantly less, allowing the potential for a portable system.

Professor Graham asked: Do you think this has the potential to displace existing PCR systems such as the LightCycler?

Mr You answered: Yes. A LightCycler still incorporates a multi-well heating and cooling block in direct contact with a high-powered Peltier and heat sink. The novelty of the LightCycler is its ability to perform real-time detection through fluorescence detection as well as melting-point analysis. Yet the method employed for thermocycling remains the same as conventional benchtop PCR thermocyclers: conduction. Because the system still relies on the heat transfer method of conduction, the minimum time for running 30 cycles in the latest model LightCycler is approximately 30 min. However, our system incorporates forced convective heat transfer along with conduction, in conjunction with a multi-chamber thermocycler design, to achieve a 6 min 50 s thermocycling time for the same 30 cycles. There is also potential to further optimize the system to less than 3 min for 30 cycles based on existing simulation data, by reducing the solution sample and optimizing the chamber temperatures. Additionally, fluorescence detection can easily be implemented into our system by incorporating LEDs and photodetectors on the printed circuit board. It is also important to mention that a Peltier module consumes significantly more power than a steady-state printed circuit board trace heating element.

Dr Brown remarked: Have you applied your methods to detect any other pathogens?

Mr You responded: The only other pathogen that we have used is *E.coli* K12.

Professor Schuhmann opened the discussion of the paper by Professor Narayan: What is the actual size of the proposed needles? Can they be fabricated so that they do not hurt while penetrating through the skin?

Professor Narayan responded: The microneedles can be of various sizes; we envision developing several microneedle sizes, which are optimized for specific applications.

The uppermost layers of the skin include the epidermis, which is approximately 100 micrometres thick, and the papillary dermis, which is also approximately 100 micrometres thick. Sensory nerve endings, including Meissner's corpuscles or "touch receptors", are located in the papillary dermis. Nociceptors or "pain receptors" enter the papillary dermis and the epidermis. Sensation and pain are minimized since the cross-section of the microneedle is significantly smaller than that of a conventional hypodermic needle. As such, microneedle–receptor interaction is minimized.

Professor Guiseppi-Elie said: Your MTT assay shows a differential response to dermal fibroblasts and human epidermal keratinocytes, can you comment on this differential cell–surface response? Is this in the range of error given the number of replicates studied and error observed?

Professor Narayan responded: Per my colleague who works on skin in the College of Veterinary Medicine, human epidermal keratinocytes are epithelial cells derived from ectoderm while fibroblasts are connective tissue cells derived from mesoderm. They are two different populations and you would expect them to be different with different receptors and reactions to compounds.

Dr Ko Ferrigno said: How do these needles compare to other microneedles, such as those recently reported by Sullivan *et al.*[1]

1 S. P. Sullivan, D. G. Koutsonanos, M. del Pilar Martin, J. W. Lee, V. Zarnitsyn, S.-O Choi, N. Murthy, R. W. Compans, I. Skountzou and M. R, Prausnitz, *Nat. Med.*, 2010, **16**, 915–920, DOI: 10.1038/nm.2182.

Professor Narayan replied: The devices described in Sullivan *et al.* are solid microneedles that contain a mixture of a biodegradable polymer and either a pharmacologic agent or a vaccine. In this microneedle type, the microneedle penetrates the stratum corneum layer of the epidermis and the biodegradable polymer releases the biologically active agent into epidermal and/or dermal tissue. The microneedles described in our paper are hollow structures, which may be used for pressure-based delivery or diffusion-based delivery of one or more biologically active agents. Agent(s) may be delivered into epidermal and/or dermal tissue over an extended period of time; as such, hollow microneedles may be used for controlled dosing of biologically active agents. It should also be noted that hollow microneedles may be used for delivery of pharmacologic agents as well as extraction of fluids.

Professor Girkin remarked: What is the surface finish that can be achieved and have you attempted to create sub-diffraction limited features due to the non-linear process involved?

Professor Narayan answered: In recent work, poly(ethylene glycol) diacrylate resins with molecular weights of 302 and 742 were processed using two photon polymerization.[1] Minimum feature sizes of 200 nm were demonstrated for both materials.

There is a trade-off between minimum feature size and processing time. Processing with thinner layers enables fabrication of structures with more intricate shapes; however, longer processing times are involved. The surfaces of the microneedles may be modified by thin film deposition. One important factor that is limiting clinical use of microneedles is the risk of infection. This risk, associated with channels formed by microneedles through the stratum corneum, may be reduced by imparting microneedles with antimicrobial properties. In recent work, antimicrobial microneedles were prepared by coating microneedles with thin silver films. These thin films were deposited on the surfaces of microneedles using pulsed laser depoistion. In this physical vapor deposition technique, a solid target is vaporized with a high energy (*e.g.*, excimer) laser. The high energies of the laser-ablated species enhance film adhesion and promote film–substrate chemical reactions. Pulsed laser deposition is ideally suited for film deposition on polymeric microneedles since coating work can be performed at room temperature. No non-uniformities were noted in the thin silver films on the coated microneedles. An agar diffusion assay was utilized to evaluate the antimicrobial activity of the silver-coated organically-modified ceramic microneedle array. An absence of *S. aureus* growth beneath the silver-coated microneedle array was noted. Furthermore, inhibition of bacterial growth was observed in the agar surrounding the microneedle array, indicating release of silver into the agar.

1 A. Ovsianikov, M. Malinauskas, S. Schlie, B. Chichkov, S. Gittard, R. Narayan, M. Lobler, K. Sternberg, K.-P. Schmitz, A. Haverich, *Acta Biomaterialia*, 2010, DOI: 10.1016/j.actbio.2010.10.023.

Professor Vadgama remarked: Silicon fabricated needles inevitably have angular geometry and sharp corners, so this is a better approach for insertion since a rounded surface is less prone to cause damage. There is no need to be defensive about the material being toxic – the cell results you are seeing are reasonable for biomaterials, though clearly there is a need to do further detailed cytotoxicity studies. Additional use of surface-immobilised heparin or polyethylene oxide would be important in order to minimise surface protein fouling and needle blockage.

Professor Narayan answered: Thanks for your comment. I agree that the cell viability profile is reasonable – it should be noted that this polymer is currently used in several Class IIa medical devices (the same class as sterile syringes). The microneedle surface may be modified to provide enhanced biological functionality (*e.g.*, antimicrobial or antifouling functionality) at the microneedle/tissue interface. Protein fouling and cell attachment can impede transport of pharmacologic agents from microneedles to nearby tissues. We have used a two photon polymerization-micromolding process to fabricate solid microneedles (microneedles without a hollow bore) out of polyethylene glycol diacrylate. Polyethylene glycol diacrylate exhibits nonspecific resistance to cell adhesion and protein adsorption. Furthermore, polyethylene glycol undergoes hydrolytic degradation on the skin surface; hydrolytic degradation may allow for gradual release of pharmacologic agents from the microneedle body. We recently fabricated polyethylene glycol diacrylate microneedle arrays, which contained microneedles with base diameters of 150 μm and lengths of 500 μm, using photopolymerization-micromolding. In an *in vitro* study, polyethylene glycol diacrylate microneedles exposed to human platelet rich plasma did not exhibit protein aggregation or platelet aggregation. One important factor that is limiting clinical use of microneedles is the risk of infection; a microneedle-fabricated pore in the stratum corneum layer of the epidermis could enable microorganisms to cause local or systemic infection. The risk of infection related to microneedle-fabricated pores may be reduced by creating microneedles that exhibit antimicrobial activity. We recently fabricated solid microneedles out of a composite material containing polyethylene glycol diacrylate and gentamicin sulfate. An agar-plating assay revealed that the gentamicin-doped polyethylene glycol microneedle array inhibited *Staphylococcus aureus* growth. Antimicrobial microneedles may also be prepared by coating microneedles with materials that exhibit antimicrobial behavior. Silver exhibits broad-spectrum antibacterial, antiviral, and antifungal properties; this activity is attributed to interruption of DNA replication and disruption of electron transport. Silver also facilitates wound healing and reduces inflammation. Coatings containing antimicrobial materials may be applied to polymer microneedle surfaces at room temperature using pulsed laser deposition. We fabricated silver-coated microneedles using two photon polymerization-micromolding and pulsed laser deposition. Solid microneedle arrays were prepared using organically-modified ceramic material by means of two photon polymerization-micromolding. Pulsed laser deposition was subsequently used to deposit silver thin films on these microneedle arrays. An agar diffusion assay was used to evaluate antimicrobial performance; an absence of *S. aureus* growth beneath the silver-coated microneedle array was noted. Inhibited growth was also noted in the region surrounding the microneedle array, which suggested silver release into the agar.

Dr Krause remarked: Is rapid prototyping suitable for mass production? What would be the price per microneedle device if the production was scaled up?

Professor Narayan responded: Increasing the two photon polymerization processing rate may be accomplished by means of (a) multibeam parallel processing and (b) increasing the speed at which the laser beam spot is translated within the resin. For example, a microlens array may be used to transform a collimated beam into multiple focus spots; this setup may be used for parallel processing of several identical structures. Indirect processes based on two-photon polymerization may also be used for creating large numbers of devices. For example, soft lithography approaches (*e.g.*, microcontact molding, microcontact printing, microtransfer molding, membrane assisted microtransfer molding, and replica molding) may be used for replication of solid microneedles as well as other two photon polymerization-fabricated microscale structures . Two photon polymerization may be used to process a variety of photosensitive resins, including acrylate-based polymers and organically-modified ceramic materials. Many of these materials may be obtained at low cost. Based on the low cost of the resin, the final price per device is anticipated to be under $0.10 US.

Dr Davies asked: There is a device, called the Dermaroller, which consists of a hand-driven cylindrical roller from which protrude steel spikes – it is available in different spike lengths and is proposed for various skin treatments. I have read that using it to perforate the skin can increase the delivery of material to the epidermis and dermis by up to 10 000-fold. I would expect there to be some indication from users of such equipment on the incidence of bacterial infection that would be relevant to microneedle dosing. Phlebotomists use an alcohol wipe before piercing with a sterile needle and this appears to produce an acceptably low level of infection, and such a procedure could be performed before use of microneedles.

Professor Narayan answered: One issue that is limiting the use of microneedles, particularly for outpatient treatment of chronic diseases, is the risk of infection. Besides its roles in sensation, ultraviolet light protection, vitamin D production, fluid regulation, and thermoregulation, the skin serves as a barrier to microorganism entry.[1] The stratum corneum layer of the epidermis prevents pathogenic bacteria and other microorganisms from entering deeper layers of the epidermis. This layer of the skin contains corneocytes as well as intercorneocyte lipids, which provide a microscopic "brick and mortar" barrier to microorganisms. In addition, the skin produces proteins that inhibit microbial growth such as cathelicidins, chemokines, defensins, neuropeptides, and protease inhibitors. Gram-positive bacteria (*e.g.*, *Staphylococcus aureus*) and other microorganisms dwell on the skin surface.[2–4] A pore in the stratum corneum created by a microneedle could allow microorganisms to enter deeper regions of the body and cause either local or systemic infection.

According to Sato *et al.*, pathogenic microorganisms may be transmitted from the skin surface to deeper skin regions by conventional hypodermic needles.[5] Ruffini *et al.* and Banga indicated that penetration of skin by microneedles and other small-scale medical devices may be associated with a risk of infection.[6,7] Ruffini *et al.* described the use of carbon nanotube arrays to evaluate electrochemical cell activity in the brain, the heart, and various other tissues. They speculated that the risk of infection with their device would be relatively low due to the small dimensions of the device. On the other hand, they noted that penetration of small microorganisms would be possible through nanotube-fabricated openings. Birchall also mentioned the formation of channels through the stratum corneum increases the risk of infection.[8] He suggested that stratum coreum damage is both transient and relatively minor; he also remarked that skin repair of microneedle damage is not fully understood. The risk of infection is of particular concern for immuno-compromised patients. Healthcare providers typically prepare the patient's skin with isopropyl alcohol or another chemical disinfectant prior to administration of a pharmacologic agent using a hypodermic needle in an inpatient setting. The insertion of medical devices under sterile conditions is considered to be the most effective method for minimizing the number of medical device-associated infections. However, it may not be possible to maintain consistent adherence to these protocols in a setting in which there are limited medical resources, in a setting in which the caregiver has limited medical knowledge, or in an outpatient setting. Incorporating antimicrobial properties within medical devices is a commonly used mechanism for providing additional protection against the development of infections.[9–12] The risk of infection associated with outpatient use of microneedles may be reduced by imparting these devices with antimicrobial properties. Pharmacologic agents are commonly incorporated within a variety of medical devices, including catheters that enter the body through the skin or the genitourinary tract, in order to impart these devices with antimicrobial functionality.[13–23] For example, central venous catheters are used by patients in an inpatient setting who require antibiotic therapy, chemotherapy, hemodialysis, or total parenteral nutrition; ~5% of these devices become infected during use.[24] Indwelling urinary catheters are medical devices that are utilized in the care of ~20% of patients during short-term hospitalizations. These devices are commonly colonized by microorganisms. Catheter-associated urinary tract infections are associated with many of the 900 000 cases of nosocomial bacteriuria that are observed in US hospitals every year.[25–29] Catheter-related infections have been associated with migration of Gram-positive organisms and other microorganisms along the external catheter surface from the surface of the skin.[30–34] Complications of these medical device-associated infections include local morbidity, bloodstream infection, systemic morbidity, death, and increased health care costs.[35–39] A variety of pharmacologic agents have been coated on catheters in order to minimize the number of catheter-related infections; antimicrobial-coated devices are currently being utilized in many countries. Meta-analyses and randomized controlled trials have shown the effectiveness of antimicrobial-coated catheters for reducing catheter-related bloodstream infections.[25,40,41] The United States Centers for Disease Control and Prevention as well as clinical guidelines described by Parker *et al.* suggest that use of antimicrobial catheters is an appropriate mechanism to decrease the catheter-related infection rate.[42,43] In addition, Drekonja *et al.* discussed consistent evidence of variable nature that antimicrobial-coated catheters limit catheter-associated infections when used for short-term procedures.[39]

1 S. H. Lee, S. K. Jeong, S. K. Ahn, *Yonsei Med. J.*, 2006, **47**, 293.
2 Y. N. Kalia, L. B Nonato, C. H Lund and R. H Guy, *J. Invest. Derm.*, 1998, **111**, 320.
3 D. L. Stulberg, M. A. Penrod, and R. A. Blatny, *Am. Fam. Physician*, 2002, **66**, 119.
4 E. Larsen, *Clin. Infect. Dis.*, 1999, **29**, 1287.
5 S. Sato, T. Sakuragi, and K. Dan, *Anesthesiology*, 1996, **85**, 1276.
6 G. Ruffini, S. Dunne, E. Farrés, J. Marco-Pallarés, C. Ray, E. Mendoza, R. Silva and C. Grau, *Sens. Actuat. A*, 2006, **132**, 34.

7 A. Banga, *Pharm. Res.*, 2007, **24**, 1357.
8 J. C. Birchall, *Exp. Rev. Med. Dev.*, 2006, **3**, 1.
9 D. Trinkaus, *Am. J. Infect. Control.*, 2004, **32**, E105.
10 N. P. O'Grady, M. Alexander, E. P. Dellinger, J. L. Gerberding, S. O. Heard, D. G. Maki, H. Masur, R. D. McCormick, L. A. Mermel, M. L. Pearson, I. I. Raad, A. Randolph and R. A. Weinstein, *Pediatrics*, 2002, **110**, e51.
11 J. W. Costerton, P. S. Stewart and E. P. Greenberg, *Science*, 1999, **284**, 1318.
12 H. Hanna, *Infect. Med.*, 2001, **18**, 38.
13 N. Safdar, *Crit. Care*, 2009, **13**, 148.
14 K. A. Halton, D. A. Cook, M. Whitby, D. L. Paterson and N. Graves, *Crit. Care*, 2009, **13**, R35.
15 J. N. Greene, *Infect. Dis. Clin. North Am.*, 1996, **10**, 255.
16 F. G. Raad II, *Cancer Treat. Res.*, 1995, **79**, 211.
17 C. J. Crnich and D. G. Maki, *Clin. Infect. Dis.*, 2002, **34**, 1362.
18 C. J. Crnich and D. G. Maki, *Curr. Infect. Dis. Rep.*, 2001, **3**, 496.
19 S. I. Blot, P. Depuydt, L. Annemans, D. Benoit, E. Hoste, J. J. De Waele, J. Decruyenaere, D. Vogelaers, F. Colardyn and K. H. Vandewoude, *Clin. Infect. Dis.*, 2005, **41**, 1591.
20 B. Renaud and C. Brun-Buisson, *Am. J. Respir. Crit. Care Med.*, 2001, **163**, 1584.
21 J. B. Dimick, R. K. Pelz, R. Consunji, S. M. Swoboda, C. W. Hendrix and P. A. Lipsett, *Arch. Surg.*, 2001, **136**, 229.
22 D. K. Warren, W. W. Quadir, C. S. Hollenbeak, A. M. Elward, M. J. Cox and V. J. Fraser, *Crit. Care Med.*, 2006, **34**, 2084.
23 I. Raad, H. Hanna, D. Maki, *Lancet Infect. Dis.*, 2007, **7**, 645.
24 D. Roe, B. Karandikar, N. Bonn-Savage, B. Gibbins and J.-B. Roullet, *J. Antimicrob. Chemother.*, 2008, **61**, 869.
25 J. R. Johnson, M. A. Kuskowski and T. J. Wilt, *Ann. Intern. Med.*, 2006, **144**, 116.
26 S. Saint and C. E Chenoweth, *Infect. Dis. Clin. North Am.*, 2003, **17**, 411.
27 J. W. Warren, *Med. Clin. North Am.*, 1991, **75**, 481.
28 D. G. Maki and P. A. Tambyah, *Emerging Infect. Dis.*, 2001, **7**, 342.
29 S. Saint, B. A. Lipsky and S. Dorr Goold, *Ann. Intern. Med.*, 2002, **137**, 125.
30 M. J. Richards, J. R. Edwards, D. H. Culver and R. P. Gaynes, *Infect. Control Hosp. Epidemiol.*, 2000, **21**, 510.
31 M. J. Richards, J. R. Edwards, D. H. Culver and R. P. Gaynes, *Crit. Care Med.*, 1999, **27**, 887.
32 J. Rello, A. Ochagavia, E. Sabanes, M. Roque, D. Mariscal, E. Reynaga and J. Valles, *Am. J. Respir. Crit. Care Med.*, 2000, **162**, 1027.
33 R. J. Sherertz, *Curr. Opin. Infect. Dis.*, 2004, **17**, 303.
34 N. P. O'Grady, *MMWR Recommendations Rep.*, 2002, **51**, 1.
35 I. Raad, R. Darouiche, J. Dupuis, D. Abi-Said, A. Gabrielli, R. Hachem, M. Wall, R. Harris, J. Jones, A. Buzaid, C. Robertson, S. Shenaq, P. Curling, T. Burke and C. Ericsson, *Ann. Intern. Med.*, 1997, **127**, 267.
36 D. G. Maki, S. M. Stolz, S. Wheeler and L. A. Mermel, *Ann. Intern. Med.*, 1997, **127**, 257.
37 R. O. Darouiche, *N. Engl. J. Med.*, 1999, **340**, 1.
38 M. L. Pearson and E. Abrutyn, *Ann. Intern. Med.*, 1997, **127**, 304.
39 D. M. Drekonja, M. A Kuskowski, T. J Wilt and J. R Johnson, *Exp. Rev. Med. Dev.*, 2008, **5**, 495.
40 D. L. Veenstra, S. Saint, S. Saha, T. Lumley and S. D. Sullivan, *J. Am. Med. Assoc.*, 1999, **281**, 261.
41 A .L. Casey, L. A Mermel, P. Nightingale and T. S. J. Elliott, *Lancet Infect. Dis.*, 2008, **8**, 763.
42 N. P. O'Grady, *MMWR Recommendations Rep.*, 2002, **51**, 1.
43 D. Parker, L. Callan, J. Harwood, D. L. Thompson, M. Wilde, M. Gray, *J. Wound Ostomy Continence Nursing*, 2009, **36**, 23.

Dr Ko Ferrigno said: Do you see any fluid coming from the punctured skin back into the needle? Could this be a way of sampling interstitial fluid, which is a biological fluid that has not been well explored?

Professor Narayan answered: In our study, water was placed on the backside of the microneedle array for use of a water immersion lens. In this case, capillary forces in the channels may have pulled some of the water through the microneedles and into the cadaveric porcine skin; this process may have resulted in dilution of the quantum dot solution. In general, hollow microneedles are an appropriate mechanism for sampling of interstitial fluid. For example, Suzuki *et al.*[1] have described the combination of a microneedle device and a microscale sensor for sampling and sensing of glucose.

1 H. Suzuki, T. Tokuda and K. Kobayashi, *Sens. Actuators, B*, 2002, **83**, 53, DOI: 10.1016/S0925-4005(01)01028-0.

Dr Mount said: It is claimed that Fig. 8b of your paper shows a broad distribution of quantum dots in the deep epidermis and dermis. However, the pattern of distribution appears counterintuitive, with the least number at the tip of the needle and a much larger concentration above half needle height and at the skin surface. Is this indicative of leakage of fluid in the cavity around the needle and how does this affect optimisation of needle length, shape and materials characteristics for optimum delivery?

Professor Narayan answered: The image demonstrates limitations of current multiphoton microscopy technology. There are two problems with multiphoton imaging of structures within tissues. First, there are difficulties associated with light penetration of tissue. Laser power may be increased to facilitate imaging of structures within tissues; however, burning of the tissue may occur. Second, there are difficulties associated with deflection of light; deflection becomes a significant issue for imaging of structures located deeply within tissues. It should also be noted that multiphoton imaging must take into account three sources of signal: (a) autofluorescence of the polymeric device, (b) autofluorescence of the skin, and (c) fluorescence of the quantum dots.

In this manuscript, imaging of the fluorophore solution several hundred micrometres within porcine tissue was shown. It is believed that imaging of structures 800–900 micrometres within porcine tissue may be possible using the current multiphoton microscopy technology. Working distance issues limit imaging beyond one millimetre.

Professor Girkin commented: When you inject into a living eye the compound frequently does follow the path of the needle when the needle is withdrawn and I am aware that this has also been seen in skin models, but I am not certain about in living human tissue in terms of skin.

Professor Narayan responded: Per my colleague who works on skin in the College of Veterinary Medicine, eye and skin are very different in terms of penetration behavior. Skin has a cornefied stratified squamous epithelium while the eye has a nonkeratinizing epithelium. The pressure behind the eye is much greater. It should be noted that the thickness of the epithelium in both tissues may be similar.

Dr Mount asked: In your paper, the stated difference between input and measured tip heights appears to be near 60% in Fig. 6a and 40% in Fig. 6b and c. What is the physical origin of this difference and are similar differences seen in base and channel diameter? Have the design rules and processing parameters been established to enable specified tip dimensions to be fabricated without iteration and are these dependent on the tip material?

Professor Narayan replied: The microneedle fabrication described in this manuscript was guided by a computer-aided design (CAD) model. In two photon polymerization (as in other rapid prototyping technologies), the CAD model surface is converted to a series of polygons; this process is known as tessellation. This series of polygons is subsequently sliced into collections of cross-sectional layers. The microneedle is then fabricated in a layer-by-layer manner. Differences between input and measured dimensions are attributed to differences in translation of the model to the physical structure. Some iteration may be necessary because we cannot predict how the model is manipulated by the software.

Professor Vadgama said: Hydrostatic pressure in the skin (dermis) is actually slightly negative to atmospheric so there would be interesting work to do to evaluate skin pressure changes in relation to physiology and pathology using percutaneous needle probes with manometric monitoring.

Professor Narayan replied: Thanks for your comment. According to one of our paper's coauthors (who completed her thesis in anatomy), the pressure would be variable because it would be dependent on anatomical features (*e.g.*, capillaries) that were penetrated by the microneedle.

Professor Graham opened a general discussion of the papers by Dr Yoon and Professor Narayan: How important is depth penetration for optical imaging when moving from mouse to human models? In mice, organs can be moved to optical windows for analysis but obviously this is harder in humans. Will we see whole body optical scanners for humans?

Dr Dhaliwal answered: I don't think so. Limited ring tomography with depth of up to 7–8 cm will likely be the limit. With established whole body molecular imaging with PET tracers, it is unlikely that whole body optical imaging will provide adequate depth penetration or sensitivity. On the other hand, endoscopic devices, hand-held perioperative devices are emerging in the clinical arena.

Dr Kendall responded: The penetration depth required for optical techniques depends on the question you are addressing. If you are diagnosing the development of epithelial cancers (most cancers are epithelial) then optical techniques with near-infrared excitation wavelengths have sufficient penetration. There are also methods for measuring Raman signals from deeper within tissue, for example, if you wish to identify whether breast calcifications are benign or malignant. Methods include: Kerr Gating, transmission Raman, SORS (spatially offset Raman spectroscopy), and SESORS (surface enhanced SORS).[1–6] A recent review has been published on pre-clinical whole-body fluorescence imaging.[7]

1 R. Baker, P. Matousek, K. L. Ronayne, A. W. Parker, K. Rogers and N. Stone, Depth profiling of calcifications in breast tissue using picosecond Kerr-gated Raman spectroscopy, *Analyst*, 2007, **132**, 48–53.

2 P. Matousek and N. Stone, Prospects for the diagnosis of breast cancer by non-invasive probing of calcifications using transmission Raman spectroscopy, *J. Biomed. Opt.*, 2007, **12**, 024008.

3 N Stone and P Matousek, Advanced Transmission Raman Spectroscopy – A promising tool for breast disease diagnosis, *Cancer Res.*, 2008, **68**, 4424–4430.

4 P. Matousek, Inverse Spatially Offset Raman Spectroscopy for Deep Noninvasive Probing of Turbid Media, *Appl. Spectrosc.*, 2006, **60**(11), 1341–1347.

5 N. Stone, R. Baker, K. Rogers, A. W. Parker and P. Matousek, Subsurface probing of calcifications with spatially offset Raman spectroscopy (SORS): future possibilities for the diagnosis of breast cancer, *Analyst*, 2007, **132**, 899–905.

6 N. Stone, K. Faulds, D. Graham and P. Matousek, Prospects of Deep Raman Spectroscopy for Noninvasive Detection of Conjugated Surface Enhanced Resonance Raman Scattering Nanoparticles Buried within 25 mm of Mammalian Tissue, *Anal. Chem.*, 2010, **82**(10), 3969–3973, DOI: 10.1021/ac100039c.

7 F. Leblond, S. C. Davis, P. A. Valdés and B. W. Pogue, Review of instruments, methods and applications, *J. Photochem. Photobiol., B*, 2010, **98**, 77–94.

Professor Girkin replied: There is frequently a trade-off in optical imaging of resolution *vs.* depth penetration. For example imaging with micron resolution up to 100 microns in many forms of tissue is possible, but it is also possible to image through a complete breast optically, though the resolution is now at the 100s of micron level. Thus whole body scanners are potentially possible, but do they provide sufficient resolution to assist at present in clinical diagnosis? At present the answer is frequently no.

Dr Davies remarked: Self-monitoring of external tissue is done wittingly and unwittingly all the time. Devices are available for the examination of some internal tissues. Breastlight^{TM} is available over the counter in some pharmacies for examination of breast tissue which is squeezed up against its intense light and the user looks for a shadow cast by more dense tissue within the breast. The Breast Sense Glove has an inner and outer layer of polyurethane separated by a thin, slippery silicon liquid which enables the hand and inner glove layer to easily slide over the squeezed breast and improve the detection of more solid tissue within the breast. Interrogation of deeper tissues is harder to envisage. It is also unfortunate for the males of our species that their "dangly bits" are anatomically ill-positioned and sensitive and so are not amenable to such enhanced self-examination.

Dr Dhaliwal said: There are clinical scanners for breast cancer based on optical imaging, *e.g.* ART (Advanced Research Technologies) commercial system, but there are three placed in world so there is limited clinical use. At £1 million per piece of kit, this may have limited penetration due to cost *vs.* mammography.

Dr Davies responded: If the performance of these scanners matches, or exceeds, the mammogram and their cost come down then I could see their use increasing. However, I think it is nicer and easier for a woman to examine herself at home with something like the BreastLight. Hopefully, in such unskilled hands the efficacy of the method will match mammography and women will act on the information it provides. Of course mammography is not a highly sensitive method, but until something better comes along it is our gold standard for mass screening.

Professor Graham the addressed the audience: Physical scientists need to engage with clinicians at an early stage to shape their research and prepare for translation from the bench to the body. Has anyone had success in commercialising a research breakthrough in a clinical market and what lessons have been learned that can be shared with everyone?

Professor Vadgama replied: We have had success in the form of a licensing arrangement on a discrete piece of technology. The underlying considerations for success I would suggest are that the technology is preferably simple (likely to be more robust) and is advanced by a collaborator who truly understands the translational challenges.

Professor Girkin replied: If you develop a radically new method/technique then it is vital to get the clinician on side as soon as possible, but this can mean having to break his reliance on a specific method which he has grown to love and trust. Sometimes the new method is compared to the old method and seems to come out badly but frequently this is because the older method is unreliable and the new method is really showing this up. When you have to prove that the accepted "gold standard" is wrong you have an up-hill battle.

Professor Graham said: There are people who have had experience and success of taking research to the clinical market. One of the major issues is compliance and regulatory issues which have to be addressed at an early stage. Scientists are talking to clinicians and with so much good science taking place the main barrier appears to be the person who can find their way from a scientific research project success to a clinical product. This is not the physical scientist or the clinician!

Dr Ko Ferrigno commented: I suggest that we should always consider taking technologies (suitably patented or protected) to established companies who have mechanisms for taking technologies to market and who may be interested in a license.

I would add that although commercialisation is often seen to be linked to an opportunity to make money, we should not ignore the altruistic motivation of simply wishing to make a difference to patients/the world. The problem is the paradigm that, for a company to take up a new idea or technology, it needs to be patented. Because patenting costs money which universities and/or academics need to recoup, the scope for giving a technology to a company to use is reduced. Thus, although patenting is seen as a prerequisite for getting a technology to market, it is also a barrier, because it creates expectations that do not always match the primary motivations of the inventor(s).

Dr Campbell remarked: To play Devil's Advocate, as academics, why should we bother with commercialisation? We are not judged by our commercial successes or failures. Maybe we should just write our work up for publication, forget translation and the end-user, and let someone else commercialise it?

Professor Girkin replied: There is some logic behind this comment, however, some return to universities would be required if this open access methodology was adopted. The issue is in fact much deeper than this: who is the research for the benefit of? The researcher (career or financial), university (status or financial), UK PLC. The drive for publication also forces researchers to constantly look for new tricks and methods just for the sake of doing this, when perhaps sometimes a standard method is just applied very well through new technology/engineering. This type of work is not well supported in the UK, and also not well viewed by our academic peers, though it might often give the best benefit in the short (5–10 year) timescale to both UK PLC and patients.

Professor Cooks answered: The well-known answer is that commercialization does not succeed without a champion. Much less commented on is the fact that driving commercialization need not be tied to personal financial enrichment.

Professor Bradley responded: Far too much lip service and not enough action is the reality of translational activity. I would suggest that we should be judged by translational success much more than we are today. Our research is paid for by the tax payer – and much research is done that has little link with reality or the real world: is this really sustainable or supportable? There always will be a number of major challenges and balancing acts that must be tempered by any academic and clearly there is a need for a balanced portfolio of research activities to ensure future translational ideas are grown and developed. It is a criminal waste of ideas, technology and science if they are not exploited and ends up with resulting damaging consequences for potential future growth of new companies and new jobs if, as scientists, we do not whole-heartedly embrace the translational agenda.

Dr Campbell asked: With reference to my previous comment, I might want to look at the situation from the opposite perspective – that by merely publishing ways of carrying out an assay without any thought of an end-user or translation we are flooding the literature with meaningless science. I would suggest that the right answer is somewhere in between these two extremes.

Dr Davies answered: I would agree that much of the material in the biosensor arena, for instance, is almost meaningless. One major problem, as someone who would like to translate from scientific paper to product, is that the biosensor people in academia come up with many clever ways of making a measurement in buffer. Humans do not run on buffer and so such experiments, while of tantalising interest, more often than not fail when evaluated against real samples. Academia and the government funders expect industry to solve the hard problem of making the sensor work in a biological sample, but industry rarely has the resource to enter this high risk domain. DeGennes took physics to dirty media with great success and it is about time someone took biosensors into real samples.

Professor Vadgama said: There is value to doing applied bio-linked research because it informs the fundamental science. If only the fundamentals are worked on then these are in danger of either becoming outdated or disconnected from practical bio-analytical needs.

Professor Cass opened the discussion of the paper by Professor Allbritton: Is it possible to estimate the number of enzyme molecules present in your sample from the amount of product that you detect?

Professor Allbritton replied: The readout provides a quantitative measure enabling us to estimate the number of moles of fluorescent PIP3 produced. To estimate the amount of enzyme present, a variety of assumptions need to be made which are not likely to be entirely accurate. These include the suppositions that PIP3 kinase can access all fluorescent substrate, no metabolism of fluorescent PIP3 is ongoing, all PIP3 kinase molecules are active, the k_{cat} measured for PIP2 is similar to that for the fluorescent substrate, and a host of other assumptions. Using the published turnover number of PI3K for nonfluorescent PIP2 ($3\ s^{-1}$),[1] we estimate the number of active enzyme molecules in our single-cell measurements to be very roughly 700 molecules. However, we believe that more careful studies targeted specifically at this question would yield a more accurate estimate.

1 J. D. Carson *et al.*, *Biochem. J.*, 2008, **409**, 519–524.

Dr Ko Ferrigno asked: Is there a need for single cell analysis in clinical diagnostics?

Professor Allbritton responded: Yes. Numerous single-cell analyses already exist in cancer diagnostics. Examples are counting the percentage of single cells with abnormal DNA content, identifying single Reed–Sternberg cells (Hodgkins Disease), immunofluorescence staining with microscopic examination at the single cell level, examination of single cells in PAP smears, *etc.* Much of the microscopy-based diagnostics currently in use for cancer involves examining specimens at the single-cell level to make a diagnosis, formulate a prognosis, or develop a therapeutic strategy. Cancer is a disease that is highly heterogeneous at the single-cell level and resistance to cancer therapeutics likely occurs initially in a single cell. Metastases also originate from single cells with unique properties that enable survival at sites distant from the primary tumor. Cancer is a disease of rapid evolution where the evolving unit is the single cell. Therapeutics needs to take into account as much of the single-cell heterogeneity as possible in order to recognize the presence of resistant cells when they are in the minority. This capability is needed to design and implement effective therapies for the individual patient. From a prognostic standpoint, it will also be important to recognize the early emergence of resistant cells during and after treatment to optimize therapy.

Dr Ko Ferrigno then asked: Following on from my previous question, in a disease such as cancer where the cells in a single tumour can be highly heterogeneous, can you be sure that you are sampling the right cell?

Professor Allbritton responded: It is clear that sampling a single cell is not sufficient, but rather the goal is to sample as many single cells as possible to develop a full picture of the tumor heterogeneity at the cellular level. The more cells sampled, the more likely one is to identify the existence of drug-resistant clones or clones that are dependent on a different set of signaling pathways compared to the bulk of the tumor. This would enable all tumor lineages or clones to be therapeutically targeted at early time points in the hopes of improving the effectiveness of therapeutic interventions. High-throughput, single-cell analysis will enable these types of diagnostic and prognostic assays to be performed.

Dr Mount said: There appears to be a requirement for high throughput measurement to enable rapid and statistically significant multiple cell characterisation. Is this an aim of your work and how are you addressing this?

Professor Allbritton replied: Developing techniques with greater throughput is a major focus of our group and we are actively improving the technology in this regard. We have recently published a platform that enables automated serial analysis of fluorescent analytes from single cells.[1] This technology development effort parallels our work to develop and validate biochemical assays such as was reported in our paper. At present, that platform has a throughput of 2 cells min^{-1}. We are currently working to enhance the rate of analysis through technical improvements on this CE-based platform and through the development of a platform based on a microfluidic chip which possesses a current throughput of approximately 30 cells min^{-1}.

1 D. Jiang, C. E. Sims and N. L. Allbritton, Microelectrophoresis Platform for Fast Serial Analysis of Single Cells, *Electrophoresis*, 2010, **31**, 2558–2565.

Dr Mount asked: Can such high throughput techniques be used profitably in combination with cell tumour location to gain information about cell variation in tumours?

Professor Allbritton responded: It is conceivable that this capillary-based assay can be utilized with a modified sampling technique to provide the ability to sample living tissue at different locations within a tumor. In this scenario, the tissue would need to be incubated with fluorescent probes followed by serial sampling or simultaneous sampling using multiplexed capillaries. In this manner, the enzyme activity of individual cells in various regions of the tumor could be assessed. Performance of such a measurement will require further technical enhancements.

Professor Schuhmann remarked: The timescale for the capillary electrophores seems to be rather long (1000–1500 s). Could you comment on this?

Professor Allbritton responded: The experiments described in this manuscript were carried out on a system which utilized a capillary of 40 cm to span the distance from the microscope stage to the fluorescence detection system, which was mounted adjacent to the microscope on this prototype system. Most analytes require a much shorter effective capillary length to achieve separation. We have developed higher throughput capillary and microfluidic systems. These platforms utilize a miniaturized fluorescence detection system mounted directly on the stage above the sample to reduce the effective length of the capillary and shorten migration times to times as short as 5 min or less. The microfluidics system delivers separation times of less than 30 s.

Dr Davies opened the discussion of the paper by Dr Campbell: Fig. 1 in your paper has controls for water and buffer but not for a non-functionalised Cy3-labelled oligonucleotide – did you do this? if you did what was the value? If the value is substantial (*i.e.* over a few 100 fluorescence units) then how do you think it attaches to the epoxy-derivatised glass? If the only chemical attachment is to the correct end of the probe then all is well, if it is to the wrong end or to a base, or bases, within the sequence then this would be expected to reduce or prevent hybridisation to probes held in that way. You do get hybridisation (Fig. 2 of your paper), but unless we are confident of the anchoring chemistry and lateral spacing of the probes then we cannot be sure that this is the best chemistry. Could it also be possible that, under the conditions used, a non-functionalised Cy3 probe physisorbs? The flip-side is that it does work, and so we can improve hybridisation. What is the purpose of the UV treatment? Under the spotting conditions the Debye screening length is below 1 nm so the oligonucleotides may adsorb by van der Waals attraction. Could UV exposure under these circumstances pin the oligonucleotide to the surface in several places?

Dr Campbell responded: In our experience, unmodified oligos will immobilise on epoxy-coated slides. I believe that this is mostly a non-covalent hydrophobic interaction and so is not particularly strong. The immobilisation of unmodified oligos can be improved by UV cross-linking. When comparing unmodified oligos with modified oligos we used a standard protocol for the unmodified oligos which included UV crosslinking because we wanted to compare novel modifications (peptide tags) with a known protocol.

Professor Bradley asked: Why do specific peptide spacers enhance hybridisation efficiency over others. For example, one might expect Lys6 would be better than Tyr6. Do you have an explanation for the optimal peptide. Also how would the acetylated-His-oligo bind to the surface if the primary amine is blocked through the side chain.

Dr Campbell responded: The optimal peptide structure is difficult to rationalise. The first aspect is why a peptide spacer improves hybridisation so much while a nucleic acid spacer does not. I suspect that having a peptide spacer moves the primary amine away from the phosphate backbone which could stabilise its protonated form, thus inhibiting nucleophilic attack.

The importance of this primary amine is reinforced by the observation that the His-oligo on which the terminal amine is acetylated does not perform as well as the non-acetylated His-oligo. The mechanism of binding between the acetylated-His-oligo and the surface is presumably through the imidazole ring. Again, I think that Tyr6 spaces the primary amine far enough from the phosphate backbone that coupling happens efficiently. While it would seem that Lys6 should also do this, I would speculate that the polycationic peptide may actually fold back and "complex" with the oligonucleotide.

Professor Bradley said: Do you think different peptides could be coupling at different rates? Does the signal relate to changes in loading at the surface or changes in the hybridization efficiency.

Dr Campbell replied: I think there is more to the effect than purely an increase in hybridization efficiency since the data in the figure suggests that the peptides either couple at different rates or pack onto the surface with different efficiencies. It would certainly be interesting to use a surface-sensitive approach to investigate the kinetics of coupling.

Professor Bradley asked: Could the peptide have a secondary structure (or even interact with adjacent peptides) to either enhance hybridization (*e.g.* α-helical formation that forces the strands apart) or lower it due to peptide aggregation?

Dr Campbell answered: I think you could go to quite extreme lengths to design peptide sequences and they would certainly improve efficiency, however I would add a note of caution: synthesising oligos with peptide conjugates is not cheap, and neither is buying them. However, using peptide tags may eventually be a more practical approach when using PNA oligos since they share the common backbone.

Dr Ko Ferrigno asked: To try and tease out the contributions of size and spacing to the improved sensitivity when using different peptide tags, have you tried using single amino acids instead of hexamers?

Dr Campbell responded: No, we haven't tried that. It's a good idea (and it might be cheaper!).

Professor Lu commented: Fig. 3 of your paper shows that the Ala_HCV displays ~50% higher mean fluorescence intensities than Lys_HCV, but Table 2 showed Ala_HCV has a ten-times worse detection limit (1 pM *vs.* 0.1 pM). While this opposite behavior may be a typical difference between sensitivity and detection limit, I am wondering when to use sensitivity to measure the performance of a method and when to use detection limit as a measure.

Dr Campbell replied: The truth is that, in my opinion, they are both important metrics. To some extent it comes down to your application – would you like to detect a very low concentration (in this case limit of detection might be important) or would you like to discriminate between different concentrations (in which case sensitivity may be more important)?

Professor Guiseppi-Elie asked: The epoxide has likewise been shown to be the most effective in our 2002 work on printed microarrays. We believed then that its effectiveness was due to the higher contrast it allowed because of better blocking of non-specific adsorption. Do you have any further insight into why the epoxide was superior to all those other surfaces evaluated?

Dr Campbell replied: Epoxide seems to have several advantages. When carrying out solution-based silanisation it seems to give more uniform coatings. In most cases spot morphology is better than (for example) amino silane and it is simple to use (doesn't need activation).

Professor Cass asked: In a complex genomic sample, the DNA could behave like an aptamer library and so have sequences that recognise the peptide tag rather than the tethered probe. Do you have any evidence for binding of DNA just to the tags?

Dr Campbell answered: That's an interesting thought. No, we have no evidence for DNA binding to His tags, but we haven't looked for it either.

Professor Graham remarked: Have you tried using pre-organised oligonucleotides in your arrays? They might help with the hybridisation rates.

Dr Campbell replied: I think that Chad Mirkin has published on this recently.[1] It seems like a good strategy, but not something that we've tried.

1 A. E. Prigodich, O.-S. Lee, W. L. Daniel, D. S. Seferos, G. C. Schatz and C. A. Mirkin, J. Am. Chem. Soc., 2010, **132**, 10638–10641.

Professor Cass said: There could be a common sequence for probes.

Professor Cass opened the discussion of the paper by Professer Motte: There have been some recent concerns over the therapeutic use of bisphosphonates; as your imaging reagents are based upon bisphosphonates do you think these concerns could translate to imaging use?

Professor Motte answered: Concerns about the therapeutic uses of bisphosphonates (HMBP) are essentially about cases of jaw osteonecrosis (also known as fossy jaw). This side effect could affect patients under radiation or cancer chemotherapy treated by nitrogen-containing HMBP. Nevertheless, it is relatively rare and is observed with free HMBP, which is used orally or intravenously in significant quantities (general medication for oral HMBP is several hundred milligrams). In our case, HMBP are used as iron chelates in quantities far less significant. Moreover, HMBP chelates (such as technetium 99m) have been routinely used as bone scintigraphy agents for many years.

Dr Ko Ferrigno asked: You mention the MIAplex option for multiplexing. How much multiplexing can be achieved?

Professor Motte responded: So far we have proven the concept with two different populations of magnetic nanoparticles. We intend to push it to four to five parameters tested at the same time by implementing a sophisticated excitation magnetic field and signal processing on one side and by using "orthogonal" magnetic nanoparticles on the other side. We expect that this technique would definitely not allow more than ten parameters.

Dr Davies remarked: In the lateral flow assay, your magnetic particles form a visible brown line. Which is the most sensitive: the magnetic measurement or a visual inspection under good lighting? By sensitive I mean which appears first, the visual or the magnetic signal, when applying increasing levels of particles to your strip with biotin sprayed at the test line?

I am confused by the sprayed biotin. Biotin binds quite deeply in the binding pocket of streptavidin (the crystal structure has been solved for this complex many times) and so I do not see how it can hold the streptavidin particles to the test strip. Did you use a biotin–protein conjugate to hold the biotin to the test line?

Professor Motte responded: The magnetic measurement is more sensitive (down to 1 ng of magnetic material) than the visual inspection. For example the deposit of 0.5 μg of magnetic nanoparticles can not be detected by visual inspection.

You are exactly right, the biotin is implemented as a BSA conjugate for staying on the strip. This has been specified in the text since the conference.

Professor Tsien asked: Did you inject systemically or into the tumor?

Professor Motte answered: To investigate the *in vivo* therapeutic efficacy of γFe_2O_3@alendronate nanoparticles, we randomly sorted mice xenografted with MDA-MB231 cells into four groups. All mice developed tumours within one week after MDA-MB-231 inoculation and the treatments were initiated at the end of the first week when tumors reached 30 mm^3. PBS solutions were injected for the control group. The three other groups were treated either with alendronate molecules or with γFe_2O_3@alendronate nanoparticles in presence or absence of a small magnet applied on the tumor for 30 min. Intracutaneous injection was not performed systemically. It was done subcutaneously at the periphery of the tumor but not into the tumor.

Professor Tsien asked: Have you tried MRI to see where the particles are going?

Professor Motte replied: These preliminary *in vivo* experiments were performed to investigate the therapeutic efficacy of γFe_2O_3@alendronate nanoparticles as antitumoral agents. Since we demonstrated the antitumoral activity of such nanosystems, a full biological study will be performed including *in vivo* MRI in order to localize the therapeutic nanoagents.

Professor Cass asked: Philips have had an extensive programme in magnetic sensing of super-paramagnetic labels using giant magnetoresistive sensors. How does you approach compare?

Professor Motte answered: Philips have tried to implement giant magnetoresistive sensors which are very sensitive as they are able to detect as little as one magnetic bead. The constraint is that such a magnetic bead must be very close to the surface of the sensor, the typical distance being one diameter of the magnetic bead. This makes it mandatory to embed the GMR inside the disposable. This generates engineering difficulties and eventually a costly solution. Philips finally abandoned the GMR sensor option. By tracking only the non-linear response of superparamagnetic particles, the Magnisense technique overcomes the background signal. It does not require the transducer (straightforward copper coils) to be as close to the magnetic beads. Once the superparamagnetic particles are inside the coil they provide an exploitable signal. Eventually, the Magnisense technique needs less engineering efforts for its implementation and allows accommodation of a variety of disposable formats.

Professor Lu commented: Since MRI of supermagnetic iron oxide nanoparticles is quite sensitive to the aggregation state of the nanoparticles, I am wondering if you have tested the stability of your multimodal supermagnetic nanoplatform *in vivo* in the absence of its target to rule out non-specific aggregation as an issue.

Professor Motte responded: We tested the stabilty of the nanoparticles *versus* pH, ionic strength and within different buffers as Hepes and PBS. The preliminary *in vivo* experiments indicate no toxicity induced by the nanoparticles since the body weight is independent of the treatment.

Professor Lu asked: Since it is difficult to test the stability of your system *in vivo*, I am wondering if you have tested its stability with respect to aggregation in serum in a test tube.

Professor Motte replied: As explained just before, we have tested the nanoparticles aggregation in test tubes *versus* pH and ionic strength by measuring the hydrodynamic diameter. The system is stable between pH 5 and 12. Precipitation appears at 0.01 mol L^{-1} NaCl of ionic concentration.

Dr Dhaliwal commented: No thorough *in vivo* tests have yet been performed. Biophosphates in lipoplexes ablate circulatory phagocytic cells (clodronate liposomes, www.clodronateliposomes.org); see a paper by Nico van Roojian *et al.* If your agent also ablates peripheral blood monocytes and tissue macrophages, it may not be a useful therapeutic. Therefore a quick assay that needs to be done is to find if there are any toxicity issues on primary monocytes and macrophages from humans and mice? As currently you have tested HUVECs/cancer cell lines.

Professor Motte answered: Thanks for your excellent comment. Clorodronate is well known to selectively deplete for macrophages. Concerning alendronate, this compound (sold as Fosamax) is clinically used for osteoporosis and several other bone diseases. In a recent study,[1] the authors show that liposomal alendronate (LA) treatments resulted in a dose-response modified biodistribution following both intravenous (IV) and intraperitoneal (IP) administrations. The biodistribution of high-dose LA (10 mg kg^{-1}), but not that of the low-dose (1 mg kg^{-1}), was similar in healthy and diseased animals. It is concluded that LA impedes its own elimination from the circulation by depleting circulating monocytes and/or inhibiting their endocytic activity in a dose-dependent manner. Both IV and IP administration of LA mediated by the partial and transient depletion of circulating monocytes effected inhibition of restenosis. Inhibition of EM was effected only by IP administration, which depleted both intraperitoneal and circulating monocytes. Thus, EM should be considered as a local inflammatory condition with systemic manifestations as opposed to restenosis, a systemic inflammatory disease. In the case of iron oxide nanoparticles coated with alendronate, we have tested the nanoplatform *in vitro* on HUVECs and MDAMB cancer cell lines. No change in cell proliferation was observed on HUVECS whereas the IC50 for MDAMB decreases from 120 to 32 micromolar comparing the free molecules to nanoplatform.

The preliminary *in vivo* experiments on xenografted nude mice show no toxicity induced by the nanoparticles since the body weight is independent of the treatment.

1 E. Haber, E. Afergana, H. Epsteina, D. Gutmana, N. Koroukhova, M. Ben-Davida, M. Schachterb and G. Golomb, *J. Controlled Release*, 2010, **148**, 226, DOI: 10.1016/j.jconrel.2010.08.030.

Professor Cass opened a general discussion of the papers by Professor Allbritton, Dr Campbell and Professor Motte: Many of the methods and devices we are discussing require us to graft some chemistry on to the underlying biology. The question is then how much of the basic biology needs to be retained or modified and where is the optimum place for the chemistry and biology to meet?

Professor Cooks responded: A follow-up comment is that we often combine spectroscopy, chemistry and biology. Mass spectrometry is often classed as a type of spectroscopy because of the method of data representation. Of course it is no such thing: it is purely a chemical (reaction) experiment and as such the intrinsic chemistry can be endlessly adapted and combined with biology as needed. A (sorry, provincial) example is reactive DESI in which the chemistry is easily recognizable because of the use of aqueous solutions as the reaction phase. However, pure unimolecular gas phase chemistry as in electron ionization mass spectrometry is still pure chemistry, not spectroscopy.

Professor Cass said: In many of our diagnostic devices there are a series of interfaces and it is engineering of these that often determines the device performance.

New ionization methods and miniature mass spectrometers for biomedicine: DESI imaging for cancer diagnostics and paper spray ionization for therapeutic drug monitoring

R. Graham Cooks,*[a] Nicholas E. Manicke,[a] Allison L. Dill,[a] Demian R. Ifa,[a] Livia S. Eberlin,[a] Anthony B. Costa,[a] He Wang,[b] Guangming Huang[a] and Zheng Ouyang[b]

Received 11th May 2010, Accepted 26th July 2010
DOI: 10.1039/c005327a

The state-of-the-art in two new ambient ionization methods for mass spectrometry, desorption electrospray ionization (DESI) and paper spray (PS), is described and their utility is illustrated with new studies on tissue imaging and biofluid analysis. DESI is an ambient ionization method that can be performed on untreated histological sections of biological tissue in the open lab environment to image lipids, fatty acids, hormones and other compounds. Paper spray is performed in the open lab too; it involves electrospraying dry blood spots or biofluid deposits from a porous medium. PS is characterized by extreme simplicity and speed: a spot of whole blood or other biofluid is analyzed directly from paper, simply by applying a high voltage to the moist paper.
Both methods are being developed for use in diagnostics as a means to inform therapy. DESI imaging is applied to create molecular maps of tissue sections without prior labeling or other sample preparation. Like other methods of mass spectrometry imaging (MSI), it combines the chemical speciation of multiple analytes with information on spatial distributions. DESI imaging provides valuable information which correlates with the disease state of tissue as determined by standard histochemical methods. Positive-ion data are presented which complement previously reported negative-ion data on paired human bladder cancerous and adjacent normal tissue sections from 20 patients. These data add to the evidence already in the literature demonstrating that differences in the distributions of particular lipids contain disease-diagnostic information. Multivariate statistical analysis using principal component analysis (PCA) is used to analyze the imaging MS data, and so confirm differences between the lipid profiles of diseased and healthy tissue types. As more such data is acquired, DESI imaging has the potential to be a diagnostic tool for future cancer detection *in situ*; this suggests a potential role in guiding therapy in parallel with standard histochemical and immunohistological methods.
The PS methodology is aimed at high-throughput clinical measurement of quantitative levels of particular therapeutic agents in blood and other biofluids. The experiment allows individual drugs to be quantified at therapeutic levels and data is presented showing quantitative drug analysis from mixtures of therapeutic

[a]*Department of Chemistry and Center for Analytical Instrumentation Development, Purdue University, West Lafayette, IN 47907, USA. E-mail: cooks@purdue.edu; Fax: (+765) 494-9421; Tel: (+765) 494-5262*
[b]*Weldon School of Biomedical Engineering and Center for Analytical Instrumentation Development, Purdue University, West Lafayette, IN 47907, USA*

drugs in whole blood. Data on cholesterol sulfate, a new possible prostate biomarker seen at elevated levels in diseased prostate tissue, but not in healthy prostate tissue in serum are reported using paper spray ionization.

1. General introduction

This study covers the implications for disease diagnostics of two new mass spectrometry experiments, both of which employ ambient ionization, *i.e.* ionization of samples in their native state in the open environment. One method, DESI, is used in the imaging mode to characterize the distribution of small molecules—especially fatty acids and phospholipids—in tissue sections. DESI imaging studies presented here focus on the detection of positively charged ions, a measurement that provides complementary information to that obtained from the more conventional negative-ion data.

The second method, paper spray ionization, is used to examine biofluids rapidly and quantitatively. Quantitation relies on the ability to add an internal standard to the sample, or more realistically to the solvent used in the experiment or the paper substrate itself. The capabilities of the methodology are tested in the case of mixtures of therapeutic drugs in blood.

Current types of mass spectrometers are not optimized for high-throughput clinical measurements of therapeutics and/or endogenous compounds in blood and other biofluids, so smaller purpose-built instrumentation is needed. The current capabilities of hand-held mass spectrometers are therefore briefly described in the last section of the paper which also deals with their adaptation to high-throughput clinical applications. These instruments, after further adaptation to improve ion transport through air from the sampling point to the (distant) mass spectrometer, also have the potential to allow real-time *in situ* analysis of tissue during surgery to provide diagnostic information that will guide the surgeon.

2. Desorption electrospray ionization imaging and disease diagnosis

2.1 Introduction to DESI imaging for disease diagnosis

Mass spectrometry imaging (MSI) is a well established technique for surface analysis which combines chemical speciation of multiple analytes with information on their spatial distribution. After the entire surface is scanned, specific compounds can be selected from the set of mass spectra collected and their spatial distribution visualized by the creation of chemical images.[1,2] Traditional methods for MSI, such as secondary-ion mass spectrometry (SIMS) and matrix-assisted laser desorption/ionization (MALDI), require a high vacuum and the addition of matrices to the sample. However, new ambient ionization technologies have been introduced in which sampling and ionization are performed outside of the mass spectrometer at atmospheric pressure without the requirement of application of a matrix or any other sample pre-treatment.[3–6]

In desorption electrospray ionization (DESI), a stream of charged droplets hits the surface creating a thin film that dissolves the analytes. Subsequent droplets splash into this film forming secondary microdroplets. The secondary microdroplets containing the analytes follow the conventional desolvation processes of electrospray ionization as they move into the mass spectrometer. This DESI mechanism known as "droplet pick-up"[3] is supported by simulations and a range of experimental evidence. DESI imaging capabilities have been demonstrated for different types of samples and surfaces.[7] For instance, the direct analysis of illicit drugs on latent fingerprints and the characterization of inks in documents of questionable provenance have been reported.[8,9] Analysis of lipids of many classes in rat brain was demonstrated after partial separation by thin-layer chromatography.[10] Recent applications to biological tissues include three-dimensional visualization of mouse brain structure by lipid analysis;[11] investigation of catecholamines in adrenal glands;[12] and analysis and imaging of rat spinal cord cross-sections.[13,14]

The application of DESI imaging for discrimination between healthy and diseased tissue using their distinct glycerophospholipid (GP) profiles has been demonstrated and the encouraging reported results indicate that DESI imaging has the potential to become a powerful tool in molecular pathology. DESI-MS is advantageous because it provides histopathological information without the requirement of sample preparation such as staining or labeling as normally used in histochemical protocols. Alterations in the composition of polar lipids, such as glycerophospholipids, are known to occur in certain malignant diseases.[14–16] There are many un-met needs in disease diagnosis and management including that for improved methods to accurately diagnose and stage a disease and to define tumor margins. It has been previously established that the lipid profile of tissues changes during disease and these changes provide information on the underlying biology of cancer that have the potential to be useful in disease diagnosis and management.[14,17–19] As medicine advances towards more personalized forms, lipids are likely to provide increasingly important information on the state of individual health.

DESI-MS imaging of biological tissue samples was initially applied systematically to studies of animal tissues, although the very first application was achieved in a human liver sample.[20] In this case, the GP profiles allowed discrimination between healthy and diseased tissues as well as the location of the border region.[20] While the potential for DESI-MS imaging to distinguish cancerous from normal tissues was clear, additional studies with a larger and more diverse cohort were needed to validate these capabilities. DESI-MS imaging has also been applied to canine bladder cancerous and adjacent normal tissues and successfully distinguished tumor from normal tissue on the basis of multiple marker lipids.[21] Following these initial studies in dog bladder transitional cell carcinoma, experiments were conducted using human bladder cancerous tissues. Tissue samples from 20 different patients encompassing tumor and adjacent normal tissue from each patient were imaged using DESI-MS.[22] Hand-selected ion images were created show-casing the differences between the tissue types as observed by the absolute and relative intensities of particular GP species recorded in the negative ion mode. These GP distributions function as molecular markers of disease and allow for a clear visual distinction between cancerous and normal tissues. Validation of DESI-MS diagnostic results was obtained by pathological examination of serial tissue sections stained with hematoxylin and eosin (H&E).

Multivariate statistical analyses were used to visualize and create a diagnosis rule for the tissue sections, using the methods of principal component analysis (PCA) and partial-least-squares discriminate analysis (PLS-DA). The statistical methods rely on the entire collection of mass spectral data acquired, not simply on selected ion images. The DESI-MS ion images correlate with the statistically generated images, which correlate in turn with the pathological diagnosis in over 80% of the cases in this one small study.[22] Human prostate cancer tissue specimens were also examined by DESI-MS imaging in the negative-ion mode. When examining prostate tissues from 34 different patients the GP profiles of the tissue sections did not provide a means by which to classify the tissue samples. Nevertheless, in this case a unique single molecule, cholesterol sulfate (CS), was found to be over expressed in precancerous lesions or cancerous tissue and undetected in normal tissue, providing a simpler approach to diagnosis.[23] Many biological roles for CS in prostate cancer progression have been hypothesized, including its ability to influence cell signaling, cellular differentiation and apoptosis.[24–26]

2.2 Methods

2.2.1 Biological sample preparation and cryosectioning. All tissue samples were handled in accordance with approved institutional review board (IRB) protocols at Indiana University School of Medicine. They were flash frozen in liquid nitrogen and stored at −80 °C until sliced into 15 μm thick sections and thaw mounted onto glass slides. The slides were stored at −80 °C; prior to analysis they were allowed to come to room temperature and then dried under nitrogen in a dessicator for approximately 20 min. Serial sections were formalin fixed and subsequently stained using H&E for pathological examination.

2.2.2 DESI-MS 2D imaging and statistical analysis. The DESI ion source used in our experiments was a lab-built prototype, configured as described previously.[27] Optimization involved obtaining a small and uniform spray spot on the sample surface. The spray solvent used for MS acquisition was methanol–water (50 : 50) with a 5 kV spray voltage applied. Acetonitrile was purchased from Sigma-Alrich (St. Louis, MO, USA) and water (18.2 MΩ cm) was from a PureLab ultra system by Elga LabWater (High Wycombe, UK). The nitrogen gas pressure was 150 psi and the solvent flow rate was 1.5 μL min^{-1}. In the imaging experiments, the tissues were scanned using a 2D moving stage in horizontal rows separated by a 250 μm vertical step until the entire sample surface had been assayed. The surface moving stage included an XYZ integrated linear stage (Newport, Richmond, CA) and a rotary stage (Parker Automation, Irwin, PA). All experiments were carried out using a LTQ linear-ion-trap mass spectrometer controlled by XCalibur 2.0 software (Thermo Fisher Scientific, San Jose, CA, USA). An in-house program allowed the conversion of the XCalibur 2.0 mass spectra files (.raw) into a format compatible with the Biomap software (freeware, htto://www.maldo-msi.org). Spatially accurate images were assembled using the BioMap software. The color scale is normalized to the most intense (100% relative intensity) peak in the mass spectra.

For statistical method development, these 20 tissue pairs were randomly divided in half with 10 samples in the training set, and with the other 10 samples in the validation set used to test the accuracy of the statistical method. For this statistical analysis, 10 representative, individual spectra were acquired from both known tumor and normal regions of the 10 samples in the training set, excluding background regions of glass slide. The validation set used the full mass spectral imaging data, with no manual manipulation, for each sample to construct synthetic images. Data were processed as described below in Matlab 2008a, The MathWorks (Natick, MA, USA).

2.3 Results and discussion

While the focus of previous work on human cancer tissue imaging using DESI has involved negative-ion mode data, mass spectrometric analysis can also be conducted in the positive-ion mode. Multiple marker lipids observed in the positive-ion mode were found to distinguish canine bladder cancerous tissue from adjacent normal tissue.[21] By imaging in the positive-ion mode important lipid species, such as glycerophosphocholines and sphingolipids are more readily detected by DESI-MS. By performing positive- and negative-mode analysis the total number of species that can be detected greatly increases, as does the confidence of the diagnosis by relying on multi-modal pattern recognition.

Paired human bladder cancer and adjacent normal tissue sections from 20 patients (40 tissues total) were imaged in the positive-ion mode using DESI-MS. By using paired tissue samples differences inherent within different patients are minimized. A series of DESI-MS ion images, showing the distribution of one particular *m*/*z* value across the sample, were hand selected to visualize the differences in relative abundances between cancerous and normal tissue. While a few ions were selected for convenience of visualization, the mass spectrum obtained for each pixel shows that the absolute intensity of a variety of ions changes between the cancerous and normal tissues. This showcases the need for multivariate statistical methods in order to base the diagnosis on all species present in the spectra. Fig. 1(a) and (b) shows

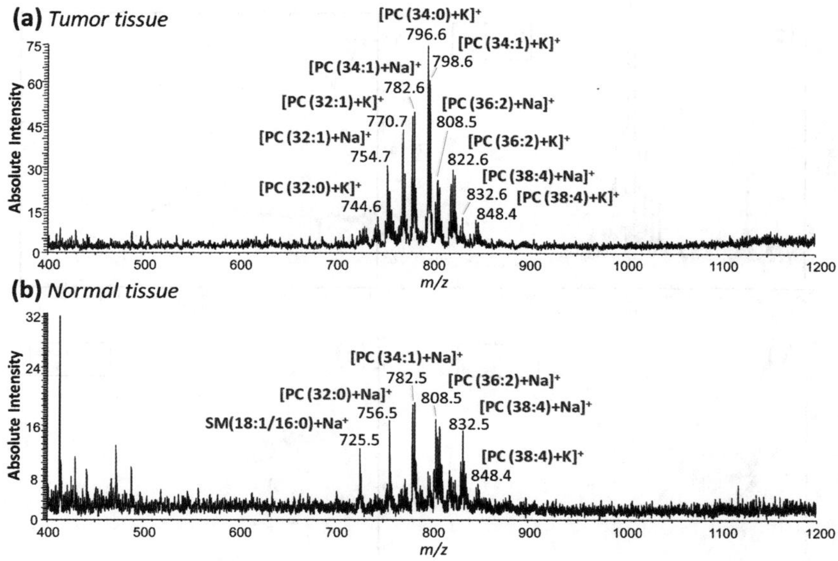

Fig. 1 Typical positive-ion mode mass spectra of human bladder (a) tumor tissue and (b) adjacent normal tissue in the *m/z* range 400–1200.

representative positive-ion mass spectra for cancerous and normal tissue, respectively. While it is not necessary to chemically identify each species contributing to the diagnostic pattern in the mass spectrum, the main GPs observed were identified based on collision induced dissociation (CID) tandem MS experiments and comparison of the generated product ion spectra with the literature.[28–30] The analyses reported here focus on glycerophosphocholines (PC) and their adducts, and while additional compounds are labeled in the spectra featured in Fig. 1, only a few selected ion images are shown. These images are compared to serial tissue sections which were stained with H&E and examined by a certified pathologist for diagnosis.

Principal component analysis (PCA) was employed as a visualization approach to the description of the disease state of the tissue. PCA is a simple unsupervised data dimensionality reduction method. In this case it is used to reduce a large library of mass spectra to small set of principal components, or linear, orthogonal combinations of DESI peaks. Spectra are acquired from known tumor and normal regions from patient samples and partitioned into a training and validation set. Each set then includes tumor and normal spectra from half the set of patients. Example average tumor and normal spectra belonging to a training set are shown in Fig. 2(a). These data originate from positive-ion mode bladder cancer section analysis, as previously described. After normalization to the median area under the curve (AUC), PCA is run on the whole training set. Typically, the first principal component [PC1, Fig. 2(b)] is found to separate tumor and normal regions of tissue, as shown in the PC1/PC2 score plot [Fig. 2(c)]. This vector is the weighted set of DESI peaks which indicate the existence of tumor. The validation set may then be projected onto the principal components to confirm the separation along the PC1 coordinate [Fig. 2(d)]. After the set of principal components is generated, full images belonging to the validation set are projected onto the training set of principal components. The projection, or loadings, of each pixel along the PC1 coordinate are plotted on an x,y cartesian coordinate system identical to that used when rastering in the DESI imaging procedure. The pixel loadings are color coded linearly from blue to red. Red pixels are then regions of tissue whose mass spectra are predicted to be tumor, based on the high contribution of the PC1 coordinate. The result is

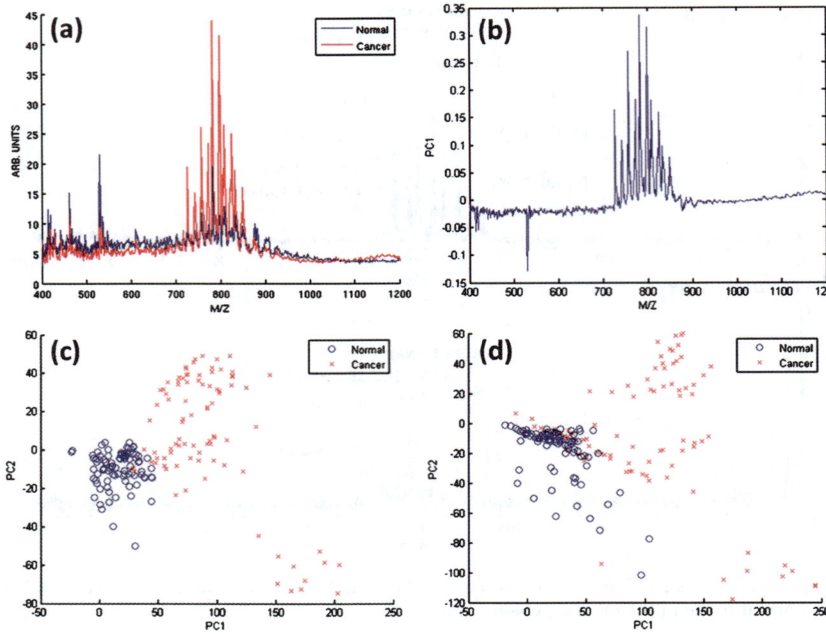

Fig. 2 Principal component analysis results. (a) Average tumor and normal spectra from the training set that were used as inputs for PCA. The spectra are resampled to unit resolution, background-corrected, and scaled to the median area under the curve for all spectra. (b) PC1 eigenvector. PC1/PC2 score plot for (c) training data set and for (d) validation data set projected onto the training set principal component eigenvectors.

a simple but powerful visual approach for the determination of likely regions of tumor on a surface in an unsupervised fashion. This method has now been applied to a variety of samples, including canine and human bladder, prostate and kidney in both negative- and positive-ion modes.

The DESI-MS ion images along with the optically scanned H&E stained tissue section and the synthetic PCA images for two bladder tissue pairs, UH00010-19 and UH0105-26, are shown in Fig. 3. The selected-ion images for both tissue samples show increased intensities in the tumor tissue for the ions at *m/z* 770.6 (PC(32 : 1) + K^+), *m/z* 782.6 (PC(34 : 1) + Na^+) and *m/z* 798.6 (PC(34 : 1) + K^+), Fig. 3(a)–(c) and (f)–(h). This correlates with the diagnosis obtained from the H&E stained tissue sections. The synthetic PCA image is shown in Fig. 3(e) for UH0010-19 and in Fig. 3(j) for UH0105-26. These PCA images also correlate with the DESI-MS ion images and the H&E stained tissue. This presentation method has advantages over hand-selected ion images and also increases the confidence of diagnosis by relying on more ion species. Interestingly these results are in line with what has been reported previously in canine bladder cancer.[21] Some of the same lipid species present at increased levels here in human bladder cancer were also found to be increased in canine bladder cancer, including PC(34 : 1) + Na^+, and PC(34 : 1) + K^+. This is not unexpected considering the similarities between canine and human bladder cancer.[31]

DESI-MS imaging is clearly a promising technique for obtaining pathological molecular information. It allows the analysis to be conducted in the ambient environment with little sample preparation and without the use of special stains or labels. DESI-MS can be used to discriminate cancerous from normal tissue in multiple human cancers in the negative-ion mode[22,23] as well as in the positive-ion mode, as

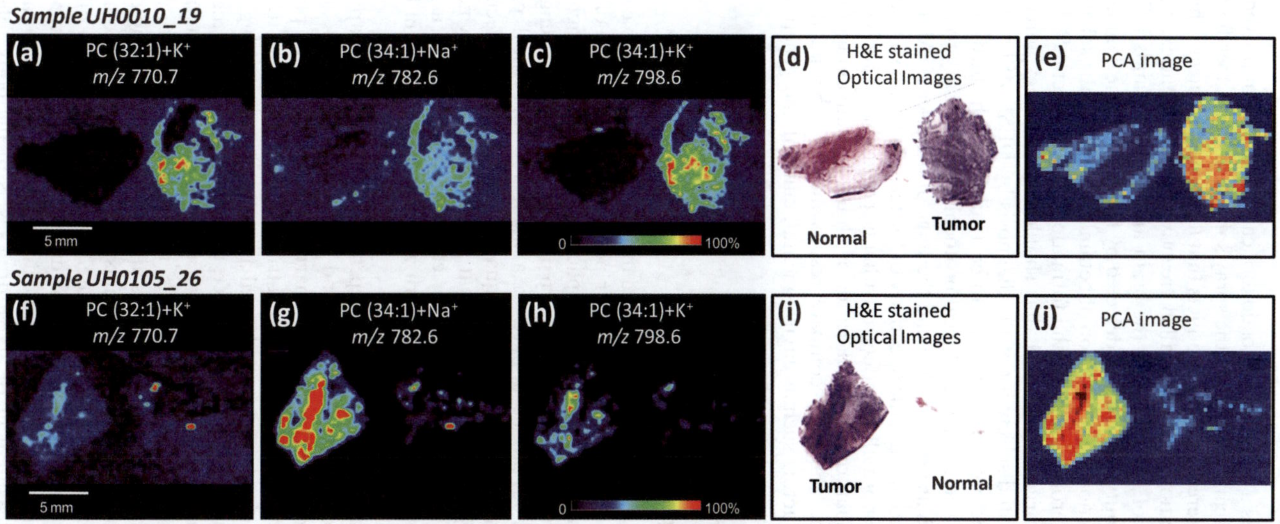

Fig. 3 Positive-ion mode tissue imaging of human bladder tissues samples including areas of cancer and adjacent normal tissue. Ion images obtained for sample UH0010_19 (tumor tissue on right and normal tissue on left) showing the distribution (a) PC(32 : 1) + K^+, *m/z* 770.7; (b) PC(34 : 1) + Na^+, *m/z* 782.6 and (c) PC(34 : 1) + K^+, *m/z* 798.6. Optical images of H&E stained adjacent sections and the PCA-based synthetic image obtained from the DESI-MS imaging data are shown in (d) and (e), respectively. Ion images obtained for sample UH0105_26 (normal tissue on right and tumor tissue on left) showing the distribution (f) PC(32 : 1) + K^+, *m/z* 770.7; (g) PC(34 : 1) + Na^+, *m/z* 782.6 and (h) PC(34 : 1) + K^+, *m/z* 798.6. Optical images of H&E stained adjacent sections and the PCA-based synthetic image obtained from the DESI-MS imaging data are shown in (i) and (j), respectively. Both samples are from the validation set, illustrating the performance of the method on future samples.

shown here. The strength of the diagnosis is demonstrated through the use of multiple marker lipids and multivariate statistical analysis.

2.4 Future of DESI imaging

The use of DESI-MSI as a disease diagnostic tool can be further improved. On the analytical side, reagents can be added to the spray solvent to cause specific reactions with the molecules of interest creating products that are easier to detect. This analytical variant on the ionization method is called "reactive-DESI".[32] One example is the detection and mapping of cholesterol in the brain and adrenal glands.[12,32] Cholesterol has low proton affinity and it is not detected by DESI using the common methanol/water solvent conditions. The addition of betaine aldehyde to the solvent makes the detection of cholesterol possible by the formation of a hemicetal.[12,32] Reactive DESI is a promising variation of DESI that should be further explored since it increases the sensitivity and selectivity of the diagnosis by DESI-MSI.

On the clinical side, before the technique can be available to the medical community for disease diagnosis and staging, clinical protocols are necessary to investigate important variables such as sex, race, age and the stage of the disease. All these variables can affect the sensitivity and the selectivity of the technique and should be evaluated. These studies are normally conducted with a large number of individuals and preferentially at more than one medical center or clinic. The advantages of multicenter trials include the ability to compare results among subpopulations with different demographic factors such as environmental, cultural and genetics factors.

3. Paper spray for bioanalysis

3.1 Introduction to paper spray capabilities

Paper spray ionization is an electrospray related method in which a sample (solution or solid) is first loaded onto paper or another porous substrate. A jet of charged droplets (*i.e.* electrospray) is formed *via* the application of solvent and high voltage, provided the tip of the substrate is cut to a sharp point to increase the strength of the electric field. The method can be applied to a number of analytical problems, although the focus thus far has been for the quantitation of small molecule drugs from blood dried on paper. We anticipate that this simple method, which could be implemented using a disposable cartridge, could be coupled with a low-cost mass spectrometer for routine clinical monitoring of therapeutic drugs in blood or other body fluids.

The successful administration of a drug depends on managing the appropriate dosing guidelines for achievement of a safe and effective outcome. This guideline is established during clinical trials where the pharmacokinetics (PK) and pharmacodynamics (PD) of the drug are studied. Clinical trials use PK-PD studies to establish a standard dose, which may be fixed or adjusted according formulae using variables like body mass, body surface area, *etc.* However, the drug exposure, *i.e.* the amount of drug circulating over time, is influenced by a number of factors that vary from patient to patient. For example, an individuals' metabolic rate, the type and level of plasma proteins, and pre-existing conditions all play a role in affecting the exposure of the drug *in vivo*.[33] Furthermore, administration of a drug in combination with other medications may also affect exposure. As a result, it is often difficult to predict and prescribe an optimum regimen of drug administration for particular patients. To address these concerns, therapeutic drug monitoring (TDM) can be employed. TDM is the measurement of active drug levels in the body followed by adjustment of drug dosing or schedules to increase efficacy and/or decrease toxicity. TDM is indicated when the variability in the pharmacokinetics of a drug is large relative to the therapeutic window, and where there is an established relationship between drug exposure and efficacy and/or toxicity.

Another requirement for TDM is that a sufficiently precise and accurate assay for the active ingredient must be available at a reasonable cost. Immunoassays and high-performance liquid chromatography (HPLC) with detection by ultraviolet–visible (UV–vis) spectroscopy or mass spectrometry (MS) are the most commonly used methods for TDM. HPLC-MS is the most flexible and powerful approach for TDM, but is labor intensive and time consuming. Due to the cost of the equipment and the training required to operate it, TDM by HPLC-MS is typically carried out in central laboratories that receive samples from numerous hospitals and specialize in such tests. Samples must therefore be shipped after collection to the central laboratory where they are prepared and analyzed. The drawbacks of this approach include the difficulty in controlling the conditions used in storage and shipment and the high costs and long turnaround times (two weeks) involved; these have been significant barriers towards widespread use of TDM as a tool in personalized medicine. A suitable device capable of TDM at the point of care would reduce the cost of clinical trials and the overall costs of drug development, and it would facilitate the practice of personalized medicine by dose optimization of drug regimen on an individual case-by-case basis. This would improve clinical outcomes and decrease drug toxicity for chemotherapeutic agents, immunosuppressive drugs, and others.

Paper spray (PS) allows mass analysis of complex mixtures without sample preparation or separation [Fig. 4(a)]. A piece of paper is cut into a triangular shape; ions are generated with the addition of a small amount (~10 μL) of solvent and application of a high voltage (3–5 kV) which generates a strong electric field at the tip of the paper, and a cone-jet of charged droplets is produced at the sharp tip [Fig. 4(b)]. The sample can be contained in the solvent or pre-deposited on the paper, for example as a dried blood spot (DBS). Observation of the spray plume together with inspection of paper spray spectra suggest that ion formation in paper spray is analogous to ESI, with some additional contributions from other mechanisms in some circumstances.

Paper spray is advantageous for the analysis of small molecules from blood for a number of reasons. First, paper is an important medium for biological fluid collection and storage. Drugs in DBSs are more stable compared with drugs in the liquid phase because enzyme enhanced hydrolysis is one of the most common reasons for drug degradation.[34,35] Also, the storage and shipment of DBS is more convenient since there is no need of a freezer or dry ice. Finally, small sample volumes are required for analysis (less than 50 μL compared to 500 μL or more for typical analyses). These numerous advantages to using paper for blood collection and collection has spurred interest in the pharmaceutical industry, with one major company using dried blood spot collection for all of its PK-PD studies in new small molecule oral drugs.[36] Paper spray uses the paper as the ionization source, whereas in the normal method, drugs are extracted from the DBS using liquid extraction and analyzed by HPLC-MS. The use of paper spray instead means that biological samples can be analyzed immediately after sampling or after storage of the dried spot. Paper spray can be used to directly ionize analytes in complex samples, such as blood, urine and other biological fluids. Moreover, paper spray has the capability to allow sample

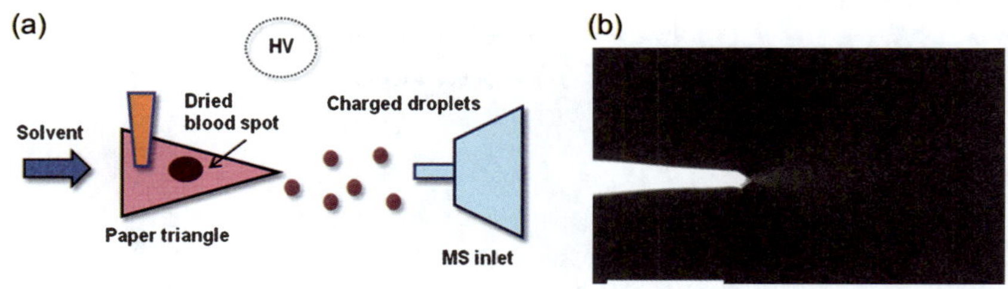

Fig. 4 (a) Schematic of paper spray ionization. (b) Photograph of a paper tip, showing the spray plume generated after application of solvent and high voltage. Adapted from the original figure by Wang *et al.*(2010) reproduced with permission from Wiley-VCH, copyright 2010.

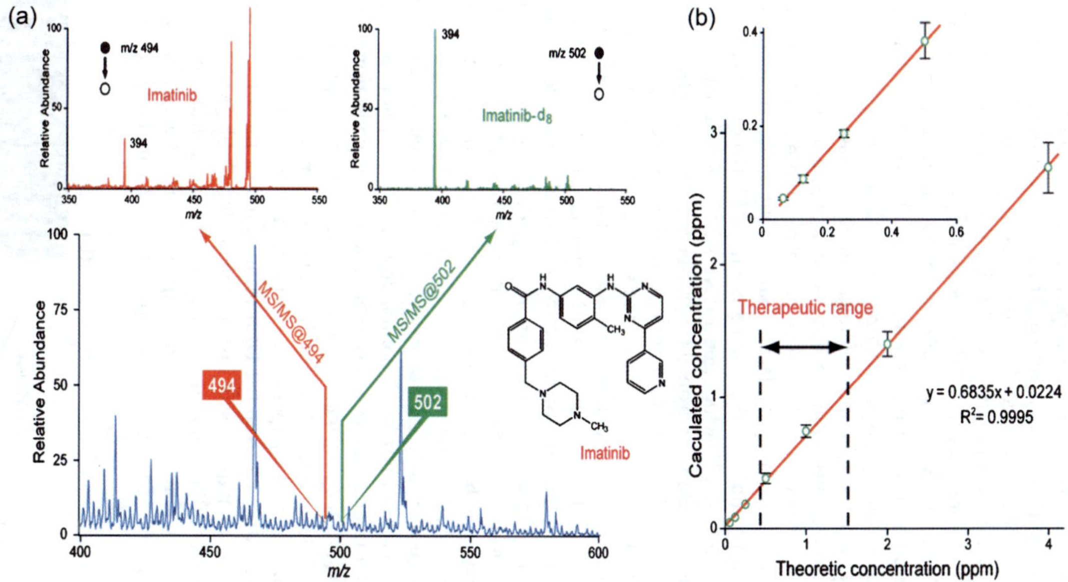

Fig. 5 (a) Full mass spectrum (bottom) and tandem mass spectra of 500 ng mL^{-1} imatinib and imatinib d_8 in blood. (b) Calibration curve for imatinib in blood with the therapeutic range set by dashed lines. Adapted from the original figure by Wang *et al.*(2010) reproduced with permission from Wiley-VCH, copyright 2010.

analysis to be performed without pneumatic assistance, which makes this method more convenient and applicable for the purpose of portable or point-of-care analysis, such as the coupling with a miniature MS. The solvent flow is driven by capillary action, eliminating the need for a syringe pump. Another advantage of paper spray is that clogging, a serious problem for conventional nanoelectrospray ion sources, is unlikely due to its multi-porous characteristics.

In our studies, drugs in various biological samples, including blood, urine and tissues, were examined by PS-MS.[37,38] After collecting the blood sample and spotting it onto paper, solvent is applied onto the paper, and an electrically charged solvent spray is generated upon application of high voltage. Fig. 5 shows the results for the analysis of dried blood containing imatinib, a therapeutic drug for leukemia, using PS-MS. The peak corresponding to the drug cannot be observed above the baseline noise in the full mass spectrum. However, upon isolating a narrow mass window and performing CID of the precursor ion, the diagnostic fragment ion at *m*/*z* 394 can be observed [Fig. 5(a)]. By using a deuterated analog as an internal standard (IS), the response of imatinib is linear from 50 ng mL^{-1} to at least 5 μg mL^{-1}, covering the entire therapeutic window of the drug [Fig. 5(b)]. In this experiment, the IS was spiked into the blood and thoroughly mixed. More realistically, the IS must be incorporated in some other way because it is not feasible to add it into the liquid blood at the point of collection. We found that acceptable precision was obtained by either pre-treating the paper with the IS or adding the IS to the dried blood spot after the blood had dried,[39] with a relative standard deviation of less than 8% for replicate measurements.

In this section, we will describe some recent progress in paper spray ionization. Detection limits for a small set of representative drugs were determined to evaluate the general applicability of paper spray for TDM. Also, new capabilities of paper spray are reported: the ability to measure therapeutic drugs and endogenous compounds from tissue and the use of electron emission from the paper substrate to analyze compounds with high electron affinity.

3.2 Methods

3.2.1 Application of paper spray to drug monitoring. All experiments were performed on a TSQ Quantum Access Max (Thermo Scientific, San Jose, CA) in the selected reaction monitoring (SRM) mode. The SRM parameters were as follows: Propranolol: *m*/*z* 260 → 183; tube lens: 108 V; Q2 offset (collision energy): 27 V/atenolol: *m*/*z* 267 → 145; tube lens: 93 V; Q2 offset: 18 V/lidocaine: *m*/*z* 235 → 86; tube lens: 94 V; Q2 offset: 18 V/amitriptyline: *m*/*z* 278 → 233; tube lens: 113 V; Q2 offset: 17 v/verapamil: *m*/*z* 455 → 165; tube lens: 150 V; Q2 offset: 27 V/imatinib: *m*/*z* 494 → 294; tube lens: 148 V; Q2 offset: 27 V.

Drug standards were a gift from AstraZeneca. Blood samples were prepared by diluting 100× stock solutions of the drug standards (Innovative Research, Novi, MI) for final concentrations of 0 ng, 250 ng, 500 ng, and 1000 ng per mL of blood. The IS used at a constant concentration of 500 ng mL^{-1}. For propranolol, amitriptyline, and lidocaine, atenolol was used as the internal standard. For verapamil, IS was imatinib. For atenolol, the IS was propranolol. Blood was spotted as 0.5 μL aliquots onto Whatman Grade 1 Chromatography paper (VWR). For each drug and concentration level, 7 replicate experiments were performed (separate blood spots, same blood sample).

For paper spray ionization, the paper was cut into a triangle of approximately 10 mm base width × 10 mm height. A pipette was used to add 10 μL of 90% methanol: 10% water (v:v) was added to the and the spray voltage was then set to 4000 V. Data was collected until the spray solvent was depleted by the spray process and evaporation (about 30 s).

For data analysis, the area under the curve (AUC) for the analyte was normalized against the AUC for the IS. This ratio was corrected by subtraction of the average blank signal, and the data was fit to a line by linear regression. The limit of detection was calculated using the formula $3s_B/m$, where s_B is the standard deviation of the blank ($N = 7$) and m is the analytical sensitivity.

3.2.2 Direct tissue analysis by paper spray. For endogenous lipid profiling, porcine adrenal gland was purchased from Pel-Freez (Rogers, AR). A metal needle was used to remove 1 mm^3 of tissue from the bulk tissue and place them onto the surface of paper triangles for PS-MS experiments. Methanol–water (1 : 1 v:v; 10 μL) was added to the paper as solvent and then 4.5 kV positive DC voltage was applied to produce the spray.

For TDM from tissue, mouse liver tissue homogenate was prepared using a wheaten tissue grinder (VWR). Atenolol was spiked into tissue homogenate, and 0.5 μL of the tissue homogenate was applied on the paper surface to form a dried spot. Ten μL of methanol–water (1 : 1 v:v) was added to the surface of the paper as solvent and then 4.5 kV positive DC voltage was applied to produce the spray.

3.2.3 Electron attachment using paper spray. Paper spray was operated in the negative-ion mode by applying −3.5 kV and 1 : 1 (v:v) methanol–water. C_{60} was introduced to the mass spectrometer by placing 50 ng of C_{60} powder 5 mm away from the paper spray tip and also 5 mm away from the mass spectrometer inlet. A fast heating film was used to increase the temperature of the C_{60} from room temperature to above 350 °C in about 4 s.

3.3 Results and discussion

3.3.1 Application of paper spray to drug monitoring in blood. A method for monitoring therapeutic drugs must fulfill several requirements. The drug must be detectable over the required concentration range, which varies depending on the drug, dose, method of administration, *etc.* It must be specific so that the drug can be distinguished from interfering species, including other drugs and drug metabolites. The measurement must meet some acceptable level of accuracy and precision, which could be set at a constant level for all drugs, although a better approach would be to tailor the acceptable amount of inaccuracy and imprecision to the therapeutic window; 20% of the therapeutic window has been previously suggested.[40] In this section, we consider the limits of detection obtained for a set of small molecule drugs (Scheme 1). This set of drugs, consisting of atenolol, propranolol, verapamil, amitriptyline, and lidocaine, have similar pK_as but differ widely in their other

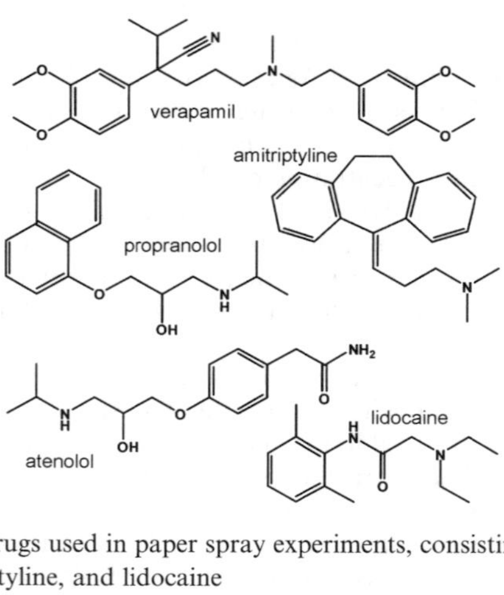

Scheme 1 The set of drugs used in paper spray experiments, consisting of atenolol, propranolol, verapamil, amitriptyline, and lidocaine

Table 1 Some molecular properties of drugs analyzed using PS-MS

Drug	MW/g mol^{-1}	CLogP[a]	tPSA/ Å^2	CMR/cm^3 mol^{-1}	H-Bond donors[b]	H-bond acceptors[c]
lidocaine	234.5	2.41	32.34	73.9	1	3
propranolol	259.34	3.35	41.49	78.3	2	3
verapamil	454.6	5.69	63.95	135.1	0	6
amitriptyline	277.4	4.63	3.24	93.4	0	1
atenolol	266.34	0.22	84.58	74.6	3	5

[a] CLopP, Topological Polar Surface Area (tPSA), and the calculate molar refractivity (CMR) were calculated in ChemDraw Ultra, which uses Crippen's fragmentation and Viswanadhan's fragmentation methods to calculate CLogP and the tPSA and Joback's fragmentation method to calculate molar refractivity. [b] The sum of all NH and OH. [c] The sum of all N and O.

molecular properties, making them a fairly representative set of small molecule drugs (Table 1). The limit of detection was calculated from the average signal intensity and standard deviation of blank blood samples and the analytical sensitivity from a three point calibration curve. A typical result is shown in Fig. 6, displaying the extracted ion chromatograms for the analyte (amitriptyline, in this case) and the internal standard (atenolol) for one concentration level [Fig. 6(a)]and the calibration curve obtained for this drug [Fig. 6(b)].

The results, summarized in Table 2, show that the detection limits for these drugs vary from 3 ng mL^{-1} for verapamil to 106 ng mL^{-1} for propranolol, but are generally in the low ng mL^{-1} range. These limits of detection lie below the therapeutic range for all of the drugs studied except for propranolol (Table 2). The approximately 20-fold difference in the limit of detection measured for the samples in this set stems from differences in both the analytical sensitivity (*i.e.* the slope of the calibration

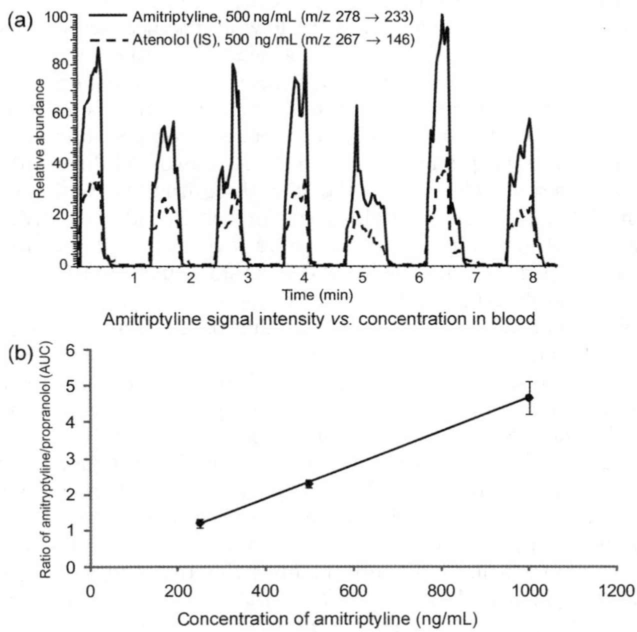

Fig. 6 (a) Total ion chromatograms for replicate measurements of 500 ng mL^{-1} amitriptyline and 500 ng mL^{-1} of the internal standard atenolol in blood. (b) AUC of amitriptyline, normalized against the AUC of the internal standard, *vs.* amitriptyline concentration in blood.

Table 2 Sensitivity, average and standard deviation of the blank signal, and limit of detection for several drugs analyzed using PS-MS

Drug	Internal standard	Sensitivity (m)/ mL ng^{-1}× 10^{-3})	$x_B \pm s_B$[b]	LOD/ ng mL^{-1}	Therapeutic range[a]/ng mL^{-1}
propranolol	atenolol	3.4	0.4 ± 0.2	106	20–300
atenolol	propranolol	1.5	0.06 ± 0.02	34	100–1000
verapamil	imatinib	6.3	0.031 ± 0.005	3	50–250
lidocaine	atenolol	18	0.06 ± 0.02	4	1000–6000
amitriptyline	atenolol	4.6	0.013 ± 0.009	6	50–200

[a] Therapeutic range given is in serum or plasma, which may differ from the concentration in whole blood. [b] x_B = average signal of the blank, s_B = standard deviation of the blank.

curve) and the standard deviation of the blank, which in turn is a function of the magnitude of the blank signal intensity because the standard deviation generally increases as the magnitude of the blank signal increases. The differences in the magnitude of the blank signal stem from the fact that blood is complex chemical mixture and, although some components are retained on the paper, a number of background chemicals from the blood or the paper itself are ionized and may overlap with the analytes. While tandem mass spectrometry further decreases interference in the spectrum, chemical noise is still present. In the case of propranolol, for example, an overlapping peak exists in the blank blood at the m/z 260 → 183 transition, which increases the magnitude, and therefore the standard deviation, of the blank.

Differences in the analytical sensitivity for each drug also contribute to the range obtained for the limits of detection. The sensitivity is affected by a variety of factors: the proton affinity of the analyte, its solubility in the spray solvent and its surface activity, and interactions with the porous substrate. Among this set of drugs, for example, atenolol and propranolol have the lowest analytical sensitivities at 0.0014 and 0.0034 mL ng^{-1}, respectively. This may be because the site of protonation in atenolol and propranolol is a secondary amine, whereas the protonation site in the other three drugs is a tertiary amine. In the gas phase, tertiary amines are stronger bases than secondary amines due the presence of more electron donating groups around the charged site. Also, propranolol and atenolol have two and three hydrogen-bond donating groups, respectively, whereas the other three drugs have either none (verapamil and amitriptyline) or just one (lidocaine). An increased ability to hydrogen bond could increase a drug's affinity for the paper substrate, and therefore decrease recovery. The set of drugs explored here is both too small and too structurally complex to precisely extrapolate general rules for predicting sensitivity.

In general, it seems reasonable to conclude that paper spray has adequate detection limits to detect many, but not all drugs at their therapeutic levels. High-potency drugs, including hormones, cytotoxic drugs, and some immunosuppressives will be difficult to detect at relevant levels. Other drugs will pose a challenge because of poor ionization efficiency. Since the detection of compounds by mass spectrometry requires ionization, drugs that do not have an easily ionized functional group or do not form adducts with counter ions easily will not be detected as sensitively. Efforts have been made previously to improve the detection of poorly ionizable compounds using on-paper derivatization, such as the use of betaine aldehyde to add charge to cholesterol.[32,41] Finally, highly polar drugs may pose a challenge because they have increased affinity for the paper and/or reduced surface activity, which is known to decrease sensitivity in electrospray ionization.

3.3.2 Direct tissue analysis by PS-MS. Rapid analysis of tissue by mass spectrometry is of interest for the diagnostic sciences. While this analysis can certainly be performed using the typical MS workflow of extraction followed by chromatography, paper spray is potentially a simpler alternative for some applications. In this experiment, a portion of tissue obtained from a surgical biopsy or from a needle aspiration biopsy is applied to the paper substrate, and chemicals from the tissue are extracted from the tissue and ionized by the application of solvent and high voltage. This method, as with the TDM of dried blood spots already presented, has the advantage of being fast and simple because of the lack of sample preparation. It has the disadvantage, naturally, of not supplying as much chemical information as can be obtained from tissue homogenization, extraction, and chromatography. This method is therefore suited for targeted analyses of a relatively small number of analytes, such as therapeutic drugs or known markers of disease state.

The pharmacological effects of a drug occur at the target tissue, not in the blood. The drug concentration in blood or plasma, which is more often measured because it is less invasive than performing a tissue biopsy, is assumed to correlate with the concentration of the drug at the active site. Variability in the drug uptake by the target tissue from patient to patient can lead to poor correlation between the plasma concentration and tissue concentration. As a result, the plasma drug concentration may correlate well with toxicity, but not with efficacy.[42] One study, for example, found that the extent of tumor perfusion had a strong effect on drug uptake by the tumor, but found no correlation between drug concentration in the plasma and in the tumor.[43] To demonstrate the feasibility of performing TDM from a needle aspiration biopsy, the drug atenolol was spiked into mouse liver homogenate, which was then applied to the paper and allowed to dry. The full mass spectrum and tandem mass spectrum shown in Fig. 7(a) demonstrate that the drug can be detected at less than 500 ng mL^{-1} concentration.

Endogenous compounds can also be detected from tissue using PS-MS. Fig. 7(b) shows the mass spectrum obtained after analyzing a section of porcine adrenal tissue in the positive-ion mode. A high number of glycerophosphocholine lipids are detected in the mass range shown, primarily as sodium and potassium adducts, although protonated species are also observed. The presumptive molecular identity of the peaks are indicated on the spectrum, but, due to the enormous complexity of the spectra obtained from such analyses, it is likely that many of the peaks actually represent two or more isobaric species. It is possible that PS-MS could be useful as a profiling tool to detect changes in the polar lipid composition at the tumor border. Another study, for example, detected different glycerophosphocholine lipid species in normal *versus* tumor prostate tissue.[44] Further study will be necessary to establish if paper spray can used to address this problem.

3.3.3 Electron attachment using PS-MS. All of the experiments presented thus far were performed in the positive-ion mode, with a spray voltage of 3.5 to 4.5 kV with methanol–water in some proportion as the solvent. When PS-MS is conducted under similar conditions in the negative-ion mode, for example −3.5 kV spray voltage with 1 : 1 methanol–water, the spray current is significantly higher (10 s of microamps *versus* less than 1 microamp in the positive mode). We have attributed this to the emission of a large number of electrons from the paper tip, concomitantly with the spray droplet formation. The electron emission might be a result of field emission or droplets carrying electrons during spraying. The presence of electrons was confirmed by the experiment shown in Fig. 8, in which 9-10-anthraquinone and Cl-9-10-anthraquinone vapors were introduced between the paper spray source and the mass spectrometer. Peaks identified as the $[M^{\bullet}]^-$ ions of both species were observed. This suggests that electrons are passing from the paper to the ground mass spectrometer, resulting in electron attachment to the neutral molecules to form negative ions.

C_{60} is representative of another group of compounds with high electron affinity. C_{60} vapor was introduced between the paper spray ionization source and the mass spectrometer by using a fast heating film to vaporize the fullerene. Ions representing

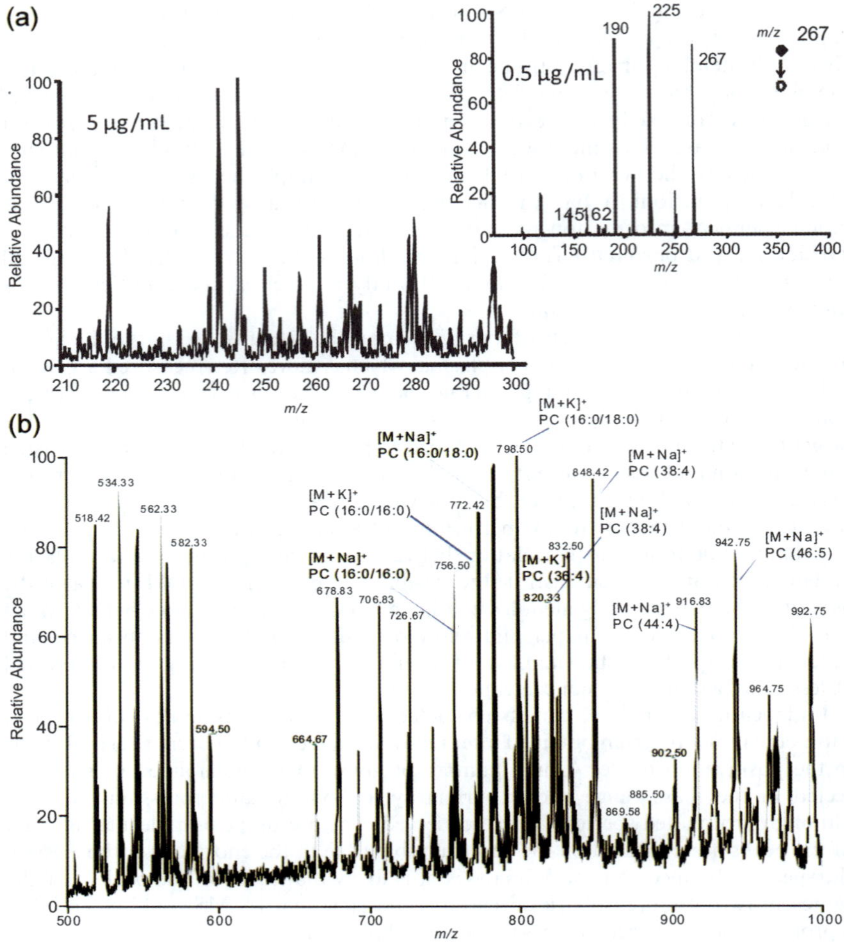

Fig. 7 (a) Analysis of atenolol spiked into mouse liver homogenate by PS-MS. (b) Phospholipids from porcine adrenal gland analyzed by PS-MS. Phospholipids such as phosphatidylcholine (PC) are identified in the spectra.

the radical anion of C_{60} were observed at m/z ~720, and the isotopic ratio was consistent with C_{60}, confirming the molecular identity of the peak [Fig. 8(b)]. Not only was the monomer observed, but also the electron-bound dimer whose isotopic peaks supported the identification as C_{120} [Fig. 8(c)]. This experiment further proved the presence of electrons during paper spray negative mode, and provides the possibility of dimer creation due to electron attachment.

In conclusion, paper spray in the negative mode can provide electrons under ambient conditions, which can be attached to either neutral molecules or charged species. This phenomenon could be useful in diagnostics for ionizing volatile chemicals expelled in the breath, for example.

3.4 Future of paper spray ionization

Paper spray ionization has shown early potential for the quantitation of small molecule pharmaceuticals. It has also shown potential in detecting endogenous compounds from serum samples, specifically detecting cholesterol sulfate, which

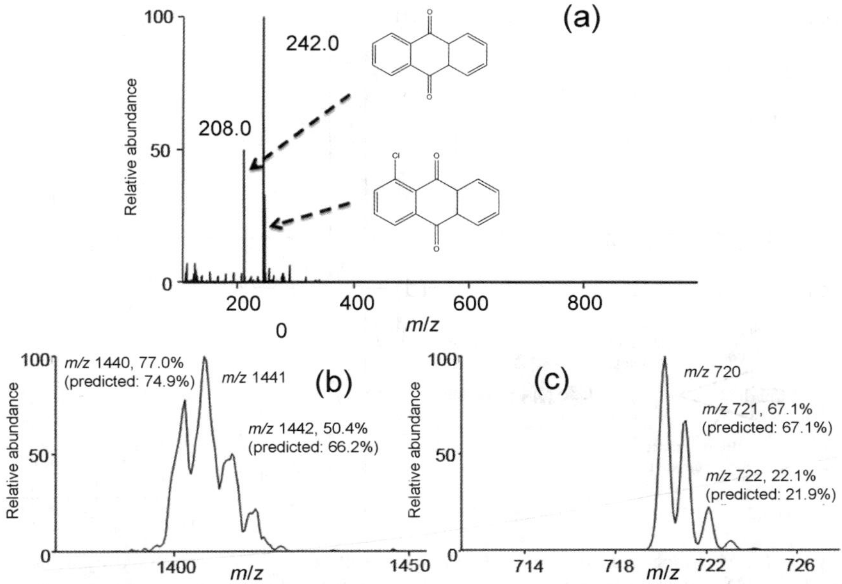

Fig. 8 (a) Electron attachment spectra of 9-10-anthraquinone and Cl-9-10-anthraquinone obtained using paper spray as the electron source. Electron attachment spectra of C_{60}: dimer (b), monomer (c).

was seen in tissue as a marker for pre-cancerous and cancerous prostate tissue.[23] A cholesterol sulfate standard compound can be detected [Fig. 9(a)] and fragmented [Fig. 9(b)] in the negative-ion mode using paper spray [Fig. 9(c)]. When applied to a serum sample from a prostate cancer patient, cholesterol sulfate was detected by PS-MS by monitoring the major fragment ion at *m/z* 97 from the parent ion at *m/z* 465 [Fig. 9(d)]. Paper-spray MS is a promising method for detection of endogenous compounds with the ability to be translated readily into a clinical setting.

The performance of the PS method is quite remarkable, considering how simple it is, consisting of only a piece of filter paper from which an electrospray plume is generated. Further development by the addition of selective methods for trapping and pre-concentrating the analyte in the substrate is anticipated to further decrease the limits of detection. This is needed for at least two reasons. First, although this method is capable of detecting drug levels below the therapeutic range for many drugs, it is likely that still lower detection limits will be required for the method to find broad applicability. It is unlikely that this method will be able successfully monitor all therapeutic drugs, but the larger the fraction of drugs that paper spray can successfully analyze, the more likely this method will successfully penetrate into clinical laboratories and have an impact on patient care. Improving the detection limit is first step in this process. Second, preconcentration and trapping the analyte(s) as it is extracted from the blood spot should decrease sources of irreproducibility introduced by the blood sample by separating the analyte from interfering chemicals in the blood. To this end, all experiments have been done on uniform blood samples. Blood collected from different individuals will naturally have different chemical content, for example in the amount of lipids or the presence of other drugs. The error introduced by the varied chemical background is partially offset by the presence of an internal standard; separation of the analyte from interfering chemicals should further reduce irreproducibility and error.

Pre-concentration of the analyte could be performed on the paper in any number of ways by borrowing from current knowledge of separations. For example, a hydrophobic or ion-exchange trapping region could be added to the paper downstream of the dried

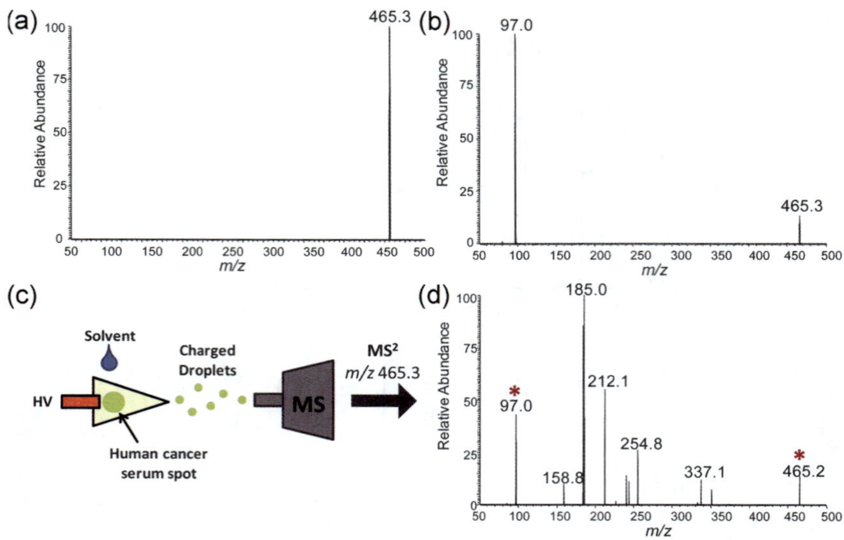

Fig. 9 (a) Mass spectrum of standard cholesterol sulfate, *m/z* 465.3, analyzed by negative-mode PS-MS. (b) Tandem MS (MS^2) of *m/z* 465.3 showing its single characteristic fragment ion *m/z* 97.0. (c) Scheme of paper spray analyses of human cancer serum and (d) MS^2 of *m/z* 465.3 from serum sample, showing parent ion and fragment ion (marked with asterisks) amongst chemical noise peaks.

blood spot. The analyte could be extracted from the blood spot and concentrated on the trapping region and then eluted with the application of a second solvent. This method could both separate at least some matrix molecules from the analyte and to serve as a method of pre-concentrating analyte molecules from a large volume of dried blood. It is important to note, however, that any method to pre-concentrate the analyte could also increase the concentration of the contaminants by an equal factor. Some additional selectivity, whether achieved through hydrophobicity, ion exchange, or another method, is needed so that the analyte can be preferentially concentrated.

The experiments described thus far have used the paper for sample collection, storage, and analysis. The use of paper or another porous substrate is also of use as a simple electrospray emitter. Typical nanospray tips suffer from problems with clogging to the small size of the capillary opening. A porous substrate, such as paper, is not subject to clogging and may be useful for coupling with microfluidic platforms with mass spectrometry. With the fast development of paper-based microfluidic devices,[45–47] paper spray is foreseen to play a major role in the integration of future microfluidic-MS systems of compact sizes.[48]

4. Miniature mass spectrometers and the future of biomedical and clinical mass spectrometry

4.1 State-of-the art in miniature mass spectrometer systems

The smallest fully functional autonomous mass spectrometer is an ion-trap-based system built at Purdue.[49,50] This wireless-controlled miniature rectilinear ion-trap MS system, weighs 5.0 kg (with batteries), consumes less than 35 W of power, and has dimensions of 22 cm in length by 12 cm in width by 18 cm in height. It is capable of tandem mass spectrometry which is essential for complex mixture analysis and has modest but adequate performance (unit mass resolution and mass range up to *m/z* 800, *m/z* 1500 in extended mass range mode). The instrument is compatible with various types of ionization sources including a glow-discharge electron impact ionization (GDEI) source for internal ionization and various atmospheric pressure ionization sources which are external to the vacuum system and connected to it through the use of a discontinuous atmospheric pressure interface (DAPI).[51,52] This interface was developed to match the rate of sample introduction to the limited pumping capacity of the miniature MS. These external sources include ESI and DESI, for solution and solids analysis, respectively. Performance characteristics of these handheld mass spectrometers have been reported.[53]

4.2 Adaptation to point-of-care drug monitoring and intrasurgical diagnostics

The availability of miniature mass spectrometers and ambient sampling ionization methods has significant potential in terms of the application of mass spectrometers in clinical point-of-care analysis and in providing real-time diagnostic information to guide surgery. The paper spray method has already been adapted to miniature mass spectrometers.[37] The next step in the development of a point-of-care clinical system will be the optimization of a disposable device in which the paper substrate is pre-mounted (and pre-supplied with appropriate internal standards and solvent). Addition of a finger prick blood spot will then be followed by ionization and recording of a mass spectrum using a miniature mass spectrometer. Miniature MS analysis systems with direct sampling ionization and tandem MS capabilities have potential as valuable new tools in high throughput point-of-care clinical analysis. These same systems should also see wide use as generally applicable, high sensitive analytical methods for applications in food safety, forensics and public safety, environmental monitoring and quality control.

The combination of a small mass spectrometer with efficient ion transfer capabilities allows one to consider the possibility that an ambient ionization source might be built into a surgeon's scalpel to provide real-time diagnostic information before tissue is excised. The geometry-independent DESI configuration[54] with coaxial charged droplet delivery and secondary droplet transport tubes is one possible configuration. The main technical problem is not the passage of ions over long distances at atmospheric pressure. This has been demonstrated for both DESI and for plasma-based ionization methods.[55,56] Rather it is maximizing the efficiency of this process when performing the experiment with a small mass spectrometer.

Acknowledgements

The authors acknowledge funding and other support from the Alfred Mann Institute for Biomedical Development at Purdue University (MIPurdue) for development of some of the technology presented in this paper. Support from the NIH (Grant 1 R21 EB009459-01) and the NSF (CHE 0848650 and 0847205) as well as support from Thermo Fisher is gratefully acknowledged.

References

1 L. A. McDonnell and R. M. A. Heeren, *Mass Spectrom. Rev.*, 2007, **26**, 606–643.
2 M. L. Pacholski and N. Winograd, *Chem. Rev.*, 1999, **99**, 2977.
3 R. G. Cooks, Z. Ouyang, Z. Takats and J. M. Wiseman, *Science*, 2006, **311**, 1566–1570.
4 G. A. Harris, L. Nyadong and F. M. Fernandez, *Analyst*, 2008, **133**, 1297–1301.
5 G. J. Van Berkel, S. P. Pasilis and O. Ovchinnikova, *J. Mass Spectrom.*, 2008, **43**, 1161–1180.
6 A. Venter, M. Nefliu and R. G. Cooks, *TrAC, Trends Anal. Chem.*, 2008, **27**, 284–290.
7 D. R. Ifa, J. M. Wiseman, Q. Y. Song and R. G. Cooks, *Int. J. Mass Spectrom.*, 2007, **259**, 8–15.
8 D. R. Ifa, L. Gumaelius, L. S. Eberlin, N. Manicke and R. G. Cooks, *Analyst*, 2007, **132**, 461–467.
9 D. R. Ifa, N. E. Manicke, A. L. Dill and G. Cooks, *Science*, 2008, **321**, 805–805.
10 G. Paglia, D. R. Ifa, C. P. Wu, G. Corso and R. G. Cooks, *Anal. Chem.*, 2010, **82**, 1744–1750.

11 L. S. Eberlin, D. R. Ifa, C. Wu and R. G. Cooks, *Angew. Chem., Int. Ed.*, 2010, **49**, 873–876.
12 C. P. Wu, D. R. Ifa, N. E. Manicke and R. G. Cooks, *Analyst*, 2010, **135**, 28–32.
13 M. Girod, Y. Shi, J.-X. Cheng and R. G. Cooks, *J. Am. Soc. Mass Spectrom.*, 2010, **21**, 1177–1189.
14 S. Shimma, Y. Sugiura, T. Hayasaka, Y. Hoshikawa, T. Noda and M. Setou, *J. Chromatogr., B: Anal. Technol. Biomed. Life Sci.*, 2007, **855**, 98–103.
15 B. Szachowicz-Petelska, S. Sulkowski and Z. A. Figaszewski, *Mol. Cell. Biochem.*, 2007, **294**, 237–242.
16 T. Utsugi, A. J. Schroit, J. Connor, C. D. Bucana and I. J. Fidler, *Cancer Res.*, 1991, **51**, 3062–3066.
17 E. O. Aboagye and Z. M. Bhujwalla, *Cancer Res.*, 1999, **59**, 80–84.
18 T. Alonso, R. O. Morgan, J. C. Marvizon, H. Zarbl and E. Santos, 1988, pp. 4271–4275.
19 S. Valitutti, P. Cucchi, G. Colletta, C. Di Filippo and D. Corda, *Cell. Signalling*, 1991, **3**, 321–332.
20 J. M. Wiseman, S. M. Puolitaival, Z. Takats, R. G. Cooks and R. Caprioli, *Angew. Chem., Int. Ed.*, 2005, **44**, 7094–7097.
21 A. L. Dill, D. R. Ifa, N. E. Manicke, A. B. Costa, J. A. Ramos-Vara, D. W. Knapp and R. G. Cooks, *Anal. Chem.*, 2009, **81**, 8758–8764.
22 A. L. Dill, L. S. Eberlin, A. B. Costa, C. Zheng, D. R. Ifa, L. Cheng, T. Masterson, M. Koch, O. Vitek and R. G. Cooks, 2010, submitted.
23 L. S. Eberlin, A. L. Dill, A. B. Costa, D. R. Ifa, L. Cheng, T. Masterson, M. Koch, T. L. Ratliff and R. G. Cooks, *Anal. Chem.*, 2010, **82**, 3430–3434.
24 C. A. Strott and Y. Higashi, *J. Lipid Res.*, 2003, **44**, 1268–1278.
25 J. I. Rearick, G. D. Stoner, M. A. George and A. M. Jetten, *Cancer Res.*, 1988, **48**, 5289–5295.
26 T. Kuroki, T. Ikuta, M. Kashiwagi, S. Kawabe, M. Ohba, N. Huh, K. Mizuno, S. Ohno, E. Yamada and K. Chida, *Mutat. Res., Rev. Mutat. Res.*, 2000, **462**, 189–195.
27 N. E. Manicke, T. Kistler, D. R. Ifa, R. G. Cooks and Z. Ouyang, *J. Am. Soc. Mass Spectrom.*, 2009, **20**, 321–325.
28 N. E. Manicke, J. M. Wiseman, D. R. Ifa and R. G. Cooks, *J. Am. Soc. Mass Spectrom.*, 2008, **19**, 531–543.
29 M. Pulfer and R. C. Murphy, *Mass Spectrom. Rev.*, 2003, **22**, 332–364.
30 F. F. Hsu and J. Turk, *J. Am. Soc. Mass Spectrom.*, 2003, **14**, 352–363.
31 D. W. Knapp, N. W. Glickman, D. B. DeNicola, P. L. Bonney, T. L. Lin and L. T. Glickman, *Urol. Oncol.*, 2000, **5**, 47–59.
32 C. Wu, D. R. Ifa, N. E. Manicke and R. G. Cooks, *Anal. Chem.*, 2009, **81**, 7618–7624.
33 G. R. Wilkinson, *N. Engl. J. Med.*, 2005, **352**, 2211–2221.
34 K. C. Waterman and R. C. Adami, *Int. J. Pharm.*, 2005, **293**, 101–125.
35 A. A. Alfazil and R. A. Anderson, *J. Anal. Toxicol.*, 2008, **32**, 511–515.
36 N. Spooner, R. Lad and M. Barfield, *Anal. Chem.*, 2009, **81**, 1557–1563.
37 J. J. Liu, H. Wang, N. E. Manicke, J. M. Lin, R. G. Cooks and Z. Ouyang, *Anal. Chem.*, 2010, **82**, 2463–2471.
38 H. Wang, J. J. Liu, R. G. Cooks and Z. Ouyang, *Angew. Chem., Int. Ed.*, 2010, **49**, 877–880.
39 N. E. Manicke, Q. Yang, H. Wang, S. Oradu, Z. Ouyang and R. G. Cooks, *Int. J. Mass Spectrom.*, 2010, DOI: 10.1016/j.ijms.2010.06.037.
40 J. H. Glick, *Clin. Chem.*, 1976, **22**, 475–483.
41 A. Saleem, R. J. Harte, J. C. Matthews, S. Osman, F. Brady, S. K. Luthra, G. D. Brown, N. Bleehen, T. Conners, T. Jones, P. M. Price and E. O. Aboagye, *J. Clin. Oncol.*, 2001, **19**, 1421–1429.
42 A. Saleem and P. M. Price, *Clin. Cancer Res.*, 2008, **14**, 8184–8190.
43 A. W. Martinez, S. T. Phillips and G. M. Whitesides, *Proc. Natl. Acad. Sci. U. S. A.*, 2008, **105**, 19606–19611.
44 H. Wang, N. E. Manicke, R. G. Cooks and Z. Ouyang, *Proceedings of the 58th ASMS Conference on Mass Spectrometry and Allied Topics*, 2010.
45 K. Abe, K. Suzuki and D. Citterio, *Anal. Chem.*, 2008, **80**, 6928–6934.
46 X. Li, J. F. Tian, T. Nguyen and W. Shen, *Anal. Chem.*, 2008, **80**, 9131–9134.
47 Z. Ouyang and R. G. Cooks, *Annu. Rev. Anal. Chem.*, 2009, **2**, 187–214.
48 R. Regenthal, M. Krueger, C. Koeppel and R. Preiss, *J. Clin. Monit. Comput.*, 1999, **15**, 529–544.
49 Z. Ouyang, R. J. Noll and R. G. Cooks, *Anal. Chem.*, 2009, **81**, 2421–2425.
50 L. Gao, A. Sugiarto, J. D. Harper, R. G. Cooks and Z. Ouyang, *Anal. Chem.*, 2008, **80**, 7198–7205.
51 L. Gao, R. G. Cooks and Z. Ouyang, *Anal. Chem.*, 2008, **80**, 4026–4032.
52 L. Gao, G. T. Li, Z. X. Nie, J. Duncan, Z. Ouyang and R. G. Cooks, *Int. J. Mass Spectrom.*, 2009, **283**, 30–34.

53 N. L. Sanders, E. Sokol, R. H. Perry, G. Huang, R. J. Noll, J. S. Duncan and R. G. Cooks, *Eur. J. Mass Spectrom.*, 2010, **16**, 11–20.
54 A. Venter and R. G. Cooks, *Anal. Chem.*, 2007, **79**, 6398–6403.
55 Z. Takáts, J. M. Wiseman and R. G. Cooks, *J. Mass Spectrom.*, 2005, **40**, 1261–1275.
56 I. Cotte-Rodríguez and R. G. Cooks, *Chem. Commun.*, 2006, 2968–2970.

Development of carbon-fluorine spectroscopy for pharmaceutical and biomedical applications

Farid Menaa,* Bouzid Menaa* and Olga Sharts

Received 27th April 2010, Accepted 19th July 2010
DOI: 10.1039/c005252c

Carbon-Fluorine Spectroscopy (CFS™), also known as Fluoro-Raman Spectroscopy (FRS™), is a relatively new patented platform technology using a family of various methods and cost-effective devices called PLIRFA™ (Pulsed Laser Isochronic Raman and Fluorescence Apparatus) developed by Fluorotronics, Inc. The key feature of this progressive and non-destructive technology is based on the discovery of a characteristic optical signature of carbon-fluorine bond(s) in the fingerprint spectral area of 500–800 cm^{-1} allowing rapid, ultra-specific and sensitive detection, characterization, imaging, and measurement of any fluoroorganics. Interestingly, the C–F bond is unique in its character and so it can be used as a molecular label. Furthermore, the C–F label is efficient, soluble, cheaper, smaller, more stable and less toxic than fluorescent dyes, nanoparticles or quantum dot materials. Thereby, C–F bonds are often incorporated into pharmaceutical, agrochemical and biological molecules in addition to polymers and nano-materials to achieve special properties (*e.g.* molecular stability, molecular tracing). In this paper, we present some of our data obtained from FRS™ applied to pharmaceuticals and biologics, and provide perspectives of FRS applications for the pharmaceutical and biomedical fields.

1. Introduction

Quantitative and qualitative analyses of compounds and molecules are an important task. Common methods of detection include molecular labelling using radioactive isotopes or fluorescent dyes. The labelled molecules can be used as tracers *in vivo*, through the body, and observed for emitted radioactivity or fluorescence respectively.[1,2] However, these types of detection have limited sensitivity, stability and/or require expensive safety and precautionary measures. In this context, we have developed a non-competitive family of compact devices called PLIRFA™ (Pulsed Laser Isochronic Raman and Fluorescence Apparatus) for multiple R&D applications across the pharmaceutical and life sciences spectrum. Among them, FRS™ (Fluoro-Raman Spectroscopy) (Fig. 1) from which FRAT™ (Fluoro-Raman Process Analytical Technology) and DELPHI™ (Definitive Express Laser Pharmaceutical Inspection) are derived. The FRAT™ provides much broader in-depth understanding of the processes, drug-drug or drug-protein interactions and can assure quality of each single pharmaceutical sample (*e.g.* pill) while DELPHI™ represents a valuable tool for differentiating legitimate and illegitimate substances (*e.g.* counterfeited drugs).

The key feature of FRS is based on the discovery of a characteristic Raman signature of carbon-fluorine bond(s) in the fingerprint spectral area of 500–800 cm^{-1} allowing reliable detection, characterization, imaging, monitoring, screening and

Fluorotronics Inc., 2453 Cades Way, Bldg C, Vista, CA, 92081, USA. E-mail: fmenaa@yahoo.com; bouzid.menaa@gmail.com

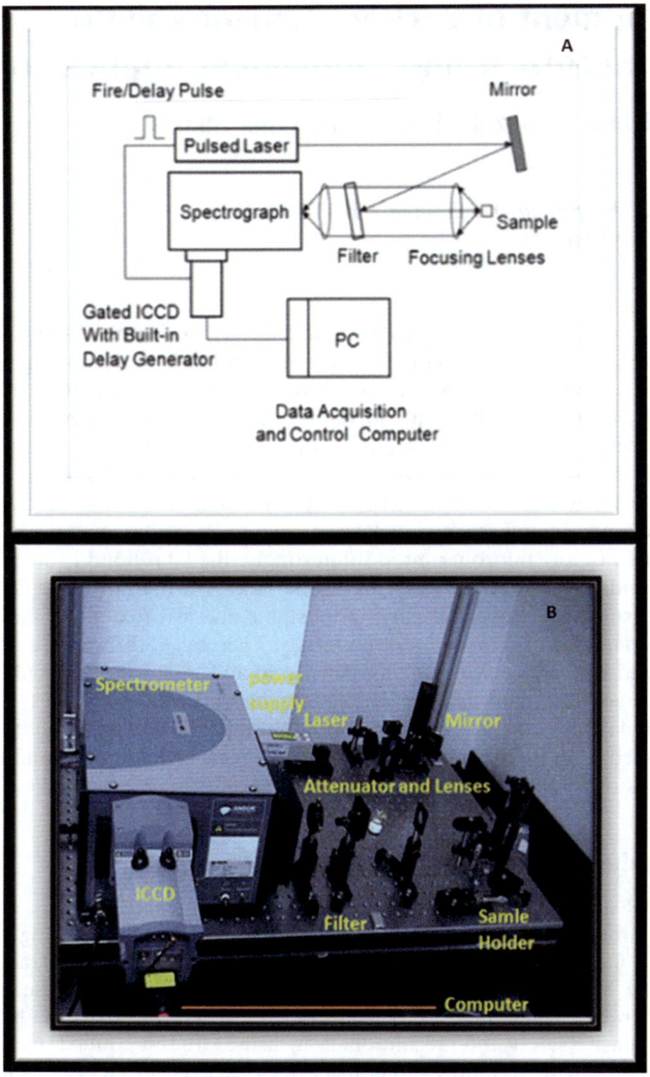

Fig. 1 **A**, Schematic representation of FRS™ (Fluoro-Raman Spectroscopy) patented apparatus (third generation). **B**, Real view of assembled FRS™. It includes mainly a laser with power unit, a spectrograph, an ICCD camera with image intensifier, a desktop computer and FRS™ software (views A and B). The upcoming generation of device is replacing the focusing lenses by fiber optics.

measurement of fluoro-nanomaterials, fluoroorganic compounds, fluoro-molecules, fluoroorganic impurities or fluoro-degradation products.[3–9] Thereby, FRS™ represents a major advance over conventional Raman spectroscopy since it allows (i) highly specific and ultrasensitive detection (ppm-ppb level or <10 cells) of C–F bonds (–C–F, $=CF_2$, $-CF_3$); (ii) detection and characterization of any other halo-organic bonds or unlabeled molecules regardless the physical state (solid, liquid or gas) which can be done *via* quartz vials, glass or polymer containers; (iii) real-time analysis and ultrafast data acquisition (10^{-8} s); (iv) quantitative determination of any fluorinated substance since the emitted C–F bond(s) signal is directly proportional to the concentration of the analyte; (v) structural characterization of molecules (*e.g.* differentiation of length of molecules with a resolution of only one carbon); (vi) preservation of the sample integrity (non-destructive, non-invasive technology); (vii) analysis with little or no sample preparation; (viii) easy use and data interpretation (*e.g.* innovative software solutions); (ix) low maintenance (*e.g.* apparatus calibration); (x) operation with one laser source which can function in 3 modes: Fluoro-Raman, general pulsed Raman and two photons excitation/time resolved fluorescence, using adequate time-gating and time-delays; (xi) the technology to be versatile, flexible and progressive (*e.g.* hybrid and portable devices such as Fluoro-Raman microscope, High Throughput Screening Raman multi-well analyser).

C–F bonds can be incorporated into pharmaceuticals (*e.g.* for drug discovery, drug safety, drug security), agrochemicals and biologics (*e.g.* for diagnostics, therapeutics, theranostics™) as well as in polymers and nano-materials to possibly achieve special properties such as (i) molecular stability; (ii) product safety using C–F as a specific/selective molecular security label; (iii) tracing a compound (*e.g.* across various stages of the pharmaceutical development and for imaging monitoring). Indeed, C–F bonds are ideal as internal (*e.g. ex vivo*, *in situ*, *in vivo*) and/or external labels (*e.g.* surface, *in vitro*) and can provide necessary improvement in productivity and efficiency,[1,2,10] since they are (i) unique and man-made; (ii) more stable than fluorescent dyes due to their covalent interactions; (iii) smaller, more soluble, and less toxic than nanoparticles or quantum dots; (iv) safe, easy to handle, avoiding logistical problems normally associated with radioactive labels; (v) efficiently incorporated into molecules; (vi) inexpensive.

Eventually, the patented broad PLIRFA™ technology platform is allowing breakthroughs in unification of the detection, measurement and characterization of fluorinated compounds, molecules and materials applied in life sciences, bio-medical, pharmaceutical and chemical industries.

Here, we present briefly some promising data and perspectives for applications in pharmaceutical and biomedical R&D fields using FRS™ as an innovative and relatively new analytical tool.

2. Materials and methods

2.1. Fluoro-Raman spectroscopy (FRS) technology

FRS™ is one of the key devices from the progressive PLIRFA™ (Pulsed Laser Isochronic Raman and Fluorescence Apparatus) technology platform along with its associated methods and applications that we have developed and patented. The apparatus has been assembled by our engineers and Raman experts in collaboration with worldwide partners, manufacturers and providers. FRS™ is composed mainly of a metal –usually copper or gold-vapor laser – with a power unit (Sierra, General Atomics, USA), a Spectrophotometer Shamrock SR-303i along with its power supply ELC-00376 (Andor, Northern Ireland), an ICCD camera DH-740-18F-33 including an image intensifier/photomultiplier (Andor, Northern Ireland), a filter RazorEdge LP03-532RU-25 (Semrock, USA), a computer and a software-time resolved Solis(t) (Andor, Northern Ireland) for data collection and processing. The global experimental setup is shown in Fig. 1. Next generations of devices are currently under development.

Briefly, the lens focused the beam into the sample and collected the scattered radiation back along the same axis. The long-pass filter Razor Edge was used to reflect the excitation to the lens, to pass the Raman scattered radiation collected by the lens and to block Rayleigh scattering from reaching the spectrograph. The ICCD camera built-in delay generator was used to acquire the Fluoro-Raman spectra.

2.2. Analysis of pharmaceuticals and biologics using FRS™ technology

3-Fluoraniline (3-FA, C_6H_6FN, 98% pure) and methanol (MeOH, CH_3OH, 99.8% pure) used as a solvent, were purchased at Acros Organics. Both 3-FA and MeOH were characterized by FRS™ (Fluorotronics, Inc.) using a flow cell of

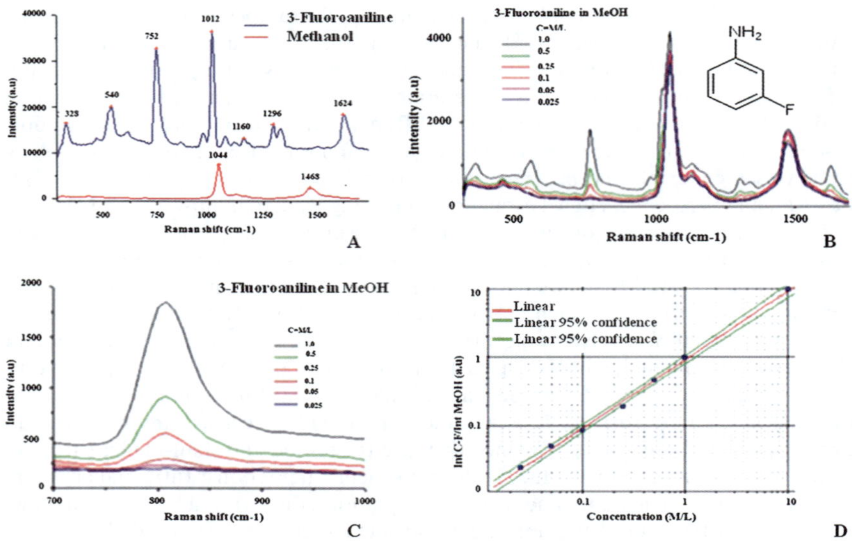

Fig. 2 Fluoro-Raman spectrum of the 3-fluoroaniline (3-FA, C_6H_6FN) *via* original glass container. Excitation at 510.6 nm with a pulsed copper-vapor laser; Average laser power 200 mW, 10 kHz repetition rate, 17 ns pulse; Recording with a double monochromatic spectrophotometer. 3-FA heights peak intensities normalized to the 1468 cm^{-1} peak intensities corresponding to methanol (MeOH) as a function of the concentration C (upper limit $C = 10.2$ M corresponds to pure, 98% of 3-FA). **A**, FRS spectra of pure 3-fluoroaniline (3-FA) and pure methanol (MeOH). C–F signal for 3-FA is observed at 752 cm^{-1}. **B**, Diluted 3-FA in MeOH using concentrations ranged from 0.025 to 1 M. C–F signal can still be detected at 752 cm^{-1} using a 3-FA concentration as low as 0.25 M (x axis: 0–1700 cm^{-1}). **C**, Sensitivity of C–F signal according 3-FA concentrations ranged from 0.025 to 1 M. After zooming (x axis: 700–1000 cm^{-1}), C–F signal can still be detected at 752 cm^{-1} using a 3-FA concentration as low as 0.05 M. **D**, Effect of 3-FA concentrations ranged from 0.0025 to 10 M on C–F intensity. The C–F bond is specifically detected at 752 cm^{-1} for 3-FA regardless to 3-FA concentration used and the characteristic C–F signal displayed, with a confidence of linearity equal to 95% and the intensity proportional to the concentration of 3-FA used. The heights peak intensities corresponding to the 3-FA were normalized to the 1468 cm^{-1} peak intensities corresponding to MeOH as a function of the concentration C (upper limit being $C = 10.2$ M and corresponds to pure (98%) 3-FA).

11 mm (Agilent). Briefly, a double monochromatic spectrometer composed of a 510.6 nm pulsed copper vapor laser (Sierra, General Atomics, USA) delivering an average laser power of 200 mW at 10 kHz repetition rate and with 17 ns pulse, was used to record the spectrum of each molecule. The C–F bond characteristic signal of 3-FA was observed at 752 cm^{-1} (Fig. 2A, 2B, and 2C). Different concentrations of 3-FA have been performed to ensure the consistency of the data (Fig. 2B and 2C). The peak intensities corresponding to the 3-FA were normalized to the 1468 cm^{-1} peak intensities corresponding to MeOH as a function of the concentration C (upper limit being $C = 10.2$ mol L^{-1} and corresponds to pure (98%) 3-FA). The results showed a confidence of linearity equal to 95% (Fig. 2D) with an intensity proportional to the concentration of 3-FA used. This experiment has served as one of the proof-of-principle validation for subsequent Raman studies.

Fluoxetine hydrochloride capsule (Prozac®, $C_{17}H_{18}F_3NO.HCl$, solid) was purchased from a drug store. The composition of the capsule is unknown other than the presence of fluoroxetine hydrochloride. 60 fg of this powdered compound representing a spot size of 100 μm was dropped on a glass surface and characterized by FRS using same parameters as for 3-FA, excepted that the laser power was set up at 100 mW. The very narrow sharp peak at 782 cm^{-1} was assigned to a symmetrical

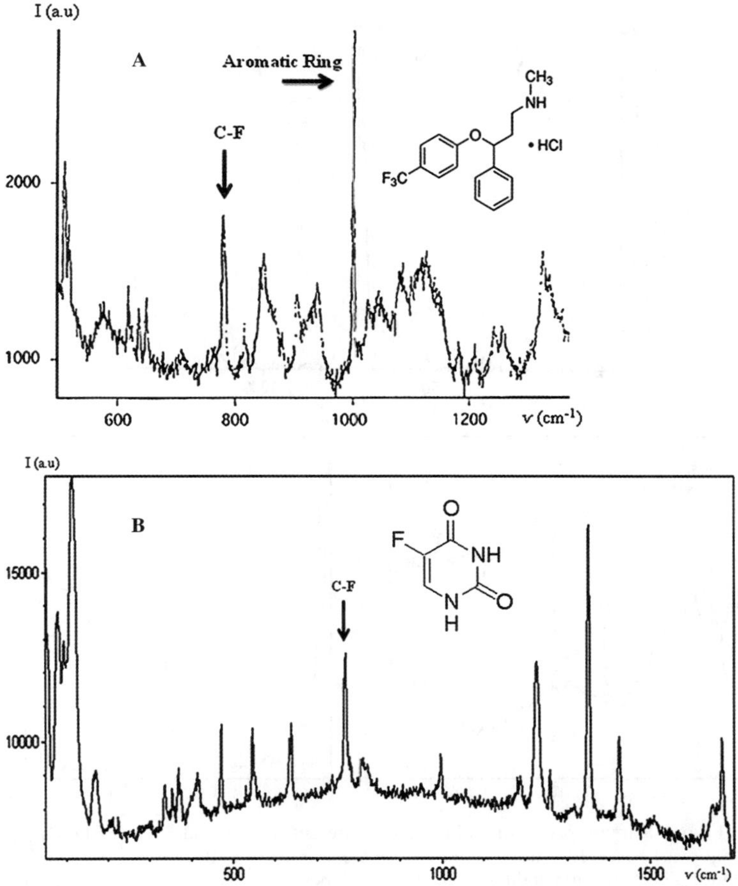

Fig. 3 Fluoro-Raman spectra of pharmaceuticals. **A**, represents fluoxetine hydrochloride (Prozac®) capsule. 60 fg of this powdered compound representing a spot size of 100 μm was dropped on a glass surface and characterized by FRS using same parameters as for 3-FA, excepted that the laser power was set up at 100 mW. The sharp peak at 782 cm^{-1} is assigned to a symmetrical vibration involving the trifluoromethyl ($-CF_3$) group. The symmetric aromatic vibration is found at 1001 cm^{-1}. **B**, represents 5-Fluorouracil (5-FU) spectrum. 1 μg ml^{-1} of this powdered compound diluted in distilled water was used for FRS analysis. Excitation at 510.6 nm with a pulsed copper-vapor laser; average laser power 100 mW, 10 kHz repetition rate, 12 ns pulse; recording with a double monochromatic spectrophotometer. The C–F signal was observed at 770 cm^{-1}.

vibration involving the trifluoromethyl ($-CF_3$) group (Fig. 3A). The symmetric aromatic vibration was found at 1001 cm^{-1}.

5-Fluorouracil (5-FU, $C_4H_3FN_2O_2$, 99% pure) was purchased from Sigma-Aldrich. 1 μg ml^{-1} of this powdered compound diluted in distilled water was used for FRS analysis using a copper-vapor laser operated at 100 mW with 12 nanosecond pulses at 10 kHz repetition rate. The C–F signal was observed at 770 cm^{-1} (Fig. 3B).

Perfluorodecalin (PFD, $C_{10}F_{18}$, 95% pure) was purchased from Sigma-Aldrich. Raman spectrum of 1 μg ml^{-1} of the compound was observed with a standard single monochromator. The 510.6 nm light source was a copper-vapor laser operated at 100 mW with 20 ns pulses at 10 kHz repetition rate. The C–F signal was observed at 692 cm^{-1}, which was assigned to a symmetrical vibration involving the difluoromethyl ($-CF_2$) group (Fig. 4A).

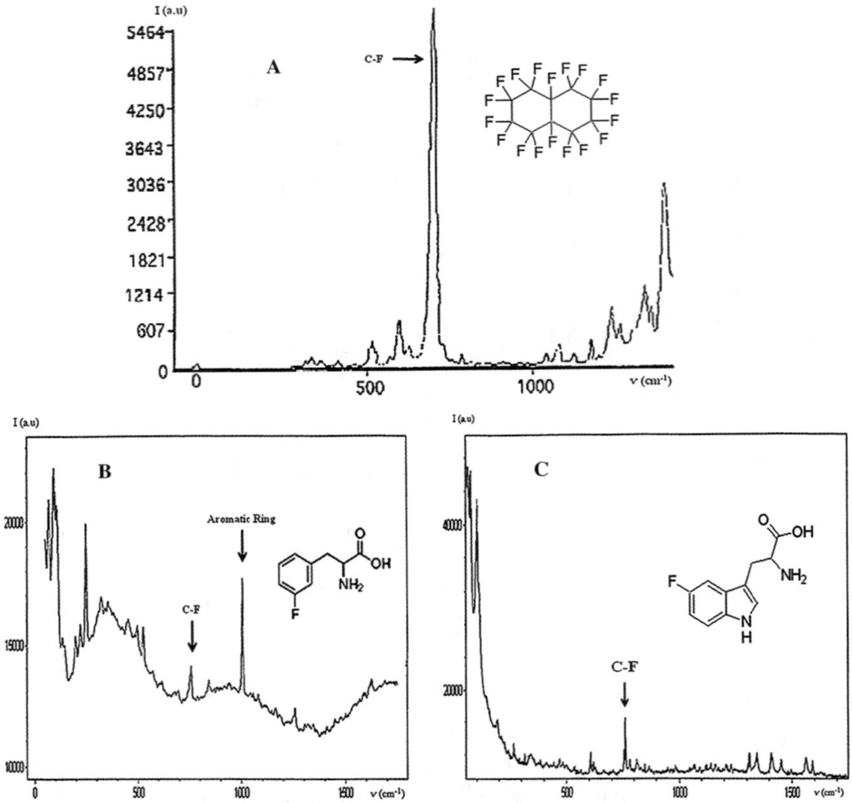

Fig. 4 Fluoro-Raman spectra of biologics. **A**, represents perfluorodecalin (PFD) Raman spectrum that was observed with a standard single monochromator, using 1 µg ml^{-1} of the compound. The 510.6 nm light source was a copper-vapor laser operated with an average power laser of 100 mW, at 10 kHz repetition rate and with 12 ns pulse. The C–F signal was observed at 692 cm^{-1}, which was assigned to a symmetrical vibration involving the difluoromethyl ($-CF_2$) group. **B**, represents 3-fluoro-DL-phenylalanine (F-Phe) spectrum. 1 µg ml^{-1} of this amino-acid diluted in distilled water was used for FRS analysis using same parameters as for 5-FU. The C–F bond characteristic signal was observed around 758 cm^{-1}. **C**, represents 5-fluoro-DL-tryptophan (F-Trp). 1 µg ml^{-1} of this amino-acid diluted in distilled water was used for FRS analysis using same parameters as for 5-FU. The C–F bond characteristic signal was observed around 768 cm^{-1}.

3-Fluoro-DL-phenylalanine (F-Phe, $FC_6H_4CH_2CH(NH_2)COOH$, ≥98%) and 5-fluoro-DL-tryptophan (F-Trp, $C_{11}H_{11}FN_2O_2$, ≥99%) were purchased from Sigma-Aldrich. 1 µg ml^{-1} of this powdered compound diluted in distilled water was used for FRS using same parameters as for 5-FU. The C–F bond characteristic signal was observed around 758 cm^{-1} (Fig. 4B) and 768 cm^{-1} (Fig. 4C) respectively.

3. Results and discussions

3.1. Presentation and mechanism of the fluoro-Raman spectroscopy

Our patented Raman instrument (Fig. 1) includes a pulsed metal-vapor laser – which presents the advantages of being cost-effective, avoiding high photodecomposition and eliminating primary fluorescence while giving enhanced sensitivity in ultraviolet (UV), visible (VIS) and near infrared (NIR) regions over continuous laser Raman spectrometers –, a mirror with dielectric and highly reflective coating to efficiently reflect a generated laser radiation, and a photodiode for collecting a laser radiation. The instrument also includes an electronic unit (*e.g.* a stroboscopic generator) and a condensing lens that focuses a laser beam on the sample to be analyzed. Condensing lenses are used to focus beams on the spectrophotometer slit after passing through the sample and an absorbance filter is used to pass a Raman emission signal (secondary radiation). Moreover, the FRS instrument includes a single or a double monochromator spectrometer (for the UV/VIS region of the spectra) equipped with a detector of secondary radiation (*e.g.* a photomultiplier), a pulse recording system (sensitive photon-counting detector system) controlled by the stroboscopic system, and, optionally, a computer for data collection, data processing and management of the spectrometer.

The general FRS system functions as follow: (i) Samples may be placed into cylindrical quartz cuvettes having parallel windows. Raman spectra are observed at 90 degrees with respect to the incident pulse by using a single or double monochromator with a single channel detection; (ii) After entering required parameters, the high energy laser pulses irradiate the fluoroorganic sample and activate a device, to form a strobe-impulse used to synchronise the pulse recording system; (iii) When irradiated, the sample emits pulsed Raman radiation and also scatters part of the incident Raman pulse. The Raman vibrational scattering emissions are detected during a narrow time internal ("gate") approximately equal to the laser pulse duration (10^{-8} s), so the Raman spectrum is not affected by fluorescence. To irradiate the samples in the UV, a doubling or a triple crystal (*e.g.* barium borate, BaB_2O_4) coupled to the laser source, can be used, allowing the observation of resonance Raman spectra (ppb level sensitivity); (iv) The absorbance filter removes scattered radiation and passes the pulsed Raman radiation to the spectrometer; (v) Signals are photomultiplied and sent to a synchronized pulse recording system; (vi) The data are collected on computer and processed using appropriate software.

Based on our experimental observations, the radiation characteristic of C–F bonds was mainly detected in the 540–785 cm^{-1} (visible) region.[3–5] For estimation purposes, a range of 500–800 cm^{-1} is reasonably used. Other radiation in the 950–1400 cm^{-1} (infra-red) region commonly associated with fluoroorganic compounds can also be observed and can sometimes be used to confirm C–F groups if these infrared absorptions are not subject to interference by other functional groups in the fluoroorganic compound.[5] We have previously shown that most of the compounds containing aromatic rings give significant enhancement of the resonance Raman emission signal due to the high self-absorbance (energy absorption at frequencies similar to those emitted) of the aromatic ring structures.[5]

3.2. Biopharmaceutical and biomedical applications

Fluoroorganic compounds are very rare in nature and the few that do occur are highly toxic. While elemental fluorine and hydrogen fluoride have such a bad reputation as hazardous, many fluorinated compounds (*e.g.* fluoro-reagents, fluoro-aromatic building blocks, fluoro-drugs, fluoro-chemicals, fluoro-biologics, fluoro-polymers, fluoro-nanomaterials) are however still on the market, continuously developed and commercialized for specific uses due to their beneficial properties. One of the earliest synthetic fluorinated drugs was the anti-neoplastic agent 5-fluorouracil (5-FU), which was first synthesized in 1957.[11] This interest in developing F-compounds can be mainly explained by the facts that (i) the fluorocompound has inherent biological activity; (ii) the introduction of fluorine into a biologically active compound generally improves its pharmacological properties (*e.g.* increased bioavailability, lipophilicity, activity against drug-resistance) and (iii) what has been described as "patent jumping".[12] Contrary to popular misconceptions, covalently bound fluorine (C–F bound length, $\sim$138 pm) has a significantly larger steric requirement than hydrogen (C–H bond length, 109 pm) and fluorinated compounds usually have a sufficient similarity in size and shape to their non-fluorinated analogues to fit a given enzyme receptor, so that they tend to have similar inherent biological activity. On the other hand, the carbon-fluorine is very strong (~485 kJ mol^{-1}, compared with C–H, ~416 kJ mol^{-1}), so the fluorinated compounds tend to be more resistant to metabolic degradation.[13] In this context, PLIRFA™ devices and methods constitute valuable process analytical tools to detect C–F bond(s), assess quality control (QC), quality assurance (QA) of any fluorinated product before (*e.g.* drug discovery or screening of molecules; drug design or chemical synthesis process) and after commercialization and can, as well, play an important role in the prevention of counterfeited products (*e.g.* drugs, cosmetic or agrochemical formulations).

As a proof-of-principle experiment for detection of organofluorine compound, we have analyzed, by FRS™, molecules such as 3-fluoroaniline (C_6H_6FN) *via* standard flow cell using methanol (MeOH) as solvent. 3-Fluoroaniline (3-FA) and derivatives are often used as intermediate for the preparation of various pharmaceuticals (*e.g.* Difloxacin or Dicural®), pesticides (*e.g.* Fluquinconazole) and dyes (*e.g.* 2.4-dinitro-5-fluoroaniline). The excitation wavelength has been performed at 510.6 nm by pulsed copper vapor laser at the average power of 200 mW. Interestingly, the sensitivity was comparable to UV spectrometric detector with the satisfactory of 10^3 dynamic range. The C–F bond was specifically detected at 752 cm^{-1} for 3-FA regardless to the concentration of the fluorinated compound but the characteristic C–F signal displayed, with a confidence of linearity equal to 95%, an intensity proportional to the concentration of 3-FA (Fig. 2). Thereby, we showed that PLIRFA™ was able to provide a fingerprint spectrum and quantify a fluoro-organic compound based on the C–F bond signature.

Thus, we have extended our investigations to several other fluoroorganic compounds. Among them, we can cite fluoro-pharmaceuticals (*e.g.* anti-depressants such Fluoxetine), blood substitutes components such as Perfluorodecalin or Perflubron), antibiotics such as fluoroquinolones, anticancer agents such 5-fluorouracil, cholesterol-lowering medicines such Atorvastatin), fluoro-agrochemicals (*e.g.* herbicides such as Fluometuron, insecticides such as Tefluthrin, pesticides such as compound 1080) and fluoro-biologicals (*e.g.* F-nucleic acids, F-aptamers, F-peptides, F-aminoacids, F-proteins, *etc.*) that we have compared with their hydrocarbon analogs.[3–5] Thereby, as examples of pharmaceuticals (Fig. 3), we report here FRS™ spectra of the antidepressant fluoxetine (Prozac®, $C_{17}H_{18}F_3NO$) (Fig. 3A) and the anticancer drug, 5-fluorouracil (5-FU, $C_4H_3FN_2O_2$) (Fig. 3B) for which the C–F signal has been respectively identified at 782 cm^{-1} and 770 cm^{-1}. In the case of fluoxetine, the C–F signal is assigned to a symmetrical vibration involving the trifluoromethyl group while the symmetrical vibration involving the aromatic group was found at 1001 cm^{-1}. Besides, we also show here that is it possible to detect by FRS™ fluoro-biologics such as the blood substitute component, perfluorodecalin ($C_{10}F_{18}$) and essential protein constituents, amino acids 3-fluoro-DL-phenylalanine and 5-fluoro-DL-tryptophan, based on the identification of their C–F characteristic signal. For perfluorodecalin, C–F signal has been observed at 692 cm^{-1} which was assigned to a symmetrical vibration involving the difluoromethyl (–CF_2) group (Fig. 4A). For 3-fluoro-DL-phenylalanine (F-Phe) and 5-fluoro-DL-tryptophan (F-Trp), the C–F signal has been detected at 758 cm^{-1} (Fig. 4B) and 768 cm^{-1} (Fig. 4C) respectively. In accordance with previous studies,[14] we were able to detect and confirm, at 1001 cm^{-1}, the characteristic aromatic ring of phenylalanine.

4. Conclusions and perspectives

Fluorinated molecules are widely used in medicinal chemistry (*e.g.* drug design) and are becoming commonly used in the biomedical field (*e.g.* fluorinated bio-probes such as F-siRNA, F-peptides which can serve as theranostics™). The number of fluorinated molecules is in constant augmentation. Thereby, in 1978, about

600 pesticides were known, but only approximately 25 (4%) contained fluorine. In 2004, fluorine-containing compounds accounted for more than 17% of all commercially available crop protection agents and many others are currently under development. In the pharmaceutical area, around 220 fluorinated drugs were on the market in 1990, representing about 8% of all synthetic drugs. Six years later, already more than 1500 fluorine-containing drugs were under development.[15] Fluorinated compounds are often vital when pharmacological properties have to be modified, special features be incorporated, or when reliable performance and stability of a compound or a formulation are needed. Indeed, fluorination, the introduction of the fluorine, can affect the molecule conformation, alkalinity, binding affinity, metabolic stability, bioavailability, solubility, lipophilicity, polarization, hydrophobicity, among other properties but may significantly reduce undesirable side effects of drug and usually offer superior pharmacological properties.[10,15]

PLIRFA™ technology platform, which includes a panel of analytical devices and methods allows rapid, sensitive and specific detection and measurement of C–F bonds without requiring detection of the entire organo-fluorine compound, which is also possible by our technology.[3–9] Consequently, our technology is currently devoted to implement methodologies (*e.g.* bioassays, imaging protocols, *etc.*) and reduce significantly the troublesome detection and time consumption associated with other routine technologies (*e.g.* chromatography, molecular imaging systems or high throughput systems).

Here, we have shown that FRS™ (Fluoro-Raman Spectroscopy), one of the device member of the PLIRFA™ family, allows not only qualitative characterization of fluoro-pharmaceuticals and fluoro-biologics using a characteristic Raman signature of the carbon–fluorine (C–F) bond but also their quantitative determination since the signature emission of the C–F bond is directly proportional to the concentration of the organofluorine compound.[3–6]

Thereby, we are routinely updating our database of spectra and currently performing other studies in the biomedical field using FRS™ as a progressive analytical tool in order to (i) understand the role of fluorine in the conformation and ellipticity of target proteins (ref. 16 and manuscript under preparation); (ii) isolate and characterize F-labelled protein biomarkers of disease from tumour cell lysates (manuscript under preparation); (iii) isolate and characterize cancer stem cells within a given tumour (manuscript under preparation); (iv) develop F-bioprobes such as F-antibodies for diagnosis and therapy purposes; (v) develop F-nanomaterials (F-nanocarbons, F-nanoparticles) as (bio-)drug delivery systems.

Thus, FRS™ may become a ‘gold standard’ technology for many biomedical and pharmaceutical applications.

Acknowledgements

The authors thank Prof. Dr Leo Avakyants and Dr Emil Djangirov for their valuable technical assistance and scientific contribution.

References

1 J. J. Pedroso de Lima, *Eur. J. Phys.*, 1998, **19**, 485.
2 C. Weir, M. Morel-Kopp, A. Gill, K. Tinworth, L. Ladd, S. Hunyor and C. Ward, *Heart, Lung and Circulation*, 2008, **17**, 395.
3 C. M. Sharts, O. Sharts and V. S. Gorelik, Method and Apparatus for Determination of Carbon-Halogen Compounds and Applications Thereof, U.S. Patent 6 307 525 B1, 2001, 1.
4 C. M. Sharts, O. Sharts and V. S. Gorelik, Method and Apparatus for Determination of Carbon-Halogen Compounds and Applications Thereof, U.S. Patent 6 445 449 B1, 2002, 1.
5 V. S. Gorelik, C. M. Sharts, O. N. Sharts, R. P. Metzger, D. F. Shellhamer and G. R. Wischnath, Detection of Carbon-Halogen Bonds, U.S. Patent 7 433 035 B2, 2008, 1.

6 C. M. Sharts, V. S. Gorelik, A. M. Agoltsov, L. I. Zlobina and O. N. Sharts, *SPIE Proceed. Photonics East*, 1998, 317.
7 V. S. Gorelik, A. V. Chervyakov, L. I. Zlobina and O. N. Sharts, *SPIE Proceed.*, 1999, 16.
8 C. M. Sharts, V. S. Gorelik, O. N. Sharts, L. I. Zlobina and A. M. Agoltsov, *SPIE Proceed.*, 1999, 309.
9 V. S. Gorelik, S. S. Verjaskin, P. P. Sverbil, A. V. Chervjakov, L. I. Zlobina and O. N. Sharts, *SPIE Proceed.*, 2001, 25.
10 D. O'Hagan, *Chem. Soc. Rev.*, 2008, **37**, 308.
11 R. Duschinsky, E. Pleven and C. Heidelberger, *J. Am. Chem. Soc.*, 1957, **79**, 4559.
12 D. Seebach, *Angew. Chem., Int. Ed. Engl.*, 1990, **29**, 1320.
13 D. O'Hagan and H. S. Rzepa, *Chem. Commun.*, 1997, 645.
14 H. I. Lee, M. S. Kim and S. W. Suh, *Bull. Korean Chem. Soc.*, 1988, **9**, 218.
15 P. Maienfish and R. G. Hall, *Chimia*, 2004, **58**, 93.
16 B. Menaa, F. Menaa, C. A. Guimarães and O. Sharts, *Int. J. Nanotechnol.*, 2010, **7**, 1.

Exploiting the diagnostic potential of biomolecular fingerprinting with vibrational spectroscopy

Catherine Kendall,*[ab] Joanne Hutchings,[ab] Hugh Barr,[bc] Neil Shepherd[bd] and Nicholas Stone[ab]

Received 19th May 2010, Accepted 20th July 2010
DOI: 10.1039/c005379a

There is immense clinical need for techniques that can detect the biochemical changes associated with pre-malignancy. The ideal diagnostic test would provide rapid, non-invasive diagnosis at the point of care with high throughput and without prior tissue processing. Over the past decade vibrational spectroscopy techniques have demonstrated their ability to provide non-destructive, rapid, clinically relevant diagnostic information. Biochemical fingerprints of tissues measured using Raman and infrared spectroscopy analysed in conjunction with advanced chemometrics have shown great potential in the diagnostic assessment of biological material. Development of Raman probes is enabling the potential of *in vivo* clinical measurements to be realised. A novel probe design has been evaluated in clinical studies to identify and classify the subtle pre-malignant biochemical changes related to the carcinogenesis process. Exciting recent developments have enabled the probing of tissue samples at depth with huge potential for breast and prostate cancer diagnostics. Furthermore, the potential of vibrational spectroscopy to provide prognostic information is tantalising. Raman spectral data acquired on oesophageal biopsy samples analysed in conjunction with patient outcome data has shown the power of spectral biomolecular fingerprinting in predicting the outcome of patients with high-grade dysplasia in Barrett's oesophagus. Raman mapping can also be used to analyse thin tissue sections on calcium fluoride slides enabling the distribution of tissue constituents to be realised. The spectral data acquired effectively enables multiplexing of digital tissue stains since a whole array of information is gathered simultaneously. Technological developments are bringing the technologies closer to the clinical reality of spectral pathology and high-throughput non-destructive measurement with high resolution.

Introduction

One in three people in the UK will be affected by cancer in their lifetime.[1] There is great clinical need for techniques that can detect the pre-malignant biochemical changes that can advance to cancer to enable early eradication of abnormal cells

[a]*Biophotonics Research Unit, Gloucestershire Hospitals NHS Foundation Trust, Leadon House, Great Western Road, Gloucester, UK. E-mail: c.kendall@medical-research-centre.com; Fax: +08454 225485; Tel: +08454 225470*
[b]*Cranfield Health, Cranfield University, Cranfield, Bedfordshire, MK43 OAL, UK*
[c]*Department of Surgery, Gloucestershire Hospitals NHS Foundation Trust, Great Western Road, Gloucester, UK*
[d]*Department of Histopathology, Gloucestershire Hospitals NHS Foundation Trust, Great Western Road, Gloucester, UK*

before systemic invasion. The majority of cancers develop in epithelial tissues which line hollow organs, and are thus exposed over a lifetime to numerous assaults from chemical and pathogenic carcinogens. Epithelial tissues can be specialized for a number of functions and depending on the organ will contain different cell architectures. Although the pathogenesis of most cancers is not fully understood, some are known to develop through a pre-malignant state.

Clinical need for novel diagnostic tools

Oesophageal cancer has been selected for this work as an example of a slowly developing cancer, which is of significance in the mortality and incidence statistics and is easily accessible with an endoscope. The incidence of oesophageal cancer has increased in recent years. Acid and bile reflux can lead to a change in the cells lining the oesophagus where the normal milky white squamous epithelium is replaced by salmon pink columnar epithelium in a condition called Barrett's oesophagus.[2] This is a metaplastic condition linked to an increased risk of oesophageal adenocarcinoma. Dysplasia, an abnormal change in the maturation of cells, is the most frequently used marker of increased cancer risk. Most patients with Barrett's oesophagus will not develop dysplasia or adenocarcinoma.[3] Reliable methods for identifying which patients with Barrett's oesophagus will progress to carcinoma do not exist at present. Therefore, techniques that enable a greater understanding of the biological processes leading to carcinogenesis will be extremely valuable.

The current gold standard for the diagnosis of cancer relies upon surveillance and screening of at-risk population groups or diagnostic tests which are carried out on presentation of suspicious symptoms. Once symptoms are present it's often too late to treat successfully, but most cancers if treated early have a good prognosis. However patients progress at different rates, therefore information on which pre-cancerous changes need to be treated and which can be monitored would be of great benefit to the clinician. Additionally some tumours respond to treatment much better than others and information to guide therapeutics would also facilitate improved prognosis.

Biopsy samples are collected, from patients with symptoms or on surveillance programmes for gastrointestinal cancers, and analysed by a histopathologist. Oesophageal cancer develops within the epithelial surface of the organ, which is approximately 500 microns thick. Often abnormal lesions are microscopic and too small to be detected by white light endoscopy alone, therefore random biopsies are collected. Even with a rigorous biopsy protocol this is a poor sampling technique akin to finding a needle in a haystack. Calculations based on collection of a 1 mm^2 biopsy at four quadrants every two cm show that less than 1% of the surface is sampled. Histopathology of excised tissue samples, is reliant on morphological information, subject to sampling errors, with significant costs in terms of time to diagnosis, and is known to be somewhat subjective.[4] However, it is likely that these architectural changes measured by histopathology are driven by early biochemical changes within tissue. The qualitative analysis of such variations provides important information in the search for a specific diagnosis, and quantitative analysis of biochemical constituents is important in measuring the extent of the disease process, in designing therapy, and in evaluating the efficacy of treatment. The ideal diagnostic test would provide an objective, rapid, preferably real-time, repeatable, non-invasive diagnosis at the point-of-care with high throughput and without prior tissue processing.

Optical diagnostics

Optical diagnostic techniques measure the interaction of light with tissue structures and component molecules to probe tissue abnormalities. Light interacts with tissue in a number of ways, including elastic and inelastic scattering, reflection of boundary layers, and absorption, leading to fluorescence and phosphorescence. Each of these techniques can be used to evaluate tissue biochemistry or morphology and lead to elucidation of the pathological state of the tissue, thus providing an additional tool in the diagnosis of disease.

Over the past decade vibrational spectroscopy techniques have demonstrated their ability to provide non-destructive, rapid, clinically relevant diagnostic information.[5] Biochemical fingerprints of tissues measured using Raman and infrared spectroscopy and analysed in conjunction with advanced chemometrics have shown great potential in the diagnostic assessment of biological material.

Raman spectroscopy

Raman spectroscopy is a powerful, non-invasive analytical technique, which has rapidly grown in its application for the interrogation of biological samples.[6] Raman spectroscopy relies on the illumination of a sample with monochromatic radiation and measurement of the inelastically scattered photons, probing the vibrational and rotational modes of molecules within the sample. A Raman spectrum is a plot of inelastically scattered intensity, as a function of the energy difference between the incident and scattered photons.

The loss (or gain) in photon energies corresponds to the difference in the final and initial vibrational energy levels of molecules involved in the interaction. The intensity of individual peaks in the Raman spectrum are linearly proportional to the concentration of the molecular constituent(s) giving rise to that peak. Therefore a precise molecular-specific fingerprint of the tissue is measured. Most biological molecules are Raman active with their own characteristic spectral fingerprint See Fig. 1 for an example of a tissue Raman spectrum with some peaks assigned; further details of assignments can be found in Table 1. Biomolecular species such as proteins, nucleic acids and lipids can all be studied.

Raman signals are inherently weak and, in addition, early diagnosis of disease requires detection of tissue molecular constituents present in low concentrations. This is accentuated by the fact that lasers with high intensity cannot be used to enhance weak signals from tissues because of the potential for sample damage. Fluorescence is another obstacle encountered in the Raman spectroscopy of biological tissue. This can be minimised by the use of near-infrared (NIR) laser excitation which generally does not induce electronic absorption in tissue chromophores; consequently, fluorescence is weak relative to the Raman signal.

Raman spectroscopy has been successfully applied to the classification of disease state by exploiting the molecular-specific fingerprint of the tissue in a number of organs[5,7] including oesophagus,[8–10] larynx,[11] bladder,[12] prostate,[13] cervix, breast,[14]

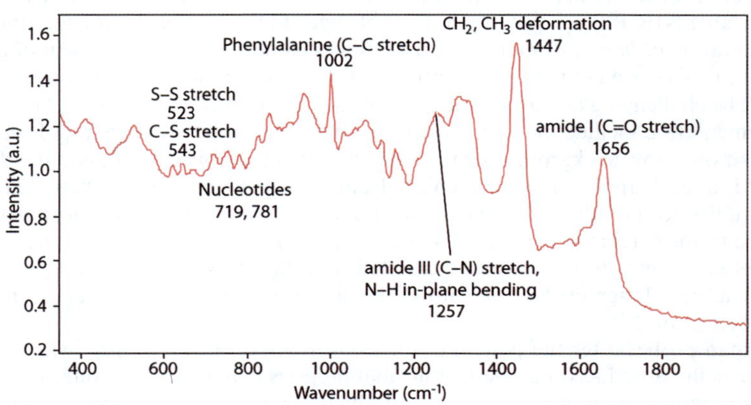

Fig. 1 Example of a Raman spectrum of oesophageal tissue with assignment of some peaks.

Table 1 Major Raman peak assignments identified in oesophageal tissue

Peak position/cm^{-1}	Assignment
490	Glycogen
524	S–S disulfide stretch proteins
621	C–C twist phenylalanine
643	C–C twist tyrosine
669	C–S stretch cystine
719	C–N (membrane phospholipid head)/nucleotide peak
755	Tryptophan
781	Cytosine/uracil
828	Out of plane ring breathing tyrosine/O–P–O stretch DNA
853	Ring breathing mode of tyrosine and C–C stretch proline ring
920	C–C stretch proline ring/glucose/lactic acid
935	C–C stretching mode of proline and valine and protein backbone (α-helix conformation)/glycogen
1001	Phenylalanine
1031	C–H in-plane bend phenylalanine
1083	C–N stretch proteins (and lipid lesser degree)
1123	C–C stretch lipids/protein C–N stretch
1155	C–C (& C–N) stretch proteins
1170	C–H in-plane bend tyrosine
1209	Tryptophan, phenylalanine
1240–1265	Amide III C–N stretch proteins, α-helix
1313	CH_3CH_2 twisting collagen/lipids
1335	CH_3CH_2 wagging collagen DNA-purine bases
1446	CH_2 bend proteins, lipids
1548	Tryptophan
1579	Pyrimidine ring, heme protein
1603	C=C in-plane bend phenylalanine tyrosine
1616	C=C stretch tyrosine tryptophan
1655	Amide I C=O stretch α-helix proteins, C=C lipid

and lymph nodes.[15] These studies indicate the potential of Raman spectroscopy as a non-invasive, real-time, diagnostic tool which measures the molecular concentrations of biochemical constituents within the tissue.

High signal-to-noise ratio (SNR) spectra can be obtained in short integration times with charge-coupled-device (CCD) detectors, single monochromators, efficient edge filters and solid-state semiconductor lasers. This has enabled the development of portable NIR Raman systems compatible with fibre optic probes facilitating implementation of *in vivo* Raman spectroscopy as a clinical tool. Development of novel Raman probes is enabling the potential of *in vivo* clinical measurements to be realised. The challenging demands placed upon the design of Raman probes for clinical implementation include the requirement for high collection efficiency, rapid spectral acquisition, a low background signal, and a well defined working distance and depth of field for each application. Customised Raman probes have been evaluated in clinical studies to identify and classify the subtle pre-malignant biochemical changes related to the carcinogenesis process in oesophageal tissue. A number of alternative probes have been developed and evaluated by other groups, but none appears to satisfy all the design criteria of a novel diagnostic tool for oesophageal pre-malignant detection.[16–19]

Raman probes offer the potential of a non-invasive, optical biopsy, with associated benefits of reduced numbers of normal biopsies being collected and processed, ensuring resection margins are clear, and obviating the need for many second visits as an immediate diagnosis could facilitate treatment at the same session.

The clinical application of Raman spectroscopy relies upon the implementation of multivariate data analysis methods to facilitate the classification of tissue samples based upon their biochemical fingerprint. Principal component analysis is an unsupervised data reduction technique which describes the major sources of independent variation in the data. Supervised classification can then be applied using information about the samples; for example the accepted gold standard information provided by the consensus opinion of expert histopathologists.[8]

Furthermore, the potential of vibrational spectroscopy to provide prognostic information is tantalising and would enable theranostics to come to the fore with customised treatment for each patient dependent on their diagnosis. The ability of Raman spectroscopy to stage transitional cell carcinoma in the bladder by measuring the surface of the lesion has been demonstrated.[20] This indicates that there are molecular changes in the tumour associated with disease progression in deeper tissues.

A further approach is to apply vibrational spectroscopy as a histopathological tool. This could eventually lead to high-throughput automated sample analysis and sorting. Normal or benign tissue samples could be excluded prior to sectioning and staining. Further samples could be analysed using unstained thin tissue sections. Raman mapping and Fourier transform infrared (FT-IR) imaging used to analyse thin tissue sections on calcium fluoride slides enables the distribution of tissue constituents to be realised.[10] The spectral data acquired effectively enables multiplexing of molecular information (which can effectively be used as digital tissue stains) since a whole array of information is gathered simultaneously. Technological developments are bringing the technologies closer to the clinical reality of spectral pathology and high throughput non-destructive measurement with high resolution.[21] Rapid Raman spectral mapping (Streamline, Renishaw) and FT-IR imaging (Spotlight, Perkin Elmer) have been used to investigate thin tissue sections of cervical, colon, oesophageal and lung tissue using multivariate statistical analysis to further enhance our understanding of the carcinogenesis process. Vibrational spectroscopic mapping of tissue sections presents increased challenges with respect to data analysis in terms of the volumes of spectral data that are generated. Care also needs to be taken in the preparation of tissue sections to ensure representative sampling with consistent section thickness and staining. Extraction of biochemical information from the spectra enables exploration of different disease states.

Experimental methods

Ethical approval for the studies was obtained from the Gloucestershire Local Research Ethics Committee. After informed consent was obtained, biopsy samples were collected from patients undergoing routine upper gastrointestinal endoscopy. Oesophageal specimens were placed on acetate paper, snap frozen in liquid nitrogen and stored at −80 °C until the Raman spectral measurements could be undertaken.

Raman probe evaluation

The novel Raman probe design has been described in detail elsewhere.[22] In summary, the probe is a miniature confocal fibre optic design, which is small enough to fit down the instrument channel of a standard medical endoscope. The optical design of the probe was optimised for the study of the carcinogenesis process in the oesophagus for application initially as a biopsy targeting tool to differentiate between normal, low-risk (Barrett's) tissue and high-risk neoplastic tissue. This would obviate the need to undertake excisional tissue biopsies and could facilitate treatment during the same endoscopy session as diagnosis.

The biopsy specimens had 7 μm frozen sections microtomed, stained with haematoxylin and eosin, and reviewed by three expert gastrointestinal pathologists blinded to each others' opinions, to act as a rigorous gold standard. The remainder of the samples were placed onto a calcium fluoride slide and passively defrosted prior to the measurement of Raman spectra on the epithelial surface using the customised probe, and a Renishaw Raman System 100 spectrometer, with 50 mW of 830 nm excitation light. Spectra were collected with 10 s and 2 s acquisition times, with 1304 Raman spectra from 123 samples measured in total.[23]

Spectra were loaded into Matlab (The Math Works, Inc.) and a fluorescent green glass energy sensitivity calibration was applied. Samples classified as homogeneous (containing only one pathology), with agreement from all three pathologists, were included in further analysis. Those with a 10 s acquisition were used to train a principal component fed linear discriminant classification model. The performance of the model was tested with the projection of the 2 s spectra on to the classification model.

Prognostic model

The aim of this study was to determine whether Raman spectroscopy can be used to detect those dysplastic cells with a higher probability of progressing to cancer.

576 Raman spectra were acquired in 30 s using a Renishaw Raman System 1000 microspectrometer with 830 nm excitation, from 25 high grade dysplasia oesophageal biopsy samples obtained from 14 patients. The spectra were analysed in conjunction with clinical follow up data collected from the patients' notes. Patients were split into four categories according to when they progressed to cancer; (1) cancer coincident at first high-grade dysplasia biopsy, (2) cancer within one year, (3) cancer at more than one year, (4) never developed cancer.

Vibrational spectroscopic mapping

10 μm thick sections of cervical tissue were microtomed onto calcium fluoride substrates. Consecutive 7 μm sections were taken above and below the mapping sections and stained with haematoxylin and eosin for histological diagnosis. Samples were defrosted prior to Raman mapping using StreamLine® technology on a customized Renishaw Raman System 1000 spectrometer coupled to a microscope fitted with a Leica 50× long working distance objective, with excitation at 830 nm (Renishaw, UK). FT-IR mapping using a Perkin-Elmer Spectrum One (Spotlight) 300 spectrometer was then undertaken. Images were acquired with 4 cm^{-1} resolution and an average of 64 scans per pixel over the wavenumber range 4000–750 cm^{-1} in transmission mode. 64 point spectra distributed across the sample were acquired in transmission mode. Mapping parameters were approximately matched between the Raman and FT-IR systems. For the Raman maps step sizes of 6.3 μm and 6.25 μm for the FT-IR maps step sizes for comparison.

The spectral mapping data was studied using principal component analysis (PCA). The Raman data was mean-centered and normalised before PCA. Saturated spectra and cosmic rays were removed using two methods; thresholding to remove outliers and zapping (interpolating between the points on either side of) the cosmic ray from the individual spectra. FT-IR data converted to absorbance and was truncated to the fingerprint region before PCA. The PC score maps and corresponding loads were plotted for comparison of the techniques and acquisition parameters.

Results

Raman probe evaluation

Fig. 2 shows that characteristic Raman spectral peaks relating to tissue biochemistry are evident in oesophageal tissue spectra measured using the probe in as little as 2 s. Raman spectral peaks seen in Fig. 1 are visible in addition to a significant background signal.

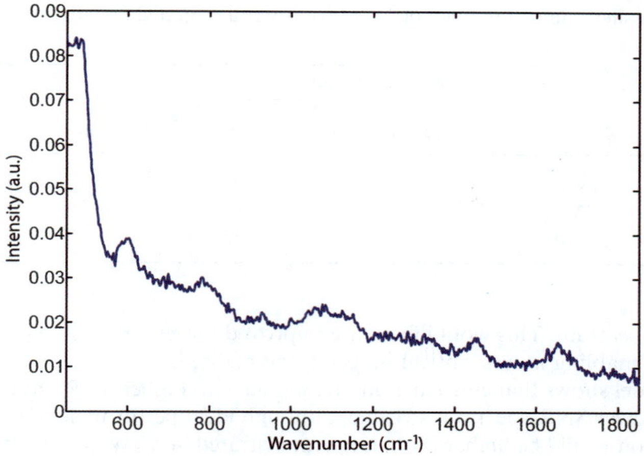

Fig. 2 Characteristic Raman probe tissue spectrum measured in 2 s.

Raman spectra have been measured from all nine pathology groups encountered in the development of oesophageal cancer. In this work the samples have been merged into normal squamous, Barrett's and neoplasia groups to demonstrate probe performance as an optical biopsy tool. Samples classified as homogeneous (containing only one pathology), with consensus agreement of all three pathologists were included in further analysis (744 spectra).

Principal component fed linear discriminant classification models were trained and tested, demonstrating an overall correct classification of 74% with sensitivity 71–88% and specificity 77–99%. Fig. 3 and Table 2 show the performance of the classification model in the separation of the three groups. This data demonstrates it is feasible to consider clinical implementation with in a clinically practical time frame

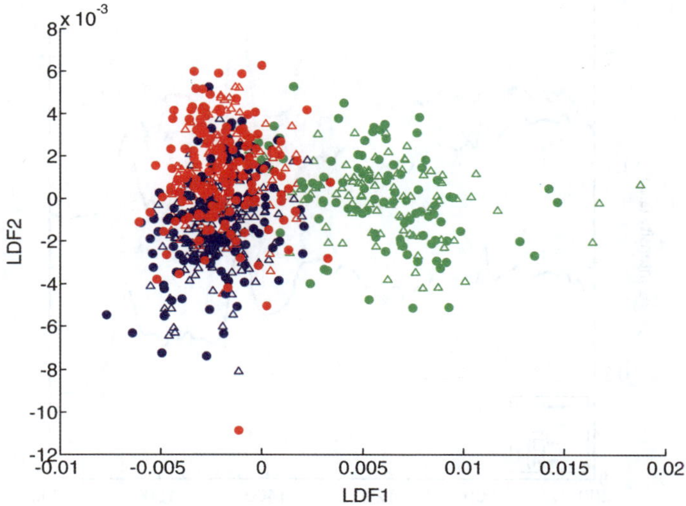

Fig. 3 Linear discriminant scatterplot of 2 s acquisition test data set (Normal = green circles, Barrett's = blue circles, Neoplasia = red circles) projected onto the training classification model (10 s acquisition: Normal = green triangles, Barrett's = blue triangles, Neoplasia = red triangles).

Table 2 Classification performance of 2 s Raman spectra projected onto the 10 s training data set

		Spectral prediction		
		Normal Squamous	Intestinal Metaplasia	Neoplasia
Pathology	Nsq	79	2	9
	IM	2	88	42
	Neo	2	42	108

of 2 s per spectrum. This would facilitate improved targeting of biopsy samples and ultimately enabling *in vivo* optical biopsy to be realised.

This model shows that construction of a model with higher SNR data and prediction with lower SNR spectra is feasible, although ideal performance for clinical implementation would be higher than that demonstrated here. Work is currently being undertaken to improve the fibre coupling between the spectrometer and probe and increase the signal-to-noise ratio of the measured tissue spectra.[22]

Prognostic model

The performance of Raman spectroscopy as a prognostic model for the determination of the time to the development of cancer from the first high-grade dysplasia biopsy was tested using PCA-fed LDA multivariate classification models. Fig. 4 shows the mean spectra for the four clinical categories. Performance of the training classification model showed 82% of samples were correctly classified, with 81–97% sensitivity and 92–99% as shown in Fig. 5.

Vibrational spectroscopic imaging

Results from one mapped cervical tissue section are shown here, Fig. 6. Discussion of results from oesophageal mapping studies can be found elsewhere.[21]

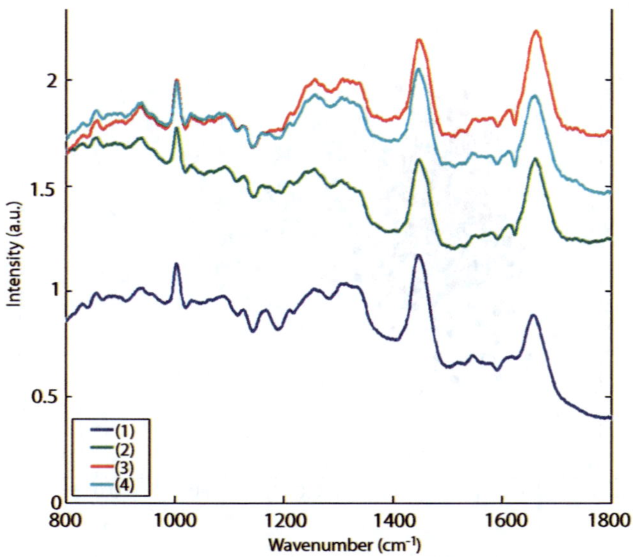

Fig. 4 Mean Raman spectra measured on high grade dysplasia biopsy samples according to the four clinical categories (1) never developed cancer (2) cancer at first biopsy (3) cancer within one year (4) cancer after one year.

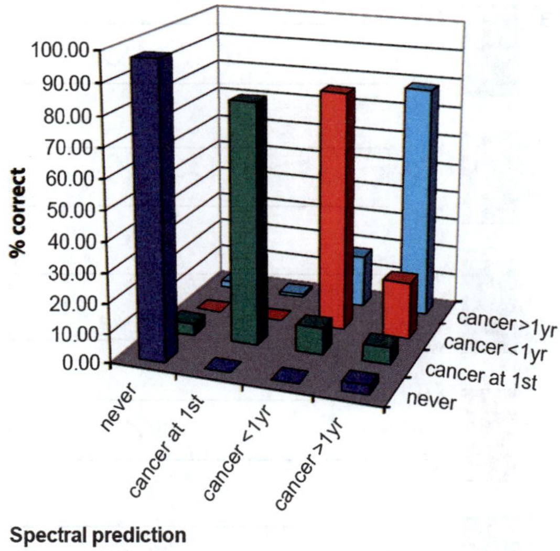

Fig. 5 Performance of a PC-fed LDA classification model demonstrating the differentiation of Raman spectra measured on high-grade dysplasia biopsies according to the time interval to the development of cancer.

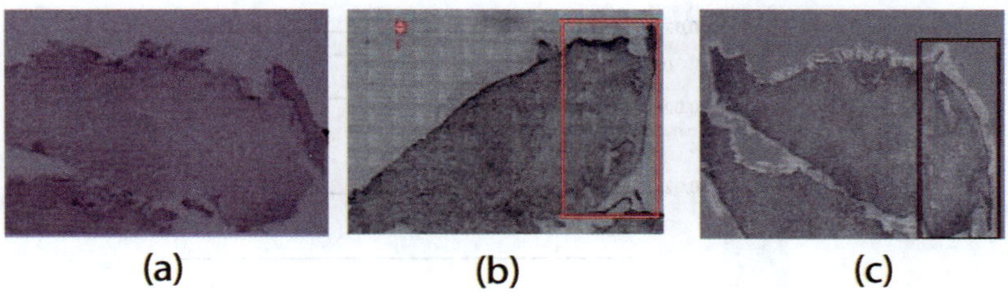

Fig. 6 (a) 7 μm thick haematoxylin- and eosin-stained ectocervical tissue section, with normal epithelium seen along the right-hand edge of the sample. Contiguous sections were used for spectral measurement. (b) White-light image of the region mapped using FT-IR (c) White-light image of the region mapped using Raman spectroscopy.

The spectral mapping data was analysed using principal component analysis (PCA). The Raman data was mean-centered and normalised before PCA. Saturated spectra and cosmic rays were removed using two methods; thresholding to remove outliers and zapping the cosmic ray from the individual spectra. FT-IR data were converted to absorbance and truncated to the fingerprint region before PCA. The PC score maps and corresponding loads were plotted for comparison of the techniques and acquisition parameters as shown in Fig. 7(a) and (b).

Evaluation of the spectral peaks in the PC loadings provides information on the biochemical constituents that are seen to delineate regions within the tissue. PC 5 on the FT-IR data set is thought to delineate the basement membrane. Cosmic rays in the Raman data sets are seen to pose a challenge to data analysis. Exclusion of spectra containing cosmic rays reduces the size of the data set and makes the score maps look speckled.

Discussion

It has been shown that tissue Raman spectra acquired in clinically practical timescales of a few seconds can be obtained using novel designed Raman probes. It

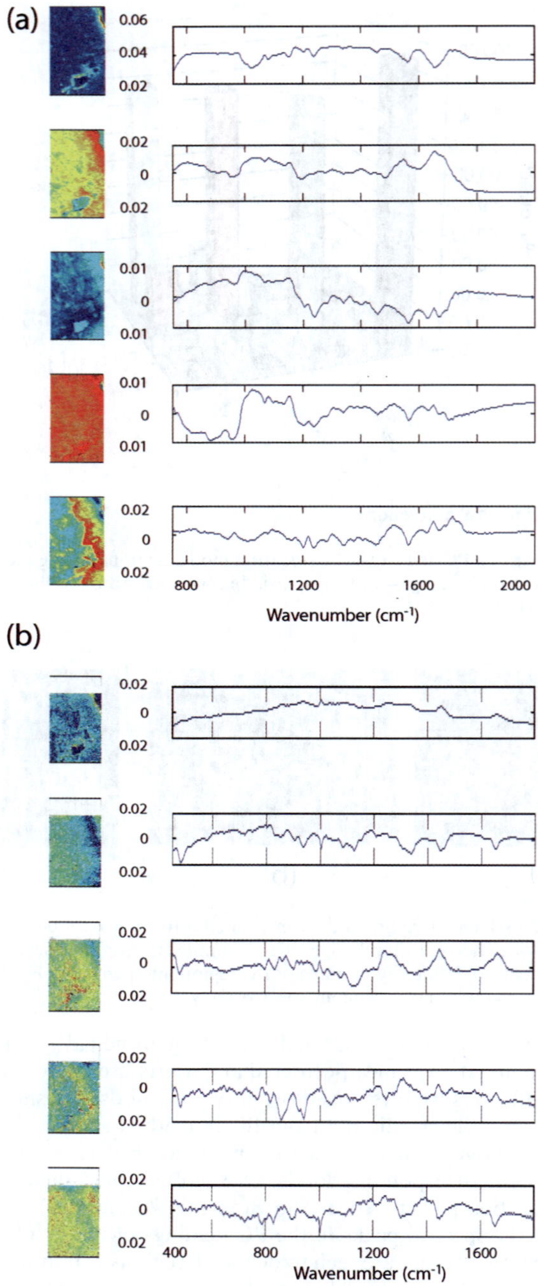

Fig. 7 (a) PCs 1 to 5 score maps and corresponding loadings for 6.25 μm step FT-IR map of 10 μm thick cervix tissue section. (b) PCs 1 to 5 score maps and corresponding loads for 6.3 μm step Raman map (7.3 s acquisition per point).

has furthermore been demonstrated that these spectra can be differentiated using multivariate statistical classification models trained using the consensus opinion of expert pathologists, the accepted gold standard. This paves the way for clinical implementation of the technique once issues related to spectrometer coupling and packaging the probes suitable for *in vivo* use are overcome. This will enable Raman spectroscopy to be used as a biopsy targeting tool during surveillance in organs which develop cancer *via* pre-cancerous lesions such as oesophageal, colorectal, cervical and laryngeal cancers. Initially, this would reduce the large number of normal biopsies that are currently collected unnecessarily.

Transferability and repeatability of tissue Raman spectra measured on different systems has been demonstrated.[24] Automated spectral analysis methods for near-real time diagnosis of the large data sets is a challenge which is currently being addressed.[25,26]

The indication from the prognostic model that the Raman spectral information can be used to predict patient outcome is exciting, although this is demonstrated here on a limited data set. This small pilot study has shown the power of spectral biomolecular fingerprinting in predicting the outcome of patients with high grade dysplasia in Barrett's oesophagus. Hopefully in the future this will lead to the *in vivo* diagnosis and prognosis of patients with dysplasia and Barrett's oesophagus.

Histological features of the tissue are clearly differentiated with both Raman and FT-IR mapping and additional information on the distribution of biochemical constituents is gleaned. Work is ongoing to correlate spectral information with molecular markers of malignant progression. Although current rates of imaging are too slow for *in vivo* measurement, ultra-low spatial resolution Raman (ULSRR) mapping using fibre probes has been demonstrated with potential for *in vivo* application for surveillance or margin assessment.[27]

Conclusions

Vibrational spectroscopy generates a wealth of both biochemical information with potential to identify markers associated with pre-malignant and carcinogenic change. There is potential for the non-invasive detection of disease, together with new information for classifying, grading, and evaluating the progression of malignant neoplasms. Analysing the biochemical state of the tissue appears to enable potentially prognostic information to be gleaned. There are exciting recent developments have enabled the probing of tissue samples at depth with huge potential for breast and prostate cancer diagnostics.[28]

Automation of vibrational spectroscopy for histopathological examination faces real challenges. Automated pathological diagnosis using FT-IR imaging and/or Raman mapping in conjunction with rapid spectral analysis has significant potential. Further understanding of the biochemical changes associated with carcinogenesis is ongoing and will help to support and validate biochemical changes interpreted from the spectral data.

Acknowledgements

The authors gratefully acknowledge funding from the UK Department of Health, Health Technology Devices Fund. National Institute for Health Research (Career Scientist Fellowship) for NS, Royal Society (Dorothy Hodgkin Fellowship) for CK, National Institute for Health Research (Post-doctoral Fellowship) for JH.

Jo Motte and Christine Braune are thanked for the preparation of tissue sections. Prof. Neil Shepherd, Dr Bryan Warren and Prof. Karel Geboes are thanked for histopathological analysis of the oesophageal biopsy specimens used, Dr Linmarie Ludeman for the cervix histopathology, Mr Simon Dwerryhouse for his assistance with the prognostic model and Prof. Hugh Barr for recruiting the oesophageal patients and for input into teaching us about the clinical need.

References

1 Cancer Research UK Incidence and Mortality Statistics (2005) http://info.cancerresearchuk.org/cancerstats/mortality/?a=5441).

2 J. A. Jankowski, R. F. Harrison, I. Perry, F. Balkwill and C. Tselepis, *Lancet*, 2000, **356**, 2079–2085.
3 B. J. Reid, P. L. Blount, C. E. Rubin, D. S. Levine, R. C. Haggitt and P. S. Rabinovich, *Gastroenterology*, 1992, **102**, 1212–1219.
4 E. Montgomery, M. P. Bronner, J. R. Goldblum, J. K. Greenson, M. M. Haber, J. Hart, L. W. Lamps, G. Y. Lauwers, A. J. Lazenby, D. N. Lewin, M. E. Robert, A. Y. Toledanao and K. Washington, *Hum. Pathol.*, 2001, **32**, 268–378.
5 C. Kendall, M. Isabelle, F. Bazant-Hegemark, J. Hutchings, J. Babrah, L. Orr, R. Baker and N. Stone, *Analyst*, 2009, **134**, 1029–1045.
6 A. Mahadevan-Jansen and R. Richards-Kortum, *J. Biomed. Opt.*, 1996, **1**, 31–70.
7 E. B. Hanlon, R. Manoharan, T.-W. Koo, K. E. Shafer, J. T. Motz, M. Fitzmaurice, J. R. Kramer, I. Itzkan, R. R. Dasari and M. S. Feld, *Phys. Med. Biol.*, 2000, **45**, R1–R59.
8 C. Kendall, N. Stone, N. Shepherd, K. Geboes, B. Warren, R. Bennett and H. Barr, *J. Pathol.*, 2003, **200**, 602–609.
9 N. Stone, C. Kendall, N. Shepherd, P. Crow and H. Barr, *J. Raman Spectrosc.*, 2002, **33**, 564–573.
10 G. Shetty, C. Kendall, N. Shepherd, N. Stone and H. Barr, *Br. J. Cancer*, 2006, **94**, 1460–1464.
11 N. Stone, P. Stavroulaki, C. Kendall, M. Birchall and H. Barr, *Laryngoscope*, 2000, **110**, 1756–1763.
12 P. Crow, N. Stone, C. A. Kendall, J. S. Uff, J. A. M. Farmer, H. Barr and M. J. P. Wright, *Br. J. Cancer*, 2003, **89**, 106–109.
13 N. Stone, M. C. Hart Prieto, P. Crow, J. Uff and A. W. Ritchie, *Anal. Bioanal. Chem.*, 2007, **387**, 1657–1668.
14 J. Smith, C. Kendall, A. Sammon, J. Christie-Brown and N. Stone, *Technol. Cancer Res. Treat.*, 2003, **2**, 4.
15 J. Horsnell, P. Stonelake, J. Christie-Brown, G. Shetty, J. Hutchings, C. Kendall and N. Stone, *Analyst*, 2010, (submitted).
16 J. T. Motz, M. Hunter, L. H. Galindo, J. A. Gardecki, J. R. Kramer, R. R. Dasari and M. S. Feld, *Appl. Opt.*, 2004, **43**, 542–554.
17 P. Crow, A. Molckovsky, N. Stone, J. Uff, B. Wilson and L.-M. Wong Kee Song, *Urology*, 2005, **65**, 1126–1130.
18 Y. Hattori, Y. Komachi, T. Asakura, T. Shimosegawa, G. Kanai, H. Tashiro and H. Sato, *Appl. Spectrosc.*, 2007, **61**, 579–584.
19 A. Nijssen, S. Koljenović, T. C. Bakker Schut, P. J. Caspers and G. J. Puppels, *J. Biophotonics*, 2009, **2**, 29–36.
20 P. Crow, J. S. Uff, J. A. Farmer, M. P. Wright and N. Stone, *BJU Int.*, 2004, **93**, 1232–1236.
21 J. Hutchings, C. Kendall, B. Smith, N. Shepherd, H. Barr and N. Stone, *J. Biophotonics*, 2009, **2**, 91–103.
22 J. C. C. Day, R. Bennett, B. Smith, C. Kendall, J. Hutchings, N. Stone, C. Born and S. Yu, *Phys. Med. Biol.*, 2009, **54**, 7077–7088.
23 C. Kendall, J. Day, B. Smith, J. Hutchings, N. Shepherd, H. Barr and N. Stone, *Analyst*, 2010, DOI: 10.1039/c0an00536c.
24 N. Stone, C. Kendall, and H. Barr, in *Vibrational Spectroscopy for Medical Diagnosis*, ed. M. Diem, J. Chambers and P. Griffiths, Wiley-VCH, Chichester, 2008.
25 M. Isabelle, K. Rogers and N. Stone, *J. Biomed. Opt.*, 2010, **15**, 026030.
26 J. Hutchings, C. Kendall, N. Shepherd, L. Ludeman, H. Barr and N. Stone, *J. Biomed. Opt.*, 2010, (in press).
27 J. Hutchings, C. Kendall, N. Shepherd, H. Barr, J. Day and N. Stone, *Proc. SPIE–Int. Soc. Opt. Eng.*, 2010, 75600N, DOI: 10.1117/12.858803.
28 P. Matousek and N. Stone, *Analyst*, 2009, **134**, 1058–1066.

Combining functionalised nanoparticles and SERS for the detection of DNA relating to disease

Duncan Graham,* Ross Stevenson, David G. Thompson, Lee Barrett, Colette Dalton and Karen Faulds

Received 26th May 2010, Accepted 10th June 2010
DOI: 10.1039/c005397j

DNA functionalised nanoparticle probes offer new opportunities in analyte detection. Ultrasensitive, molecularly specific targeting of analytes is possible through the use of metallic nanoparticles and their ability to generate a surface enhanced Raman scattering (SERS) response. This is leading to a new range of diagnostic clinical probes based on SERS detection. Our approaches have shown how such probes can detect specific DNA sequences by using a biomolecular recognition event to 'turn on' a SERS response through a controlled assembly process of the DNA functionalised nanoparticles. Further, we have prepared DNA aptamer functionalised SERS probes and demonstrated how introduction of a protein target can change the aggregation state of the nanoparticles in a dose-dependant manner. These approaches are being used as methods to detect biomolecules that indicate a specific disease being present with a view to improving disease management.

1 Introduction

The development of new techniques capable of providing useful diagnostic or clinical markers is an area of research growth. Issues of cost, both of consumables and equipment, ease of use and robustness dominate the uptake of use of any new technologies. However, the advent of new instrumentation coupled with complex optics are leading to new, more sensitive diagnostic techniques capable of providing a genuine clinical benefit. Here we report on how we use metal nanoparticles to detect specific DNA sequences when combined with SERS.

Coinage metal nanoparticles (1–100 nm) have been used as both delivery vehicles for biomolecular uptake into cells and as detection probes for a variety of targets. They are easily synthesised by a variety of methods, can be stored long-term and consequently have low associated costs. Gold[1,2] and silver[3] are the most commonly used nanoparticles and their small size (comparable to mid-large proteins) leads to unique physical characteristics not observed with bulk materials or molecular salts.[4–6] Nanoparticles are stable, have a large surface area to mass ratio, can be functionalised with a choice of ligands, and their optical brightness and massive scattering efficiency makes them ideally suited for *in vitro* and *in vivo* applications.[7,8] The characteristic colour of gold and silver nanoparticle suspensions is due to the localised surface plasmon resonance (LSPR) of the cloud of electrons associated with the surface of the nanoparticle. The absorbance maximum of the LSPR of a colloidal suspension is dependant upon a range of factors including the metal

Centre For Molecular Nanometrology, WestCHEM, Department of Pure and Applied Chemistry, University of Strathclyde, 295 Cathedral Street, Glasgow, UK G1 1XL. E-mail: duncan.graham@strath.ac.uk

used, size, shape, and the dielectric constant of the medium.[4] The LSPR is also dependant upon the aggregation state of the nanoparticles with larger aggregates red-shifting the LSPR band and leading to spectroscopic (Raman/SERS) and visual colour changes. SERS is an optical spectroscopic technique that provides molecularly specific fingerprints directly related to the vibrational and stretching modes of a target analyte. Developed in 1974, it offers researchers a significant advancement over conventional Raman spectroscopy with sensitivity typically 10^6 greater than the classical technique.[9] Raman spectroscopy is an inherently weak process relying on the collection of inelastically scattered photons. With an estimated 1 in every 10^8 photons scattered inelastically, the technique requires large, expensive equipment and is predominantly a fringe technique. The observation that adsorption of the analyte on a roughened metallic surface leads to significant signal enhancement led to a renaissance in Raman spectroscopy and the technique is now accepted as one of the most sensitive and specific detection methods available. Use of gold and silver nanoparticles in SERS has been extensively studied and enhancement of the Raman signal due to the interaction between the surface adsorbed analyte and the LSPR of the nanoparticle. Further enhancement can be achieved by the controlled formation of electromagnetic 'hot spots' between 2 or more aggregated nanoparticles.[10] SERS gives rise to exquisite sensitivity with single molecule detection reported.[11] Further, SERS offers excellent opportunities in the detection of multiple analytes as the narrow, sharp, fingerprint spectra lend themselves to a multicomponent detection technique.[12–14]

Gold and silver nanoparticles can be functionalised with a range of chemical moieties, commonly achieved by coupling surface atoms to a thiol group. Oligonucleotide-gold nanoparticle (OGN)[15,16] and oligonucleotide-silver nanoparticle (OSN)[17,18] conjugates provide robust and versatile detection systems. Synthesis of OGN and OSN conjugates is relatively simple. A thiol-modified oligonucleotide is introduced to a suspension of gold or silver nanoparticles and is salt aged over a short period (2–48 h). When oligonucleotide-nanoparticle conjugates are used for detection, two different methodologies can be applied: (1) using the nanoparticle as a label for an analyte of interest and (2) inducing the assembly of the nanoparticles to create a measurable visual or spectroscopic response.

Nanoparticle assembly can be controlled to such an extent that a biological interaction, such as DNA hybridisation, can be detected by the changes caused in the LSPR.[19,20] An assembly-based assay using OGN or OSN conjugates typically uses two batches of nanoparticle probes functionalised with different, non-complementary oligonucleotides. Addition of a target sequence, complementary in part to both conjugates, can cause hybridisation to occur. The nanoparticle assembly ensures individual nanoparticles are brought into close enough proximity to red-shift the LSPR band and allow DNA detection solely through colorimetry

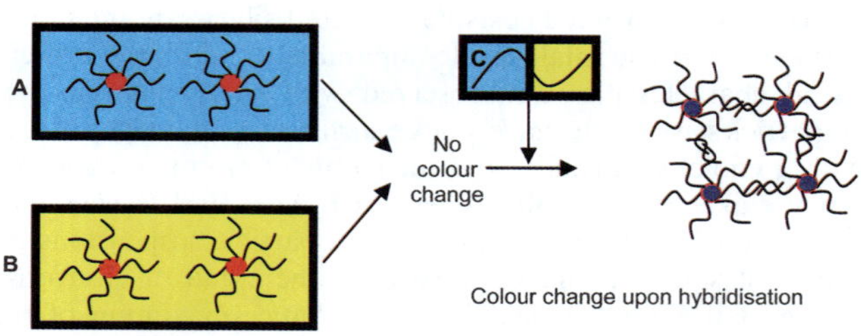

Fig. 1 Schematic representation of an OGN/OSN conjugate assay. Nanoparticle conjugates functionalised with two different, non-complementary oligonucleotides (A + B) hybridize upon addition of target oligonucleotide (C). The molecular recognition event and the formation of structured nanoparticle assemblies cause a visual colour change in the nanoparticle suspension.

(Fig. 1). OGN and OSN conjugates have been used to detect a wide variety of biologically relevant targets including genomic DNA,[21] bacterial DNA,[22] and duplex and triplex binding sequences[23] and have been coupled with optical,[24] electrochemical[25] and vibrational[26] detection systems.

Protein detection is also possible using OGN conjugates in the form of an aptamer sequence adsorbed to the nanoparticle surface. Aptamers are nucleic acid ligands selected to bind a target molecule with high selectivity and specificity. Typically short sequences of both ssDNA and RNA of 20–80 nucleotides (~6–26 kDa) have been successfully selected to interact with a range of analytes including proteins, peptides and small molecules. Aptamers form unique tertiary folded structures that present grooves and clefts capable of binding targets with affinities that often rival if not surpass an antibody alternative. First developed in 1990,[27,28] most aptamers are selected using the 'systematic evolution of ligands by exponential enrichment' (SELEX) process. SELEX is a reiterative amplification technique that screens a starting library of up to 10^{15} unique sequences to a final optimized pool of 10–100 candidates. The real strengths of the technique stem from the diversity of the starting pool, which can lead to highly sensitive aptamers being selected with extremely high specificity. An example being the ability of a theophylline aptamer to molecularly discriminate between its target and caffeine.[29] These small molecules differ by a single methyl group, and the theophylline aptamer has a greater than 10 000-fold preference for its target over caffeine.[29]

The properties of aptamers are commonly compared to antibodies due to their molecular targeting abilities. Both species can be used for sensitive and specific detection though aptamers have a number of further advantages over antibody alternatives. Aptamers are selected with a relatively easy and effective selection technique. Once selected, they can be reproduced in sizeable amounts using fast, cheap synthetic approaches with little batch to batch variability and where simple, diverse modifications are possible. Further advantages include their relatively small size, their low toxicity and their slow degradation kinetics compared to any protein-based alternative. A sign of their success and potential as clinical drug candidates can be illustrated in that in less than two decades since their introduction to the scientific community, the FDA approved pegaptanib (or Macugen®, marketed by Pfizer) as an aptamer drug for use in the treatment of age-related macular degeneration (AMD).[30] More recently, researchers have looked to functionalize nanoparticles with DNA and aptamer sequences.

Aptamer functionalised nanoparticles have been reported for a diverse range of targets including prostate specific membrane antigen (PSA),[31] platelet derived growth factor (PDGF),[32] and cocaine.[33] The cocaine aptamer is specially noteworthy as it utilises a unique 'reverse molecular beacon' approach. In the absence of target, the aptamer adopts an almost linear conformation. Addition of target causes the aptamer to fold around the drug and brings the 3′ end of the oligonucleotide into close contact with the surface of the nanoparticle. When the 3′ end is functionalised with a fluorescent dye, the conformational change on binding switches on the SERS and the cocaine can be detected by the SERS signal.

Here we show the development of oligonucleotide-nanoparticle conjugates with enhanced stability for the assembly-based detection of the *mecA* gene of methicillin-resistant *Staphylococcus aureus* (MRSA) and the use of aptamer functionalised nanoparticles and SERS for the detection of thrombin.

2 Results and discussion

Development of a MRSA detection oligonucleotide-nanoparticle system

Functionalised nanoparticles offer excellent opportunities as SERS active DNA probes. SERS from both OGN and OSN conjugates has been used to detect specific oligonucleotide sequences at low concentrations (Fig. 2).[34–37]

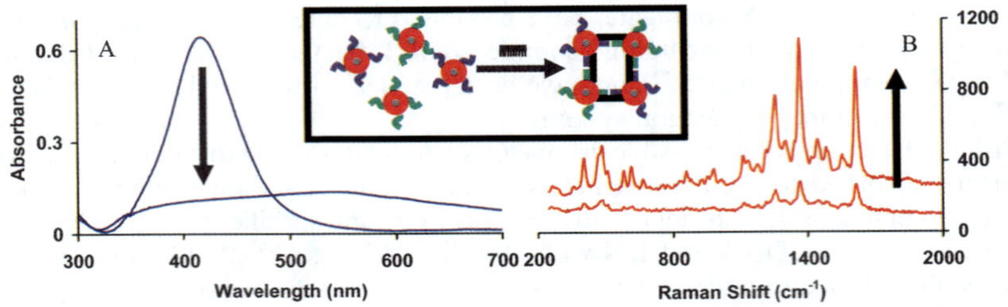

Fig. 2 SERS and LSPR analysis of sequence specific DNA nanoparticle probes. Combined diagram showing the simultaneous decrease in LSPR (A) and increase in SERS intensity (B) caused by the hybridisation of dye-coded DNA functionalised silver nanoparticles to an unmodified complementary target oligonucleotide (inset).

The first approach used dye-coded oligonucleotide-silver nanoparticle (DOSN) conjugates to detect an unmodified target oligonucleotide.[34] Two batches of silver nanoparticles were coated with different surface-seeking Raman reporter molecules.[38] Each batch was then functionalised with probe oligonucleotides non-complementary to each other, yet antisense to half a target sequence. When exposed to the target, the probe sequences hybridise, causing aggregation of the nanoparticles, changing the LSPR and increasing the intensity of the SERS response. Initial work discriminated between targets that were either fully complementary or non-complementary, but this was later refined to discriminate single base mismatches in the target sequence at room temperature.[39] Although performed using model sequence oligonucleotides, it was assumed that extending this system to sequences of interest would present no significant problem. However, attempts to move to biologically relevant targets found that far from being universally suited to all sequences, the stability and ability of the conjugates was found to be highly dependant on the sequence used. Short-term and long-term conjugate stability is vital for a successful assembly-based assay; short-term stability to ensure a robust control as the response from any non-complementary sequence should not change and long-term stability to ensure repeatable results over time. The reasons assumed for the

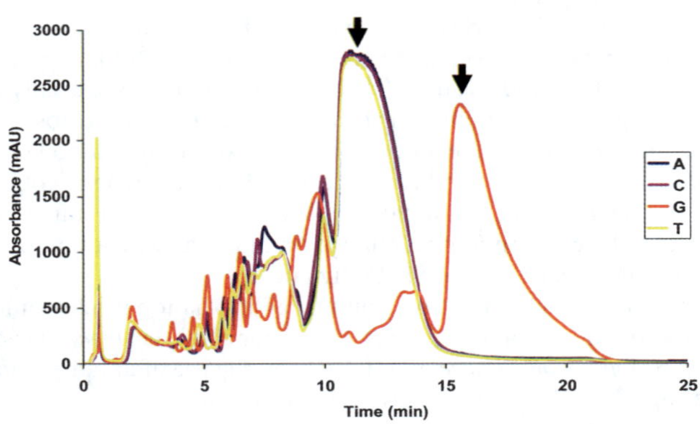

Fig. 3 HPLC analysis of four synthetic oligonucleotides that differ by a single base. The traces of the final synthesis products (designated with arrows) of four thiol-modifed oligonucleotide sequences from their failure sequences are illustrated. The sequences were (5′–3′): A, thiol-AAAAAAAAAAGGACAACCT; C, thiol-AAAAAAAAAAGGACCACCT; G, thiol-AAAAAAAAAAGGACGACCT; T, thiol-AAAAAAAAAAGGACTACCT. Note that the sequences only differed by a single base. The ten adenine bases act as a 'tether', distancing the sequence of interest from the nanoparticle surface, minimising any steric effects that could affect hybridisation efficiency.

sequence dependence are varied. Sequences that contain a high degree of self-complementarity are unsuitable as the conjugates self-hybridise. Sequences that take some form of secondary structure can also be unsuitable (shown in Fig. 3).

Fig. 3 contains the HPLC traces from the purification of four oligonucleotide sequences that differ only by a single base. Traces were obtained using ion-exchange chromatography, separating DNA sequences based upon the number of negatively charged phosphate residues present. The data indicates that the sequence containing guanine at the midpoint is retained longer than the others and suggests formation of a secondary structure or the possible incorporation of a second oligonucleotide. Significantly, oligonucleotide-nanoparticle conjugates could be successfully created using all except that sequence containing the mid-point guanine. This issue suggests applications that require specific sequences for disease detection may need further surface stabilisation or passivation to enable a SERS probe to work to the best of its ability. The secondary-structure dependence could also impact the creation of aptamer-nanoparticle conjugates which can form highly convoluted structures.

For this reason we have undertaken extensive research to define suitable sequences as well as developing a new, more stable form of oligonucleotide attachment. Although linkers containing multiple thiol groups can be used to improve the stability of oligonucleotide-nanoparticle conjugates, 5′ mono-thiol oligonucleotides are currently the easiest, cheapest and therefore most commonly used approach for the attachment of DNA to a nanoparticle. Multiple thiol attachment groups have been shown to greatly increase the stability of OGN and OSN conjugates to higher salt concentrations, displacement molecules and high temperatures.[17,40,41] Our developments have led to the commercialisation of multiple thiol anchor groups based on thioctic acid.[41] In that particular study model sequences of both OGN and OSN conjugates were used throughout. Fig. 4 shows the absorbance data from OGN conjugates created using single- or dual-thiol attachments. The approach is significant as it used a sequence complementary to part of the *mecA* gene of MRSA, a relevant biological target.

Fig. 4 also illustrates the stability of OGN conjugates when treated with DTT, a small molecule that displaces thiol moieties from the nanoparticle surface. Fig. 4A and 4B compare the increase in absorbance at 675 nm over time and

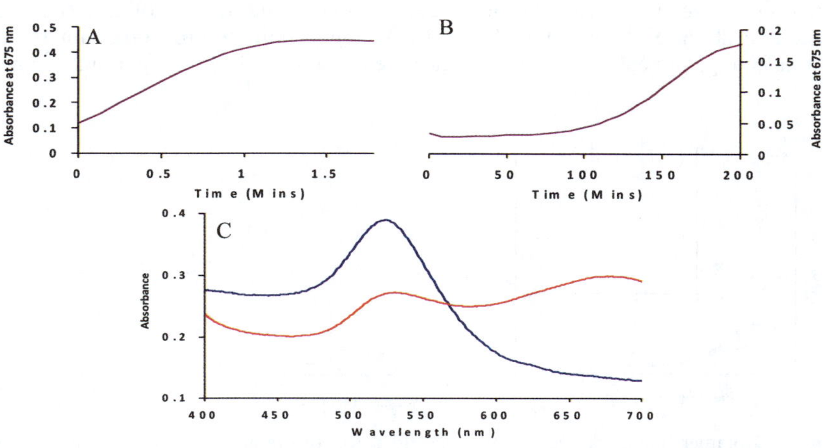

Fig. 4 Analysis of thiol-nanoparticle stability. UV/vis absorbance analysis of oligonucleotide probe conjugates before and after the addition of dithiothreitol (DTT). The absorbance of mecA-OGN conjugates monitored at 675 nm with (A) monothiol attachment and (B) multiple thiol attachment after the addition of DTT. The UV-Vis spectra of OGN conjugates (C) before (blue) and after (red) the addition of DTT. Note the formation of a second maximum at 675 nm.

correspond to the aggregation of the nanoparticles. The data indicates the mono-thiol attachment is displaced much quicker than the dithiol anchor thus providing a less stable linker molecule. Conjugates created with thioctic acid, containing the *mecA* complement sequence, were more stable than those created with the single thiol attachment This shows that surface attachment chemistry is crucial in producing probes suitable for use in diagnostic applications.

Development of a thrombin nanosensor

As demonstrated with the *mecA* OGNs, nanoparticles offer facile, robust and specific SERS detection platforms. DNA-DNA interactions are well studied and the sense-antisense attraction provides researchers an excellent platform to aggregate nanoparticles. Less well studied are DNA-protein interactions and specifically the use of aptamer-functionalised nanoparticles as a sensitive protein detection assay. Our approach looked at the feasibility of using aptamer-functionalised nanoparticles to detect thrombin. Thrombin protein plays a key role in homeostasis, as well as regulating many inflammation processes and aiding tissue repair.[42] In blood, thrombin is typically found at low-nanomolar levels, whereas significantly lower levels can often be indicative of a blood coagulation abnormality. The thrombin aptamer was one of the earliest published, and is one of the most studied DNA aptamers.[42] X-Ray crystallographic analysis of the DNA 15-mer interacting with the protein found that the aptamer forms a G-quartet, stabilized by Hoogstein hydrogen bonding.[43] The Bock aptamer[42] interacts at two distinct regions on the thrombin molecule, the fibrinogen exosite and the heparin binding site.[44] Here we show how the LSPR and Raman enhancement effects observed when nanoparticles are brought into close proximity can be exploited by using SERS active silver nanoparticles functionalized with thrombin DNA aptamers.

Mixed monolayer Raman nanoparticle probes were prepared using the same method as DOSN conjugates but with a 5′ thiol-modified version of the Bock aptamer (5′ - (thiol) TTT TTT TTT TTT TTT GGT TGG TGT GGT TGG). Thrombin was added to the colloidal suspension and UV spectra were recorded periodically over a 90 min period (shown in Fig. 5A). The data indicates that molecular recognition by the aptamer probes induces a change in the LSPR. This suggests that the aptamers remain functional when adsorbed to the nanoparticle surface. SERS analysis was then performed to observe the Raman signal from probes in the presence and absence of thrombin (Fig. 5B). Although both spectra look similar, the subtle change in peak intensities is indicative of nanoparticle aggregation. Further,

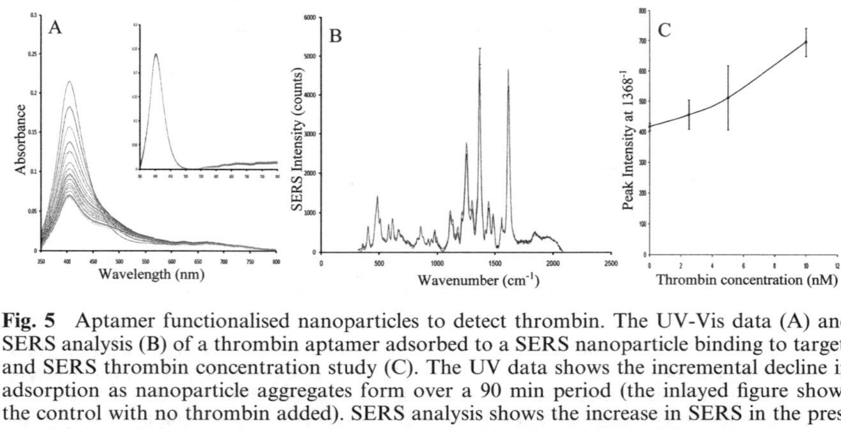

Fig. 5 Aptamer functionalised nanoparticles to detect thrombin. The UV-Vis data (A) and SERS analysis (B) of a thrombin aptamer adsorbed to a SERS nanoparticle binding to target, and SERS thrombin concentration study (C). The UV data shows the incremental decline in adsorption as nanoparticle aggregates form over a 90 min period (the inlayed figure shows the control with no thrombin added). SERS analysis shows the increase in SERS in the presence (pink) and absence of thrombin (blue). Figure C shows the effect of thrombin concentration on the change in intensity of the peak at 1368 cm^{-1}.

a concentration-dependant Raman response was observed (Fig. 5C) where the peak height increases with an increase in concentration of added thrombin. Comparison of DNA and thrombin detection using DOSN conjugates show that the magnitude of the response for unmodified oligonucleotide and protein are quite different, with the oligonucleotide detection yielding a much greater change in intensity (Fig. 2 and Fig. 5B). The recognition of the protein shows a modest SERS response which is probably due to the large size of the thrombin. This means that although active and bound to the target, the aptamer nanoparticles may be too far apart to allow a plasmon overlap, hindering electromagnetic SERS enhancement and leading to the modest SERS improvements over the monodisperse colloidal suspension.

The turning on of SERS through the interaction of the aptamer nanoparticles with the protein target shows that this concept can be extended from a purely nucleic acid system and offers opportunities in terms for protein detection related to disease.

3 Conclusion

DNA-functionalised SERS probes offer excellent opportunities in molecular recognition. The use of mixed monolayers on nanoparticles, containing a Raman dye and specific oligonucleotide sequences, can lead to a novel range of detection systems capable of screening both nucleotide sequences and proteins. The approaches developed show how DNA-functionalised nanoparticle probes can be used in the detection of DNA sequences related to disease and in clinically relevant protein targets. Although work is still in the preliminary stages, it is possible to envisage how such systems could offer real opportunities to sensitively detect the expression of multiple genes and/or proteins in a cellular environment. Further, the ability of SERS to detect two very different recognition events should allow for the multiplexed detection of biological interactions including protein-protein and protein-small molecule. Recent developments in *in vitro* and *in vivo* nanoparticle delivery, coupled to such a highly sensitive spectroscopic technique, may yield a highly sensitive and specific technique of use in clinical diagnosis.

4 Experimental

All reagents were purchased from Sigma Aldrich, UK, apart from those used in oligonucleotide synthesis which were purchased from Link Technologies, UK. Aptamer sequences were purchased from Europhins MWG Operon, Germany.

Oligonucleotide-nanoparticle synthesis

All conjugates were synthesised by previously published methods.[18,34] Both gold[45] and silver[46] nanoparticles were synthesised *via* citrate reduction methods. Briefly, a 10 μM solution of thiol or thioctic acid modified DNA prepared in 60 mM phosphate buffer pH 8.5 was desalted using a Sephadex superfine desalting column fitted to a Dionex HPLC using distilled water as eluent directly onto a 3 mL solution of concentrated gold nanoparticles or as prepared silver nanoparticles. For the synthesis of DOSN conjugates the silver nanoparticles were first exposed to a Raman reporter molecule for 24 h at a final concnetration of 1×10^{-5} M. The solutions were then left in the dark for 24 h before gradual salt aging. The conjugates were then centrifuged twice at 7000 rpm for 20 min with resuspension in 0.3 M PBS pH 7.

SERS

A Leica DM/LM microscope equipped with an Olympus 20×/0.4 long working distance objective was used to collect 180° backscattered light from a standard cuvette. The spectrometer system was a Renishaw inVia and a Spectra Physics Argon ion laser (514.5 nm) was used as the excitation source with a power output measured to be 6 mW at the sample. Dielectric edge filters were used to reject Rayleigh scattered light.

DOSN assay

For SERS analysis, a 30 pM solution of DOSN conjugates was prepared. An unmodifed complementary oligonucleotide was added to this solution in a 100× excess. For UV-Vis analysis the DOSN conjugate solution concentration was 10 pM. For both techniques, scans were taken immediately after target addition and 30 min later.

mecA-OGN stability

2 nM of OGN conjugates were prepared in a quartz glass cuvette. To this an aliquot of DTT was added at a final concentration of 10 mM. The change in absorbance was then monitored using a Cary Win 300 UV-Visible spectrometer.

HPLC

HPLC traces were obtained using a Dionex P680 pump fitted with a UVD 170U detector and a 1 mL ResourceQ column with sodium perchlorate used as the counter ion.

Thrombin aptamer nanoparticles

DOSN conjugates were prepared using the method previously stated apart from that a buffer containing 0.21 M NaCl, 0.01 M KCl, 9 mM Na phosphate buffer pH 7.5 was used throughout rather than 0.3 M PBS.

UV-vis analysis: Thrombin (20 nM) was added to OSN conjugates (10 pM) and the change in absorbance was recorded every 5 min over a 90 min period using a Cary Win 300 UV-Visible spectrometer. The control study repeated the work in the absence of thrombin.

SERS: Thrombin (5 nM) was added (or buffer for control) to DOSN conjugates (50 pM) and allowed to incubate for 45 min. An extended scan was performed (300–2100 cm^{-1}) using a 10 s accumulation time. Analysis was performed in triplicate. The concentration analysis used functionalised nanoparticles (10 pM) incubated with thrombin (0, 2.5, 5, 10 nM) for 45 min prior to SERS analysis.

References

1 M. Faraday, *Philos. Trans. R. Soc. London*, 1857, **147**, 145.
2 J. Turkevich, P. C. Stevenson and J. Hiller, *Discuss. Faraday Soc.*, 1951, **11**, 55.
3 P. C. Lee and D. Meisel, *J. Phys. Chem.*, 1982, **86**, 3391.
4 G. Schmid and U. Simon, *Chem. Commun.*, 2005, 697.
5 M. C. Daniel and D. Astruc, *Chem. Rev.*, 2004, **104**, 293.
6 Y. Sun and Y. Xia, *Analyst*, 2003, **128**, 686.
7 J. Yguerabide and E. E. Yguerabide, *Anal. Biochem.*, 1998, **262**, 157.
8 J. Yguerabide and E. E. Yguerabide, *Anal. Biochem.*, 1998, **262**, 137.
9 M. Fleishman, P. J. Hendra and A. J. McQuillan, *Chem. Phys. Lett.*, 1974, **26**, 163.
10 K. Faulds, R. E. Littleford, D. Graham, G. Dent and W. E. Smith, *Anal. Chem.*, 2004, **76**, 592.
11 S. Nie and S. R. Emory, *Science*, 1997, **275**, 1102.
12 K. Faulds, R. Jarvis, W. E. Smith, D. Graham and R. Goodacre, *Analyst*, 2008, **133**, 1505.
13 K. Faulds, F. McKenzie, W. E. Smith and D. Graham, *Angew. Chem., Int. Ed.*, 2007, **46**, 1829.
14 A. MacAskill, D. Crawford, D. Graham and K. Faulds, *Anal. Chem.*, 2009, **81**, 8134.
15 A. P. Alivastos, K. P. Johnsson, X. Peng, T. E. Wilson, C. J. Loweth, M. P. Bruchez and P. G. Schultz, *Nature*, 1996, **382**, 609.
16 C. A. Mirkin, R. L. Letsinger, R. C. Mucic and J. J. Storhoff, *Nature*, 1996, **382**, 607.
17 J. S. Lee, A. K. R. Lytton-Jean, S. J. Hurst and C. A. Mirkin, *Nano Lett.*, 2007, **7**, 2112.

18 D. G. Thompson, A. Enright, K. Faulds, W. E. Smith and D. Graham, *Anal. Chem.*, 2008, **80**, 2805.
19 J. J. Storhoff, R. Elghanian, R. C. Mucic, C. A. Mirkin and R. L. Letsinger, *J. Am. Chem. Soc.*, 1998, **120**, 1959.
20 R. C. Mucic, J. J. Storhoff, C. A. Mirkin and R. L. Letsinger, *J. Am. Chem. Soc.*, 1998, **120**, 12674.
21 J. J. Storhoff, A. D. Lucas, V. Garimella, Y. P. Bao and U. R. Muller, *Nat. Biotechnol.*, 2004, **22**, 883.
22 H. D. Hill, R. A. Vega and C. A. Mirkin, *Anal. Chem.*, 2007, **79**, 9218.
23 A. K. R. Lytton-Jean, M. S. Han and C. A. Mirkin, *Anal. Chem.*, 2007, **79**, 6037.
24 T. A. Taton, C. A. Mirkin and R. L. Letsinger, *Science*, 2000, **289**, 1757.
25 S. J. Park, T. A. Taton and C. A. Mirkin, *Science*, 2002, **295**, 1503.
26 Y. C. Cao, R. Jin and C. A. Mirkin, *Science*, 2002, **297**, 1536.
27 C. Tuerk and L. Gold, *Science*, 1990, **249**, 505.
28 A. D. Ellington and J. W. Szostak, *Nature*, 1990, **346**, 818.
29 R. D. Jenison, S. C. Gill, A. Pardi and B. Polisky, *Science*, 1994, **263**, 1425.
30 S. A. Vinores, *Curr. Opin. Mol. Ther.*, 2003, **5**, 673.
31 O. C. Farokhzad, S. Jon, A. Khademhosseini, T.-N. T. Tran, D. A. LaVan and R. Langer, *Cancer Res.*, 2004, **64**, 7668.
32 C.-C. Huang, Y.-F. Huang, Z. Cao, W. Tan and H.-T. Chang, *Anal. Chem.*, 2005, **77**, 5735.
33 J. Chen, J. Jiang, X. Gao, G. Liu, G. Shen and R. Yu, *Chem.–Eur. J.*, 2008, **14**, 8374.
34 D. Graham, D. G. Thompson, W. E. Smith and K. Faulds, *Nat. Nanotechnol.*, 2008, **3**, 548.
35 F. McKenzie, K. Faulds and D. Graham, *Nanoscale*, 2010, **2**, 78.
36 X. Qian, X. Zhou and S. Nie, *J. Am. Chem. Soc.*, 2008, **130**, 14934.
37 D. Graham, K. Faulds, D. G. Thompson, F. McKenzie, R. J. Stokes, C. Dalton, R. Stevenson, J. Alexander, P. Garside and E. McFarlane, *Biochem. Soc. Trans.*, 2009, **37**, 697.
38 G. McAnally, C. McLaughlin, R. Brown, D. C. Robson, K. Faulds, D. R. Tackley, W. E. Smith and D. Graham, *Analyst*, 2002, **127**, 838.
39 D. G. Thompson, K. Faulds, W. E. Smith and D. Graham, *J. Phys. Chem. C*, 2010, **114**, 7384.
40 R. L. Letsinger, R. Elghanian, G. Viswanadham and C. A. Mirkin, *Bioconjugate Chem.*, 2000, **11**, 289.
41 J. A. Dougan, C. Karlsson, W. E. Smith and D. Graham, *Nucleic Acids Res.*, 2007, **35**, 3668.
42 L. Griffin, G. Tidmarsh, L. Bock, J. Toole and L. Leung, *Blood*, 1993, **81**, 3271.
43 K. Padmanabhan, K. P. Padmanabhan, J. D. Ferrara, J. E. Sadler and A. Tulinsky, *J. Biol. Chem.*, 1993, **268**, 17651.
44 K. Padmanabhan and A. Tulinsky, *Acta Crystallogr., Sect. D: Biol. Crystallogr.*, 1996, **52**, 272.
45 K. C. Grabar, R. G. Freeman, M. B. Hommer and M. J. Natan, *Anal. Chem.*, 1995, **67**, 735.
46 C. H. Munro, W. E. Smith, M. Garner, J. Clarkson and P. C. White, *Langmuir*, 1995, **11**, 3712.

Mesoporous silicon photonic crystal microparticles: towards single-cell optical biosensors†

Bin Guan,[a] Astrid Magenau,[b] Krisopher A. Kilian,[a] Simone Ciampi,[a] Katharina Gaus,[b] Peter J. Reece[c] and J. Justin Gooding*[a]

Received 12th May 2010, Accepted 11th June 2010
DOI: 10.1039/c005340f

In this paper we demonstrate the possibility of modifying porous silicon (PSi) particles with surface chemistry and recognition molecules (antibodies) such that these devices could potentially be used for single-cell identification or sensing. This is achieved by modifying PSi Rugate filters *via* hydrosilylation with surface chemistry that serves firstly, to protect the silicon surfaces from oxidation; secondly, renders the surfaces resistant to nonspecific adsorption of proteins and cells and thirdly, allows further functionality to be added such as the coupling of antibodies. The surface chemistry remained unchanged after sonication of the PSi to form PSi microparticles. The ability to monitor the spectroscopic properties of microparticles, and shifts in the optical signature due to changes in the refractive index of the material within the pore space, is demonstrated. The particles are shown to remain stable in physiological buffers and human blood for longer than one week. Finally, the modification of the PSi particles with functional antibodies is achieved.

1 Introduction

Porous silicon (PSi) is a particularly versatile material having found applications in photonic devices,[1] electrodes,[2] sensor systems[3] and drug-delivery devices.[4,5] Hence PSi also has potential to be applied in theranostics. Mesoporous silicon, where pores are in the size range of 2–50 nm, is the most relevant material for these applications. It is most commonly prepared by anodically etching the silicon in hydrofluoric acid (HF) in, effectively, a corrosion reaction,[2] where the silicon is first oxidised and the resultant SiO_2 is then dissolved away in the HF. Remarkably, under many conditions the corrosion commences at a nucleation site on the silicon surface and proceeds linearly through the material to result in a columnar pore. Importantly for the above-mentioned applications, the size of the pore, and hence the extent of porosity, can be easily tuned by controlling the current density, and hence the dissolution rate,[6] during formation. That is, the exquisite control over reactivity that electrochemistry provides is directly translated into the same exquisite control over the material's structure. It is this ability to control and alter porosity, and pore size, with nanoscale resolution as the PSi is being fabricated that allows one-dimensional

[a]*School of Chemistry, The University of New South Wales, Sydney, NSW, Australia. E-mail: justin.gooding@unsw.edu.au; Fax: +61-2-937585 6141; Tel: +61-2-9385 5384*
[b]*Centre for Vascular Research, The University of New South Wales, Sydney, NSW, Australia*
[c]*School of Physics, The University of New South Wales, Sydney, NSW, Australia*

† Electronic supplementary information (ESI) available: Chemical modification schemes, quantification of particle size distributions, additional FTIR spectra and stability studies of particles with optical signatures in the near-IR are presented. See DOI: 10.1039/c005340f

photonic crystals, with complicated multilayer structures, to be produced. High-quality optical structures such as Bragg reflectors,[7] microcavities[8] and Rugate filters[9] have been produced from PSi.

It is the optical structures formed from PSi that sees this material applied to sensing. The optical signature of the photonic crystals is sensitive to the average refractive index of the structure. Therefore, in sensing, the adsorption of material to the pore walls will cause a red-shift (to longer wavelengths) if it has a higher refractive index than the material it replaces because the average refractive index of the structure increases. Similarly, if this material desorbs from the pore walls, the average refractive index decreases, and there is a concomitant blue-shift (to shorter wavelengths). By modifying the pore walls with the appropriate surface chemistry, PSi photonic crystals can be turned into label-free optical detectors with the required selectivity for the analyte of interest.[3]

Apart from being a label-free system, PSi photonic crystals are particularly attractive for biosensing as the optical signature can be tuned, *via* changes in the porosity and periodicity of the multilayer photonic crystals, over the entire visible and infrared regions of the electromagnetic spectrum. Tuning the optical signature to the near-IR between 700–1000 nm is particularly attractive as transmission through tissue is then possible[10] and hence PSi may be amenable to implantable devices for non-invasive interrogation. Another aspect of PSi that makes it attractive for use in biological systems, such as *in vivo* applications, is that PSi has been shown to degrade in aqueous solution to orthosilicic acid by the following mechanisms:[11]

$$Si(s) + 2H_2O(l) \rightarrow SiO_2(s) + 2H_2(g)$$

$$SiO_2(s) + 2H_2O(l) \rightarrow Si(OH)_4(l)$$

As orthosilicic acid is the natural form of silicon found in the body it can be readily excreted *via* the kidneys[12] and has been shown to elicit a limited foreign body response that is comparable to titanium surfaces when implanted in rats.[4,13] Hence implanted devices will eventually degrade away and not require surgical removal. It is the degradation that sees PSi used as a material for drug delivery of poorly solubilised drugs.[4,5] With the drug incorporated into the pore space of the PSi, the effective solubility of poorly soluble drugs is increased while the degradation of the PSi in biological fluids enables controlled release of the drug. The issue here is, however, that the degradation of freshly etched PSi is rapid (within hours).[14] To slow the degradation down requires chemical modification of the PSi.

There are two main surface chemistries employed for the chemical modification of the PSi. One strategy is to thermally oxidise the PSi and then use classical silane chemistry.[3,15] This chemistry however does not effectively passivate the PSi from degradation in aqueous solutions and is prone to hydrolysis itself. Further, it is difficult to determine whether monolayers or multilayers have been formed.[3] The other common strategy is to exploit the fact that the freshly prepared PSi presents a hydrogen-terminated silicon surface. Linford and Chidsey[16,17] showed that alkenes and alkynes could be reacted with such surfaces to give very stable monolayers on silicon without an intervening oxide layer (Scheme 1). This surface chemistry does not suffer from the drawbacks of silane chemistry but it is challenging to modify silicon with this surface chemistry without some oxide forming.[18] Any oxide present will then cause the PSi to degrade in aqueous solution.[14,19,20] With regard to stabilising PSi for drug delivery, Canham and co-workers[21] showed that using the hydrosilylation of dodecene to modify PSi the material showed little degradation over four weeks. Note however, that as dodecene renders the material hydrophobic, part of the reason for the excellent stability enhancement is that little

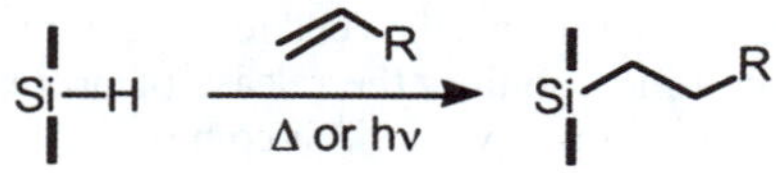

Scheme 1 A general representation of the hydrosilylation reaction used to modify hydrogen-terminated silicon surfaces with alkenes or alkynes with heat or UV light.

or no water penetrates into the pore space. More recently, we have shown that degradation is much more rapid in plasma if the modifying layer formed *via* hydrosilylation is hydrophilic.[14,19] However, the use of hydrophilic surface chemistry that resists protein adsorption, enhances the stability of PSi in plasma and biological buffers markedly. Modifying a PSi photonic crystal using undecenoic acid followed by the addition of an hexa(ethylene oxide) antifouling moiety allowed optical signatures from the photonic crystals to be recorded even 60 days after immersion in biological buffer.[14] Furthermore, we have shown that the degradation rate of PSi can be controlled by subtle variations in the coverage of the organic molecules attached to the PSi.[18]

Surface chemistry similar to that described above has also been crucial in our efforts to turn PSi photonic crystals into biosensors.[3] For biosensors, the surface chemistry is required not only to stabilise the device but also to give the photonic crystal the selectivity for the target analyte. Initial studies into using PSi for biosensing focussed on affinity sensors for the detection of DNA,[22,23] proteins[22,24,25] and microorganisms.[26] However, more recently there has been a trend to use PSi photonic-crystal-based biosensors for detecting enzymatic activity, specifically protease enzyme activity.[20,27–32] The focus of interest on protease enzymes was motivated by the importance of this class of enzyme in natural processes such as cell migration and tissue remodelling, as well as acting as a marker of many pathologies and infection, where early detection is often critical.[33]

Our main aim is to detect protease activity and hence cell activity for the development of responsive cell chips for diagnostic and theranostic purposes. Macrophage and other immune cells respond to environmental stimuli, such as chemical cues, chemokines, and the presence of foreign species, by releasing granules containing a high concentration of protease enzymes. The enzymes remodel the surrounding extracellular matrix, often leading to substantial tissue damage. Hence, monitoring the activity, and not just the presence of, protease enzymes serve as an excellent extracellular marker of the response of cells to their environment, of immune responses and cell activities that underlies a pathology.[34–36] To develop a chip that

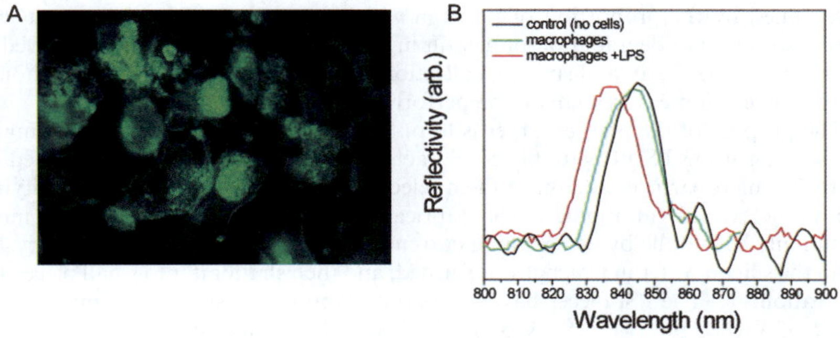

Fig. 1 PSi photonic crystals detect secretion of gelatinase enzymes. A) Human monocyte derived macrophages stained for actin adhere uniformly to gelatin coated PSi. B) After 24 h of lipopolysaccharide (LPS) stimulation the photonic resonance of the Rugate filter blue-shifts 8 nm, indicating digestion of the pore-encapsulated polypeptides.

can monitor protease release from specific cells requires a material that can capture the cells and, at the same time, monitor the release of protease enzymes from cells with the required sensitivity. This is where PSi comes in.

Using PSi we have 1) developed the surface chemistry to stabilise this material in biological buffers and plasma,[14,19,20] 2) shown that we can detect protease activity by attaching peptide substrates, for the protease enzymes, to the pore walls of the PSi,[20,27] 3) demonstrated that cell adhesion ligands can be attached onto the exterior of a PSi photonic crystal while different chemistry can be attached to the internal pores[37,38] and 4) that the release of protease enzymes from live cells captured onto the PSi surface could be detected *via* degradation of a gelatin substrate immobilised inside the pores of the PSi.[28,39] The types of data obtained using these four developments are shown in Fig. 1.

Here we used Rugate filters. Rugate filters are a class of photonic bandgap structures with a periodic sinusoidal refractive index profile. They are formed in PSi by modulating the porosity of the film on the sub-micrometre scale during etching. The structures are characterised by a spectral band of high reflectivity where the central wavelength (λ) is given by $\lambda = nd$, where n is the average refractive index and d is the physical thickness of each period. The width of the reflectivity is proportional to the amplitude of the refractive index modulation, and for small variations, the reflectivity band can be very sharp; of the order of 10 nm. Importantly, the corresponding porosity modulation means there is very little variation in porosity across a photonic crystal. For example, with the photonic crystals used in Fig. 1, the porosity varied between 64 and 66%, with pore diameters of 50 nm.[27] This essentially means the pore size is virtually unchanged through the structure and hence there will be little variation in the rate of mass transport of enzymes throughout the structure.

Fig. 1 shows the power of using these engineered PSi photonic crystals for cell chips. The adherence of human monocyte derived macrophages (HMDMs) to the PSi substrate is shown in Fig. 1A where cells are well spread on the substrate. Fig. 1B shows the change in the optical signature of the photonic crystals as a result of the release of matrix metalloproteases (MMPs) from the HMDMs. These optical responses are obtained from an estimated 1500 cells.[39] The blue-shift in the optical signature is a result of degradation of gelatin from the pores of the PSi, with lower-refractive-index water ($n \sim 1.3$) replacing the higher-refractive-index gelatin ($n \sim 1.4$). An increased response was seen as a larger blue-shift when the cells were exposed to LPS (a lipopolysaccharide found on the exterior of gram-negative bacteria) because the LPS triggers the release of MMP-2 and MMP-9 which are both gelatinases.[40,41] The number of cells that give the response shown in Fig. 1B was a consequence of the size of the footprint of the light source used to acquire the data rather than the sensitivity of the device. The measured sensitivity is actually determined by the proportion of pores in which gelatin was degraded and it was demonstrated that degradation of gelatin in only 1 in 30 pores is easily observed.[28] Hence, reducing the area of spectral collection down to the micrometre range should allow single-cell measurements to be performed.

The purpose of the present paper is to outline the progress made towards single-cell sensing using PSi Rugate filters. To achieve this in cell culture, we decided to form PSi microparticles and make them selective for specific cell types by modifying the particles with antibodies. The fabrication of PSi microparticles has most commonly been done by using a high current pulse to lift-off a PSi photonic crystal from the silicon wafer in which it was formed, and then shatter it into small pieces *via* sonication.[42–44] The PSi pieces have a broad distribution of sizes in the micrometre range of 4 to 80 μm (Figs. S1 & S2 of the ESI†). Extending our previous work on detecting protease activity in single-cell sensors requires a number of questions to be answered. These are: 1) Can we measure optical signatures from single particles and how reproducible are they? 2) How do we modify particles with surface chemistry to stabilise them and incorporate selectivity for a target analyte? An important part of this question is to work out the sequence of fabrication steps *e.g.* is the protective surface chemistry best performed before the high current pulse in HF that is necessary to cause the lift-off from the silicon crystal, after lift-off but before formation of particles or after the formation of the particles? 3) How stable will the particles be in aqueous solution and biological buffers and how will the surface chemistry influence this? and 4) Can the particles be modified with antibodies such that their binding affinity is retained? The success of these developments is outlined below.

2 Experimental section

2.1 General

All chemicals, unless noted otherwise, were of analytical grade and used as received. Chemicals used in surface modification procedures were of high purity (≥ 99%). 1,8-Nonadiyne (Alfa Aesar, 97%) was redistilled from sodium borohydride (Sigma-Aldrich, 99+%) under reduced pressure (79 °C, 8–9 Torr) and collected over activated molecular sieves (Fluka, 3 Å pore diameter, 10–20 mesh beads, dehydrated with indicator), and stored under a dry argon atmosphere prior to use. MilliQ™ water (>18 MΩ·cm) was used to prepare solutions and for chemical reactions. Dichloromethane, ethanol, and ethyl acetate for surface cleaning, chemical reactions and purification procedures, were redistilled prior to use. Dry acetonitrile was obtained from PureSolv MD 7 Solvent Purification System (Innovative Technology, Inc., Galway, Ireland). The azide compounds, 1-azido-octane and 11-azido-3,6,9-trioxaundecan-1-ol were synthesized as described previously.[38] Prime grade single-side polished silicon wafers, 100-oriented (<100> ± 0.05°), p-type (boron), 350 ± 25 μm thick, 0.069–0.077 Ω·cm resistivity, were obtained from M.M.R.C Pty. Ltd. (Malvern, VIC, Australia).

The antibodies, Cy3 AffiniPure donkey anti-rabbit IgG (H + L), Cy3 AffiniPure donkey anti-goat IgG (H + L) and Cy3 AffiniPure donkey anti-mouse IgG (H + L) were all purchased from Jackson ImmunoResearch (West Grove, PA, USA). Mouse-hosted FITC anti-human CD3 primary antibody was obtained from eBioscience (San Diego, USA).

2.2 Fabrication of porous silicon lift-off films

Rugate filters with 60 periods of low and high refractive index layers were prepared in a custom-made electrochemical cell with 1 : 1 by volume solution of 48% hydrofluoric acid/99% ethanol as described in ref. 9. A current density varying between 42 mA·cm^{-2} and 124 mA·cm^{-2} was applied periodically to afford a porous structure with a porosity variation from 54% to 64% and an average pore size of 20 nm (see Fig. S4†). After etching, a high current pulse (11.3 mA·mm^{-2}) was employed to lift the PSi structure off the substrate (but with the circumference of the PSi remaining attached to the substrate) with 1 : 2.3 by volume solution of 48% hydrofluoric acid/99% ethanol. The lift-off wafer was then rinsed with ethanol and pentane and dried with argon.

2.3 Chemical modification of porous silicon lift-off films

Stepwise modification of the PSi Rugate filter lift-off films followed a previously reported procedure.[38] Briefly, freshly etched filters were modified with neat 1,8-nonadiyne in a dry argon atmosphere at 170 °C for 3 h to form acetylene monolayers on the surface. The filters were then placed in a typical "click" reaction mixture consisting of: 1) the azide compound (10 mM, ethanol/water, 1 : 1 v/v); 2) copper(II) sulfate pentahydrate (1 mol % relative to the azide); 3) sodium ascorbate (25 mol % relative to the azide) for further modification. In the absence of a Cu(I)-stablizing ligand, *N*,*N*,*N*′,*N*′-tetramethylethane-1,2-diamine (TMEDA), the lift-off samples were functionalized with 11-azido-3,6,9-trioxaundecan-1-ol for 19 h to form tetra(ethylene glycol) (EO_4) terminated surfaces on the exterior only. Modification of the interior pore surface with either 1-azido-octane or 11-azido-3,6,9-trioxaundecan-1-ol was followed an analogous procedure, but with 0.5% (v/v) TMEDA in the "click" mixture to afford either an octyl- or an EO_4-terminated interior. After each modification step, the PSi films were rinsed consecutively with ethanol and ethyl acetate and blown dry under a gentle stream of argon before further reaction.

2.4 Engineering of porous silicon microparticles

The PSi lift-off samples with different functionalities on the exterior and interior were floated off the substrate and placed in a 2-mL eppendorf tube with 1 mL ethanol for sonication. PSi microparticles in the size range of 4–80 μm were engineered in a Misonix Sonicator 2000 with a Microtipt™ probe (Qsonica, LLC. USA) at the energy of 266 J (amplitude 30) for 1 min with one repeat. The particles were dried under vacuum and stored under argon for further use.

2.5 Porous silicon microparticle stability test

Porous silicon microparticles at the size of around 10 μm were incubated in phosphate buffered saline (1× PBS, pH 7.4), Dulbecco's Modified Eagle Medium (DMEM) or a white blood cell concentrate from human blood (composed of mostly leukocytes and platelets and referred to as 'buffy coat') (diluted 1 : 5 v/v with 1× PBS) at 37 °C in 5% CO_2, while reflectivity measurements and epi-illumination micrographs were taken over time in solution.

2.6 Antibody functionalities on PSi particles

Tetra(ethylene glycol) EO_4-terminated PSi microparticles were placed in a reaction tube containing: a) *N*,*N*′-disuccinimidyl carbonate (DSC, 0.1 M) and b) *N*,*N*-dimethylaminopyridine (DMAP, 0.1 M) in dry acetonitrile for 20 h in the presence of activated molecular sieves (3 Å). The succinimidyl-terminated particles were rinsed with ethyl acetate and dichloromethane and then transferred to an antibody solution (1 : 500 v/v) in 1× PBS for 1 h at room temperature. Antibody-coupled PSi particles were then rinsed with MilliQ™ water and 1× PBS washing buffer with 0.5% (v/v) Tween 20 (Sigma-Aldrich) and stored in 1× PBS for antibody functional tests.

Antibody functionality tests were conducted in two ways: 1) to test the ability of antibodies to recognise protein-modified particles. Here, PSi microparticles were coated with mouse anti-human CD3 IgG-FITC and then incubated with either anti-goat IgG-Cy3 (1 : 500 v/v in 1× PBS) or with anti-mouse IgG-Cy3 (1 : 500 v/v in 1× PBS), respectively, for 1 h at room temperature; 2) to explore whether antibody-modified particles selectively bind to protein markers. PSi microparticles were coated with anti-rabbit IgG-Cy3 or with anti-mouse IgG-Cy3. The modified PSi particles were then incubated with mouse anti-human CD3 IgG-FITC-(1 : 500 v/v in 1× PBS). The particles were then rinsed twice with 1× PBS washing buffer (0.5% (v/v) Tween 20), followed by 1× PBS and stored in 1× PBS at 4 °C for fluorescence imaging.

2.7 Optical and surface characterisation

Optical reflectivity spectra of the microparticles were measured in the visible and near-IR at normal incidence using a custom-built optical arrangement on an inverted Leica DM IL LED microscope (Leica Microsystems Pty Ltd, USA). In this set-up a USB2000+ miniature fiber optic spectrometer (Ocean Optics Inc., USA), having a spectral resolution of 1 nm, was coupled to a multimode fibre with the collecting facet mounted in the conjugate image plane of the microscope. The microscope was equipped with a N PLAN L 40×/0.55 air objective, an epi-illumination brightfield slider and a ProgRes® CFscan camera (JENOPTIK Laser, Optik,

Systeme GmbH, Germany) for capturing the reflectance images of particles. The collecting aperture of the fibre-coupled spectrometer had a diameter of 100 μm which corresponded to an effective collecting spot of approximately 3 μm at the sample plane. Reflectivity spectra were recorded and processed using the SpectraSuite spectroscopy software platform (Ocean Optics Inc., USA).

Fluorescence images of particles in antibody function tests were obtained with the same Leica microscope, fitted with a mercury short-arc reflector lamp. FITC-fluorescence was detected with a excitation filter: BP470/40, a dichroic mirror: 500 and a emission filter: 525/50. Cy3 fluorescence was detected with a excitation filter: BP545/50, a dichroic mirror: 565 and a emission filter: 610/75. The images were recorded with the ProgRes Capture Pro 2.7 software (JENOPTIK Laser, Optik, Systeme GmbH, Germany) and processed with ImageJ 1.42q (National Institutes of Health, USA).

The Spotlight 400 Fourier transform infrared spectroscopy (FTIR) microscope system was from PerkinElmer and was used to record transmission FTIR spectra from single PSi particles. The particles were vacuum dried and mounted on a KCl disc. The spectra were then collected at the resolution of 2 cm^{-1} and an average of 128 scans with a light aperture of 45 × 45 μm.

Scanning electron microscopy (SEM) images were taken using a Hitachi S900 SEM with a cold-field emission source (4 kV). PSi microparticles were mounted on a brass sample base using carbon tape.

3 Results and discussion

3.1 Formation of porous silicon nanoparticles and their reflectivity spectra

Rugate filters were fabricated in single crystal, p-type, Si(100) with 60 layers and a pore size of 20 nm. After preparation, a high current pulse was used to cause a fracture between the crystalline silicon and the PSi sample. However, the lateral edges of the PSi remain attached to the silicon crystal. It is this partially attached PSi sample that is modified prior to particle formation according to scheme S1† as discussed below. After surface modification, the PSi lift-off sample was completely removed from the silicon crystal and converted into particles. After particle formation, it was important to demonstrate that reflectance spectra of individual particles could be monitored and that each particle from a batch exhibited a similar reflectivity spectrum. To take spectra of single particles, a normal epi-fluorescent microscope was

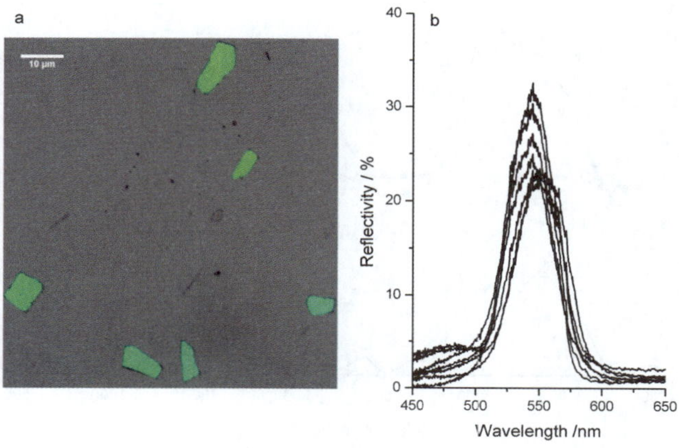

Fig. 2 a) Epi-illumination bright-field images of ethylene glycol modified PSi particles after sonication and b) reflectivity spectra of PSi particles from the same modified PSi film, featuring a high reflectivity peak at 547 ± 3 nm (mean ± SD).

employed where a collecting optical fibre mounted at the conjugate image plane in the camera port was used to record spectral data. The types of epi-illumination images of the particles and the reflectivity spectra of these particles are shown in Fig. 2.

As is evident from Fig. 2 there is some variation in the colour and spectra of these particles. This variation arises from slight differences in the fabrication of the Rugate filter across the sample due to uneven distribution of electrical field lines during the fabrication of the PSi. Importantly however, the optical properties are not significantly altered during sonication of the lift-off PSi sample to form particles.

3.2 Modification of porous silicon microparticles *via* hydrosilylation

Rather than first forming the particles and then modifying them, it was decided to use a high current pulse to partially detach the PSi from the silicon crystal on which it was formed and then modify the PSi before sonication, according to Scheme S1 of the ESI.† The reason for this was ease of handling but it did mean 1) that there would be unmodified silicon where the silicon was fractured and 2) that the surface chemistry was required to be capable of surviving the high energy of the sonication. As shown in Scheme S1 the PSi structure is first modified with the dialkyne, 1,8-nonadiyne, in an inert atmosphere at 170 °C for three hours. Subsequently, further functionality is added by reacting an azide-bearing moiety with the distal alkyne in the presence of a Cu(I) catalyst *via* the Huisgen 1,3-dipolar cycloaddition reaction; the archetypal 'click' reaction. It has been shown previously that this surface chemistry provides excellent protection of the underlying silicon surface from oxidation, even under anodic potentials,[45] with both flat silicon[46] and PSi.[47] We note that in our hands it was much easier to produce good protection of PSi from oxidation using this 1,8-nonadiyne than the chemistry we used previously that employed undecenoic acid.[3] Furthermore, although unnecessary on flat silicon, for the 'click' reaction to proceed within mesoporous silicon requires the complexation of the Cu(I) catalyst with a *N*,*N*,*N*′,*N*′-tetramethylethane-1,2-diamine ligand for the reaction to proceed inside the pores.[47] This observation was used to modify PSi with completely different chemistry inside the pore space relative to the exterior surface[38] as shown using optical reflectivity, X-ray photoelectron spectroscopy (XPS) and transmission Fourier transform infrared (FTIR) spectroscopy.[2] Using this strategy, the exterior surface was modified with an azido tetra(ethylene glycol) to render the external surface resistant to nonspecific protein and cell adsorption but also to allow further attachment

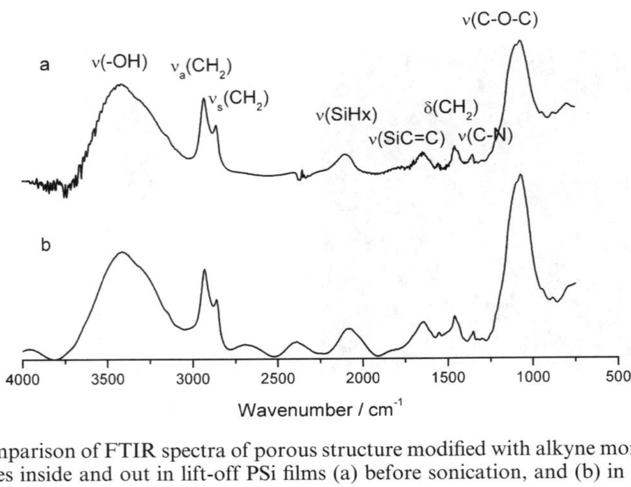

Fig. 3 Comparison of FTIR spectra of porous structure modified with alkyne monolayers and EO_4 moieties inside and out in lift-off PSi films (a) before sonication, and (b) in PSi particles after sonication.

of biomolecules to the alcohol terminus of this molecule as described previously.[48] The internal pore surfaces were then modified with the same tetra(ethylene glycol) species to make the internal surface hydrophilic, such that water will easily penetrate the pores, and to allow attachment of the gelatin substrate required by matrix metalloprotease enzymes secreted from macrophage cells (MMP-2 and MMP-9). Alternatively the internal pores were modified with an octane species to make the internal volume hydrophobic.[37] The reason for this modification was to explore whether better stability of particles in biological media was achieved if the ingress of aqueous solution into the material was limited.

After modifying the PSi with the base 1,8-nonadiyne layer followed by clicking on different functionality to the internal and external surfaces, the PSi disk was cut from the silicon crystal and sonicated in ethanol to give micrometre-size particles. To show the surface chemistry was not altered by the highly energetic conditions of sonication transmittance FTIR was performed before and after the sonication step (Fig. 3).

Fig. 3a shows the asymmetrical and symmetrical stretching of the CH_2 group at 2925 cm^{-1} and 2852 cm^{-1} remains unchanged. The broad OH stretch in the range 3500–3200 cm^{-1} and C–O stretching vibration at 1150 cm^{-1} support the presence of ethylene glycol moieties on the surface. Also present are the bands at 1630 cm^{-1}, 1455 cm^{-1} and 1350 cm^{-1}, ascribed to Si–C=C stretching, CH_2 scissoring and C–N bond respectively, strongly supporting a complete surface modification on PSi particles. Significantly, there is no broad band around 1150 cm^{-1} which would be indicative of any silicon oxide, thus indicating the effectiveness of the surface chemistry at modifying the silicon without any silicon oxides being formed. Most importantly however, Fig. 3a and b show no discernable difference indicating the sonication did not cause any harm to the surface chemistry. Similar results are

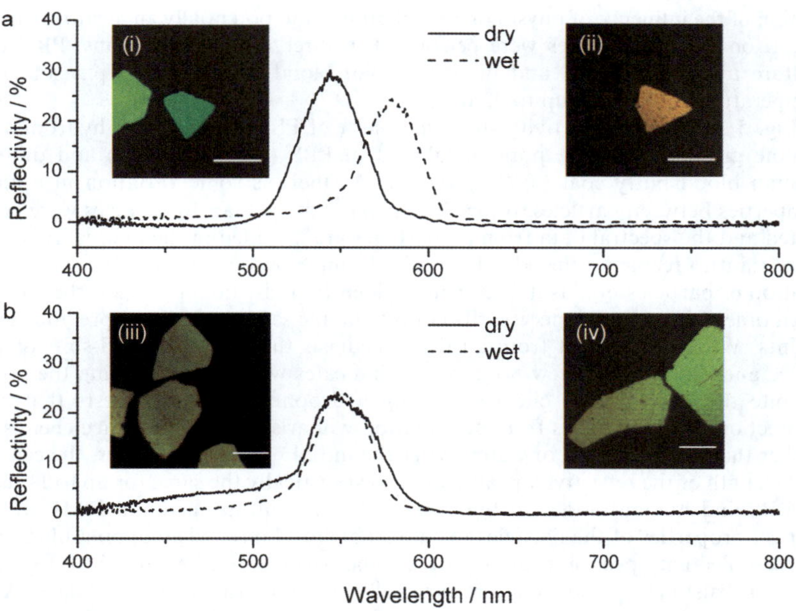

Fig. 4 PSi particles with hydrophilic and hydrophobic internal functionalities in dry/wet conditions. (a) Reflectivity spectra of a single hydrophilic particle in air and in phosphate-buffered saline (PBS). A color change from green (i) to red (ii) in the same hydrophilic particle can be observed. (b) Reflectivity spectra of a hydrophobic particle in air and in PBS. The insets (iii) and (iv) show the illumination images of hydrophobic particles in dry and wet conditions, respectively. Scale bar equals 50 μm.

observed when the internal surface chemistry is terminated with octyl moieties (see Fig. S3 of the ESI†).

The effectiveness of the surface modification in controlling the way the particles interacted with aqueous solution, and hence a demonstration of the ability to place different surface chemistry on the inside and outside of the PSi particles, was demonstrated using reflectivity measurements. The particle with internal octane-terminations are expected to exclude water from penetrating into the pores, while with a hydrophilic group, *e.g.* ethylene glycol, the pores are expected to be filled with water. As water has a higher refractive index than the air it replaces, the penetration of water into the structure will result in a red-shift in the reflectivity peak to longer wavelengths. Photonic crystals with reflectivity bands in the visible region were employed for this purpose so the changes could also be seen visually (Fig. 4). Fig. 4a shows a 30 nm spectral shift in the reflectivity spectrum caused by water infiltration and the two inset graphs [(i) & (ii)] display an abrupt color change of the same particle. No significant spectral shift in Fig. 4b suggests a well-passivated hydrophobic surface in the pores of PSi particles, which resist infiltration of water into the pore spaces.

3.3 Stability of PSi particles under physiological conditions

The next step was to evaluate the different particles, those with hydrophilic modification inside the pores and those with hydrophobic surface chemistry, for their stability in physiological media. We have performed such studies for similar hydrophilic surface chemistry with porous silicon still incorporated within the single crystal silicon and showed that PSi structures gave recognisable optical signatures for up to 60 days incubation in PBS buffer at 37 °C.[14,19] The particles however may have significantly lower stability because the fabrication of the particles exposed unmodified silicon that could then act as the site where degradation begins. The evaluation of the influence of physiological condition, and potentially an *in vivo* environment, on the PSi particles were performed in three different solutions: PBS, cell culture medium DMEM and diluted human blood buffy coat at physiological temperature and pH for up to 23 days.

Fig. 5 shows the reflectivity stop-band shift of PSi particles with hydrophobic octane groups inside the pores incubated in PBS (a), DMEM (b) and diluted human blood buffy coat (c), respectively. As there is some variation in optical properties between particles, for each time point more than 10 particles were evaluated and the spectral data from all particles are included in the figures. It is also important to recognise that the data at each time point come from the same population of particles but, as it is difficult to identify individual particles, the data at each time point does not necessarily come from the same particles as previous time points. What is apparent from all three media is that there is a red-shift of the reflectance band after a few hours which indicates water is penetrating the pores despite the pore surface chemistry being hydrophobic. This suggests that any protection of the particles from degradation will arise from the surface chemistry rather than the exclusion of water. After the initial red-shift, however, the central wavelength of the reflectivity peak remained statistically the same for up to 11 days in PBS and human blood. In the case of the cell culture medium, DMEM, the optical properties of the particles remain stable for 7 days whereupon a blue-shift in the reflectivity peak appears to suggest the particles are beginning to degrade, although distinct optical signatures are still easily discernable after 23 days. Also shown in Fig. 5b are optical micrographs of particles at different times during their incubation in the DMEM to visually illustrate the change in optical properties of particles.

The PSi particles functionalized with hydrophilic ethylene glycol moieties were also evaluated for stability in the same three media: PBS, DMEM and human blood buffy coat (Fig. 6). Unlike the hydrophobic particles there was no initial red-shift in

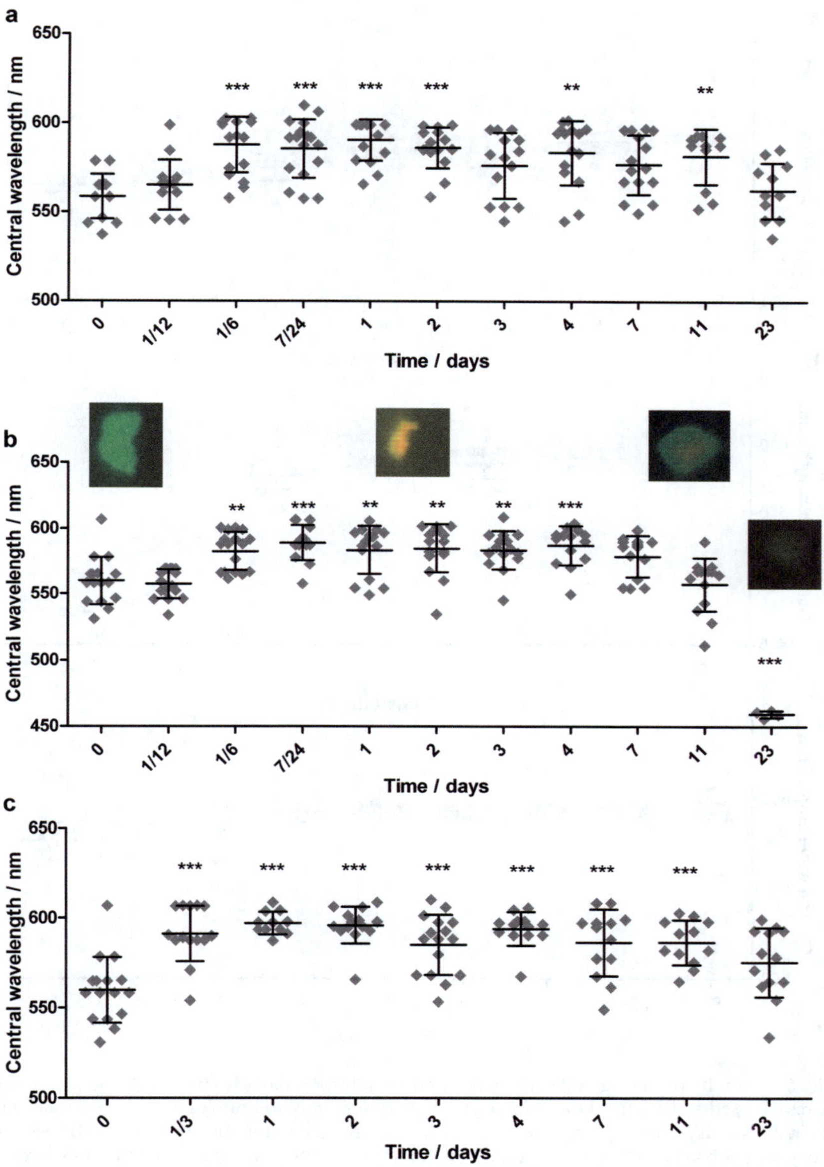

Fig. 5 Stability of the high reflectivity peak position for PSi particles with hydrophobic internal pores exposed to PBS (a), DMEM (b) and diluted human blood buffy coat(c) over time. The asterisks stands for the significance level of variances relative to the starting point (time 0) based on one-way ANOVA analysis: *** (99.9% confidence), ** (99% confidence) and * (95% confidence). The insets are illumination images of PSi particles at different incubation times in DMEM.

the reflectivity as the pore spaces of the particles were already filled with water. A gradual blue-shift in the position of the reflectivity band, presumably due to oxidation of the silicon and hence a degradation of the particles, was observed after two-day incubation in PBS (Fig. 6a). Similar results were shown for incubations in DMEM in Fig. 6b with the blue-shift becoming statistically significant after four

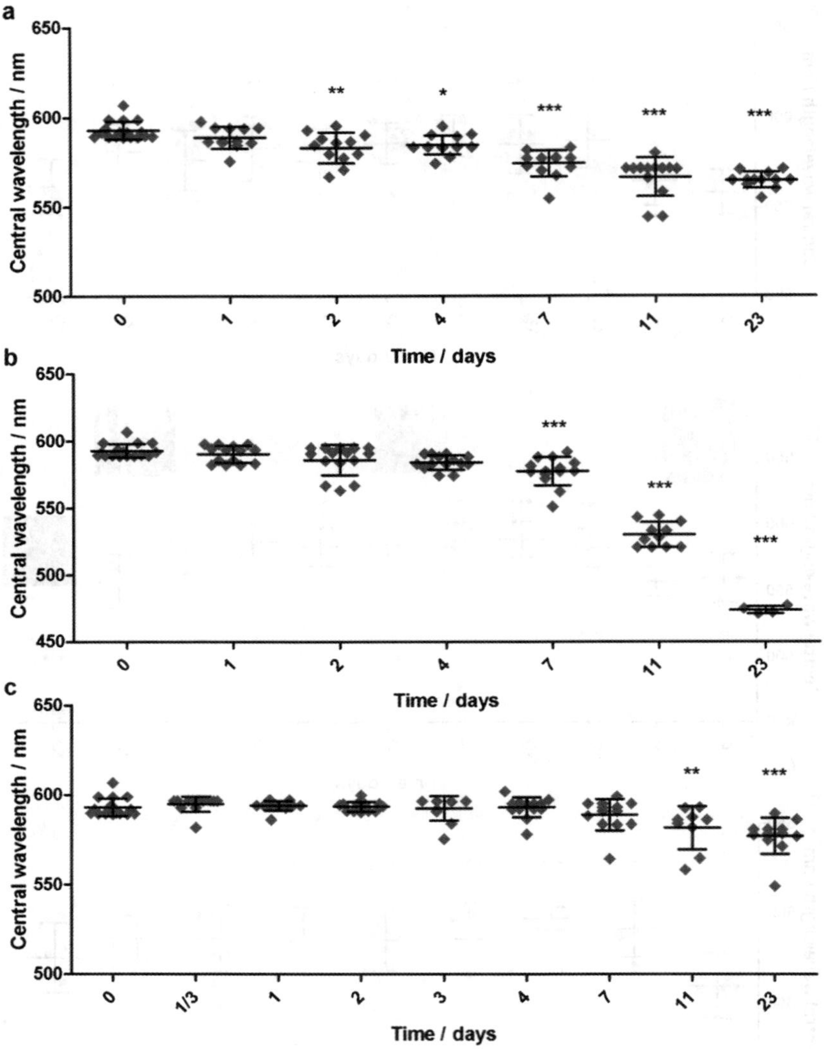

Fig. 6 Stability of the high reflectivity peak position for PSi particles with hydrophilic internal pores exposed to PBS (a), DMEM (b) and diluted human blood buffy coat (c) over time. The asterisks stands for the significance level of variances related to the start point (time 0) based on a one-way ANOVA analysis: *** (99.9% confidence), ** (99% confidence) and * (95% confidence).

days. In blood, Fig. 6c, the blue-shift, indicative of particle degradation, only became significant after more than one week. For comparison, the same stability tests were also performed on particles with hydrophilic interiors and exteriors, as in Fig. 6, but for particles with optical signatures tuned in the near-IR (the 'tissue window'). The near-IR particles showed virtually identical behaviour to those in Fig. 6 above (see Fig. S4 of the ESI†).

The stability studies were very encouraging as they show there is no major degradation of particles within the first few days of exposing them to physiological media which is well beyond the expected operation time for most diagnostic and theranostic applications. Furthermore, it is clear that the particles will still provide some optical information for 23 days after application.

3.4 Antibody immobilization on PSi microparticles

The final questions explored in this study were whether particles could be modified with antibodies on the external surface and whether the antibodies maintained their functionality? Initial attempts to modify the exterior surface of the PSi with antibodies prior to sonication of the PSi lift-off sample to form the microparticles resulted in destruction of the antibodies. As a consequence, modified PSi particles were first prepared with a hydrophobic internal pore surface chemistry and a tetra (ethylene glycol) species on the external surface as shown in Scheme 2. The lift-off PSi was then sonicated to form particles; a process shown above not to affect the surface chemistry. The distal alcohol moieties on the tetra(ethylene glycol) were converted to succinimide esters using DSC/DMAP,[48] which then allows covalent coupling of amines such as those found on lysine residues of antibodies (Scheme 2). A fluorescence labeled antibody, cyanine 3 affiniPure donkey anti-mouse IgG (H + L) was chosen to test the antibody immobilization by examining its fluorescence under the Leica microscopy. The immobilization steps were further monitored using a transmission-mode FTIR microscope that was able to measure spectra from each single microparticle.

FTIR spectra obtained from ethylene glycol-terminated PSi particles, DSC-activated and antibody-coupled particles are shown in Fig. 7. The presence of $\nu_a(CH_2)$, $\nu_s(CH_2)$, $\delta(CH_2)$, ν(C–N) and ν(C–O–C) modes at 2925 cm^{-1}, 2852 cm^{-1}, 1455 cm^{-1}, 1350 cm^{-1} and 1150 cm^{-1}, respectively, in Fig. 7a verifies the PSi particles have been modified with ethylene glycol moieties, as shown before. The appearance of IR stretching at 1750 cm^{-1} in Fig. 7b, ascribed to C=O, suggests successful activation of the distal hydroxyl group of tetra(ethylene glycol) to a succinimide ester. After incubating the activated particles in the anti-mouse IgG antibody solution for 2 h, the stretching of the C=O bond disappeared. Instead, two broad peaks at 3380 cm^{-1} and 1680 cm^{-1} appeared pointing to the presence of polypeptides, indicative of anti-mouse IgG coupling on the PSi particles. In detail, there were two main peaks related to peptides around the wavenumber of 3400 cm^{-1}: one at 3454 cm^{-1} which is from the free N–H stretch vibration and another peak at 3330 cm^{-1} from the hydrogen-bonded N–H stretch vibration.

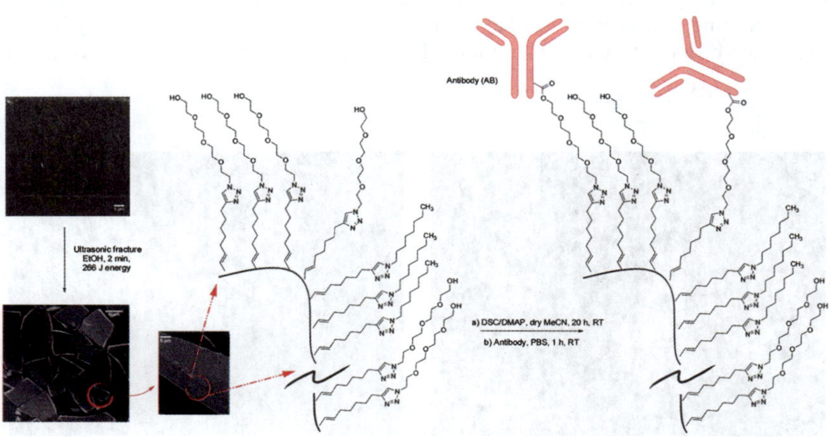

Scheme 2 The formation of PSi Rugate filter microparticles and the surface chemistry used to modify the microparticles with antibodies on the exterior surface.

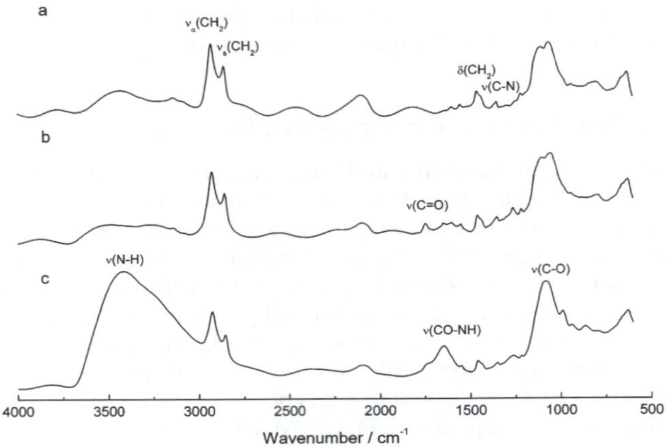

Fig. 7 Transmission-mode FTIR spectra of PSi microparticles at each step of immobilization. (a) PSi particles with distal ethylene glycol groups on the exterior and distal octane groups inside the pores. (b) DSC/DMAP-activated PSi particles. (c) Cy3-labeled antibody-coupled PSi particles.

Fluorescence images of PSi particles in Fig. 8 also support the successful immobilization of Cy3-labeled antibody through covalent bonding. The micrographs in Fig. 8a and b were taken from the same particles, but under different imaging modes: fluorescence and white light epi-illumination, respectively. The fluorescence images (Fig. 8a) show Cy3-labelled antibody across the entire PSi surface although it is noted that there seems to be a higher density of antibody at the particle edges. The particle edges are bare silicon, where no nonspecific protein adsorption resistant chemistry is located, and hence the high fluorescence intensity may reflect nonspecific adsorption of the antibodies.

Fig. 8 illustrates that antibody is immobilised onto the PSi microparticles but provides no evidence of antibody functionality. To assess the binding capabilities of surface-bound antibodies, the binding of free secondary antibody (AB^2) to recognize primary antibody-modified particles was examined here. In the method 1 of antibody functionality test, mouse FITC-labeled anti-human CD3 primary antibody was first covalently coupled onto PSi particles by the same chemistry approach in Scheme 2. The particles were then incubated with two types of secondary antibody (AB^2): cyanine 3 affiniPure donkey anti-mouse IgG that should selectively bind to the surface-bound, mouse anti-human CD3 primary antibodies and Cy3 AffiniPure

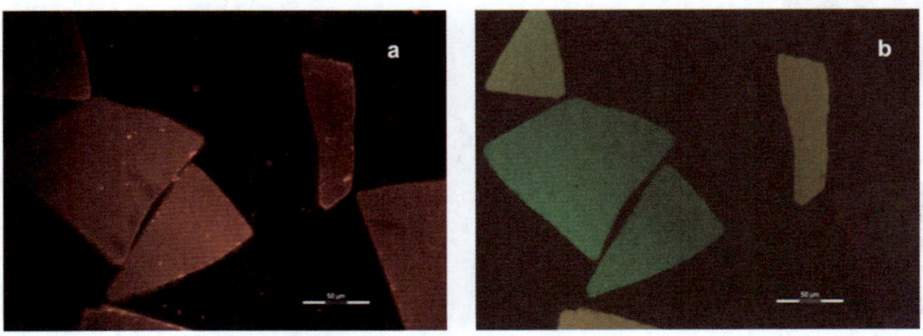

Fig. 8 (a) Fluorescence images of anti-mouse Cy3-labeled particles and (b) epi-illumination images of the anti-mouse Cy3-labeled particles.

donkey anti-goat IgG as a control that should have no specific affinity for the surface-bound antibodies.

Fluorescence images of the particles incubated with two different AB^2 were taken using a Leica DM IL microscope with a Cy3 filter and were quantified to compare the fluorescence intensities, shown in Fig. 9. Both anti-goat and anti-mouse AB^2-incubated particles show some fluorescence (Fig. 9a and b), indicating some nonspecific binding of anti-goat AB^2 onto the PSi particles. However, based on a one-way ANOVA analysis, anti-mouse AB^2-bound particles have significantly higher fluorescence intensity than the control particles with anti-goat AB^2 (Fig. 9c), in accordance with the fact that only anti-mouse AB^2 can specifically bind to the mouse-hosted anti-human CD3 primary antibody. This result serves as a demonstration that the binding epitopes of proteins immobilised on the PSi particles are preserved for binding to other proteins, an essential criterion for targeting specific cell types.

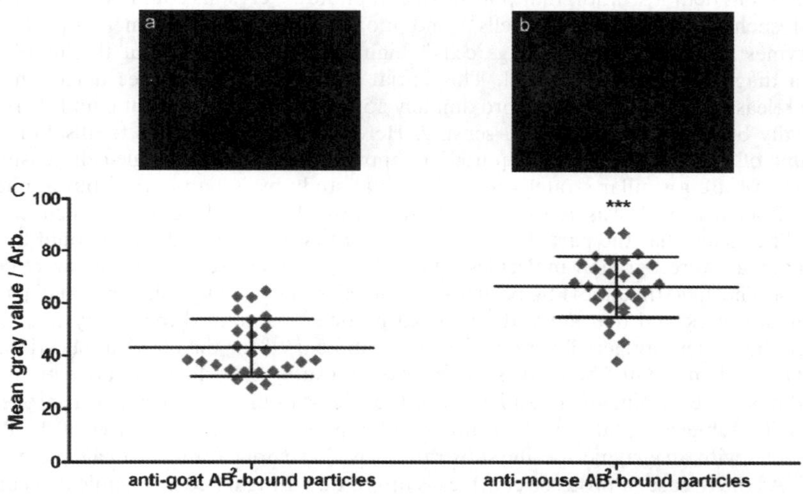

Fig. 9 Comparison of the fluorescence intensities from anti-goat AB^2-bound PSi particles and anti-mouse AB^2-bound particles. (a) and (b) fluorescence images of PSi particles with anti-goat AB^2 and anti-mouse AB^2, respectively. (c) Statistical analysis of fluorescence emissions by antibody-bound particles. The asterisks represent significant difference based on a one-way ANOVA analysis, while *** stands for significance at 99.9% confidence interval.

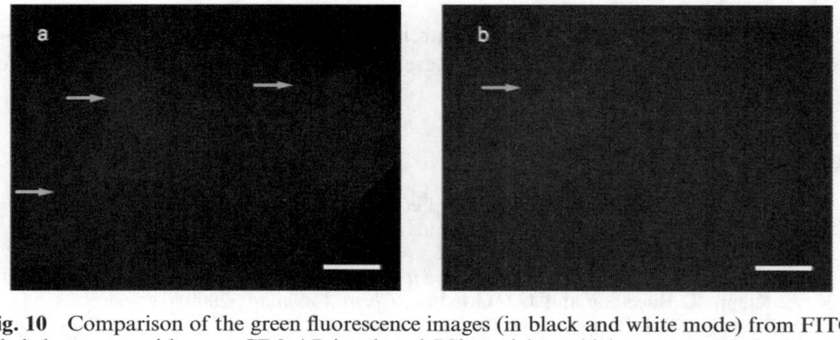

Fig. 10 Comparison of the green fluorescence images (in black and white mode) from FITC-labeled mouse anti-human CD3 AB-incubated PSi particles, which were modified with anti-mouse AB^2 (a) and anti-rabbit AB^2 (b), respectively. Arrows display the position of particles. Scale bar equals 50 μm.

To illustrate that antibodies bound to the particles remain capable of binding to specific epitopes of cells, a second antibody test was performed. In method 2, secondary antibodies, either Cy3-labeled anti-mouse AB^2 or Cy3-labeled anti-rabbit AB^2 were immobilised on the PSi particle surface first for selective binding to free primary antibodies in the solution. Fig. 10 shows the fluorescence micrographs of two different AB^2-modified PSi particles after incubating with FITC-labeled mouse anti-human CD3 primary antibodies. With anti-mouse AB^2, which specifically binds the FITC-labeled anti-human CD3 primary antibody, the particles showed brighter fluorescence than the reference ones with anti-rabbit AB^2 which have no specific affinity for the FITC-labeled anti-human CD3 primary antibody. Hence, this supported that the active binding sites on surface-bound antibodies maintain their functionality.

4. Conclusions

The recent demonstration that porous silicon photonic crystals could be functionalised such that they capture cells[32] and monitor the release of metalloprotease enzymes from living macrophage cells[26] highlighted the potential of this material as a diagnostic/theranostic tool. This initial work showed label-free detection of the release of enzymes from approximately 1500 cells but the concept could therotetically be applied to single-cell sensing. Here we have outlined the realisation of many of the developments required to apply these ideas to single-cell sensing. Here, we use a similar strategy as the original study by Kilian *et al.*[34] but applied it to PSi particles. It was necessary to show that the PSi particles could be chemically modified such that the particles were stable in physiological media and could have functional biorecognition molecules attached to give them selectivity for target cell types. The fact that the PSi particles remained stable in physiological media for over one week and that antibody-modified particles maintained the ability to selectively bind their antigen illustrates that they can be used for the target application.

The study presented here takes us towards being able to perform selective binding of PSi particles to single cells and monitor the release of protease enzymes from these cells. To achieve the latter will require modifying the internal pore spaces of the PSi particles with an enzyme-specific substrate, as we have done for macroscale pieces of PSi. As the release of protease enzymes is up- or down-regulated by chemical agents, such as therapeutics, these PSi particles could be used for monitoring the impact of specific therapeutics on cell behaviour. This capability could then be used for monitoring drug efficacy and could be tailored to the cells of individuals, thus providing patient specific information of the drug action and dosage requirements; the very essence of theranostics.

Acknowledgements

The authors thank the Australian Research Council (project number DP1094564) and the National Health and Medical Research Council (project grant 630545) for support.

References

1 M. Gal, P. J. Reece, W. H. Zheng and G. Lerondel, in *Photonics: Design, Technology, and Packaging*, ed. C. Jagadish, K. D. Choquette, B. J. Eggleton, B. D. Nener and K. A. Nugent, SPIE, Bellingham, 2003, pp. 9–16.
2 G. X. Zhang, *Electrochemistry of silicon and its oxides*, Kluwer Academic, New York, 2001.
3 K. A. Kilian, T. Boecking and J. J. Gooding, *Chem. Commun.*, 2009, 630–640.
4 C. A. Prestidge, T. J. Barnes, C. H. Lau, C. Barnett, A. Loni and L. Canham, *Expert Opin. Drug Delivery*, 2007, **4**, 101–110.
5 E. J. Anglin, L. Y. Cheng, W. R. Freeman and M. J. Sailor, *Adv. Drug Delivery Rev.*, 2008, **60**, 1266–1277.

6 L. Canham and Editor, *Properties of Porous Silicon*, IEE, London, 1997.
7 W. H. Zheng, P. Reece, B. Q. Sun and M. Gal, *Appl. Phys. Lett.*, 2004, **84**, 3519–3521.
8 P. J. Reece, G. Lerondel, W. H. Zheng and M. Gal, *Appl. Phys. Lett.*, 2002, **81**, 4895–4897.
9 S. Ilyas, T. Bocking, K. Kilian, P. J. Reece, J. Gooding, K. Gaus and M. Gal, *Opt. Mater.*, 2007, **29**, 619–622.
10 S. J. Matcher, M. Cope and D. T. Delpy, *Appl. Opt.*, 1997, **36**, 386–396.
11 L. T. Canham, *Adv. Mater.*, 1995, **7**, 1033–1037.
12 D. M. Reffitt, R. Jugdaohsingh, R. P. H. Thompson and J. J. Powell, *J. Inorg. Biochem.*, 1999, **76**, 141–147.
13 A. Rosengren, L. Wallman, M. Bengtsson, T. Laurell, N. Danielsen and L. M. Bjursten, *Phys. Status Solidi A*, 2000, **182**, 527–531.
14 K. A. Kilian, T. Bocking, K. Gaus, M. Gal and J. J. Gooding, *Biomaterials*, 2007, **28**, 3055–3062.
15 B. E. Collins, K.-P. S. Dancil, G. Abbi and M. J. Sailor, *Adv. Funct. Mater.*, 2002, **12**, 187–191.
16 M. R. Linford and C. E. D. Chidsey, *J. Am. Chem. Soc.*, 1993, **115**, 12631–12632.
17 M. R. Linford, P. Fenter, P. M. Eisenberger and C. E. D. Chidsey, *J. Am. Chem. Soc.*, 1995, **117**, 3145–3155.
18 O. Seitz, T. Bocking, A. Salomon, J. J. Gooding and D. Cahen, *Langmuir*, 2006, **22**, 6915–6922.
19 K. A. Kilian, T. Bocking, S. Ilyas, K. Gaus, W. Jessup, M. Gal and J. J. Gooding, *Adv. Funct. Mater.*, 2007, **17**, 2884–2890.
20 T. Böcking, K. A. Kilian, K. Gaus and J. J. Gooding, *Adv. Funct. Mater.*, 2008, **18**, 3827–3833.
21 L. T. Canham, C. L. Reeves, J. P. Newey, M. R. Houlton, T. I. Cox, J. M. Buriak and M. P. Stewart, *Adv. Mater.*, 1999, **11**, 1505–1507.
22 V. S. Y. Lin, K. Motesharei, K.-P. S. Dancil, M. J. Sailor and M. R. Ghadiri, *Science*, 1997, **278**, 840–843.
23 S. Chan, P. M. Fauchet, Y. Li, L. J. Rothberg and B. L. Miller, *Phys. Status Solidi A*, 2000, **182**, 541–546.
24 K.-P. S. Dancil, D. P. Greiner and M. J. Sailor, *J. Am. Chem. Soc.*, 1999, **121**, 7925–7930.
25 H. Ouyang, L. A. DeLouise, B. L. Miller and P. M. Fauchet, *Anal. Chem.*, 2007, **79**, 1502–1506.
26 S. Chan, S. R. Horner, P. M. Fauchet and B. L. Miller, *J. Am. Chem. Soc.*, 2001, **123**, 11797–11798.
27 K. A. Kilian, T. Boecking, K. Gaus, M. Gal and J. J. Gooding, *ACS Nano*, 2007, **1**, 355–361.
28 K. A. Kilian, L. M. H. Lai, A. Magenau, S. Cartland, T. Bocking, N. Di Girolamo, M. Gal, K. Gaus and J. J. Gooding, *Nano Lett.*, 2009, **9**, 2021–2025.
29 M. M. Orosco, C. Pacholski, G. M. Miskelly and M. J. Sailor, *Adv. Mater.*, 2006, **18**, 1393–1396.
30 L. Z. Gao, N. Mbonu, L. L. Cao and D. Gao, *Anal. Chem.*, 2008, **80**, 1468–1473.
31 L. Gu, M. Orosco and M. J. Sailor, *Phys. Status Solidi A*, 2009, **206**, 1374–1376.
32 M. M. Orosco, C. Pacholski and M. J. Sailor, *Nat. Nanotechnol.*, 2009, **4**, 255–258.
33 C. Chang and Z. Werb, *Trends Cell Biol.*, 2001, **11**, S37–S43.
34 C. T. N. Pham, *Int. J. Biochem. Cell Biol.*, 2008, **40**, 1317–1333.
35 J. D. Raffetto and R. A. Khalil, *Biochem. Pharmacol.*, 2008, **75**, 346–359.
36 G. A. Rosenberg, *Lancet Neurol.*, 2009, **8**, 205–216.
37 K. A. Kilian, T. Bocking, K. Gaus and J. J. Gooding, *Angew. Chem., Int. Ed.*, 2008, **47**, 2697–2699.
38 B. Guan, S. Ciampi, G. Le Saux, K. Gaus, P. J. Reece and J. J. Gooding, *Langmuir*, 2010, submitted.
39 K. Kilian, A. Magenau, T. Böcking, K. Gaus, M. Gal and J. J. Gooding, *Proc. SPIE*, 2009, **7397**, 739703.
40 P. J. Bergin, S. Wen, Q. Pan-Hammarström and M. Quiding-Järbrink, *FEMS Immunol. Med. Microbiol.*, 2005, **45**, 159–169.
41 B. Xie, Z. Y. Dong and I. J. Fidler, *J. Immunol.*, 1994, **152**, 3637–3644.
42 J. L. Heinrich, C. L. Curtis, G. M. Credo, K. L. Kavanagh and M. J. Sailor, *Science*, 1992, **255**, 66–68.
43 J. R. Link and M. J. Sailor, *Proc. Natl. Acad. Sci. U. S. A.*, 2003, **100**, 10607–10610.
44 M. J. Sailor and J. R. Link, *Chem. Commun.*, 2005, 1375–1383.
45 S. Ciampi, P. K. Eggers, G. Le Saux, M. James, J. B. Harper and J. J. Gooding, *Langmuir*, 2009, **25**, 2530–2539.
46 S. Ciampi, T. Böcking, K. A. Kilian, M. James, J. B. Harper and J. J. Gooding, *Langmuir*, 2007, **23**, 9320–9329.
47 S. Ciampi, T. Böcking, K. A. Kilian, J. B. Harper and J. J. Gooding, *Langmuir*, 2008, **24**, 5888–5892.
48 T. Böcking, K. A. Kilian, T. Hanley, S. Ilyas, K. Gaus, M. Gal and J. J. Gooding, *Langmuir*, 2005, **21**, 10522–10529.

Development of smart nanoparticle–aptamer sensing technology†

Haiyan Zhang,[a] Peter G. Stockley[b] and Dejian Zhou*[a]

Received 18th May 2010, Accepted 19th July 2010
DOI: 10.1039/c005373b

Quantum dots (QDs) are excellent donors in Förster resonance energy transfer (FRET) based sensors because of their broad absorption and narrow symmetric emission. However, the strict requirement of a short donor–acceptor distance to achieve high FRET (hence sensitivity) has posed a significant challenge for QD-FRET-based sensors due to challenges associated with the preparation of QD conjugates that are both compact and highly stable. Consequently, most robust QD-FRET sensors are often too bulky to produce FRET efficiently, especially at low target-to-QD copy numbers. They have largely relied on increasing the target:QD ratio to achieve high FRET, making them undesirable and inefficient in situations of low target:QD copy numbers. Herein we report our work on the preparation of stable, compact and water-soluble QDs and their subsequent use in making compact, functional QD-DNA-based smart nanoparticle sensors for labelled and label-free DNA and protein detection. We have developed two strategies to prepare QD-DNA sensors: 1) *via* QD-thiolated DNA self-assembly, and 2) *via* covalent coupling between DNA and a QD surface ligand functional group. We found that thiolated DNA (fluorophore labelled) can self-assemble onto a 3-mercaptopropionic acid-capped QDs to produce highly efficient FRET (~80%) at a DNA:QD ratio of 1 : 1. However, this system suffers from strong non-specific adsorption and the self-assembled single-stranded (ss) DNA target is unable to hybridise to its complementary probe. More recently, we found that a dihydrolipoic acid-capped QD-ssDNA self-assembled system can hybridise to a labelled complementary probe to produce efficient FRET that can be exploited for labelled DNA probe quantification. Further, incorporating an anti-thrombin DNA aptamer to this system leads to a QD-DNA aptamer sensor that can specifically detect a 10 nM unlabelled protein probe (thrombin). The non-specific adsorption problem can be eliminated by introducing a poly(ethylene glycol) (PEG) linker to the QD capping ligands or by capping the QD with a chelating dendritic ligand. The resulting QD-DNA sensors can specifically detect 1 nM unlabelled or 35 pM labelled DNA probes using QD-sensitised dye FRET signals on a conventional fluorimeter. Extension of the DNA target to other functional DNAs or DNA/RNA aptamers should allow the development of a multi-functional QD-DNA platform suitable for biosensing, disease diagnosis and therapeutic applications.

[a]*School of Chemistry and the Astbury Centre for Structural Molecular Biology, University of Leeds, Leeds, LS2 9JT, UK. E-mail: d.zhou@leeds.ac.uk; Fax: +0113 343 6565; Tel: +0113 343 6230*
[b]*Astbury Centre for Structural Molecular Biology, University of Leeds, Leeds, LS2 9JT, UK*

† Electronic supplementary information (ESI) available: The synthesis and spectroscopic characterisation of the chelating dendritic ligand ethyl ester and supporting figures. See DOI: 10.1039/c005373b

Introduction

Aptamers are short DNA or RNA oligonucleotides artificially generated against specific targets with binding affinities often comparable to or better than antibodies.[1] They have several advantageous properties over the more widely used antibodies as ligand binding reagents, including a wider target choice; higher ligand specificity with comparable binding affinity (nM to pM); production by totally *in vitro* protocols, *e.g.* by chemical synthesis, eliminating batch-to-batch variation; are more robust against thermal and chemical denaturation and are able to undergo several rounds of denaturation/renaturation without losing activity. They are very useful for developing sensors of a range of targets: environmental pollutants, DNA, proteins, street drugs, viruses, and even cancerous cells.[2]

Fluorescent semiconductor nanocrystals, also known as quantum dots (QDs), have been of considerable research focus over the past two decades due to their unique, size-dependent, bright and extremely stable emissions.[3,4] They have been widely used in a broad range of bio-related applications, such as imaging, cell tracking/trafficking, biosensing, multiplexed sensors *etc.*[4] In particular, their broad absorption and narrow symmetric emission make them especially attractive in Förster resonance energy transfer (FRET) based sensors, because such spectral characteristics allow the use of a wide range of excitation wavelengths to reduce the direct acceptor excitation, and the choice of narrow bandpass filters for the effective separation of donor–acceptor emissions, leading to greatly reduced background.[5–7] The FRET efficiency (E) is strongly dependent on the donor–acceptor distance (r) and decreases dramatically with the increasing r: $E = 1/[1 + (r/R_0)^6]$, where R_0 is the Förster radius of a given donor–acceptor pair under which $E = 50\%$. It is thus important to reduce the r value to achieve high E (hence sensitivity), *i.e.* a compact QD bioconjugate is essential to highly sensitive QD FRET based sensors. The strict requirement of small r for high sensitivity has posed a significant challenge for developing QD FRET based sensors because the QD fluorescent cores are relatively large, with radii varying from 1.6 to 2.9 nm for 490 to 620 nm emission CdSe–ZnS core–shell QDs.[8] Furthermore, the cores are capped with a layer of stabilising ligands, whose thickness can vary from 0.5 to 15 nm, depending on the capping methods used. Thus the overall radius of the QD donor alone prior to bioconjugation is often comparable or even greater than the R_0 of most QD-dye FRET systems, which is typically in the range of 4–7 nm. This has been a major limiting factor for high sensitivity in QD FRET based sensors.[4–7] Although a high E can also be achieved by increasing the number of acceptors attached to each QD (*i.e.* increasing the target:QD copy number),[7] such systems are inefficient and often unsuitable for low target:QD copy number situations.

Another key requirement for QD FRET based sensors is that the QDs should be stable and soluble in aqueous media. Although compact, water-soluble core QDs (*e.g.* CdTe, CdS *etc.*) can be prepared directly from aqueous media,[9] these core-only QDs are sensitive to their environment, are liable to chemical and photo-oxidation, and often lose their fluorescence significantly on bioconjugation using standard conjugation chemistries,[10] limiting their usefulness and robustness as biosensors. In contrast, core–shell (*e.g.* CdSe–ZnS) QDs, where the fluorescent core is protected by an inorganic shell, are robust and much less environmentally sensitive, and well-suited for sensing applications. In fact, most QD FRET based sensors reported so far have been based on CdSe–ZnS core–shell QDs.[4–7] Since most high-quality core–shell QDs are prepared by an organometallic route,[11] capped with hydrophobic ligands, they are therefore water-insoluble. Consequently they have to be made water soluble prior to use. Several strategies have been developed to make these hydrophobic QDs water soluble, including wrapping with amphiphilic polymers,[12] capping with PEGylated phospholipid micelles,[13] coating with silica shells,[14] and *via* ligand exchange with hydrophilic thiols.[5,6,15] Although QDs capped by polymers or phospholipids (as is the case for most commercial water-soluble QDs) are highly stable and water soluble, their resulting large sizes (*e.g.* hydrodynamic diameters, HDs, for T1 and T2 Evitags are ~40 and 25 nm)[8] limit their sensitivity significantly. Ligand-exchange on the other hand, is easy and cheap to conduct.[6,15,16] Moreover it can provide a compact QD (*e.g.* HD <10 nm) which can be critical for high sensitivity. Further, by coating the QDs with chelative ligands, the stability of the resulting QDs can be significantly enhanced.[16] Besides, the surface chemistry linking the target biomolecules (*e.g.* DNA aptamer) and QDs is also important. For high sensitivity, it is important to use as short a linker as possible, which often leads to non-specific adsorption and poor sensor stability. Whereas a long linker can provide sufficient steric stabilisation for high stability and prevention of non-specific adsorption, it however, often leads to low sensitivity. Thus it is important to fine tune the delicate balance between sensitivity, specificity and stability for QD FRET based sensors as these requirements are often incompatible.

In this paper, we report our recent advance in combining QDs and DNA (aptamers) to develop specific, sensitive FRET-based QD-DNA (aptamer) sensors. We found that thiolated double-stranded (ds) DNA (dye-labelled) can self-assemble onto an MPA-capped QD to produce efficient FRET (*ca.* 80%) at a DNA:QD copy number of 1 : 1 from single-molecule analysis.[17] However, attempts to use an equivalent ssDNA-QD system for detection of a complementary probe was unsuccessful, due to strong non-specific adsorption. The self-assembly of a ssDNA on a chelative ligand-capped QD can hybridise to its complementary probe, producing efficient FRET that is suitable for labelled DNA quantification. Incorporating an anti-thrombin DNA aptamer to the self-assembled QD-DNA system resulted in a QD-aptamer sensor able to detect 10 nM unlabelled protein target (thrombin), although it did not completely eliminate non-specific adsorption. We have developed two further strategies to overcome the non-specific adsorption problem: 1) by capping the QD with a novel chelating dendritic ligand (CDL); 2) by introducing a PEG linker to the QD surface capping ligand.[18] We found that the resulting QD-DNA sensors can specifically detect 35 pM labelled or 1 nM unlabelled complementary DNA probes at low probe:QD copy numbers with a conventional fluorimeter.

Experimental section

Materials and reagents

All CdSe–ZnS core–shell QDs (first exciton peaks, 535, 545 and 585 nm, and emission peaks (EM) ~550, 560, and 600 nm) were purchased from Evident Technologies Inc (Troy, USA). They were supplied in toluene solutions capped with trioctylphosphine oxide (TOPO) ligands. All other chemicals and reagents, *N*-(3-dimethylaminopropyl)-*N*-ethyl-carbodiimide hydrochloride (EDC, 99%), *N*-hydroxysuccinimide (NHS, 98%), thrombin (from bovine plasma), sodium bicarbonate, 3-mercaptopropionic acid (MPA, 99%), DL-thioctic acid (>98%), sodium borohydride, Tris base, DNAse and RNAse free NaCl, HPLC grade ethanol, methanol and chloroform were from Sigma-Aldrich (Dorset, UK). All DNA used was purchased from IBA GmbH (Göttingen, Germany). The 5′-C_6 amine- and C_6 thiol- modified capture DNA and unlabelled DNA probes were of HPLC grade, and all dye-labelled DNA was double HPLC purified by the supplier. 10 × PBS buffer (100 mM phosphate, 1.50 M NaCl, pH 7.2), phosphate buffer (PB, 20 mM phosphate, pH 6.0), Tris buffer (10 mM Tris, HCl, 100 mM NaCl, pH 7.4), and sodium bicarbonate buffer (50 mM $NaHCO_3$, pH 9.0) were prepared with ultra-pure MilliQ water (resistance >18 MΩ.cm).

Synthesis of functional thiol capping ligands

Scheme 1 gives the chemical structures of the functional thiols used in this report. Dihydrolipoic acid (DHLA) was prepared by reducing DL-thioctic acid with sodium borohydride following a literature procedure.[16] The synthesis of EG_3OH and

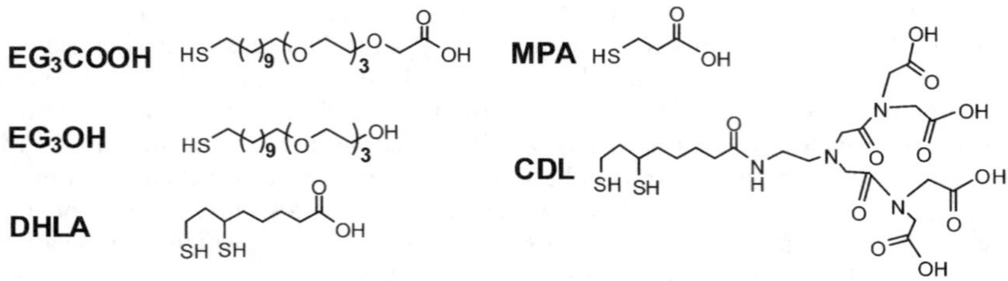

Scheme 1 Chemical structures of the functional QD surface capping ligands.

EG_3COOH was done by following a literature procedure and their spectroscopic characterisation has been reported previously.[18,19] The synthesis and spectroscopic characterisation of the chelating dendritic ligand ethyl ester (CDL-ester) are given in the ESI.†

Preparation of water-soluble QDs by ligand exchange

The ligand exchange of the QDs with functional thiols were carried out following our earlier procedures.[18] Briefly, the TOPO-capped CdSe–ZnS core–shell QDs were precipitated using ethanol, followed by centrifugation and discarding of clear supernatant three times to remove free TOPOs. The resulting QD pellet was dissolved in $CHCl_3$, to which the functional thiol ligands and tetramethylammonium hydroxide or sodium hydroxide (2 mole ligand equivalent) in ethanol was added. The resulting solution was refluxed for 4 h under N_2. After which the solvent was removed and the solid was dissolved in ethanol, and $CHCl_3$ was added to precipitate the QDs followed by centrifugation and removal of supernatant. This process was repeated three times to remove free ligands. The pellet was air dried and dissolved in MilliQ water to make the QD stock solution.

A modified procedure was used to make the CDL-capped QDs. After the initial ligand exchange, the resulting solid was dissolved in $CHCl_3$, precipitated with hexane and the process was repeated twice to obtain the CDL-ethyl ester-capped QDs. The ethyl ester groups on the QD were hydrolysed by dissolving the pellet in a mixed solution of ethanol and 1 M K_2CO_3 (1 : 1 v/v), followed by refluxing for 2 h. The resulting precipitate was collected, dissolved in the minimum amount of water, precipitated with ethanol and centrifuged. The process was repeated three times to remove free ligands. Finally the pellet was dissolved in water to make the QD stock solution. The concentrations of the QD stocks were calculated from their absorbance at their exciton peaks using their respective extinction coefficients.[8] Fluorescence quantum yields (QYs) of the QDs were estimated using Rhodamine-6G ethanol solution (excitation at 480 nm, QY = 95%) as standard.[17]

Preparation of QD-DNA conjugate sensors

Self-assembled QD-ssDNA conjugates. Self-assembled QD-ssDNA conjugates were prepared by mixing TBA (50 μl, 200 μM) with DHLA-QD (3.4 μM, 590 μl) in PBS overnight under a fixed QD:DNA ratio of 1 : 5 to form the self-assembled QD-TBA conjugates. Then different concentrations of complementary DNA probes (labelled with Atto-647N fluorophore) were introduced to make a series of samples and were allowed to hybridize at RT for 12 h.

Preparation of EG_3-QD-DNA sensors by EDC/NHS coupling[18]. EDC (30 mg) and NHS (15 mg) were dissolved in 200 μL of a 1 : 1 (v/v) EtOH–H_2O mixture, to which phosphate buffer (20 mM, 200 μL, pH 6.0) was added. Then EG_3OH/EG_3COOH-capped QDs (4 μM, 250 μl) were added and the resulting solution was briefly sonicated and allowed to stand at RT for 1 h, after which it was centrifuged. The pellet was carefully washed with water (2 × 200 μL), and then 175 μL MeOH–H_2O (1 : 1 v/v), C_6-amine modified DNA-T (440 μM, 25 μL, DNA-T:QD ratio = 11 : 1) and 100 μL $NaHCO_3$ buffer (50 mM, pH 9.0) were added. The mixture was sonicated to break up the pellet, and then allowed to react at 4 °C overnight. The sample was centrifuged and the clear supernatant was carefully removed. The pellet was washed with 200 μL MeOH–H_2O (1 : 1 v/v). An absorption spectrum was taken of the combined supernatant and the absorbance at 260 nm was used to calculate the amount of DNA not bound to the QD. The number of DNA-T molecules attached to each QD was estimated to be 2.2. The pellet was dissolved in 0.5 mL of pure water to obtain a clear stock solution and stored in the dark at 4 °C.

Hybridisation of DNA probes. The total volume of the hybridisation reaction solution was kept constant at 500 μL, with a fixed final QD concentration in PBS. The reaction was carried out in batches under identical conditions, where 50 μL of the 10× PBS, the calculated amount of water, (ethidium bromide where necessary), DNA probes, and the required amount of the QD-DNA conjugate were sequentially added and thoroughly mixed. The hybridisation reaction was carried out at RT.

Fluorescence measurements and FRET calculations

Fluorescence spectra were recorded on a Cary Eclipse or a Fluormax-3 fluorescence spectrophotometer.[17,18] The emission spectra (460–750 nm range) were recorded under a fixed excitation wavelength of 450 nm (to minimise direct excitation of the dyes, Alexa 594, Alexa 647, and Atto 647N). For ethidium bromide (EB) intercalation experiments, the sample was excited at 378 nm to minimise EB direct excitation. The spectra were corrected for direct dye excitation using the same dye-labelled DNA as a reference. The fluorescence intensity referred to here is the integrated fluorescence. The FRET signals were obtained by using the integrated fluorescence of the whole spectrum (after correction for direct dye excitation) subtracting the QD fluorescence. The calculation was based on assuming that the QD fluorescence maintained the same shape as the QD-DNA conjugate only sample, and the integrated intensity is proportional to the height of the QD fluorescence peak (all acceptors do not fluoresce at their respective QD fluorescence peaks). The apparent FRET efficiency E was estimated using: $E = I_{dye}/(I_{dye} + I_{QD})$, where I_{QD} and I_{dye} are integrated QD and dye fluorescence intensities respectively.

Results and discussion

Self-assembled QD-DNA systems

Self-assembled MPA-QD-dsDNA system[17]. The most effective way to reduce r is to directly attach the target biomolecule to the QD *via* an appended affinity tag (*e.g.* a His_6 tag) that can bind to the QD crystal surface.[6] Since alkyl thiols are widely used in ligand exchange to make water-soluble QDs,[5–7] we envisaged that thiolated DNA may also be able to bind to the QD surface directly (*via* coordination to the Zn^{2+} ions on the ZnS shell). To test this idea, we mixed a thiolated dsDNA (labelled with Alexa 594) with a MPA-capped CdSe–ZnS core–shell QD (MPA-QD, EM = 560 nm) at different molar ratios in Tris buffer and studied their FRET (Fig. 1).[17]

Fig. 1(a) shows that with the increasing DNA:QD ratio, the QD fluorescence, I_{QD}, at ~560 nm is progressively quenched, meanwhile the Alexa 594 (acceptor) fluorescence, I_{dye}, at 618 nm is enhanced. This result clearly indicates the successful DNA-QD self-assembly that brings the dye to the close proximity to the QD since FRET only happens at small r (*e.g.* <10 nm). The apparent E taken from the spectra (bulk) increases with the increasing DNA:QD ratio initially, but quickly levels off and saturates at a DNA:QD ratio of only 0.5, giving an E of ~50% [Fig. 1(b)]. In contrast, single-molecule (SM) analysis of the samples reveals a different trend, where E increases continuously, reaching ~88% at a DNA:QD ratio of 2 : 1. The E obtained

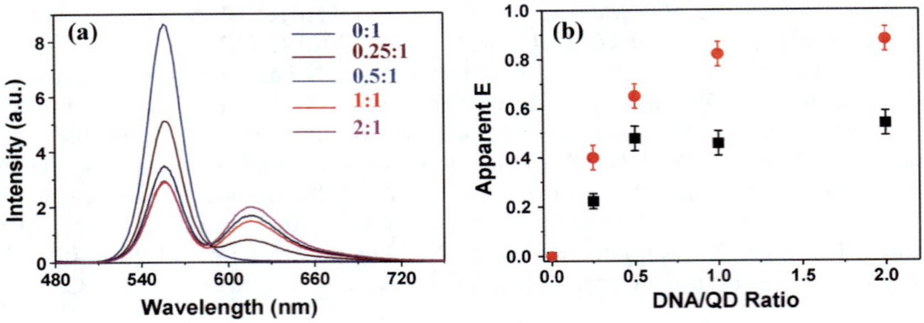

Fig. 1 (a) Fluorescence spectra of the self-assembled MPA-QD-dsDNA conjugate at different DNA:QD ratios (C_{QD} = 100 nM). The duplex DNA sequences are: HSC_6H_{12}-5′-CAT AAA AGA GCT CCA TAT CCA ACC TGC ACG-3′/3′-GTA TTT **T**CT CGA GGT ATA GGT TGG ACG TGC-5′, the base **T** shown in bold is labelled with Alexa 594. (b) A plot of the apparent *E versus* the DNA:QD ratios. The *E* values obtained from the fluorescence spectra (bulk) and single-molecule analysis are shown as black squares and red dots respectively. This figure is taken from ref. 17 after modification.

from SM study is higher than that of the bulk measurement. This has been attributed to the fact that SM measurement only detects bright QD-DNA conjugates (a threshold of 50 counts per ms bin was used to differentiate SM bursts from the background), whereas the bulk method measures all QD-DNA conjugates, bright and dim. This shows that thiolated DNA readily self-assembles with the MPA-QD to form compact a QD-DNA conjugate with a short *r* value. As a result, efficient QD-sensitised dye FRET (~80%) can be achieved at a DNA:QD ratio of 1 : 1 from SM analysis. This compact QD-DNA conjugate appears to be very attractive for highly sensitive QD FRET based sensors. However, attempts to use this system for specific DNA detection were unsuccessful, presumably because the ssDNA which initially binds to the QDs *via* their thiol groups adsorb strongly and non-specifically to the QDs, making them unavailable for specific hybridisation.

Self-assembled DHLA-QD-DNA aptamer systems. The MPA used above contains only a single thiol which can not bind to the QDs very strongly. It is thus likely that some of the MPA may be displaced by the incoming DNA (the phosphate and amine groups of the DNA can also bind to the QD surface Zn^{2+} ions), leading to the DNA wrapping around the QD and so becoming unavailable for hybridisation. We thought that if the QDs were capped by strong, chelative ligands then the incoming DNA may not be able to displace them, ensuring the DNA chain is extended away from the QD surface and available for hybridisation. To verify this, a DHLA-capped QD (EM =550 nm) was prepared where each DHLA can chelate strongly to the QD surface *via* two thiolate groups. To verify that thiolated DNA can self-assemble with the DHLA-QD, we mixed the QD with labelled thiolated dsDNA (TBA/TBA-C: TBA = HSC_6H_{12}-5′-ACA TTT TT*A GTC CGT GGT AGG GCA GGT TGG GGT GAC T*-3′, which contains a 29-mer anti-thrombin aptamer sequence[20] shown in italic; and TBA-C = Atto-647N-3′-T CAG GCA CCA TCC CGT CCA ACC CCA CTG A-5′, which is complementary to the anti-thrombin (TB) aptamer) at different molar ratios in PBS. The resulting fluorescence spectra clearly show that as the DNA:QD ratio increases, the QD fluorescence at ~550 nm is progressively quenched, meanwhile the Atto-647N fluorescence at ~670 nm is greatly enhanced, confirming the successful QD-dsDNA assembly. The apparent *E* initially increases rapidly with the increasing DNA/QD ratio, but the rate of the increase gradually levels off and saturates at a DNA:QD ratio of 7 : 1, giving an *E* of ~ 55% (see Fig. S1 of the ESI†).

Having established that thiolated dsDNA can self-assemble to the DHLA-QD, we then explored the use of this self-assembled QD-ssDNA system as a sensor for

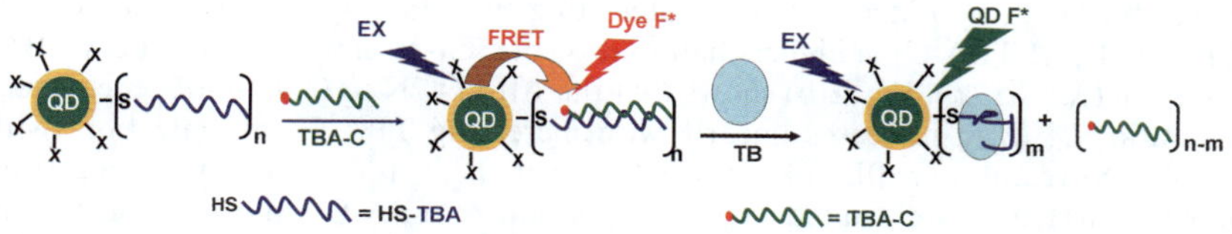

Scheme 2 Schematic of using the self-assembled QD-ssDNA aptamer system for the detection of labelled DNA probe and label-free protein targets.

labelled DNA probe and unlabelled protein detection. Scheme 2 shows our strategy schematically. First, a QD-TBA conjugate is prepared by self-assembly, then a labelled complementary probe (TBA-C) is introduced. Hybridisation of TBA-C to the QD-TBA conjugate will bring the Atto-647N dye in close proximity to the QD, producing Atto-647N FRET on excitation of the QD for labelled DNA probe detection. Finally, an unlabelled protein target, TB, is introduced to the QD-TBA/TBA-C system. Since the TBA strand contains a 29-mer anti-TB DNA aptamer sequence that has high affinity ($K_d \sim 0.5$ nM) for TB,[20] the TB will compete with TBA-C for TBA binding. The formation of a TBA/TB complex will detach DNA-C from the QD-TBA conjugate, reducing the QD to dye FRET, leading to QD fluorescence recovery that may be exploited for label-free TB sensing.

Fig. 2(a) shows the fluorescence spectra of a series of samples, each containing 20 nM QD-TBA and different concentrations of TBA-C. It is apparent that as the TBA-C concentration increases, the QD fluorescence at ~550 nm is quenched whilst the Atto 647N fluorescence at ~670 nm is greatly enhanced. Since the Atto-647N is not excited at 450 nm (also corrected for direct dye excitation), the observed Atto-647N fluorescence must come from the QD-sensitised FRET, confirming the

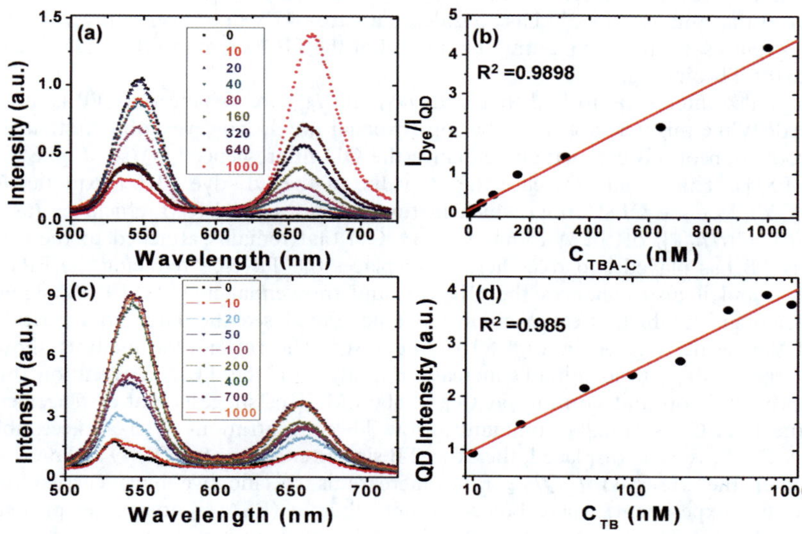

Fig. 2 (a) Fluorescence spectra of a series of the self-assembled QD-TBA sample (TBA:QD = 5 : 1, final $C_{QD} = 20$ nM) after hybridisation to different concentrations of the complementary TBA-C (Atto-647N labelled) in PBS. (b) A plot of the fluorescence intensity ratio of Atto-647N to QD, I_{dye}/I_{QD} [obtained from (a)] *versus* the TBA-C concentration. (c) Fluorescence spectra of self-assembled QD-TBA/TBA-C1 conjugates (final $C_{QD} = 10$ nM) after reaction with different concentrations of TB in PBS. (d) A plot of the QD fluorescence intensity shown in (c) *versus* the TB concentration.

successful DNA hybridisation. It is interesting that the apparent E achieved here (81%, at 1 μM TBA-C) is higher than that of the self-assembled QD-dsTBA/TBA-C system (*ca.* ~55%). Fig. 2(b) shows that the Atto-647N to QD fluorescence intensity ratio, I_{dye}/I_{QD}, increases linearly with increasing TBA-C concentration (with a linear correlation coefficient, $R^2 = 0.99$) over the 1–1000 nM range, suggesting that this system is well-suited for quantification of TBA-C. Another noteworthy point is the ratiometric signal here that it is effectively insensitive to instrument noise and signal fluctuations, allowing accurate and reliable measurement.

To investigate the potential of such self-assembled QD-DNAs for TB detection, a QD-TBA conjugate (TBA:QD ratio = 5 : 1) was prepared, and then hybridised to TBA-C1 to form QD-TBA/TBA-C1 conjugate (QD:TBA-C1 ratio = 1 : 5, TBA-C1 = 5′-AGT CAC CC**T** AAC CTG CC**A** TAC CAC GGA CT-3′-Atto 647N, which contains double base mismatches, shown in bold, to the anti-TB aptamer sequence to tune the stability of the TBA/TBA-C1 duplex, making TBA-C1 more readily displaced by TB binding).[21] The QD-DNA conjugate was divided into aliquots which were allowed to react with different amounts of TB overnight (final C_{QD} = 10 nM), and the corresponding fluorescence spectra are given in Fig. 2(c).

Fig. 2(c) clearly shows that I_{QD} is significantly enhanced by addition of TB. I_{QD} initially increases rapidly with increasing TB concentration, C_{TB}, but gradually levels off and finally saturates at ~400 nM TB, after which I_{QD} remains broadly constant throughout the concentration range studied. This observation is consistent with the expectation that the TB can bind to the QD-TBA to form a QD-TBA/TB complex, leading to the detachment of TBA-C1 (with the Atto-647N label). This reduces the QD quenching *via* FRET to Atto-647N, leading to I_{QD} recovery at ~550 nm. When all of the accessible TBAs on the QD surface have bound to TB, further increasing the amount of TB can not lead to any further TBA-C1 displacement, the I_{QD} reaches saturation. It is interesting to find that the I_{QD} increases roughly linearly with the log(C_{TB}) up to 700 nM [$R^2 = 0.985$, Fig. 2(d)], suggesting that this system is suitable for label-free TB detection and quantification. We also noticed that this system still has considerable Atto-647N fluorescence present after the I_{QD} has saturated, suggesting that not all of the TBA bound to the QD is accessible for TB binding.

It is also interesting to find that both I_{QD} and I_{dye} are enhanced as TB is introduced. While this may appear to be controversial, we do however, observe that the introduction of TB can actually enhance the QD fluorescence QY (see Fig. S2 of the ESI†). This should increase the R_0 value of the QD-dye FRET pair (as $R_0 \propto (QY)^{1/6}$ of the QD donor) which in turn increases the FRET efficiency (as $E = 1/[1 + (r/R_0)^6]$) of the Atto-647N (TBA-C1) that remains attached to the QD. Thus TB has played two roles here. It displaces the TBA-Cs reducing the FRET signal, and it also enhances the QD QY and thus enhancing the FRET signal. To distinguish which effect dominates, we have analysed the data ratiometrically and the result is given in Fig. S3 of the ESI.† The ratiometric analysis shows that the I_{QD}/I_{dye} ratio initially increases rapidly with the TB concentration, but quickly levels off and saturates at $C_{TB} \sim 300$ nM. This suggests that displacement of the TBA-C1 is initially dominant at low TB concentration. After all accessible TBA-C1s have been displaced, the extra TB simply enhances the QD QY. To verify whether the observed I_{QD}/I_{dye} ratio increase is specific, we have carried out a control experiment, where bovine serum albumin (BSA, a non-target protein, 250 nM) was added to the QD-TBA/TBA-C1 system under identical conditions. Addition of BSA has been shown to enhance the DHLA-QD fluorescence (see Fig. S2†), however, it does not cause significant increase of I_{QD}/I_{dye} ratio (<10%). In contrast, the same TB concentration led to almost doubling of the I_{QD}/I_{dye} ratio. This result highlights the importance of the ratiometric approach for sensor reliability and specificity. Despite problems such as incomplete DTA-C1 displacement and limited sensor stability, the results presented here confirm that the self-assembled QD-TBA/TBA-C1 system can be exploited for specific label-free TB detection with 10 nM sensitivity.

Covalently coupled QD-DNA systems

Although self-assembled DHLA-QD-DNA systems have shown good promise in protein and DNA sensing, problems such as non-specific adsorption and lack of long-term stability have limited their practical application. Since each DHLA ligand contains only a single alkyl chain, it can not pack densely on the highly curved QD surface to prevent non-specific DNA adsorption [Fig. 3(a)]. A dendritic ligand should provide a better match for the requirement of a dense, entangled hydrophilic coating for enhanced stability and better resistance to non-specific adsorption [Fig. 3(b)].[22] Based on this ideal, we have designed and synthesised a novel chelating dendritic ligand (CDL, Scheme 1) which combines both chelative binding and

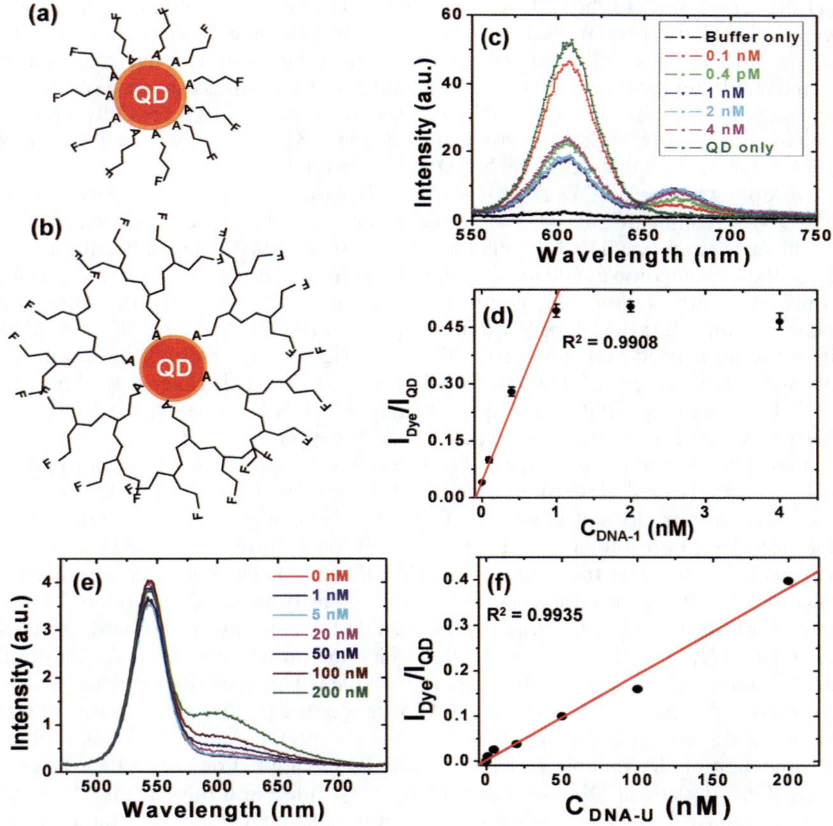

Fig. 3 Schematics of a QD capped with DHLA ligands (a) and with chelating dendritic ligands (b). A = chelate group (DHLA), F = functional group (COOH). (c) Fluorescence spectra of a series of CDL-QD-DNA-0 samples after hybridisation to different concentrations of DNA-1 (final C_{QD} = 0.2 nM) in Tris buffer. (d) A plot of the intensity ratio of the Alexa 647 (650–750 nm) to QD, I_{dye}/I_{QD}, *versus* the DNA-1 concentration (the two most concentrated data points are masked from the fit, error bars show standard deviations of three parallel samples). (e) Fluorescence spectra of a series samples, each containing 100 nM EG_3-QD-DNA-T conjugate, 2 μM EB and different concentrations of DNA-C in PBS. (f) A plot of the integrated intensity ratio of EB to QD, I_{dye}/I_{QD}, *versus* the concentration of DNA-C. Graphs (e) and (f) are taken from ref. 19 after modification.

dendritic packing (see the ESI for detailed synthetic procedures†). Each ligand contains a chelate moiety (DHLA) for strong QD binding and a dendritic moiety with four carboxylic acids to create a dense, compact hydrophilic coating on the QD surface for water-solubility and high stability. We prepared CDL-capped QDs (CDL-QD, EM ~ 600 nm) by ligand exchange. The resulting water-soluble CDL-QD was found to be very stable: extended storage (1 year) in a fridge did not cause precipitation or significant fluorescence decrease. More importantly, it can eliminate the DNA non-specific adsorption completely: mixing the CDL-QD with DNA-1 (Alexa 647-3′-TAG CGT TCT TAA GGT ACT GGT TGG ACG TGA-5′, 10 mole equivalent) did not produce any observable Alexa 647 FRET signal. Moreover, thiolated DNA-0 (HS–C_6H_{12}-5′-ATC GCA AGA ATT CCA TGA CCA ACC TGC ACT-3′) was unable to self-assemble on the CDL-QD, confirming that the CDL ligands form a stable and dense coating around the QD surface. We then explored using the surface carboxylic acid (COOH) groups of the CDLs for attaching the target DNA. Attempts to couple an amine-modified DNA-0 (sequence identical to thiolated DNA-0, with the 5′-end modified with a $H_2NC_6H_{12}$-linker) to the CDL-QD *via* EDC/NHS-mediated coupling were unsuccessful. An alternative strategy that first transformed some of the surface COOH groups into amines, followed by reacting with a hetero-functional cross linker (4-maleimidobutanoic acid NHS ester) to introduce maleimide groups, and finally by reaction with the thiolated DNA-0, was developed and successfully used to prepare the CDL-QD-DNA-0 conjugate (Scheme S2†). The average number of DNA-0 molecules attached to each QD was estimated to be ~5.

To verify whether the QD-DNA-0 system is functional, fluorescence spectra of the QD-DNA-0 conjugate (50 nM) were recorded before and after addition of a labelled complementary probe, DNA-1 (500 nM) in PBS [Fig. S4(a)†]. A significant quenching of the QD emission at 600 nm and simultaneous enhancement of Alexa 647 emission at 670 nm is observed, confirming the successful DNA hybridisation. As a control, an Alexa 647 labelled non-complementary DNA, DNA-NC, added in the same way, produced no Alexa 647 FRET signal [Fig. S4(b)†], indicating no DNA-NC attachment to the QD-DNA-0 conjugate. The experiments confirm that the FRET signal is specific for the complementary DNA probe, so the CDL-QD-DNA-0 system can be used for specific DNA detection.

Having confirmed the specificity of the QD-DNA-0 system, we went on to investigate its sensitivity. A series of samples, each containing 0.2 nM of QD-DNA-0 and different concentrations of DNA-1 in Tris buffer were prepared. The resulting fluorescence spectra are shown in Fig. 3(c). As DNA-1 is introduced, the QD fluorescence is quenched and the Alexa 647N FRET signal is enhanced, confirming the successful DNA hybridisation. A plot of the intensity ratio of integrated fluorescence Alexa 647 to QD, I_{dye}/I_{QD}, *versus* the DNA-1 concentration reveals a linear relationship ($R^2 > 0.99$) up to 1 nM, after which the ratio remains broadly constant throughout the DNA-1 concentration range studied. This is consistent with the estimation of ~5 copies of DNA-0 on each QD, suggesting that all DNA-0s on the QD are functional and available for hybridisation. The FRET signal saturation is likely an indication of all available DNA-0s on the QD having hybridised to their targets. Thereafter, any added DNA-1s can no longer hybridise and thus remain "free" in solution. Since FRET only happens over short distances (<10 nm), these "free" DNA-1s do not contribute to the observed FRET signal. This is a particular advantage for FRET based sensors, which only detect species in close proximity, any free molecules not bound to the donor (*e.g.* QD) do not interfere with the detection. This result also shows that this QD-DNA-0 sensor system is well suited for specific, labelled DNA probe quantification. The detection limit for the QD-DNA-0 system here is estimated to be 35 pM for DNA-1 on a conventional fluorimeter, which is about 2–3 orders of magnitude higher than earlier reports under similar conditions.[23]

Introduction of a PEG group to surface coatings has shown to be very effective in resisting non-specific adsorption of biomolecules.[24] Here we have extended this strategy to the highly curved QD surfaces.[18] We synthesised two functional alkyl thiols, EG_3OH and EG_3COOH (see Scheme 1), and used them in ligand exchange to make a water-soluble EG_3OH/EG_3COOH (2 : 1 molar ratio) capped QD (EG_3-QD, EM ~ 550 nm). The hydrophilic hydroxyl (OH) and COOH terminal groups are spaced out from the QD surface by the EG_3 spacer, enhancing the QD stability and water solubility. This surface coating can resist DNA non-specific adsorption: mixing of an Alexa 594 labelled DNA-C (Alexa 594-3′-GTA TTT TCT CGA GGT ATA GGT TGG ACG TGC-5′) with the EG_3-QD did not produce observable dye FRET signal [Fig. S5(a)†]. To use this QD as a DNA sensor, we attached an amine-modified target DNA, DNA-T ($H_2NC_6H_{12}$-5′-CAT AAA AGA GCT CCA TAT CCA ACC TGC ACG-3′) to the EG_3-QD using EDC/NHS-mediated coupling to form a QD-DNA-T conjugate, where on average, each QD has 2.2 DNA-T molecules attached.

To confirm that the QD-DNA-T conjugate is functional, we mixed the QD-DNA-T conjugate (100 nM QD) with different ratios of the labelled complementary probe, DNA-C, in PBS and the corresponding fluorescence spectra are shown in Fig. S6†. It is clear that hybridisation of DNA-1 to the QD-DNA-T quenches the QD emission at 550 nm, whilst enhancing the emission of Alexa 594 at 618 nm. The apparent E increases approximately linearly with the increasing DNA-1:QD ratio initially, and then levels off at a ratio of just over 2 [Fig. S6(b)†], in excellent agreement with the estimation that there are only 2.2 DNA-T molecules attached to each QD. This confirms that all the DNA-T molecules on the QD are available for hybridisation. To confirm the sensor specificity, an Alexa 594 labelled control DNA-NC (3′-AAT CAG GGA TTT ACG TGC ACG ACA CAC ACT-5′-Alexa 594, whose sequence is non-complementary to DNA-T) was also added but produced no detectable Alexa 594 FRET signal [Fig. S5(b)†]. This confirms that the FRET signal is specific for the complementary DNA, suggesting that the QD-DNA-T system is suitable for specific detection of labelled complementary DNA probes.[18]

To extend the QD-DNA-T sensor for label-free DNA detection, a novel strategy based on EB intercalation of dsDNA was developed: hybridisation of an unlabelled DNA probe to the QD-DNA-T conjugate leads to the formation of a QD-dsDNA conjugate and simultaneous EB intercalation. Thus excitation of the QD will produce an EB FRET signal suitable for unlabelled DNA probe detection. The experiments were carried out on a series of samples, each containing 100 nM QD-DNA-T conjugate, 3 μM EB and different concentrations of an unlabelled complementary DNA-U (sequence identical to DNA-C, but without the Alexa 594 label). The resulting fluorescence spectra [Fig. 3(e)] clearly show that the EB fluorescence increases with the increasing DNA-U concentration. The observed EB FRET signal is DNA sequence-specific, since a non-complementary unlabelled DNA (500 nM) did not produce any observable EB FRET signal, presumably because EB binds strongly only to dsDNAs but not to ssDNAs. Therefore only the formation of dsDNA leads to significant EB intercalation and hence FRET signal. A plot of the integrated intensity ratio of the EB FRET signal (575–750 nm) to that of the QD (I_{dye}/I_{QD}) *versus* the DNA-U concentration shows a good linear fit [R^2 = 0.994, Fig. 3(f)], suggesting this system is well-suited for quantification of unlabelled complementary DNA probes. The detection limit is estimated to be ~1 nM DNA-U on a conventional fluorimeter without optimisation, which is comparable or better than other widely used electrochemical or surface plasmon resonance (without signal amplification) based label-free methods for short unlabelled DNA probes (whose sensitivity typically ranges from 10 to a few 10s of nM).[25] In addition, the ratiometric signal used here can provide accurate and reliable measurement, highlighting the great potential of the QD-DNA conjugate sensors for both label and label-free DNA target detection. Extension of the DNA target to other aptamers and incorporation of multiple types of aptamers into a single QD-DNA system should allow the development of a general multiplexed QD-DNA sensing platform with broad applications in biosensing and disease diagnosis.

Conclusion

In summary, we have presented two different strategies that have been successfully used to prepare functional QD-DNA (aptamer) sensors, 1) *via* QD-thiolated DNA self-assembly, and 2) *via* coupling to QD surface functional groups. The first strategy is simple and easy to conduct, involving no surface coupling steps, and can achieve highly efficient FRET (80%) at low target:QD copy numbers (*e.g.* 1 : 1), making it very attractive for high-sensitive QD-FRET based sensors. However, problems such as non-specific adsorption and lack of long-term stability are likely to be significant limiting factors for their practical applications. Nevertheless, we have demonstrated that a self-assembled DHLA-QD-DNA system can specifically detect 10 nM of an unlabelled protein target, thrombin, by incorporating an anti-thrombin aptamer sequence to QD-DNA sensor. In the second strategy, we have shown that DNA non-specific adsorption can be completely eliminated by coating the QD surface with a novel chelating dendritic ligand or incorporating a PEG linker to the QD surface coating ligands. We have prepared functional QD-DNA sensors by coupling the target DNA to QD surface ligand functional groups covalently. The resulting compact QD-DNA conjuagte sensors can quantify specific 35 pM labelled or 1 nM unlabelled complementary DNA probes *via* QD-sensitised dye FRET signals at low probe:QD copy numbers using a conventional fluorimeter. An advantage here is the ratiometric signal, which is less liable to instrumental noise and signal fluctuations, ensuring accurate and reliable measurement. We are currently exploring several strategies (*e.g. via* multi-thiolated DNA-QD self-assembly, reducing the alkyl linker length, developing PEGylated CDL ligands, and improving DNA conjugation efficiency) to make the QD-DNA sensors more robust and sensitive. We are also incorporating other functional DNA molecules (*e.g.* a pH-responsive DNA molecular machine,[26] micoRNA targets[27]) and other DNA/RNA aptamers specific for disease markers or viral surface proteins[2] to the QD-DNA system to develop a multi-functional platform that may be used for simultaneous sensing, diagnosis and therapeutic applications. Studies on these aspects of applications are currently underway in our laboratory.

Acknowledgements

We would like to thank Prof. David Klenerman and Prof. Chris Abell from University of Cambridge and Dr Liming Ying from Imperial College London for their contributions to some of the earlier results described in this paper. We also thank the Food and Environment Research Agency (FERA), the Royal Society (RG 2009R1) and the University of Leeds for funding this project.

Notes and references

1 A. D. Ellington and J. W. Szostak, *Nature*, 1990, **346**, 818; C. Tuerk and L. Gold, *Science*, 1990, **249**, 505.
2 D. H. J. Bunka and P. G. Stockley, *Nat. Rev. Microbiol.*, 2006, **4**, 588; G. Mayer, *Angew. Chem., Int. Ed.*, 2009, **48**, 2672; J. Liu, Z. Cao and Y. Lu, *Chem. Rev.*, 2009, **109**, 1948.
3 A. P. Alivisatos, *Science*, 1996, **271**, 933.
4 A. P. Alivisatos, *Nat. Biotechnol.*, 2004, **22**, 47; X. Michalet, F. F. Pinaud, L. A. Bentolila, J. M. Tsay, S. Doose, J. J. Li, G. Sundaresan, A. W. Wu, S. S. Gambhir and S. Weiss, *Science*, 2005, **307**, 538; M. Y. Han, X. Gao, J. Z. Su and S. Nie, *Nat. Biotechnol.*, 2001, **19**, 631; J. M. Klostranec and W. C. W. Chan, *Adv. Mater.*, 2006, **18**, 1953; S. C. Somers, M. G. Bawendi and D. G. Nocera, *Chem. Soc. Rev.*, 2007, **36**, 579; I. L. Medintz and H. Mattoussi, *Phys. Chem. Chem. Phys.*, 2009, **11**, 17.
5 A. R. Clapp, I. L. Medintz, J. M. Mauro, B. R. Fisher, M. G. Bawendi and H. Mattoussi, *J. Am. Chem. Soc.*, 2004, **126**, 301; L. Shi, V. D. Paoli, N. Rosenzweig and Z. Rosenzweig, *J. Am. Chem. Soc.*, 2006, **128**, 10378; B. Bakalova, Z. Zhelev, H. Ohba and Y. Baba, *J. Am. Chem. Soc.*, 2005, **127**, 11328; I. Potapova, R. Mruk, C. Hübner, R. Zentel, T. Basché and

A. Mews, *Angew. Chem., Int. Ed.*, 2005, **44**, 2437; S. F. Lee and M. Osborne, *J. Am. Chem. Soc.*, 2007, **129**, 8936; D. J. Zhou, A. Bruckbauer, C. Abell, D. Klenerman and D.-J. Kang, *Adv. Mater.*, 2005, **17**, 1243.
6 I. L. Medintz, A. R. Clapp, H. Mattoussi, E. R. Goldman, B. Fisher and J. M. Mauro, *Nat. Mater.*, 2003, **2**, 630; I. L. Medintz, J. H. Konnert, A. R. Clapp, I. Stanish, M. E. Twigg, H. Mattoussi, J. M. Mauro and J. R. Deschamps, *Proc. Natl. Acad. Sci. U. S. A.*, 2004, **101**, 9612; T. Pons, I. L. Medintz, X. Wang, D. S. English and H. Mattoussi, *J. Am. Chem. Soc.*, 2006, **128**, 15324.
7 C. Y. Zhang, H. C. Yeh, M. T. Kuroki and T. H. Wang, *Nat. Mater.*, 2005, **4**, 826; E. Oh, M. Y. Hong, D. Lee, S. H. Nam, H. C. Yoon and H. S. Kim, *J. Am. Chem. Soc.*, 2005, **127**, 3270; S. Hohng and T. Ha, *ChemPhysChem*, 2005, **6**, 956; M. Levy, S. F. Cater and A. D. Ellington, *ChemBioChem*, 2005, **6**, 2163.
8 Product specification for Evidots nanomaterials and Evitags from Evident Technologies Inc., see: http://www.evidenttech.com/products/evidots.html.
9 N. N. Mamedova, N. A. Kotov, A. L. Rogach and J. Studer, *Nano Lett.*, 2001, **1**, 281; S. J. Byrne, S. A. Corr, T. Y. Rakovich, Y. K. Gun'ko, Y. P. Rakovich, J. F. Donegan, S. Mitchell and Y. Volkov, *J. Mater. Chem.*, 2006, **16**, 2896; Y. Zheng, S. Gao and J. Y. Ying, *Adv. Mater.*, 2007, **19**, 376.
10 H. Y. Zhang and D. J. Zhou, unpublished results. We found that an MPA-capped CdTe core only QD completely lost its fluorescence during activation by EDC/NHS in a phosphate buffer (20 mM sodium phosphate, pH 6.0).
11 C. B. Murray, D. J. Norris and M. G. Bawendi, *J. Am. Chem. Soc.*, 1993, **115**, 8706; Z. A. Peng and X. Peng, *J. Am. Chem. Soc.*, 2001, **123**, 183.
12 X. Wu, H. Liu, J. Liu, K. N. Haley, J. A. Treadway, J. P. Larson, N. Ge, F. Peale and M. P. Bruchez, *Nat. Biotechnol.*, 2003, **21**, 41; W. W. Yu, E. Chang, J. C. Falkner, J. Zhang, A. M. Al-Somali, C. M. Sayes, J. Johns, R. Drezek and V. L. Colvin, *J. Am. Chem. Soc.*, 2007, **129**, 2871.
13 B. Dubertret, P. Skourides, D. J. Norris, V. Noireaux, A. H. Brivanlou and A. Libchaber, *Science*, 2002, **298**, 1759.
14 M. Bruchez Jr., M. Moronne, P. Gin, S. Weiss and A. P. Alivisatos, *Science*, 1998, **281**, 2013.
15 W. C. W. Chan and S. Nie, *Science*, 1998, **281**, 2016; Q. Wang, Yang Xu, X. Zhao, Y. Chang, Y. Liu, L. Jiang, J. Sharma, D. K. Seo and H. Yan, *J. Am. Chem. Soc.*, 2007, **129**, 6380; W. Liu, H. S. Choi, J. P. Zimmer, E. Tanake, J. V. Frangioni and M. Bawendi, *J. Am. Chem. Soc.*, 2007, **129**, 14530.
16 A. R. Uyeda, I. L. Medinitz, J. K. Jaiswal, S. M. Simon and H. Mattoussi, *J. Am. Chem. Soc.*, 2005, **127**, 3870; K. Susumu, B. C. Mei and H. Mattoussi, *Nat. Protoc.*, 2009, **4**, 424; F. Dubois, B. Mahler, B. Dubertret, E. Doris and C. Mioskowski, *J. Am. Chem. Soc.*, 2007, **129**, 482; S. Pathak, S. K. Choi, N. Arnheim and M. E. Thompson, *J. Am. Chem. Soc.*, 2001, **123**, 4103.
17 D. J. Zhou, J. D. Piper, C. Abell, D. Klenerman, D.-J. Kang and L. M. Ying, *Chem. Commun.*, 2005, 4807.
18 D. J. Zhou, L. M. Ying, H. Xin, E. A. H. Hall, C. Abell and D. Klenerman, *Langmuir*, 2008, **24**, 1659.
19 K. L. Prime and G. M. Whitesides, *J. Am. Chem. Soc.*, 1993, **115**, 10714; D. J. Zhou, X. Z. Wang, L. Birch, T. Rayment and C. Abell, *Langmuir*, 2003, **19**, 10557.
20 Y. Kim, Z. Cao and W. Tan, *Proc. Natl. Acad. Sci. U. S. A.*, 2008, **105**, 5664.
21 H. Y. Zhang and D. J. Zhou, unpublished results. We found that in addition to TBA-C1 which contains double base mismatches to TBA, the fully complementary TBA-C can also be displaced by TB binding, although the level of signal increase was found to be weaker than TBA-C1.
22 Y. A. Wang, J. J. Li, H. Chen and X. Peng, *J. Am. Chem. Soc.*, 2002, **124**, 2293.
23 W. R. Algar and U. J. Krull, *Anal. Chim. Acta*, 2007, **581**, 193.
24 K. L. Prime and G. M. Whitesides, *Science*, 1991, **252**, 1164; D. Zhou, A. Bruckbauer, L. M. Ying, C. Abell and D. Klenerman, *Nano Lett.*, 2003, **3**, 1517; D. S. Liu, A. Bruckbauer, C. Abell, S. Balasubramanian, D.-J. Kang, D. Klenerman and D. J. Zhou, *J. Am. Chem. Soc.*, 2006, **128**, 2067.
25 X. H. Lin, P. Wu, W. Chen, Y. F. Zhang and X. H. Xia, *Talanta*, 2007, **72**, 468; A. Erdem, F. Sayar, H. Karadeniz, G. Guven, M. Ozsoz and E. Piskin, *Electroanalysis*, 2007, **19**, 798; L. He, M. D. Musick, S. R. Nicewarner, F. G. Salinas, S. J. Benkovic, M. J. Natan and C. D. Keating, *J. Am. Chem. Soc.*, 2000, **122**, 9071.
26 D. S. Liu and S. Balasubramanian, *Angew. Chem., Int. Ed.*, 2003, **42**, 5734; W. X. Wang, Y. Yang, E. J. Cheng, M. C. Zhao, H. F. Meng, D. S. Liu and D. J. Zhou, *Chem. Commun.*, 2009, 824; E. J. Cheng, Y. Z. Xing, P. Chen, Y. Yang, Y. W. Sun, D. J. Zhou, L. J. Xu, Q. H. Fan and D. S. Liu, *Angew. Chem., Int. Ed.*, 2009, **48**, 7660.

27 R. C. Lee, R. L. Feinbaum and V. Ambros, *Cell*, 1993, **75**, 843; C. D. Johnson, A. Esquela-Kerscher, G. Stefani, N. Byrom, K. Kelnar, D. Ovcharenko, M. Wilson, X. W. Wang, J. Shelton, J. Shingara, L. Chin, D. Brown and F. J. Slack, *Cancer Res.*, 2007, **67**, 7713.

General discussion

Dr Campbell opened the discussion of the paper by Professor Cooks: In your paper spray technique, do you use the paper for separation (*i.e.* are you doing paper chromatography)?

Professor Cooks answered: No, not in the experiments so far reported. It is possible to do this but not necessary in the systems we have examined, including the whole blood analysis we have done.

Dr Davies remarked: Collection of blood on filter paper is a well accepted procedure for some tests, *e.g.* neonate heel prick for PKU testing, where the blood is dried onto a Guthrie card and sent off for extraction and assay for hyperthyroidism, cystic fibrosis, phenylketonuria and sickle cell disease. Have you compared dried versus non-dried bloods for the recovery of different molecular species (proteins, lipids, carbohydrates, nucleic acids and small molecules) to see how versatile this technique is?

If you needed to add a preservative and perhaps a surfactant to facilitate release of adsorbed species, how well would mass spectroscopy cope with these?

Professor Cooks responded: Our normal procedure in the paper spray experiment is to allow the pinprick of blood (*ca.* 0.4 microlitres) to dry on the filter paper. This takes a few seconds and facilitates quantitation which is then based on the area of the spot not the volume of the blood and on signal comparison with an isotopically labeled internal standard preloaded onto the paper. We have not made the comparison with non-dried blood. Nor have we done systematic studies (which would include quantitation) on any class of molecules other than small drugs. I do not expect that the presence of a surfactant or preservative would be problematical for mass spectrometry. Preservatives are more likely to be used in experiments with cards shipped to remote labs than in the POC analysis which we are attempting. Surfactants have been found to improve performance in the related ambient ionization method of DESI.[1] Another point of note is that DESI has been used to examine dried bloods.[2]

1 See: A. Badu-Tawiah and R. G. Cooks, Enhanced Ion Signals in Desorption Electrospray Ionization using Surfactant Spray Solutions, *J. Am. Soc. Mass Spectrom.*, 2010, **21**, 1423–1431.

2 See: J. M. Wiseman, C. A. Evans, C. L. Bowen and J. H. Kennedy, Direct analysis of dried blood spots utilizing desorption electrospray ionization (DESI) mass spectrometry, *Analyst*, 2010, **135**(4), 720–725.

Professor Bradley asked: Was it based on an evolving library of samples that allowed correlation of cancerous state with specific markers (by MS)?

Professor Cooks responded: If the question concerns the chemical identification of specific marker compounds, then the answer is that this information is not necessary for diagnostic purposes. However, there are many other reasons to be interested in the nature of the marker compounds and a library of MS/MS product ion spectra was used in making these identifications. If the question concerns a library of tissue types for which DESI mass spectra were recorded, then the answer is that H&E staining and pathological examination of tissue was the reference method used when classifying tissue sections (or portions of tissue sections) as cancerous or healthy. This allowed certain patterns of ion abundance features in the mass spectral data to be associated with disease state. Formal methods of recognizing the presence of these features in particular spectra were developed using principal component analysis (PCA). In later work, multivariate statistical analysis using orthogonal projection to latent structures treated partial least square discriminate analysis (PLS-DA) was used to classify the data as corresponding to diseased tissue or not. This was achieved by dividing the samples into a training set to recognize these key features and then testing performance using a validation set. This training set functioned as our initial tissue library associating particular features with the cancerous or normal states. We plan to expand this initial library using a larger set of samples.[1]

1 See: A. L. Dill, L. S. Eberlin, A. B. Costa, C. Zheng, D. R. Ifa, L. Cheng, T. A. Masterson, M. O. Koch, O. Vitek and R. G. Cooks, Multivariate Statistical Classification of Human Bladder Carcinomas using Ambient Ionization Imaging Mass Spectrometry, *Chem.–Eur. J.*, 2010, accepted; Allison L. Dill, Livia S. Eberlin, Cheng Zheng, Anthony B. Costa, Demian R. Ifa, Liang Cheng, Timothy A. Masterson, Michael O. Koch, Olga Vitek and R. Graham Cooks, "Multivariate statistical differentiation of renal cell carcinomas based on lipidomic analysis by ambient ionization imaging mass spectrometry" Anal. & Bioanal. Chem., 2010, 398, 2969-2978, DOI: 10.1007/s00216-010-4259-6.

Professor Cass remarked: Paper can be a highly variable material, do you see much batch to batch variation in the papers that you use?

Professor Cooks answered: Actually high quality paper, such as that used for laboratory filtration is subject to strict quality control and we see little variation in performance. Different papers have different amounts of intrinsic compounds associated with the production process and characteristic ions in the mass spectra result. These are known and reproducible so do not represent a problem. It is worth noting that other polymeric materials might well displace paper in the paper spray experiment. Certainly for hydrophilic drugs porous PVF's for example might well be superior (after selecting an appropriate solvent).

Professor Bradley asked: What are the *in vivo* practicalities?

Professor Cooks replied: This short question requires a long answer. This answer is provided in full in a forthcoming paper by Nathalie Agar *et al.* in an issue of *Analyst* on Grand Challenges in Analytical Chemistry.[1] Briefly, ambient imaging MS techniques are poised to deliver technology that can be applied routinely in clinical and surgical settings, offering enhanced spatial resolution and chemical specificity not currently available. The required invasive nature of these imaging methods is necessarily satisfied in surgery. Moreover, because ambient ionization techniques have characteristic ease and speed of execution, we believe the approach might allow *in situ* analysis of tissue to assist in intraoperative surgical decision making. However, to meet these needs progress must be made in instrumentation, in understanding the underlying mechanisms of these new ionization techniques and of the biological basis of the ion images. As clinical applications advance, more diverse sample populations will be necessary for validation and with these come the demands of data reduction and data processing. With all opportunities come challenges but one can be cautiously optimistic that these will be overcome allowing imaging MS to become widely used in clinical practice.

1 R. G. Cooks, A. Dill, L. Eberlin, Z. Takats and J. M. Wiseman, Mass Spectrometry in Surgery: Challenge and Opportunity, *Analyst*.

Dr Mount remarked: What limits the spatial resolution of DESI?

Professor Cooks replied: In principle, the difficulty of focusing a "beam" of charged droplets is a fundamental limitation. Of course one can aperture the spray down to a small spot accepting the associated loss of signal. Spot sizes of 40 microns diameter have been shown.[1] In tissue diagnostics, however, the spatial resolution of DESI is not the limiting factor. It is the time taken to record the image. The commonly used spot size of 200 microns requires 16 times less time to record the same sized image as does a 50 micron spot. This is true of all microprobe imaging including almost all reported imaging MS.

1 V. Kertesz and G. J. Van Berkel, Improved imaging resolution in desorption electrospray ionization mass spectrometry, *Rapid Commun. Mass Spectrom.*, 2008, **22**(17), 2639–2644.

Dr Mount asked: How much variation in quantitation do you find across different sample matrices (*e.g.* blood, tissue, *etc.*) and is there a limitation or effect of solvent choice on clinical measurement?

Professor Cooks replied: Quantitation of analytes in blood and in tissue use distinctly different technologies. Quantitation of small molecule analytes in blood using paper spray is proving to be satisfactory in terms of relative standard deviations and accuracy both of which are typically about 10% or less at levels from 1 ng mL^{-1} upwards. This information is based on a relatively small number of drugs in blood with internal standard added to the paper or to the blood, the former being preferred. The use of an isotopically labeled internal standard in paper spray has been shown to eliminate relative matrix effects across blood from different donors. Quantitation of small molecule analytes in tissue using DESI imaging is much more difficult because an internal standard cannot be intimately mixed. These results are semi-quantitative at best. However, the relative amounts of related compounds, for example different phospholipids, can be measured with good accuracy. It is these relative signals that provide the basis for disease diagnostics. These relative signals are also reproducible between serial tissue sections cut from the same sample and also within disease groups, allowing for distinction of diseased tissue from normal tissue. Between serial tissue sections the percentage relative standard deviation (%RSD) of the ratio of two intense peaks in the mass spectra is less than 10% (unpublished data). Within a single uniform tissue section the %RSD of the ratio of two intense peaks in the mass spectra is less than 10% (unpublished data). In tissue, we have observed that the solvent choice affects the compounds that are desorbed from the sample. However, for clinical applications, standardized solvent mixtures are used for the analysis of both normal and disease states, so that no variability is added.

Professor Bradley commented: I have some basic questions of the abilities of ion to travel from the site of ion-generation to the mass spec. How far can it travel before the ions are lost? How long does this take (time delay)?

Professor Cooks responded: An assumption many people seem to make is that ions will be lost by reactions with components of air or that they will be neutralized by charge exchange reactions. This might be the case for radical ions (*e.g.* ions generated by electron ionization) but the species generated by electrospray and the ambient ionization methods (DESI and paper spray, for example) are typically protonated or deprotonated species and their thermochemistry disallows proton transfer reactions with the components of air (the proton affinities and gas phase acidities of the components of air make these reactions endothermic). So the ions are stable and losses are due only to scattering. Early experiments using electrospray showed ion transport over distances of tens of centimetres in open air with scattering dispersion of the "beam" as visualized chemically.[1] Later experiments have demonstrated transport of ions generated by DESI over distances of several metres, with substantial losses but with improved signal to noise as background ions in the mass spectrometer were removed more rapidly than the even-electron species generated by DESI.[2] The possibility that transport was successful only for droplets containing analyte rather than for the analyte ions themselves was eliminated by experiments done directly on solids using a low temperature plasma ion source.[3] This study and others have also shown that losses of ions during transport through air can be minimized using supplementary pumping and that once laminar flow is established further losses are negligible over distances of up to 6 m.

1 See: Z. Takáts, J. M. Wiseman, B. Gologan and R. G. Cooks, Electrosonic Spray Ionization. A Gentle Technique for Generating Folded Proteins and Protein Complexes in the Gas Phase and for Studying Ion–Molecule Reactions at Atmospheric Pressure, *Anal. Chem.*, 2004, **76**(14), 4050–4058.
2 See: I. Cotte-Rodríguez, C. C. Mulligan and R. G. Cooks, Non-Proximate Detection of Small and Large Molecules by Desorption Electrospray Ionization and Desorption Atmospheric Pressure Chemical Ionization Mass Spectrometry: Instrumentation and Applications in Forensics, Chemistry, and Biology, *Anal. Chem.*, 2007, **79**, 7069–7077.
3 See: G. Huang, W. Xu, M. A. Visbal-Onufrak, Z. Ouyang and R. Graham Cooks, Direct Analysis of Melamine in Complex Matrices using a Handheld Mass Spectrometer, *Analyst*, 2010, **135**, 705–711.

Professor Bradley opened the discussion of the paper by Dr Menaa: What is the main market for this technology?

Dr Menaa answered: Biomedical, agrochemical, biopharmaceutical, nanotechnology, food technology, energy and environment, and toxicology with bio-hazards detection, among others. Pharmaceutical companies, health care centres, food department agencies, research laboratories both in academia and industries are concerned.

Professor Cooks commented: One notes differences in peak positions in your spectra, what information does this provide on the molecular environment?

Dr Menaa replied: Different factors have to be taken into account for the peak positions and possible shifting of the same molecule depending on its molecular environment. For example, to take the same molecule, if we work in solution, the C–F bond signal can be shifted based on the presence and nature of solvents (*e.g.* 3-fluoroaniline in methanol). Comparing different fluorinated biomolecules, neighbouring atoms or groups (such as carbonyl, vinyl, allyl groups and so on) can also have an effect on C–F vibrations. The peak position can further depend on the number of fluorine atoms present at the carbon. We are building databases of those parameters and others. In addition, our technology is not only able to recognize specifically and sensitively C–F bonds but also can provide full information about the molecular structure *via* the vibrational assignments of neighbouring functional groups in the main spectral regions. Another interesting aspect is the molecule interactions such as protein–DNA or receptor–ligand interactions that can be probed as function of the peak position of C–F bond, comparatively to the same non-interacting molecules. Such interactions are easy to determine and allows the identification of a specific molecule and its possible molecular interactions.

Professor Cooks asked: To follow up then, is their information on the type of functional group present if there is not useful information on molecular recognition.

Dr Menaa responded: As stated before, in addition to the possibility to molecular identification from the different assignments of the molecular vibrations for C–F bond but also for neighbouring functional groups, we can also use this technique for molecular recognition. For instance, the C–F bond signal can shift as function of interactions of modified amino acids, proteins, fluorinated antigens and antibodies or other biomolecules, *ex vivo*, *in vivo* and *in vitro*. Thereby, from the fluoro-Raman spectrum of fluorinated antigen and its specific antibody, we can monitor easily *in situ* the interactions *via* the C–F bond shift. So, we can use this method for molecular recognition.

Professor Bradley asked: Can the technique pick up fluorine materials attached to a cell or a receptor *in vivo*?

Dr Menaa responded: Yes, it is possible to measure and detect molecule–molecule interactions. For example, we could observe a shift in the C–F Raman signal when a molecule (*e.g.* ligand, substrate, epitope, or non-enzymatic protein) is interacting with another molecule (*e.g.* receptor, enzyme, antibody, or nucleic acid). It is also possible to reverse the interaction by dissociation assays, so we can pick up fluorine materials attached to cells (*e.g.* *via* receptors).

Dr Mount said: Is it possible to use this approach for deep tissue imaging?

Dr Menaa replied: Yes. It is known that with an excitation wavelength below 600 nm (which is usually employed in current technologies) fluorescent biomolecules or fluorescent biomarkers do not re-emit well through tissues. Our technology has different modes and, among others, a two-photon excitation time fluorescence mode is capable of excitation wavelength >800 nm which will enable deep tissue imaging. If we consider the possibility of the hybrid device with an imaging module, then the C–F Raman spectroscopy technique has great potential. Our progressive technology can then be applied for non-invasive deep-tissue molecular imaging for an acceptable signal-to-noise ratio in a living subject using fluoro-nanoparticles. We aim to demonstrate as soon as possible the use of our technology for whole-body images or *in vivo* tissue targeting (*e.g.* tumor) taking into consideration the pharmacokinetics of fluoro-nanomaterials (*e.g.* nanoparticles).

Dr Mount asked: How is data analysis envisaged to be carried out in the final clinical system?

Dr Menaa answered: Data can be obtained quite rapidly with our multi-functional CFS technology (a.k.a. C–F Raman spectroscopy). It can help to diagnose rapidly and efficiently a particular disease, provide information about clinical evolution, and eventually choose for the best individualized therapeutic option. Thus, our technology aims to carefully follow up the patients. A specific software tool for data analysis adapted to clinicians and other health professionals is intended to be developed.

Dr Campbell commented: It seems that your technology is in competition with fluorescence. What are the advantages of your technology over this established technique which will entice users to use carbon fluorine spectroscopy.

Dr Menaa replied: C–F Raman spectroscopy (CFRS) can be used in fluoro-Raman mode or in fluorescence mode. This is useful when you want to compare detection or screen a molecule with only one device. Indeed, you can label a molecule (*e.g.* antibody) with C–F or using a fluorescent dye and detect it on specific cells or tissues using CFRS or CFRS coupled to a confocal microscope in the case of molecular tracking. For medical diagnostics, fluorescence spectroscopy has been preferred to classic Raman studies. However, in medical applications it is often desired to gain more information than it is possible to extract from fluorescence, and then Raman spectroscopy seems to be an attractive alternative. Our technology overcomes the problems encountered in sample preparation, and also about the signal resolution seen with the classic Raman spectroscopy and, in the mean time, our technology can be associated with fluorescence: (i) The instrumentation itself operates both in

Raman and fluorescence mode. We can also tune the wavelength excitation by switching the desired mode and make measurements on tissues with two photon excitation life time fluorescence and excitation wavelengths >800 nm. (ii) The set up of the instrumentation is valuable to get strong fluorescence signals and Raman signals just by switching the modes and in a way that the fluorescence noise which can be encountered does not interfere with our Raman signals. (iii) We can detect specifically and sensitively C–F molecules but also fluorescently labeled molecules (*i.e.* molecules labeled with tags such as Fluorescein). Finally, some of the greatest advantages of using C–F Raman spectroscopy (CFRS) over other existing technologies reside in its multiple features in one apparatus. Thereby, it represents a sensitive and non-destructive tool which can (i) quantify molecular dynamics in biophysics and biochemistry, such as protein–protein interactions, protein–DNA interactions, and protein conformational changes; (ii) to detect, track, screen and quantify rapidly and specifically C–F labeled molecules which are more stable than molecules stained with fluorescent dyes (which are usually light sensitive, toxic, too short half-time for some *in vivo* experiments or to use as a security label to prevent counterfeited drugs, generate background problems, *etc.*); (iii) be used in fluorescence mode; (iv) represents a progressive technology (*e.g.* portable device which can be coupled to other devices to implement some features).

Dr Davies remarked: For this technology to be taken up by researchers there needs to be a readily available library of fluorine compounds to label biomolecules. Are such compounds available, and how easy are they to use?

Dr Menaa answered: We have used several and various fluoro-based materials as supports for analysis using our fluoro-carbon Raman technology such as those described in the paper. We are building our database based on fluoro- and non-fluoro-biomolecules. Among the fluoro-biomolecules, we can cite a few such as, for example, fluoro-antigen/antibodies, fluoro-therapeutics, fluoro-proteins, fluoro-amino-acids (*e.g.* tryptophan, fluoro-tyrosine), nucleic acids such as fluoro-DNA, fluoro-RNA, fluoro-cells (prokaryotes and eukaryotes), fluoro-peptides but also fluoro-nanomaterials as solid host matrices such as mesoporous nanoparticles, polymers, sol–gel glasses for bio-encapsulation and adsorption for drug delivery system applications using biologically active fluoro- or non-fluoro-biomolecules. Recently, we probed hydrophobic surfaces (functionalized with fluoro-groups) of nanoporous sol–gel glasses in which proteins were encapsulated, enhancing the stability and protein helical structure. We are also welcoming new fluoro-biomolecules, fluoro-therapeutics from researchers (both for industrials and academics) to widen our database.

Professor Bradley commented: There are a range of materials used in F-MRI based imaging – could these therefore be used to supplement FSC?

Dr Menaa replied: Yes, but our goal and the advantage of our technology is also to avoid using isotopes, so one of our current goals is to combine both powerful technologies CFS-MRI for imaging purposes (cell migration monitoring, *etc.*), without the need to use the isotope ^{19}F, in order to label cells *ex vivo* using F-molecules (such as the novel perfluoropolyether (PFPE) nanoparticle). We intend to do the same with using a combination of CFS and PET. Indeed, PFPE is not degraded *in vivo* and has no known intracellular biological reactivity.

Dr Perney communicated: After re-reading your paper, I still not fully understand the process by which fluorescence is eliminated? Are the spectra presented in Fig. 3 showing raw data or are they corrected for background?

Dr Menaa communicated in reply: Our technology involves a pulsed metal vapor laser which avoids high photodecomposition and eliminates primary fluorescence, while giving enhanced sensitivity in UV/VIS/NIR and the determination of the C–F signal. For instance, experiments with continuous laser Raman spectrometer (conventional Raman spectroscopy) have not shown such effects. Thereby, C–F characteristic signal could not be revealed using conventional Raman spectroscopy due to the continuous wavelength laser excitation. The spectra presented in Fig. 3 have been obtained with a pulsed copper vapor laser without the need to subtract the background.

Professor Bradley opened the discussion of the paper by Dr Kendall: Is it possible to actually deconvolute the spectra to say what are the major individual Raman markers that make up the majority of the "finger-print" of the samples?

Dr Kendall answered: Yes. Molecular bonds from different species are associated with the characteristic Raman spectral peaks. The peaks can be assigned to the molecular species, which are listed extensively in the literature, and the main tissue constituents can be identified (protein, lipids, DNA, glycogen, *etc.*) (see Fig. 1 and Table 1 of our paper, and ref. 1). Further, spectra of biochemical constituents have been measured and mathematically fitted to tissue spectra to demonstrate the relative distribution of these constituents through the tissue in relation to the disease state.[2]

1 N. Stone, C. Kendall, N. Shepherd, P. Crow and H. Barr, Near-infrared Raman spectroscopy for the classification of epithelial pre-cancers and cancers, *J. Raman Spectrosc.*, 2002, **33**(7), 564–573.
2 G. Shetty, C. Kendall, N. Shepherd, N. Stone and H. Barr, Raman spectroscopy: Elucidation of biochemical changes in carcinogenesis of oesophagus, *Br. J. Cancer*, 2006, **94**, 1460–1464.

Professor Cooks asked: Can you say more about the predictive capabilities of the experiment; the number of patients and length of time followed?

Dr Kendall replied: The prognostic model presented here is preliminary data collected from 14 patients. 576 Raman spectra measured from 25 high grade dysplasia samples were correlated with patient outcome over a period of two years. It is appreciated that this is a small model in terms of numbers with short follow-up compared to the natural history of oesophageal cancer. Work is currently underway to review the notes of these patients over what is now an 8 to 9 year follow-up. It is hoped this will permit correlation of the Raman spectra of Barrett's oesophagus samples with patient outcome.

Dr Zhou enquired: How repeatable is the Raman spectroscopy diagnosis for tumour tissues? Are the Raman spectra affected by extended storage and/or different hydration state of the sample?

Dr Kendall answered: Classification models built with data collected over periods of several months and tested with with cross validation or independent data show high sensitivity and specificity. A number of studies have been undertaken to look at the effect of storage duration and tissue preparation, indicating that extended storage at −80 °C does not adversely affect the spectral quality.

Dr Carr asked: To what degree are foreign materials which could coat the oesophagus such as saliva, reflux from the stomach, phlegm, food residue, *etc.* interfering with typical measurements obtained by Raman spectroscopy?

Dr Kendall replied: The Raman spectra of bile, mucus, spittle, blood and endoscope lubricant have been measured and compared to oesophageal tissue spectra (measured without washing of the sample) indicating that these potential confounding factors do not appear to compromise the Raman spectra of oesophageal tissue.[1]

1 N. Stone, C. Kendall, N. Shepherd, P. Crow and H. Barr, Near-infrared Raman spectroscopy for the classification of epithelial pre-cancers and cancers, *J. Raman Spectrosc.*, 2002, **33**(7), 564–573.

Dr Davies asked: People have been taking various silver colloid preparations for years in the belief that they give a health benefit. They do little harm, and no good. A puff of silver particles onto the oesophagus might enhance the Raman signals shown in Fig. 2 of your paper, particularly from extracellular molecular species. The penetration depth of the radiation used will be many cell layers and perhaps this use of silver powder could enable one to say whether the unique signals seen in Barrett's oesophagus were intracellular or extracellular?

Have you any idea what the Raman spectra are "seeing" in oesophageal cancer? Is it reasonable to expect it to pick up, for example, an increase in the concentration of proteins in a sea of proteins? Ge Zhou *et al.*, working with Petrioin, found in oesophageal cancer some 58 spots on a 2D gel were upregulated >3-fold and 107 were downregulated >3-fold in a fairly constant background of about 1050 spots.[1]

1 G. Zhou *et al.*, *Mol. Cell. Proteomics*, 2002, **1**, 117–124.

Dr Kendall answered: Silver particles could be used to enhance the Raman signal. However enhancement would be on the epithelial surface where there may be contamination as mentioned above. The Raman signals collected described in our paper are averaged over several cells and contain signals from both intracellular and extracellular components. A significant amount of confocal Raman work has been undertaken on cells to characterise the signals further.

The differences in the Raman spectra with pathology have been characterised using peak assignments in the literature and measurement of biochemical constituents. This indicated changes in protein conformation, lipid, DNA, glycogen *etc.* The intensity of the Raman signal is proportional to the concentration of the constituents within the tissue. Raman has been shown to differentiate between the relative concentrations of different proteins in tear fluid.[1]

1 J. Filik and N. Stone, Investigation into the Protein Composition of Human Tear Fluid using Centrifugal Filters and Drop Coating Deposition Raman Spectroscopy, *J. Raman Spectrosc.*, 2009, **40**, 218–224.

Dr Mount asked: It is clear in principle that it is the case in many applications, but has it been established in practice in your clinical application that a combination of Raman and IR spectroscopy gives additional essential clinical information over either technique separately?

Dr Kendall replied: For *in vivo* application there are issues with strong water absorption in infrared spectroscopy, therefore we have not undertaken comparative studies on oesophageal tissue for this reason. However such combination of Raman and FTIR spectroscopy has been undertaken on oesophageal lymph nodes[1] and by Krafft *et al.* on lung[2] and brain[3] tissue.

1 M. Isabelle, N. Stone, H. Barr, M. Vipond, N. Shepherd and K. Rogers, Lymph Node Pathology Using Optical Spectroscopy in Cancer Diagnostics, *Int. J. Spectrosc.*, 2008, **22**(2–3), 97–104.
2 C. Krafft, D. Codrich, G. Pelizzo and V. Sergo, Raman and FTIR imaging of lung tissue: Methodology for control samples, *Vib. Spectrosc.*, 2008, **46**(2), 141–149.
3 C. Krafft, S. B. Sobottka, G. Schackert and R. Salzer, Raman and infrared spectroscopic mapping of human primary intracranial tumors: a comparative study, *J. Raman Spectrosc.*, 2006, **37**(1–3), 367–375.

Dr Mount said: Unlike IR, Raman has the potential for the variation of excitation wavelength to interrogate samples to different depths. Have you or others tried this and does optimisation or variation of this give additional information?

Dr Kendall responded: The use of different excitation wavelengths has been evaluated.[1] Near-infrared wavelengths of 830 nm and 785 nm have been shown to be optimal for Raman spectroscopy of tissue. Near-infrared radiation gives improved depth of penetration of the light into the tissue. In addition, at visible wavelengths significant fluorescence is excited by fluorophores within the tissue which floods the Raman spectrum. It terms of the tissue optics it would be beneficial to use longer wavelengths to penetrate more deeply than 830 nm but the sensitivity of the detectors drop off beyond this wavelength, and therefore compromise is required. In addition for epithelial cancer diagnostics it is important to probe the epithelial layers (1–500 microns deep, optimum 100–200 microns in the oesophagus).

1 N. Stone, *Raman spectroscopy of biological tissue for application in optical diagnosis of malignancy*, PhD thesis, Cranfield University, 2001.

Professor Bradley asked: Are they labels that can be attached to various bio-molecules (*e.g.* aptamers) that would allow the technique to be used only as an alternative to fluorescence imaging?

Dr Kendall replied: Yes, indeed it has been shown that functionalised metallic nanoparticles can be used for Surface Enhanced Raman Spectroscopy (SERS) and Surface Enhanced Resonance Raman Spectroscopy (SERRS) to target specific sequences. These assays are molecularly specific and are more sensitive than those using fluorescence, and have the ability to identify and determine several analytes in a mixture.[1–4]

1 D. Graham, B.J. Mallinder, D. Whitcombe, N. D. Watson, W. E. Smith, Simple Multiplex Genotyping by Surface Enhanced Resonance Raman Scattering, SERRS, *Anal. Chem.*, 2002, **74**(5), 1069–1074.
2 B. D. Moore, L. Stevenson, A. Watt, S. Flitsch, N. J. Turner, C. Cassidy and D. Graham, Rapid and Ultra-sensitive Determination of Enzyme Activities Using Surface-enhanced Resonance Raman Scattering, *Nat. Biotechnol.*, 2004, **22**(9), 1133–1138.
3 D. Graham, D. G. Thompson, W. E. Smith and K. Faulds, Control of enhanced Raman scattering using a DNA-based assembly process of dye-coded nanoparticles, *Nat. Nanotechnol.*, 2008, **3**, 548–551.
4 G. Sabatte, R. Keir, M. Lawlor, M. Black, D. Graham and W. E. Smith, Comparison of surface-enhanced resonance raman scattering and fluorescence for detection of a labeled antibody, *Anal. Chem.*, 2008, **80**, 2351–2356.

Professor Guiseppi-Elie asked: The current standard of care is currently histopathology. How do you propose to develop the predictive capability against the gold standard, which is itself inherently fraught with error?

Dr Kendall answered: This is indeed a challenge. Right from our early work using Raman spectroscopy we identified this as an issue and demonstrated the Raman classification was affected by the quality of the gold standard used to train the model. We showed that improved sensitivity and specificity of the spectral diagnosis was achieved with the use of the consensus opinion of three expert pathologists and that there was considerable inter-observer disagreement between the pathologists.[1] The prognostic model presented in this paper provides weight to the argument that Raman spectroscopy can probe the biochemical basis of the disease and provide useful information to the clinician. Further work to demonstrate the prognostic value of spectral measurements will be beneficial in this respect. We also need to further explore the basis for the spectral discrimination of disease by correlation with other accepted diagnostic tools such as immunohistochemistry and molecular markers.

1 C. Kendall, N. Shepherd, B. Warren, K. Geboes, H. Barr and N. Stone, Raman spectroscopy a potential tool for the objective identification and classification of neoplasia in Barrett's oesophagus, *J. Pathol.*, 2003, **200**, 602–609.

Professor Cass remarked: What is the attitude of the regulatory authorities (FDA, MHRA, EMEA) towards diagnostics based around 'fingerprints' rather than defined molecular species?

Dr Kendall replied: I don't know. I am trying to find out more from MHRA.

Professor Girkin asked: What is the spatial resolution of the system? What are the issues associated with sampling only certain areas and the relatively slow collection of each spectrum? Added to this, how might a Raman system be linked with an OCT system which might do the original searching before the Raman was used for a more detailed chemically specific evaluation?

Dr Kendall answered: It is accepted that the work presented to date is a first generation probe and as such is a point measurement device. The probe is designed with a 100 μm core diameter collection fibre and objective lens to enable collection efficiency comparable to a typical ×10 microscope objective on a free space system and samples tissue 147 μm deep.[1] The current gold standard of histopathology involves the collection of biopsy material and even with a rigorous biopsy protocol collecting 4 quadrant biopsies every 2 cm through the Barrett's region is a poor sampling technique (0.5–1% of the surface). To be able to probe these spots non-invasively using Raman would be a step forward. Faster acquisition times would facilitate the measurement of more tissue. There are a number of ways of improving the optical throughput of the probe design or indeed increasing the incident laser power. Further, the use of multivariate statistical classification models permit the classification of spectra with lower signal-to-noise than that demonstrated in the paper. There are also ways of addressing the sampling issue, either to use some method of scanning the tissue to build up a map as used with other optical techniques[2,3] or use a multi-modality approach in combination with other optical imaging techniques such as fluorescence[4] or OCT to pinpoint the area of interest to be subjected to detailed biochemical inspection using Raman spectroscopy.[5] A number of groups have been working on methods of combining technologies.

1 J. C. C. Day, R. Bennett, B. Smith, C. Kendall, J. Hutchings, G. M. Meaden, C. Born, S. Yu and N. Stone, A miniature confocal Raman probe for endoscopic use, *Phys. Med. Biol.*, 2009, **54**, 7077–7087.
2 J. Hutchings, C. Kendall, N. Shepherd, H. Barr, J. Day and N. Stone, Ultra-low spatial resolution Raman mapping using a novel fibre optic probe, *Proc. SPIE*, 2010, 7560, DOI: 10.1117/12.858803.
3 L. Qui, D. K. Pleslow, R. Chuttani, E. Vitkin, J. Leyden, N. Ozden, *et al.*, Multispectral scanning during endoscopy guides biopsy of dysplasia in Barrett's esophagus, *Nat. Med.*, 2010, **16**, 603–606, DOI: 10.1038/nm.2138.
4 M. C. Grimbergen, C. F. van Swol, R. J. A. van Moorselaar and N. Stone, Feasibility of Raman spectroscopy *in vitro* after 5-ALA based fluorescence diagnosis in the bladder, *Proc. SPIE*, 2006, 6078, B-51.
5 C. A. Patil, N. Bosschaart, M. D. Keller, T. G. van Leeuwen and A. Mahadevan-Jansen, Combined Raman spectroscopy and optical coherence tomography device for tissue characterization, *Opt. Lett.*, 2008, **33**, 1135–1137.

Professor Cooks commented: It is interesting to compare mass spectrometry with Raman/IR. Mass spectrometry has a much greater information density (greater number of resolution elements); perhaps the strength of Raman/IR lies in the reproducibility of the relative intensity data.

Dr Kendall answered: It depends on the level of information required. MS has more resolution elements but Raman/IR will provide reproducible characteristic tissue spectra which may be more readily interpreted. Raman also permits non-invasive *in situ in vivo* measurement of tissue biochemistry.

Dr Mount addressed Professor Cooks and Dr Kendall: What is the prospect of PCA being used to discriminate normal from abnormal cells in clinical applications? How large would the training set need to be for this and what is an acceptable rate of false positives and negatives for the application? Are there other more appropriate analysis techniques?

Professor Cooks answered: The analysis dealt with in our paper is not at the cellular level. Information is obtained on pixels on the order of 1/25 mm^2. In practice, information is likely to be obtained on the 1 mm^2 scale, a scale that is certainly suited to surgical applications. In prospect is the implementation of DESI imaging as an adjunct to normal pathology. Pathology measurements during surgery are not common although with certain diseases like brain cancer they could become common if (i) the new methodology is first extensively tested against established methods and (ii) if information such as tumor grade and extent is provided by the methodology. As already noted, an attempt to provide a fuller answer to this question is given in a paper shortly to be submitted to *Analyst*.

Regarding the question of the size of the training set needed to validate tissue diagnostics, some dozens of tissue section pairs (diseased and adjacent normal for example) is satisfactory in the early stages of research. I have no special information on false positive and false negative rates, since PCA does not allow for a formal classification. For this purpose, multivariate statistical analysis using orthogonal projection to latent structures treated partial least square discriminate analysis (PLS-DA) was used to classify the tissues and give detailed misclassification rates. In bladder cancer, using the negative ion mode data, the misclassification rate was 12%.[2]

1 R. G. Cooks, A. Dill, L. Eberlin, Z. Takats, J. M. Wiseman, Mass Spectrometry in Surgery: Challenge and Opportunity, *Analyst*, to be submitted for an issue on Grand Challenges in Analytical Chemistry.

2 See: A. L. Dill, L. S. Eberlin, A. B. Costa, C. Zheng, D. R. Ifa, L. Cheng, T. A. Masterson, M. O. Koch, O. Vitek and R. G. Cooks, Multivariate Statistical Classification of Human Bladder Carcinomas using Ambient Ionization Imaging Mass Spectrometry, *Chem.–Eur. J.*, 2010, accepted; Allison L. Dill, Livia S. Eberlin, Cheng Zheng, Anthony B. Costa, Demian R. Ifa, Liang Cheng, Timothy A. Masterson, Michael O. Koch, Olga Vitek and R. Graham Cooks, "Multivariate statistical differentiation of renal cell carcinomas based on lipidomic analysis by ambient ionization imaging mass spectrometry" Anal. & Bioanal. Chem., 2010, 398, 2969-2978, DOI: 10.1007/s00216-010-4259-6.

Dr Kendall replied: PCA performs well as a method of multivariate statistical analysis. It is an unsupervised method and enables a review of loadings to identify the sources of variation in the dataset and validate the biochemical basis for separation. It can be used in conjunction with linear discriminant analysis to develop classification models, with projection of an independent test set onto the training dataset. We have demonstrated classification with high levels of sensitivity and specificity in classification models developed with excised tissue from a number of organs in the body. PCA-fed LDA is a widely accepted technique for pathology classification of Raman spectral data. Other multivariate techniques such as cluster analysis and artificial neural networks have been used for Raman and FT-IR spectroscopic analysis. Cluster analysis cannot be used to explicitly generate diagnostic algorithms as independent data cannot be projected onto the model. ANNs have high computation demands and the variables used by the discriminating algorithms are not transparent in the way that PC loadings can be viewed. Support vector machines have recently demonstrated enhanced performance as an alternative analysis method for Raman data.[1]

The acceptable level of false positives and negatives depends on how Raman is being used clinically. As a screening tool the requirement is very few false negatives so all (pre)-malignant changes are picked up. However as a diagnostic tool to guide invasive patient management, for example indicating the need for an oesophagectomy on finding high grade dysplasia in Barrett's oesophagus, the requirement is no false positives.

1 M. Sattlecker, C. Bessant, J. Smith and N. Stone, Investigations of support vector machines and Raman spectroscopy for lymph node diagnosis, *Analyst*, 2010, **135**, 895–901.

Professor Bradley addressed Professor Cooks, Dr Menaa and Dr Kendall: What are the strengths and weaknesses of your techniques? What are the hurdles for *in vivo* application and the timeframes that the technology and equipment will require for *in vivo*/clinical application?

Dr Menaa responded: The strengths: the key feature of this progressive and non-destructive technology is based on the discovery of a characteristic optical signature of carbon–fluorine bond(s) in the fingerprint spectral area of 500–800 cm^{-1} allowing rapid, ultra-specific and sensitive detection, characterization, imaging, and measurement of any fluoroorganics. Interestingly, the C–F bond is unique in its character and so it can be used as a molecular label. Furthermore, the C–F label is efficient, soluble, cheaper, smaller, more stable and less toxic than fluorescent dyes, nanoparticles or quantum dot materials. Thereby, C–F bonds are often incorporated into pharmaceutical, agrochemical and biological molecules, in addition to polymers and nano-materials to achieve special properties (*e.g.* molecular stability, molecular tracing). FRS represents a major advance over conventional Raman spectroscopy since it allows highly specific and ultrasensitive detection (ppm to ppb level or 800 nm), adequate time-gating and time-delays. The technology is versatile, flexible and progressive (*e.g.* hybrid and portable devices such as F-Raman microscope, HTS Raman multi-well analyser).

The weakness at this stage is that we cannot currently use it for *in vivo* imaging experiments as a single apparatus, which is the main limitation at this time, but we are implementing our technology (coupling devices such as integrated microscope and performing adapted CFS set-up).

Professor Cooks replied: I take the question to refer to both the paper spray methodology for therapeutic drug analysis and to the use of DESI to image tissue for disease diagnosis. In regard to therapeutic drug analysis, the paper spray method is rapid, sensitive and molecularly specific. It is successful as a quantitative method. It is minimally invasive (pin-prick). The major weakness is that some drugs are used in doses that are too low for direct analysis, so method modification to allow preconcentration (in some form) will be necessary in these cases. In regard to tissue imaging for disease diagnostics, the disadvantage is the invasive nature of mass spectrometry. However, for applications in surgery the method faces only the usual hurdles involving demonstration of value in clinical practice, safety and regulatory compliance. The advantage is that the experiment is conducted in the ambient environment, is relatively fast and provides chemically specific information.

Dr Kendall answered: Strengths: Multiplexed biochemical analysis in near-real time. Raman can be used for *in vivo* and *in vitro* diagnostics. Recent developments in imaging/rapid mapping technology enables rapid characterisation of thin tissue sections at micron scale spatial resolution. SERS will enable highly specific measurements and SESORS could enable deep readout of these signals from multiple targets of interest. Weaknesses: *In vivo* probe measurements at present are point spectra, therefore with poor sampling. Raman generates a wealth of data but this then needs to be understood and analysed. Developments in computing power, computational methods (chemometrics) and user interfaces are helping in this regard, but much more needs to be done to enable Raman to be used for real-time diagnosis.

Hurdles to *in vivo* application: Raman has been applied *in vivo*,[1] but acceptance by the wider medical community will require larger multi-centre studies and accessible user interfaces. Demonstration of improvements in diagnosis and cost benefit with respect to the current gold standard is required.

1 E. B. Hanlon, R. Manoharan, T.-W. Koo, K. E. Shafer, J. T. Motz, M. Fitzmaurice, J. R. Kramer, I. Itzkan, R. R. Dasari and M. S. Feld, Prospects for *in vivo* Raman spectroscopy, *Phys. Med. Biol.*, 2000, **45**(2), R1.

Dr Mount asked Professor Cooks, Dr Menaa and Dr Kendall: How much of your motivation in this work is to get to an understanding of the biomolecular basis which gives rise to the distinct clinical signature you measure, as compared to the development of a viable and robust clinical diagnostics device?

Dr Menaa replied: In addition to the development of a promising fluoro-therapeutics field that we are focusing on, our technology is actually meeting the needs for competitive biomedical diagnostic tool. Understanding the biomolecular interactions, biocompatibility with the factors directing protein folding, molecular recognition with fluoro-labelled antigen–antibody, development of fluoro-therapeutics owing to the different C–F optical signature (vibrations, wavelength shifts, *etc.*) will undoubtedly give rise to information for the use of therapeutics but also to map in a rational manner processes in order to have a clinical signature. The method is of high sensitivity, high spatial and temporal resolution, high multiplexing capacity, low cost, and high-throughput and so offers advantages over other imaging methods such as positron emission tomography (PET), single photon emission computed tomography, magnetic resonance imaging, computed tomography, optical bioluminescence and fluorescence, high frequency ultrasound, and several other emerging modalities.

Professor Cooks replied: Initially our motivation is to do good by developing a device that is viable and robust. Understanding the basis for such performance seems to me more appropriately placed later in time.

Dr Kendall responded: Both apply. The main aim is to develop a viable robust clinical diagnostic tool, but to do so it is essential to understand the basis for the differentiation between disease states which in turn may contribute to an improved understanding of the carcinogenesis process. The research team is multidisciplinary in nature, with scientists and clinicians working together, to achieve both of these aims.

Dr Campbell asked: Going beyond the instrumentation required to carry out near patient Raman spectroscopy or mass-spectrometry, what are the requirements of the user interface? Is there a challenge in designing a user interface where either reciprocal centimetres or mass/charge ratios do not need to be interpreted by a clinical end user.

Professor Cooks answered: This is a considerable task but one that has been achieved in other mass spectrometry applications, notably those involving public safety and explosives detection. Small instruments have been developed by several companies which include automated library comparison and the identification of compounds present in a sample on the basis of a match between the sample mass spectrum (or MS/MS spectrum) and one of those in a small library.

Dr Kendall replied: The end user should not be presented with the raw data, the clinician is interested in understanding what spectral information tells them about their patient without having to interpret the spectra. In conjunction with our probe development work we have developed a prototype user interface which uses a coloured bar chart to present the percentage likelihood of the sample being normal, pre-cancerous (Barrett's) or cancerous (Neoplasia) for any measured spectrum, based upon the results of the classification model (see Fig. 1). The clinician can then use this information to inform his clinical judgement. Automated spectral analysis for near-real time diagnosis is a challenge in particular for the large datasets acquired with Raman and FTIR mapping/imaging systems despite continual improvements in computing capabilities. We are exploring new chemometric methods for rapid spectral analysis.

Professor Graham commented: Two comments on the use of Raman spectroscopy. The first is that instrumentation is changing rapidly and chemically multiplexed rapid mapping of tissue is possible as is using handheld spectrometers at low cost for *in vivo* imaging. It depends on the use of the technique but the instrumentation is generally available. The second point is that for the end user to make full use of the technique the software has to be simple and easy to use. Again, this is happening but it is still behind the fluorescence community in this regard.

Dr Davies opened the discussion of the paper by Professor Graham: Can you explain the physics of enhancement in the aggregates? I presume this is more than you would get by increasing the thickness of the Raman label layer.

From my delvings into the literature on colloidal aggregates I would expect your aggregates to be of a range of sizes and shapes and to be fairly open, like a 3D self-avoiding random walk. However, I would still expect there to be "size" above which there is no further enhancement of signal. The photon capture cross-section for silver is about 10× the particle cross section and so this would reduce the penetration depth into the aggregate, making those particles below this "skin" layer optically silent. So, as the aggregates form you would expect to see an increase to a plateau in the Raman signal and then, on a slower timescale, the signal would fall as the aggregates stick together so that one aggregate acts to shield another from the incoming light? (Gentle mixing may prevent sedimentation without disruption of the aggregate, especially if the interaggregate bonds are strong and multiple.) Do you see this?

If there is, either theoretically or empirically, an optimal particle "size" then assuming they could be prefabricated as labels to which one would attach a biorecognition element (antibody/DNA/aptamer/*etc.*) would this improve upon the few Raman labels that have been reported in the literature?

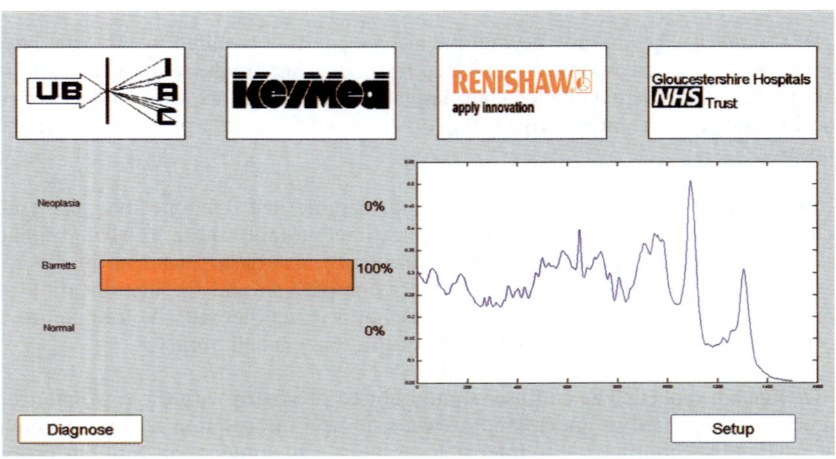

Fig. 1 Example snapshot of prototype user interface for clinical application of *in vivo* Raman spectroscopy using a probe system at endoscopy.

Professor Graham answered: Single nanoparticles give a surface enhancement which is dependent on excitation wavelength and plasmon resonance, which in turn is affected by size, shape, *etc.* of the nanoparticle. By choosing an excitation frequency that matches that of the aggregate rather than the single nanoparticle we can tune into the aggregate. There is also a significant increase in the electromagnetic field when the plasmons couple as two or more particles get close enough to interact. We have looked at a number of different experimental conditions to try and understand the relationship between aggregation and enhancement. We and many others have found that dimers give excellent SERS compared to single nanoparticles and there are a number of papers now on isolating the dimers and the functionalising them with an active targeting species. We also find that not all single nanoparticles give SERS and that to ensure any cluster gives SERS 100% of the time we found empirically it had to contain 13 or more nanoparticles! Adding more reporter molecule will only work effectively up to monolayer coverage and then multilayer effects and distance from the surface reduce the degree of enhancement relative to the monolayer species so aggregation does give more signal return. As aggregation occurs we do see a plateauing and then reduction of signal and it is all dependent on the conditions used to aggregate. This means there tends to be an optimal window for measurement of the maximum signal which can be several hours. Hopefully this has answered all of your points!

Dr Mount said: Detection is dependent not only on the amount of aggregation. The nature and amount of the plasmon interaction (which should control, for example, the spectral shift) should also depend on the nanoparticle separation distance. This should lead to a measured dependence on probe/target sequence length and a maximum detectable target sequence length. Has this been observed and what is this maximum length?

Professor Graham replied: We are in the process of carrying out a systematic study on this very aspect of the DNA hybridisation induced aggregation and from previous studies we know we can see plasmon interactions up to at least the radius of the nanoparticle used. Does this mean we see SERS though? The answer to that question is still under investigation.

Professor Lu said: It is very nice to see that your method can be extended from detection of DNA to detection of protein thrombin. As you have pointed out, the thrombin aptamer has special properties that most other aptamers do not have, *i.e.* the thrombin aptamer interacts at two distinct regions with thrombin so that one thrombin molecule can bind to two thrombin aptamers, making it possible for aggregation-based detection. I am wondering if your method can be extended to other aptamers for other targets, including non-protein targets.

Professor Graham answered: The short answer is yes, although there are always considerations. If the target offers two binding sites as with thrombin then aggregation will work, however if it is 1 : 1 then alternative strategies are needed. For instance, splitting the aptamer sequence and using the binding to bring it back together again should work but we have not done this yet.

Dr Mount asked: What are the issues with undesired nanoparticle aggregation and how does this affect detection in clinical samples?

Professor Graham responded: We have not got to clinical samples with our nanoparticle aggregation system yet, but we are wary of undesired nanoparticle aggregation. We have performed a number of experiments and have developed robust chemistry to produce the DNA-functionalised nanoparticles to minimise undesired aggregation and so far have been successful!

Professor Cass said: Have you tried assembling small nanoparticles on the surface of larger nanoparticles to generate 'hotspots'?

Professor Graham responded: We have and we have tried mixing the types and sizes of the metals too. We have not quantified the difference from sizes but have found that silver gives greater enhancement than gold (as expected) and are currently investigating size effects with a view to understanding the relationships between enhancement and particle size.

Dr Mount asked: What is your assessment of the advantages and disadvantages of using OGN and OSN conjugates in electrochemical, optical and vibrational detection modes?

Professor Graham answered: We have not embarked on any electrochemical studies so have no data to use in this type of assessment. In terms of the optical and vibrational, we find the optical is very sensitive and more so with OSNs due to their higher extinction coefficients. However, the vibrational gives us much more of an understanding about what is interacting as we can identify the specific conjugates involved without separating those which do not hybridise. The disadvantage is that they can be tricky to make and sometimes there is a high attrition rate before stable conjugates can be isolated. Apart from that, they are pretty good for both optical and vibrational detection.

Dr Dhaliwal enquired: Are you ready to detect thrombin *in vivo* with your SERS thrombin approach?

Professor Graham replied: No. We do not have thrombin detection on our critical path as the changes in the SERS are still quite subtle *in vitro*, so before we go *in vivo* we need to get more discrimination between bound and free.

Dr Brown requested: Can you provide more detail about information you have obtained using SAXS in terms of how your data relate to structural models of your system?

Professor Graham answered: Its very preliminary data and we could obtain particle size and some crude interaction data but it does need more work before more can be said.

Dr Brown asked: Do you have any evidence to suggest that your samples are polydisperse, by SAXS or other methods? If so then how do you intend to analyse the data?

Professor Graham replied: Data from SEM and plasmon measurements indicate our nanoparticles are fairly mono-dispersed following synthesis. Our very preliminary SAXS experiments showed this to be true too and allowed measurements to be made but needs more work before we can speak definitively about the data.

Dr Mount opened the discussion of the paper by Professor Gooding: How can precise control of dissolution and orthosilicic acid excretion be achieved (*i.e.* prolonged stability followed by rapid dissolution and excretion when not required)?

Professor Gooding responded: Precise control of the dissolution of the porous silicon, and hence rate of excretion of orthosilicic acid, can be achieved by controlling the coverage of the surface chemistry. We have shown this previously,[1] where the coverage of the surface chemistry, as determined by the shift in optical reflectivity and using IR, directly correlated to the stability of the porous silicon structures in physiological fluids. That is, the higher the density of molecules in the monolayer attached to the porous silicon, the greater the stability and hence the slower the rate of orthosilicic acid secretion.

1 T. Böcking, K. A. Kilian, K. Gaus and J. J. Gooding, The monolayer surface coverage in mesoporous silicon optical filters determines the stability degree of functionalization for biosensor and drug delivery applications, *Adv. Funct. Mater.*, 2008, **18**, 3827–3833.

Dr Krause asked: When the porous silicon is broken into smaller particles, unmodified silicon becomes exposed. Does this affect the stability of the particles?

Professor Gooding responded: Yes. As the base surface chemistry layer is attached to the porous silicon before it is fractured into particles the edges of the particles will be exposed silicon. As this exposed silicon will be the most likely site of degradation of the particles, we felt it was important to evaluate the stability in biological solutions as reported in the paper. The particles do not show significant degradation within the first week, but as expected the particles do not have quite as good long term stability as porous silicon still attached to a wafer as we reported previously.[1]

1 K. A. Kilian, T. Böcking, K. Gaus, M. Gal, J. J. Gooding, Si–C Linked Oligo(ethylene glycol) Layers in Silicon-based Photonic Crystals: Optimization for Implantable Optical Materials, *Biomaterials*, 2007, **28**, 3055–3062.

Dr Krause said: Can the porous silicon be modified after it has been broken into smaller particles?

Professor Gooding answered: Modification of the porous silicon after it has been broken into pieces is certainly possible and we have done this. In fact, in our paper we attached the antibodies to the particles after formation because the antibodies did not withstand the sonication required to produce the particles. The main reason we did not do the entire modification after sonication is for ease of handling. We found it significantly easier to modify the porous silicon prior to particle formation simply because it was a single macroscale sheet. In the end the stability of the particles exceeds what is required for our measurements, so having the small amount of exposed silicon, and the reduced stability that results, does not influence our ability to use the particles for their intended application.

Dr Krause asked: Why does the modification of hydrogen-terminated silicon surfaces with dialkynes result in better blocking of the surface than modification with alkenes?

Professor Gooding replied: We believe the better stabilisation of the hydrogen-terminated silicon surfaces when modified with dialkynes, rather than alkenes, is because the distal moieties with the dialkynes have an affinity for each other rather than a repulsion. So with the alkenes that we used previously, the distal moiety was a carboxylic acid. Hence a close packed monolayer formed from undecenoic acid will result in electrostatic repulsion between the distal carboxylic acids. In contrast, with the dialkyne, π–π bonding can occur between the distal alkynes. After further modification of the dialkyne *via* the 'click' reaction, the distal end of the monolayer possesses a triazole ring and so π–π bonding can still occur. Evidence to support this hypothesis comes from a previous paper[1] where we diluted the dialkynes with monoalkynes. In these experiments, the monoalkynes spaced the dialkynes apart and hence reduced the amount of π–π interactions at the distal surface of the monolayer. In this case there was significantly less effective protection of the underlying silicon from oxidation.

1 S. Ciampi, P. K. Eggers, G. Le Saux, M. James, J. B. Harper and J. J. Gooding, Silicon Electrodes Resistant to Oxidation in Aqueous Solutions: An Unexpected Benefit of 'Clicking' Azides to Surface Acetylene Moieties, *Langmuir*, 2009, **25**, 2530–2539.

Dr Krause asked: Is the triple bond in alkynes essential for getting good blocking of the silicon surface?

Professor Gooding replied: It is not the triple bond that we think is important in giving good stabilisation of the silicon per se but the affinity between the molecules towards the distal end of the monolayer. This is what we think the distal triple bond provides. We have, however, not tested this hypothesis other than diluting the distal alkynes as mentioned above.

Dr Krause said: Why is complexation of the Cu(I) catalyst required for the click reaction to occur inside the pores of mesoporous silicon?

Professor Gooding responded: We have a working hypothesis for why a ligand to complex the Cu(I) catalyst is required for the click reaction to occur within the pores. This is that in the absence of the ligand, the triazole rings of already proceeded click reactions complex the copper and prevent further copper entering the pores. The ability of the triazoles to bind copper has been discussed by Chan *et al.*[1]

1 T. R. Chan, R. Hilgraf, K. B. Sharpless, V. V. Fokin, *Org. Lett.*, 2004, **6**, 2853–2855.

Dr Mount commented: You have claimed that the "click" reaction leads to denser monomolecular surface films than other functionalising strategies, at least in part due to the neutral state of the reacting species minimising intermolecular repulsion. Given the requirement for the ingress, egress and participation of copper catalyst, can the preservation of neutral species be assumed throughout this reaction?

Professor Gooding replied: The key point we were trying to make does not relate to charge *versus* neutrality per se. We believe the higher density of monolayers using the dialkyne modification compare with monolayers containing distal charged moieties, such as carboxylic acids, and hence the better stabilisation of the porous silicon structures, is that the distal moieties have an affinity for each other rather than repulsing each other. If copper was complexed within the monolayer, as long as the π–π interactions were operating we do not think the copper would have a deleterious effect on the stabilisation.

Dr Dhaliwal asked: Regarding your end *in vivo* suggested application, what is the relevance of the eye model?

Professor Gooding responded: The eye model is primarily to provide a point of entry to using these materials for *in vivo* sensing and teaching us how to monitor the particles when they are *in vivo*. Clinicians we are working with believe the particles may have applications for early detection of Age-Related Macular Degeneration.

Dr Davies said: You have a rigid diffraction grating which will behave in a similar fashion to the volume holograms formed by depositing layers of elemental silver in a soft matrix, *e.g.* a gelatin photographic film, and commercialised by Smart Holograms in Cambridge, UK (see www.smartholograms.com). Soft holograms will respond to changes in pH and ionic strength because the polymer network supporting them will expand and shrink and so alter the spacing of the silver planes. One advantage of the grating is that it will not suffer from such environmental effects. Do you see other advantages as far as sensing is concerned with the grating over the soft hologram?

Professor Gooding replied: In addition to the above-mentioned benefit of environmental stability, the highly developed silicon modification chemistry allows a much greater possibility for incorporating other types of functional elements, *e.g.* different types of enzyme-selective polymers, anti-fouling layers, *etc.* Another benefit of de-coupling the transducer from the sensing element is that the optical response of the Rugate filter may still be detected even if substrate is completely removed. In the case of the soft hologram the transducer degrades as the polymer is digested.

Dr Davies said: The volume fraction of air in the porous silicon, how it is distributed and its depth will determine both the optical properties (*e.g.* sensitivity in terms of shift in central wavelength with solution refractive index, finesse or half-height width of the reflected central wavelength, and signal/noise) and the kinetic response of the grating. Did you optimise the performance of the porous silicon on the Rugate filter for monitoring protease activity? If you did, what factors were important? Could you follow the fall in refractive index as the gelatin was broken down and diffused out of the silicon over a period of a few seconds?

Is it known whether a Rugate filter can respond to the small increases of refractive index occurring when an antibody binds to it and displaces water? SPR sensors, the resonant mirror and thin film interferometers can pick up less than 10 micrograms of protein binding per square metre of surface.

For your cell protease work, you broke up the porous silicon into pieces of various sizes and shapes. How quickly do they report the proteolytic action and do you see a quicker response as the particle size or shape varies – one would expect a thin, flat flake to respond faster than a sphere of equivalent volume.

Professor Gooding answered: Optimising the performance of the Rugate filter for detecting protease enzymes was crucial to our earlier studies (compare the Rugate filter in ref. 1 with an earlier paper[2]). The main issues considered were the average porosity of the structure being chosen such that the pore sizes would allow enzymes to freely diffuse through the structures. The porosity modulation was chosen to be small to create sharp spectral features. This also had the benefit of providing minimal disturbance to the pore morphology, which would otherwise influence the enzyme diffusion kinetics. As for monitoring the gelatin breakdown in a matter of seconds, this is possible but we did not do this because, rather than sitting at one wavelength, we scanned over a broader wavelength range, and because the release of enzymes from living cells was on a slower time scale.

Rugate filter can in principle be used to detect binding events under low concentration as the sensitivity (wavelength shift per refractive index unit change) is comparable. The important points to realise from these experiments are that (i) the protease activity produces a blue-shift in spectral features which is more readily distinguished from non-specific binding events, and (ii) the activity of the protease on the polymer substrate will lead to a much larger optical response than the actual protein binding events.

In all cases we use thin flat plates rather than spheres. For the work where we have actually monitored the release of protease enzymes from living macrophage cells we used a porous silicon disk. So we have yet to use the particles for detecting protease activity. So we cannot yet definitively answer the question as to whether size and shape will be important in determining the response but we suspect not. The reason for this belief is that all particles are the same thickness, they are just disks of smaller or larger area, and we know that what determines the response to protease activity is the percentage of pores in which gelatin is degraded.

1 K. A. Kilian, T. Böcking, K. Gaus, M. Gal and J. J. Gooding, Peptide Modified Optical Filters for Detecting Protease Activity, *ACS Nano*, 2007, **1**, 355–361.

2 K. A. Kilian, T. Böcking, S. Ilyas, K. Gaus, W. Jessup, M. Gal and J. J. Gooding, Forming Antifouling Organic Multilayers on Porous Silicon Rugate Filters towards in vivo/ex vivo Biophotonic Devices, *Adv. Funct. Mater.*, 2007, **17**, 2884–2890.

Professor Cass asked: Could you measure the production of extracellular matrix (ECM) by cells if the ECM penetrated the pores?

Professor Gooding replied: In principle this would be relatively easy. What the porous silicon photonic crystals would tell you is that material released from the cells has gone down the pores. It would be much more difficult to configure the system to give specific chemical information, such as whether it was ECM proteins or other proteins released by the cells or what proteins were released.

Dr Mount said: Arguably, the development of methods for the reproducible functionalisation of surfaces is crucial to reproducible detection in clinical measurement. Are optical measurements on PSI photonic crystals a good method for assessing the efficacy of and developing novel methods for silicon surface functionalisation?

Professor Gooding responded: The optical shifts observed in porous silicon photonic crystals when the coverage refractive index of the material changes are very useful in characterising the surface chemistry but are just one in a suite of techniques we use including XPS and FTIR. Where the reflectivity shifts are most useful is the magnitude of the optical shift provides an indication of the coverage of the organic monolayers, and the extent of coupling for subsequent layers, while any changes in the full width half maxima of the peaks provide some information on whether the surface chemistry has modified the entire surface or only a portion. As the reflectivity shift provides no chemical information, chemical characterisation techniques are also vital.

Dr Wark asked: Is the reflectivity peak position sensitive to refractive index changes at the outer surface of the PSi microparticles as well as responding to changes within nanopores?

Professor Gooding replied: No, we see no shift in the reflectivity peak when the cells first attach to the surface. It is only once material is released into the pores. The cells on the surface can, however, scatter light which may influence the reflectivity but we do not observe this.

Dr Wark requested: Could you please comment on the uniformity and the range of pore sizes that can be achieved.

Professor Gooding answered: The size, shape and uniformity of pores produced is highly variable and highly dependent on the etching conditions, doping type and levels of the silicon and current densities used during the etching process. A good example of the types of different pore sizes and structures can be found in the paper by Fauchet and co-workers,[1] but there are many other studies that also demonstrate the scope of pore sizes (spanning 1–2 nm to more than 100 nm and hence materials can be microporous, mesoporous or macroporous), shapes (linear and branched) and uniformity (from highly uniform columnar pores to very ill-defined) that can be achieved. In our study the etching conditions were such that highly uniform columnar pores were achieved.

1 H. Ouyang, M. Christophersen, R. Viard, B. L. Miller and P. M. Fauchet, *Adv. Funct. Mater.*, 2005, **15**, 1851–1859.

Dr Krause asked: Could porous silicon be replaced with TiO_2 nanotubes, the diameter and length of which can be very tightly controlled?

Professor Gooding responded: From a perspective of forming well-defined optical structures with very high optical performance (narrow reflectivity peaks in the present case), what is important is to be able to precisely control the pore size through the depth of the material. In the case of the Rugate filters used in this study, this means a sinusoidal variation in porosity over many layers. I am not aware that TiO_2 nanotubes can be as easily or precisely modified in this way.

Professor Lu opened the discussion of the paper by Dejian Zhou: It is very nice to see that the QDs capped with your dendritic ligands worked so well. You have shown selectivity over non-complementary DNA. I am wondering if you have tried to test the system to differentiate DNA with mismatches, such as DNA with a single-base mismatch.

Dr Zhou replied: This is a good point. We have not tried to test our QD sensor for single-base mismatch DNA target yet but will do that in due course. According to our previous experience, it would be difficult to differentiate single-base mismatch 30-mer normal DNA (used in this paper) from its perfect match target using standard DNA hybridization without heat treatment. The use of a shorter target, PNA or locked-DNA, should enhance the selectivity and may achieve single-base mismatch differentiation.

Dr Mount said: QDs are arguably a type of optical label. How does the demonstrated DNA detection sensitivity and specificity in this work compare to the current gold standard of traditional fluorophore labelling?

Dr Zhou responded: The sensitivity (35 pM) of the QD-DNA FRET sensors described in this paper is comparable to traditional fluorophore labelling used in DNA microarray. We should state, however, that two different detection techniques (conventional fluorescence spectroscopy in our case and microarray reader for the later) are used, making it difficult to directly compare the sensitivity. It has been shown that if a QD-labelled DNA is used as the probe in DNA micoarray detection, it can offer 1–2 orders of magnitude higher sensitivity compared to the traditional fluorophore labelled probe.

Dr Mount asked: Can you tune the excitation wavelength to enable near-infrared through tissue sampling?

Dr Zhou answered: Yes, but a different material has to be used to make the QD. The QDs used in this paper are CdSe/ZnS core/shell, where the longest emission wavelength obtainable is ~640 nm. By using lower energy bandgap materials, *e.g.* InAs or PbS QDs, the QD exitation and emission wavelength can be extended to the near-infrared region, *e.g.* 900–1300 nm. Another feature for QD is its broad absorption: it can be excited by any photos with wavelength shorter that its exciton peak, and the shorter the wavelength, the stronger the absorption. Therefore you have a wide range of excitation wavelengths to choose from.

Dr Mount commented: Your work uses steady-state emission measurements. Would not other modes of QD measurement such as fluorescence lifetime imaging measurement and/or time-resolved fluorescence anisotropy[1] potentially offer more information-rich modes of imaging and sensitive, specific detection?

1 See our recent work: G. Giraud, H. Schulze, T. T. Bachmann, C. J. Campbell, A. R. Mount, P. Ghazal, M. R. Khondoker, A. J. Ross, S. W. Ember, I. Ciani, C. Tlili, A. J. Walton, J. G. Terry and J. Crain, *Int. J. Mol. Sci.*, 2009, **10**(4), 1930–1941; G. Giraud, H. Schulze, T. T. Bachmann, C. J. Campbell, A. R. Mount, P. Ghazal, M. R. Khondoker, S. W. J. Ember, I. Ciani, C. Tlili, A. J. Walton, J. G. Terry and J. Crain, *Chem. Phys. Lett.*, 2010, **484**(4–6), 309–314.

Dr Zhou replied: I totally agree on this point. Currently we do not have access to good instruments that can do such measurements and I am very interested in collaborating with you to carry out these measurements, if possible.

Dr Mount asked: How mature is the production and supply of QDs compared to traditional fluorescent labels?

Dr Zhou responded: The QDs are not yet as mature as the traditional fluorophore labels. We did notice considerable batch-to-batch variations in terms of both stability and quantum yield for commercial QDs even from the same supplier, Evident Technologies. This is unfortunately one of the limitations currently facing the QDs field. This is due to the fluorescence properties of QDs being strongly affected by the size, shape, ligand coating and surface defects, which are very difficult to be controlled exactly the same every batch. Whilst the quality of the commercial QDs are improving, they are still quite some way from perfect yet.

Dr Mount said: What is the toxicity of these QDs? How rapidly will they be cleared from the body and will they able to be used for *in vivo* applications?

Dr Zhou replied: The toxicity is strongly dependent on the QD materials and its surface coating. At cellular level, the CdSe/ZnS core/shell QDs (such as the ones used in this paper) did not show any cytotoxicity over a period of 1–2 days,[1] but this is not certain over a longer period. QDs have also been used to image tumours in animal models. The main toxicity is believed due to the leaking of the heavy metal ions such as Cd^{2+}. The toxicity issue can be overcome by using other QDs, such as InP or Mn-doped ZnS QDs, which do not contain any toxic heavy metals. Therefore these QDs should be able to used for *in vivo* applications.

1 See: V. Bagalkot, L. Zhang, E. Levy-Nissenbaum, S. Jon, P. W. Kantoff, R. Langer and O. C. Farokhzad, *Nano Lett.*, 2007, **7**, 3065.

Professor Gooding addressed Professor Graham and Dr Zhou: The modification chemistry you have both used to modify your nanoparticles has a common motif of a thioctic acid-type moiety and, Professor Graham, in your paper you discuss the reason for this being related to the instability of the metal–thiolate bond, especially on the highly curved surfaces of nanoparticles. This clearly improves the stability of the modification chemistry, although it is utilises to the same basic bonding of sulfur to a noble metal. So my question is, is this surface chemistry good enough or would you ideally like more stable surface chemistry if it was available for these types of surfaces?

Dr Zhou answered: The chelating dendritic ligand-based surface coating is good enough for our currrent *in vitro* sensing applications, but may not so if we want to extend it to *in vivo* applications because of the competition from intracellular glutathione, which is typically in the mM range. More stable surface chemistry will be certainly very welcome if it was readily available.

Professor Graham responded: In our applications we have found that the stability offered by the thioctic acid modification has been sufficient and have observed excellent temperature stability in some systems. The packing around the nanoparticle is also very stabilising, so good surface attachment and robust packing seems to work.

Dr Mount opened a general discussion of the papers by Professor Graham, Professor Gooding and Dr Zhou: What are the barriers to the adoption of detection technologies based on nanoparticles for *in vivo* assays? Are these likely ever to be approved for use by the regulatory authorities?

Professor Vadgama replied: As with all discreet foreign bodies entering the circulation or tissue space, nanoparticles can be taken up by the body's scavaging cell system (especially liver or spleen) and so may not be able to reach the intended tissue destination, even if functionalised with a specific targeting molecule. Alternatively, they could reach tissues where their long term presence may cause harm; the classical example here is the known ability of nanoparticles to penetrate the blood/brain barrier and so enter the central nervous system. This adds to the technical challenge of achieving stable sensing/reporting by nanoparticles in the face of surface fouling agents in biological fluids. Regulatory constraints regarding safety will not currently allow the use of nanoparticles for detection unless it is established that renal or other excretion allows for removal, and of course detailed nanotoxicity studies are available; there is an analogy here with the introduction of new medicines. However, for the future, one good route to approval might be to develop nanoparticles that are degradable and that the degradable products are established as being non-toxic. This overcomes the major problem with nanoparticles in the body, their internal longevity and therefore their ability to elicit a continuous, adverse body reaction.

Professor Graham responded: The main barrier to using nanoparticles for *in vivo* assays will be FDA approval. Some approaches using nanoparticles for drug delivery are being proposed as commercial ways forward, so FDA will have to be approached for these materials. The diagnostic approaches will use similar nanoparticles so there could be a way forward for this in due course.

Dr Mount addressed Professor Graham and Professor Gooding: Which of the detection technologies discussed are approaching or have reached the requirements for single cell measurement? What sensitivity is required for single cell measurement for the application discussed and why is single cell measurement important?

Professor Gooding replied: The porous silicon photonic crystals described based on the release of protease enzymes from macrophage cells have yet to reach the single cell measurement level. However, this is well within the capabilities of the technique and we expect to make these measurements within the next year. The detection limit of the porous silicon photonic crystals is not based on the number of cells but the proportion of pores in which the enzyme substrate is removed. So in our proof-of-concept work the number of cells interrogated was defined by the size of the laser spot from which the reflectivity spectrum was taken. We estimate that reflectivity shifts can be easily measured if 1 in 30 pores have the gelatin substrate removed.[1] In the paper presented we show we can now detect reflectivity spectra from single cell-sized porous silicon photonic crystals and we can also measure shifts in these microscale photonic crystals. Therefore single cell measurements should be achievable. Single cell measurements are important because they take us towards learning how the bulk tissue response is made up of the multiplicity of single cell responses. Therefore we could learn whether the bulk tissue response is composed of a single type of cell behaviour or several different types of cell behaviours. Single cell measurements also provide the potential to identify whether malfunctioning/disease cells may be present in a sample and potentially what that condition is.

1 K. A. Kilian, L. M. H. Lai, A. Magenau, S. Cartland, T. Böcking, N. Di Girolamo, M. Gal, K. Gaus and J. J. Gooding, Smart Tissue Culture: In Situ Monitoring of Cellular Secretion With Nanostructured Photonic Crystals, *Nano Lett.*, 2009, **9**, 2021–2025.

Professor Graham responded: This is a tricky one to answer as it depends what you are measuring in the single cell! Most methods for use in single cells will do as expected, however it is when you want to push the boundaries that new approaches are needed. Is single molecule detection important? Probably, so the ultimate in sensitivity is single molecule and it is important when looking for new interactions. So single molecule detection in real time is important, but also the interactions of molecules. Moving beyond single cells though, it is important not to get fixed (excuse the expression) on single cells as it is very often the interaction between cells in a supported matrix, *i.e.* tissue, that is important to study but with the same resolution as possible with single cells!

Professor Gooding mentioned: A comment about the nanoparticle-type systems and their ability to perhaps monitor single molecule binding events, is that it is not clear, at least to me, what this information would actually mean. What I mean by this comment is that we usually think about most affinity diagnostic systems in terms of thermodynamics, which relates to large ensembles of binding events. With single molecule binding we are really in the domain of statistical thermodynamics and this may guide us in relating the single molecule information obtained with the analytical questions we are trying to answer.

Professor Graham replied: I agree, and the single molecule is the ultimate in sensitivity, but it has to be remembered that each measurement has to be taken in context and has to be meaningful. Single molecule is fine for fundamental studies but how relevant and meaningful is it for clinical diagnosis?

Professor Cass communicated: When developing new diagnostic devices or assays, at what point in the development process should we engage with the regulatory authorities?

Professor Vadgama replied: Certainly it is beneficial to engage the end user at the earliest stage of diagnostic device development, because it informs the research itself. Contact with regulatory authorities, however, is a different type of interaction. For *in vitro* diagnostics it is useful to define the overall planned system as the regulator can offer experience in end user needs and indicate potential weaknesses, *e.g.* relating to safety, practical handling, calibration and materials disposal (contaminated or otherwise). These are overall processing issues and not referable to any individual chemistry that might be being developed. For *in vivo*, it is again a different matter, and definitely beneficial to contact the regulator at the beginning of the development; the safety and reliability challenges are much greater and the invasiveness of such devices demand more early consideration of what is or is not acceptable medically.

Concluding remarks

Pankaj Vadgama

Received 2nd November 2010, Accepted 3rd November 2010
DOI: 10.1039/c0fd90001j

Clinical context and fundamental advances: the need to build linkages

Innovation in analytical science has been a key factor in many of our advances towards the mechanistic understanding of disease processes. This contribution has been such that it is now taken for granted in 21st century medicine. The evolution to healthcare diagnostics of many of the core techniques has, moreover, made for an analytical repertoire that is both substantial and seamless within the overall patient management process. The influence of chemistry is evident in all of this right across the spectrum of prevention, diagnosis and therapy. At one extreme, the highly automated central hospital diagnostic laboratory delivers daily assay numbers running into the thousands. Here, integration with reliable automation and processing technologies has been critical, the majority of such advances inevitably coming from industry. However, the required robust chemistries for high accuracy and precision, core to these advances, have also been realized and are a testament to the inventiveness of the researcher undertaking basic analytical science research. Without the latter, many of the groundbreaking advances would have been impossible. The integration of complex chemistries with engineering and automation systems has itself created a quite unique set of challenges. Moreover, there are 'real world' outcomes, evident to all, such as instrument reliability and method failure, which can put a rather different complexion on what is, or is not, a significant advance.

The integration of bulk solution chemistry methods with automation and low sample volume handling has also kept remarkable pace with the rapid throughput needs of contemporary medicine. Indeed, the current state of the art is such that multiple assays can now be carried out on just microlitres of sample, something that the neonatologist is well aware of. The increased need for expensive biological reagents, such as enzymes and antibodies, as the analytical repertoire becomes more sophisticated, brings additional pressures to shrink assay volumes down to microlitre volumes, and yet, at the same time, to ensure no erosion of assay performance.

So what is to be the next evolutionary phase of diagnostics development? It is the scoping of this future landscape that was the essential subject of this Faraday Discussion. The opportunity for extended discussion allowed for detailed and critical examination of the many nascent capabilities that were presented in the formal papers. The timing and clinical focus was especially welcome given the need to touch fundamentals as a basis for the next phase of more complex and integrated diagnostics. A particular benefit was the bringing together of established exponents of a broad range of physical and chemical methodologies, normally not achieved within a single meeting; mass spectroscopy would not typically be discussed alongside bio- and chemical sensors. The discussion also gave an airing to the developing concept of theranostics, increasingly considered in healthcare circles as a potentiator of the patient management process. This latter, new vision diagnostics, sooner or later, is also going to allow for clinical assessment that will be sufficiently individualised to ensure truly bespoke therapies. The outcomes could well be both

IRC in Biomedical Materials, School of Engineering and Materials Science, Queen Mary University of London, London, UK

economically and therapeutically significant with, for example, higher drug dosage efficiency, fewer side effects and a reduced dependence on large scale clinical trial work. All this may yet be some way off, but better diagnostics is a good place to start. This Faraday Discussion certainly highlighted the art of the possible in all this, both for laboratory and extra-laboratory applications.

Whilst the Faraday Discussion covered many recently developed analytical and technological platforms, it clearly could not be fully comprehensive, or representative, as is the nature of any such gathering. So the inclusion of the above theranostic paradigm was rather more implicit than explicit; it is too early to have expected more. What was brought out in good measure, however, was the unifying challenge of moving analysis from the research 'comfort zone' of simple solution assays to real biological measurements and environments. The formal sessional breakdown was: (i) systems and devices to inform therapy, (ii) towards real-time clinical measurement, (iii) high-throughput methods of analysis, and (iv) physical techniques for diagnostics. These are certainly well matched to what is happening in this area. Moreover, the resulting broad discussions identified some challenges from the perspective of the end user, unfortunately not often given sufficient prominence in traditional analytical science meetings. From the Discussion it would seem that addressing the latter does not have to lead to a dilution of the science or its rigour. Quite the contrary, the feedback of applications into basics is healthy for generating new conceptual and fundamental science advances.

On a specific, practical front, without contextual awareness, it is hard to see how adaptation to clinical analysis can be secured properly. It has certainly been the model for analytical science developments in Clinical Biochemistry right from the start. Candidate diagnostics, like candidate drugs, need to encounter at a quite early stage real biological systems to allow for a more accelerated assessment of feasibility. Some of the presentations at this Faraday Discussion showed the accelerating effect of early stage biomedical links.

Along with the formal presentation and discussion, it was the informal exchanges around the wide ranging set of posters that especially helped to give this meeting its fully interactive element. The engagement of early stage researchers also added considerably to the outreach value of the Faraday Discussion.

Devices to inform therapy

If indeed there is an already existing, potent analytical capability at the disposal of clinical diagnostics, then what might be the need for further refinement? Perfection has certainly not been achieved. Also, while there is no single system problem that allows for a unified resolution, a set of common end user issues do emerge from out of all the detail. Despite the automation and industrialisation of measurement processes, there remains a dependence on soluble reagents, cumbersome whole blood separation, and repeated, expensive, calibration. Miniaturisation is as yet incomplete despite an era of extreme micro-nanotechnology, and reliable continuous monitoring remains an elusive challenge. In truth, there is greater pressure to resolve such generic issues in Point of Care Testing (POCT) than in an automated central laboratory, and undoubtedly the former smaller area is a growing market, so our capabilities do need to change. This opens up pathways for just the type of technologies presented at this Faraday Discussion; it is estimated that while currently 15% of the world *in vitro* diagnostic market is taken up by POCT, a remarkable doubling of this is anticipated by 2014.[1]

A primary medical consideration is whether qualitative or quantitative measurement is needed and whether a system needs to be single or multiple use. For continuous operation, quite a different toolset is needed; reagent binding reversibility, a notable problem for antibody-based systems, non-intrusive design, sterilisation, *etc.* There is, further, a need to integrate any diagnostic element into the patient management pathway. This is especially the case for rapid, *qualitative* diagnostics

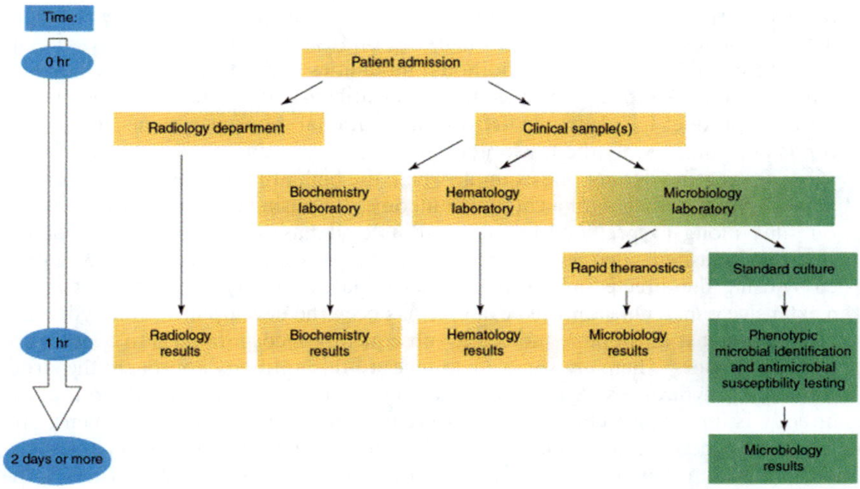

Fig. 1 Diagnostic pathway for microbiology *vs.* other disciplines showing the delay of traditional testing methods (from ref. 2 with permission).

in infectious disease management where early antibiotic therapy may be critical to success. Microbial diagnostics are unfortunately uniquely slow in that they need organism growth for amplified detection of specific proteins or metabolic end points. To overcome the limitations of protein-based diagnostics and the time needed for microbial culture, there has, accordingly, been a move to molecular biology diagnostics.[2] The reliability and immediacy of the latter is set to reverse the 1–2 day delay associated with our current microbiology methods, essentially refined versions of those used in the 19th century. By bringing microbial testing in line with other diagnostics, including radiology (Fig. 1), it will be possible to streamline the patient pathway through the sequence of admission, triage, primary evaluation, history/physical, provisional diagnosis, laboratory assay and therapy.

Sensing strategies

There are well-known advantages to using monolithic, solid-state devices for rapid diagnostics. Whether there is a biological molecule incorporated, as in the classical biosensor embodiment, or whether a purely artificial chemical construct is used, the advantages are significant. Thus, an on-board receptor/reagent eliminates the need for further reagent requirement, potentially avoids a sample dilution and separation step, and through direct readout goes a long way towards deskilling the entire analytical process. What the clinical gains are has been well-established in the case of home use pregnancy test kits and glucometers.[3,4] Now, with the advent of improved micro- and nanofabrication technologies, both further miniaturisation and complex sensor multiplexing become practical, reducing operator time and delivering performance reliability. What does remain a constraint is the affinity window of the binding reagent to trigger the entire sensing cascade. Too low an affinity and the target molecule cannot be detected, too high an affinity and sample dilution becomes obligatory. Weak binding is certainly a barrier to our reaching down to the low levels at which hormones and drugs are present in biological fluids.

Methods for modifying affinity usually takes a molecular biology route aimed at altering the binding region of some artificially expressed construct. Predictable success here has not been forthcoming and reliance is really on empiricism since we do not fully understand which structures define any given affinity property for practical purposes. However, examples do exist, such as in the case of a higher affinity cholinesterase,[5] where, through use of fusion protein constructs, it has also been possible to introduce additional modalities such as tags for labelling and immobilisation. Antibodies offer a highly versatile affinity system, but the lack of a reactive product has demanded effort on additional reporter systems, a less easy proposition than simply the tracking of an enzyme reaction.

Catalytic antibodies are one example where the high affinity capability of an antibody can be combined with a catalytic outcome. By raising an antibody to a transition state along a specific reaction coordinate, it has been possible to generate a degree of quasi-enzymic activity. The resulting catalytic action can be further modified by using site-directed mutagensis and candidate antibody selection can be from an extensive panel. One reported example has been the development of a hydrolytic antibody against an organic molecule, subsequently incorporated into a potentiometric biosensor.[6] There are some advantages to this antibody approach: the structural lability, typical of enzymes, is avoided, the standard 'Y' structural motif of the antibody is uniformly achieved, facilitating uniform immobilisation regimens, and unique catalytic variants can be harvested through *de novo* immunisation and selection, rather than having to depend on a naturally found enzyme. Whilst this is the theory, in practice target transition state design has been difficult, good catalytic efficiency elusive and versatility in target conversion not a feature of progress here.

Such catalytic antibodies were not featured in this Faraday Discussion; however, what they do provide is a timely reminder from an earlier field that the highly emphasised aptameric systems of the Faraday Discussion will have similar challenges to face as artificially generated receptors. The aptamer work described, of course, reflected the interest of the particular delegates, but it is also likely that these will play a more substantial part in future biorecognition systems. Certainly their huge structural diversity from a few, simple, sub-units, and new accelerated platforms for production will make for improved discovery of tailored receptors. The now well-honed 'evolution of ligands by exponential enrichment' (SELEX) gives rapid convergence to a set of affinity candidates and the speed, automation and ease of approach is likely to be further advanced with new engineering methods. Given this, and the attraction of aptamer structural stability, at least over the more floppy peptide and protein equivalents, this growing interest in aptamers is a logical progression; the four-fold increase in reported citations between 2002 and 2008[7] is a clear manifestation of this.

Specifically, this Faraday Discussion illustrated innovations in both the design and testing of labelled aptamers, including use of the ferrocene model as a nominally faithful reporter of binding and aptamer near-surface organisation.[8] Incorporation of additional sequences on an aptamer that allowed for its capture and subsequent optical readout at a solid surface was described, and also peptide aptamers were shown as alternatives, conveniently retained in 3D scaffolds.[9] Aspects that will still need more fundamental consideration are the basics of aptamer conformation and conformational flexibility, the effects of target binding and solid surface proximity on this and how some element of predictability can be instituted regarding performance.

As with so many immobilised bioreagents, we still do not have a sufficient handle on the effects of the immobilisation surface on the original bulk solution properties. So for designing future aptamers, it would now be timely to incorporate molecular modelling as part of the selection process; a mechanistic understanding would seem to be a necessary way to extend the aptamer repertoire to a fuller molecular range. The achievement of aptameric selectivity for mercury and other ions at this Faraday Discussion does, however, show how new niches can still arise.[10] Scale-up of candidate selection is, of course, not limited to nucleic acids, and the availability of phage display libraries as a central resource could allow for rapid antibody selection for more traditional diagnostics; detection of pathogens by this route was demonstrated at this Faraday Discussion.[11]

A counterpart to any selective binding is the surface tailoring of the transduction surface. Amperometry has been a mainstay of sensing, but impedimetric analysis offers a further tool that is less often used, and which may be better suited to an era of smart and reactive materials. The example at this Faraday Discussion was oral pathogen protease detection by 'hard wiring' target peptides into polymer gel films;[12] impedimetric tracking of gel disruption then became possible as an index of enzyme activity. Microbial enzymes do, certainly, allow for selective detection through their characteristic indicator cascades, and further effort on relevant indicator end points using solid state chemistries offers a convenient route to POCT. Whilst here molecular biology can accelerate the diagnostic detection process, the instrumentation needed is more difficult to apply outside the laboratory.

Whatever the analytical convenience, any solid sensing interface used for detection is admittedly subject to a range of extraneous surface reactions, and these are not always easy to extract from the true response. Such non-specific interactions, though rather indeterminate and difficult to predict, warrant much more attention than has been the case hitherto. We now have unprecedented abilities to create designer, architectured surfaces and sensing arrays, including variously through the use of nanoparticle-based modification, layer-by-layer polymer deposition and molecular self assembly.[13–15] The baseline surface control these now offer make it opportune to examine non-specific interactions at surfaces and to use this added understanding to tackle issues of sensing reliability in biological fluids. Approaches with model surfaces are already well advanced in biomaterial research, but the jump to sensing and analytical science has been limited. Organothiol bridging to gold is one relatively straightforward methodology that has extended our understanding of surface biocompatibility.

Biomaterials research[16] thus has a useful part to play in our evolution of viable sensors. By combining molecular reactivity work with materials research, it will be easier to rationalise the differences between bulk solution and surface behaviour of bioaffinity systems. Obvious places to start would be the examination of the double layer at the surface and the altered organisation of surface and near-surface water, both potent effectors of macromolecular conformation change and of the overall colloid adsorption process from a biomatrix. Of course, single use sensors that only need to survive the biomatrix for a matter of seconds are not at risk of adventitious surface reactions, but the paucity of our understanding of most such surface events creates uncertainty over the correct choice of sensing surface even here, and will certainly constrain future efforts at extended period measurement.

Real-time measurement

Biological systems have an inherent rhythmicity and show constant change in various time domains. At one level, the study of such phenomena can be regarded as chronobiology.[17] However, there is always a stimulus-response process underlying much of biochemistry that provides us with a signature of a disease state and which may also condition the final responsiveness to therapy. The archetypical example of this is diabetes, where control is evident in the fluctuations of glucose and other metabolites.[18] Any device purporting to follow relevant metabolite changes needs to survive the sample matrix to a much greater degree than is ever needed for single point assay. A more critical problem is also faced if the device is *in vivo*; none more so than if micro- and nanoparticles are to be injected. Typical use has been for radiological and ultrasound imaging.[19] However, with chemical functionalisation using fluorescent and other ligands, selective chemistries are possible, together with a convenient cell entry route, as shown in this Faraday Discussion.[20]

Provided safety and biocompatibility is acceptable, there are additional benefits to be accrued using the 'intelligent', targeted mobility of carrier cells such as leucocytes. pH, O_2 and cation responsive dyes within capsular structures provide a further means for localised monitoring.[21] However, the possibility of extrapolation to clinical use is uncertain given the lack of understanding of the eventual fate and sequestration of such particles in the body. Programmed self-destruction or biodegradation would be useful areas to research to ameliorate any toxicity outcomes. At least now nanotechnology is furnishing us with greater structural elegance coupled with a facility for improved tissue targeting and more precise optical interrogation, quantum dots and nano-shell constructs[22] being examples of the latter. Overall, parallel work on nanoparticle toxicity, which has kept such structures out of clinical use, would repay the investment,[23] and the tissue mobility of nanoparticles may provide early insights into their ultimate fate in the body.[24]

Though we are a long way from dynamic monitoring of nucleic acid changes in the body, undoubtedly this will be possible eventually. It is a useful start to at least establish diagnostics based on rapid molecular biology, and to combine the multi-processing needs for nucleic acid amplification and detection; these are not easy partners on a single analytical platform.

The example in this Faraday Discussion of an integrated cycler/detector for SNP identification provided an example of routes to such integration,[25] though commercial developments are also moving apace to facilitate one-step POCT molecular biology.[26] There is always room for engineering improvement, and new chemistries are not the sole route to diagnostic advance; this was indicated by the ultrafast droplet-sample thermal cycler system presented for rapid influenza A virus detection.[27]

High-throughput methods

Array devices are set to be compressed in size and made more analytically convenient, most likely with present day 2D arrays, though 3D architectures are likely in the future with new fabrication technologies. Microfluidics will provide a new adjunct to this. The Faraday Discussion showed another counterpart to this: extreme miniaturisation, in this case sample separation and assay for single cell biochemistry,[28] where capillary electrophoresis bridged the sampling and assay modality. One basic advantage of such an approach for biology would be the assessment of the heterogeneity of cell populations; this would amount to a biochemical histology if it could be realised. Regardless of whether a detector surface is 2D or is an 'inside out' system using, say, nanoparticles inside the biomatrix, the theme of improving and modifying the detector surface and its biofunctionalisation, vital for both, was evident here, and showed the quantitative impact of even minor changes in a surface motif.

Physical techniques

Techniques that are the domain of physical and basic sciences are increasingly being tested and refined for clinical assay. Much of the development resides in the sample handling and interfacing aspects. There are also possibilities for miniaturisation which could allow eventual extra-laboratory use. Mass spectrometry is one of these core techniques against which so many other assays are benchmarked. However, whatever its analytical sensitivity and specificity, the various sample preparation and presentation challenges, at the very least, make it a difficult technique for routine service use. There is a need for sample clean up, separation and careful instrument management by skilled operators. The Faraday Discussion touched on the clinical potential, provided some of these basic constraints can be relaxed. Thus, it appeared to be possible achieve ionisation outside the vacuum confines of the instrument and to simplify the delivery of sample ions with sample presentation under ambient conditions.[29] Once this is operationally achieved, it becomes possible to adapt the technique to both chemical speciation and chemical imaging of tissue. There are already important insights generated using secondary ion mass spectrometry (SIMS) for tissue imaging.[30] Other examples of technique adaptation discussed were Raman spectroscopy for C–F bonds, which can enable molecular fingerprinting and which may have use in pharmaceuticals where there is increased development of fluoro-therapeutic agents as well as in biomedical research, though impact in clinical diagnostics is less likely.

The encounter of spectroscopic techniques with tissue complexity can be partly, but not entirely, optimised on the basis of biological understanding, for it is also heavily dependent on the physical domains of signal handing and chemometric routines in order to extract meaningful information. Raman spectroscopy of precancerous epithelial tissue was a considered example at the Faraday Discussion, which could, in principle, meet many of the needs of early diagnostic screening. However, tissue diagnostics remains the only means of establishing unequivocal diagnosis; as yet neither automated cytology screening nor screening for biochemical markers escape significant false positive/negative rates.

With the free distribution of nanoparticles in bulk solution, it becomes possible to apply a variety of colloid science methods, especially to detect binding and aggregation. So provided the surface function is controlled and designed to be biospecific, numerous non-contact transduction methods can be applied. One embodiment, presented at the Faraday Discussion, utilised surface-enhanced Raman scattering (SERS) as an especially versatile demonstrator for quantitative aggregation monitoring after biodetection.[31] It was also shown at the Discussion that quantum dot optical donors could be used to identify binding proximity to an optical acceptor molecule as reporter;[32] porous silicon with its high internal area offered another novel and versatile particle-based sensing platform.[33] The transferability of such models to biological fluids and to cell suspensions would be a next logical stage that would then require critical decisions on sample preparation needs for such methods and the feasibility of automated platforms.

Prospects

Multi-disciplinarity has become a byword for much medically related research. It is clear that the scale of the challenges are no longer such that they can be readily accommodated within a mono-discipline, even one as broad as chemistry. So while so many enabling ideas have come from chemistry, their fate is rather more often determined by gearing to the end user, and other discipline support. At the very least, the risks are reduced by taking the multi-discipline route, a facet well demonstrated at the Discussion, and which also indicated the commonality of many of the challenges faced by the new generation of clinical diagnostics and theranostics.

Acknowledgements

Pankaj Vadgama expresses his thanks to the EPSRC (CO/2444432) for generous support during the preparation of this manuscript.

References

1 *Point of Care Testing Application of Microfluidics Technologies*, Yole Devéloppement, Lyon, 2010.
2 F. J. Picard and M. G. Bergeron, *Drug Discovery Today*, 2002, **7**, 1092.
3 C. M. Estes, J. Ramierez, L. Tiezzi and C. Westhoff, *Contraception*, 2008, **77**, 40.
4 B. Harrison, R. Markes, P. Bradley and I. A. Ismail, *Clin. Biochem.*, 1996, **29**, 521.
5 G. Istarnboulie, S. Andreescu, J. L. Marty and T. Noguer, *Biosens. Bioelectron.*, 2007, **23**, 506.
6 G. F. Blackburn, D. B. Talley, P. M. Booth and C. N. Durfor, *et al.*, *Anal. Chem.*, 1990, **62**, 2211.
7 G. Mayer, *Angew. Chem., Int. Ed.*, 2009, **48**, 2672.
8 A. E. G. Cass and Y. Zhang, *Faraday Discuss.*, 2011, DOI: 10.1039/c005487a.

9 K. Abe, D. Ogasawara, W. Yoshida, K. Sode and K. Ikebukuro, *Faraday Discuss.*, 2011, DOI: 10.1039/c005359g.
10 S.-F. Torabi and Y. Liu, *Faraday Discuss.*, 2011, DOI: 10.1039/c005404f.
11 O. Quazi, M. Rani, A. J. Gnanam, T. W. Cullent, C. M. Stead and H. Kensing, *et al.*, *Faraday Discuss*, 2011, DOI: 10.1039/c005422b.
12 X. Zheng, J. P. Cook, M. Watkinson, S. Yang and I. Douglas, *et al.*, *Faraday Discuss.*, 2011, DOI: 10.1039/c005364c.
13 M. Gebala, G. Hartwich and W. Schuhmann, *Faraday Discuss.*, 2011, DOI: 10.1039/c005365a.
14 M. Compas and C. O'Sullivan, *Anal. Lett.*, 2003, **36**, 2551.
15 J.-Y. Park, Y.-S. Lee, B.-Y. Chang, B.-H. Kim and B. Hyean, *et al.*, *Anal. Chem.*, 2010, **82**, 8342.
16 P. Thevenot, W. J. Hu and L. P. Tang, *Curr. Top. Med. Chem.*, 2008, **8**, 270.
17 E. Haus, *Adv. Drug Delivery Rev.*, 2007, **59**, 985.
18 C. B. Landersdorfer and W. J. Jusko, *Clin. Pharmacokinet.*, 2008, **47**, 417.
19 D. J. Bharali and S. A. Mousa, *Pharmacol. Ther.*, 2010, **128**, 324.
20 K. Dhaliwal, L. Alexander, G. Ercher and A. Unciti-Broceta, *et al.*, *Faraday Discuss.*, 2011, DOI: 10.1039/c005410k.
21 S. M. Buck, Y. E. L. Koo, E. Park, H. Xu, M. A. Philbert, M. A. Brasul and R. Kopelman, *Curr. Opin. Chem. Biol.*, 2004, **8**, 540.
22 N. G. Portney and M. Ozkan, *Anal. Bioanal. Chem.*, 2006, **384**, 620.
23 C. F. Jones and D. W. Grainger, *Adv. Drug Delivery Rev.*, 2009, **61**, 438.
24 S. D. Gittard, P. R. Miller, R. D. Boehm, A. Ovsianikov and B. N. Chichkov, *et al.*, *Faraday Discuss.*, 2011, DOI: 10.1039/c005374k.
25 J. M. Girkin, M.-I. Mohammed and E. M. Ellis, *Faraday Discuss.*, 2011, DOI: 10.1039/c005271j.
26 S. Raja, J. Ching, L. Q. Xi, S. J. Hughes and R. Chang, *et al.*, *Clin. Chem.*, 2005, **51**, 882.
27 D. J. You, P. L. Tran, H.-J. Kwon, D. Patel and J.-Y. Yoon, *Faraday Discuss.*, 2011, DOI: 10.1039/c005326k.
28 D. Jiang, C. E. Sims and N. L. Allbritton, *Faraday Discuss.*, 2011, DOI: 10.1039/c005362g.
29 R. G. Cooks, N. E. Manicke, A. L. Dill and D. R. Ifa, *et al.*, *Faraday Discuss.*, 2011, DOI: 10.1039/c005327a.
30 P. Malmberg, H. Nygren, K. Richter and Y. Chen, *et al.*, *Microsc. Res. Tech.*, 2007, **70**, 828.
31 D. Graham, R. Stevenson, D. G. Thompson and L. Barrett, *et al.*, *Faraday Discuss.*, 2011, DOI: 10.1039/c005397j.
32 H. Zhang, P. G. Stockley and D. Zhou, *Faraday Discuss.*, 2011, DOI: 10.1039/c005373b.
33 B. Guan, A. Magenau, K. A. Kilian and S. Ciampi, *et al.*, *Faraday Discuss.*, 2011, DOI: 10.1039/c005340f.

Poster titles

Systematic analysis of IgG antibody immune response against varicella zoster virus (VZV) using a self-assembling protein microarray, **A Ceroni*, J LaBaer, J Haas and C J Campbell,** *University of Edinburgh, UK*

Statistical differentiation of multiple human cancers based on lipidomic analysis by ambient ionization imaging mass spectrometrym, **Allison Dill*, Livia Eberlin, Cheng Zheng, Anthony Costa, Demian Ifa, Liang Cheng, Timothy Masterson, Michael Koch, Olga Vitek and R Graham Cooks,** *Purdue University, USA*

A fully-automated CMOS fluorescence lifetime biosensor, **B. R. Rae*, J Yang, J McKendry, D Li, Zheng Gong, E Gu, MD Dawson and RK Henderson,** *University of Edinburgh, UK*

New sensors for imaging redox-mediated biological processes with subcellular resolution, **Craig Auchinvole*, Hamish McNab and Colin J. Campbell,** *University of Edinburgh, UK*

On-chip real-time quantitative nucleic acid sequence-based amplification (NASBA) for RNA detection and amplification, **M-N Tsaloglou*, M. Bahi, M Mowlem and H Morgan,** *University of Southampton, UK*

Miniaturising protein microarrays using Dip-pen nanolithography for alternative disease detection methods, **Eleanore Irvine*, Aaron Hernandez-Santana, Stacey Laing, Karen Faulds and Duncan Graham,** *University of Strathclyde, UK*

Quantification of DNA hybridization using post-labelling with a biotin-modified intercalator for the detection of *Staphylococcus aureus,* **Magdalena Gebala*, Gerhard Hartwich and Wolfgang Schuhmann,** *Ruhr-Universität Bochum, Germany*

Flourescence lifetime imaging of DNA microarrays: quantum dot labelling and SPAD array detection, **G Giraud*, H. Schulze, T. T. Bachmann, C. J. Campbell, A. R. Mount, P. Gazal, M. R. Khondoker, I. Ciani, C. Tlili, A. J. Walton, J. Terry, J. Crain,** *University of Edinburgh, UK*

A planar format screenable fluorescent assay with confocal readout, **Mithun Parambath*, Homanaz Ghafari, and Quentin S Hanley,** *Nottingham Trent University, UK*

Enzymatic on-chip enhancement of genotyping DNA microarrays, **Holger Schulze*, Gerard Giraud, Holli Vase, Shiromi Baier, Timo Barl, Till T Bachmann,** *University of Edinburgh, UK*

On-chip integrated antibody-labelling, dissociation and detection of primary tumour cells, **Jane Woods*, P. T. Docker, Charlotte E Dyer, Stephen J Haswell and John Greenman,** *University of Hull, UK*

Novel ExoSERRS detection of DNA, **Jennifer A Dougan*, Duncan Graham, Karen Faulds,** *University of Strathclyde, UK*

Using ubiquitous personal glucose meters for portable, low-cost and quantitative detection of a broad range of targets based on novel functional DNA sensors, **Yi Lu*, Yu Xiang,** *University of Illinois at Urbana-Champaign, USA*

Enhanced detection through enzymatic signal amplification for impedimetric DNA biosensors, **M Kaatz*, H Schulze, I Ciani, F Lisdat, AR Mount, TT Bachmann,** *University of Edinburgh, UK*

Magnetic LOAC for the isolation and analysis of circulating tumor cells, **O Henry*, C K O'Sullivan, L Laggae, T Stakenborg, O Nilsson, C Fermér, K S Drese, T Roeser, M Ritzi, S Hauch, W Albert, E Borgen, N Laddach, A O H Nygren,** *Universitat Rovira i Virgili, Tarragona, Spain*

A simple screening marker for prostate cancer, **Robert Pal* and David Parker,** *Durham University, UK*

Single molecule detection in fast flow: potential for rapid, ultrasensitive diagnostic applications, **Rohan T. Ranasinghe*, Haitao Li, Jung-uk Shim, Richard W. Clarke, Chris Abell and David Klenerman,** *University of Cambridge, UK*

Dielectrophoretic studies and enrichment of human embryonic stem cells, **S Velugotla*, A Menachery, S Smith, S Pells, P De Sousa, R Pethig,** *University of Edinburgh, UK*

High frequency dielectrophoresis, **C Chung*, A Menachery, R Pethig and S Smith,** *University of Edinburgh, UK*

NIR-fluorescent nano- and micrometer-sized particles for biomedical and diagnostical applications, **Thomas Behnke, Katrin Hoffmann, C Würth, J Pauli, Julia Mathejczyk, Frauke Alves, and Ute Resch-Genger*,** *BAM Federal Institute for Material Research and Testing, Germany*

Potentiometric gene sensor based on EMF generation from redox-labelled oligonucleotide probe pair, **Koji Nakano*, Takayuki Kimura, Hisao Yoshinaga, Nobuaki Soh and Toshihiko Imato,** *Kyushu University, Japan*

Multistep enzyme-DNA reactions monitored using multidimensional flourescence imaging and microfluidics, **T Robinson*, M A A Neil, G S Baldwin, A J de Mello and P M W French,** *Imperial College London, UK*

Optical classification of human embryonic stem cells, **N M B Perney*, P Horak, J S Wilkinson, T Melvin, J Hughes and N Hanley,** *University of Southampton, UK*

Identification of common fungal pathogens in a single sample, **Kate Dempsey, Julie Redden*, Karen M McCarney, Alastair Ricketts, W Ewen Smith, Mark Woolfrey and Ian Bell,** *Renishaw Diagnostics Ltd, UK*

The Skinner Prize for the best poster was awarded to Craig Auchinvole of the University of Edinburgh for his poster on new sensors for imaging redox-mediated biological processes with subcellular resolution.

List of participants

Dr K. Abe, *Japan*
Professor N. Allbritton, *University of North Carolina, USA*
Professor M. Ashfold, *University of Bristol, United Kingdom*
Mr C. Auchinvole, *University of Edinburgh, United Kingdom*
Dr T. Bachmann, *University of Edinburgh, United Kingdom*
Miss L. Ballamy, *Convatec Ltd., United Kingdom*
Professor P. Bartlett, *University of Southampton, United Kingdom*
Professor R. Beynon, *University of Liverpool, United Kingdom*
R. Bircham, *Royal Society of Chemistry, United Kingdom*
Professor M. Bradley, *University of Edinburgh, United Kingdom*
Dr K. Brown, *University of Texas At Austin, USA*
Mr J. Butement, *University of Southampton, United Kingdom*
Dr N. Cade, *SELEX Galileo Ltd, United Kingdom*
Dr C. Campbell, *University of Edinburgh, United Kingdom*
Dr R. Carr, *Ingenza Ltd, United Kingdom*
Professor A. Cass, *Imperial College London, United Kingdom*
Mr A. Ceroni, *University of Edinburgh - Division of Pathway Medicine, United Kingdom*
Mr C. Chung, *The University of Edinburgh, United Kingdom*
Professor G. Cooks, *Purdue University, USA*
Dr J. Counsell, *Royal Society of Chemistry, United Kingdom*
Dr T. Dafforn, *University of Birmingham, United Kingdom*
Dr R. Davies, *SPD Development Company Ltd, United Kingdom*
Dr E. Dempsey, *Institute of Technology Dublin, Republic of Ireland*
Dr K. Dhaliwal, *University of Edinburgh, United Kingdom*
Ms A. Dill, *Purdue University, USA*
Dr J. Dougan, *University of Strathclyde, United Kingdom*
Dr K. Faulds, *University of Strathclyde, United Kingdom*
Dr E. Fletcher, *Scottish Enterprise, United Kingdom*
Dr N. Freeman, *University of Manchester, United Kingdom*
Dr M. Gebala, *Ruhr-Universität Bochum, Germany*
Dr G. Giraud, *University of Edinburgh, United Kingdom*
Professor J. Girkin, *Durham University, United Kingdom*
Professor J. Gooding, *University of New South Wales, Australia*
Professor D. Graham, *University of Strathclyde, United Kingdom*
Prof Dr A. Guiseppi-Elie, *Clemson University, USA*
Dr O. Henry, *Universitat Rovirал Virgili, Spain*
Mr J. Hopkins, *Waters, United Kingdom*
Miss E. Irvine, *University of Strathclyde, United Kingdom*
Miss M. Kaatz, *University of Edinburgh, United Kingdom*
Dr C. Kendall, *Gloucestershire Hospitals NHS Foundation Trust, United Kingdom*
Dr P. Ko Ferrigno, *Leeds Institute of Molecular Medicine, United Kingdom*
Dr S. Krause, *Queen Mary, University of London, United Kingdom*
Prof Dr Y. Lu, *University of Illinois at Urbana-Champaign, USA*
Dr B. Menaa, *Fluorotronics, USA*
Professor L. Motte, *Universit Paris 13, France*
Dr A. Mount, *University of Edinburgh, United Kingdom*
Prof Dr K. Nakano, *Kyushu University, Japan*
Professor R. Narayan, *UNC-NCSU Joint Department of Biomedical Engineering, USA*

Dr R. Pal, *Durham University, United Kingdom*
Mr M. Parambath, *Nottingham Trent Unversity, United Kingdom*
Dr N. Perney, *University of Southampton, United Kingdom*
Dr M. Platt, *University College Dublin, Republic of Ireland*
Dr B. Rae, *University of Edinburgh, United Kingdom*
Dr R. Ranasinghe, *University of Cambridge, United Kingdom*
Miss J. Redden, *Renishaw Diagnostics, United Kingdom*
Dr U. Resch-Genger, *BAM Federal Institute for Materials Research and Testing, Germany*
Mr T. Robinson, *Imperial College London, United Kingdom*
Professor W. Schuhmann, *Ruhr-Universitt Bochum, Germany*
Dr H. Schulze, *University of Edinburgh, Division of Pathway Medicine, United Kingdom*
Miss S. Syed, *University of Edinburgh Medical School, United Kingdom*
Dr M. Townsend, *Royal Society of Chemistry, United Kingdom*
D. Trenzinger, *Waters UK Ltd, United Kingdom*
Dr M. Tsaloglou, *University of Southampton, United Kingdom*
Professor R. Tsien, *University of California, San Diego, USA*
Professor P. Vadgama, *Queen Mary & Westfield College, United Kingdom*
Mr S. Velugotla, *The University of Edinburgh, United Kingdom*
Dr A. Wark, *University of Strathclyde, United Kingdom*
Mrs J. Woods, *University of Hull, United Kingdom*
Mr D. You, *University of Arizona, USA*
Dr D. Zhou, *University of Leeds, United Kingdom*

Index of contributors*

* The page numbers in **bold** type indicate papers submitted for discussions.

